Solvatochromic Probes and Their Applications in Molecular Interaction Studies—a Themed Issue to Honor Professor Dr. Christian Reichardt

Solvatochromic Probes and Their Applications in Molecular Interaction Studies—a Themed Issue to Honor Professor Dr. Christian Reichardt

Editors

William E. Acree, Jr.
Franco Cataldo

Basel • Beijing • Wuhan • Barcelona • Belgrade • Novi Sad • Cluj • Manchester

Editors
William E. Acree, Jr.
Department of Chemistry
University of North Texas
Denton
United States

Franco Cataldo
Actinium Chemical Research
Institute
Rome
Italy

Editorial Office
MDPI
St. Alban-Anlage 66
4052 Basel, Switzerland

This is a reprint of articles from the Special Issue published online in the open access journal *Liquids* (ISSN 2673-8015) (available at: www.mdpi.com/journal/liquids/special_issues/M981Y02RGI).

For citation purposes, cite each article independently as indicated on the article page online and as indicated below:

Lastname, A.A.; Lastname, B.B. Article Title. *Journal Name* **Year**, *Volume Number*, Page Range.

ISBN 978-3-7258-1038-3 (Hbk)
ISBN 978-3-7258-1037-6 (PDF)
doi.org/10.3390/books978-3-7258-1037-6

© 2024 by the authors. Articles in this book are Open Access and distributed under the Creative Commons Attribution (CC BY) license. The book as a whole is distributed by MDPI under the terms and conditions of the Creative Commons Attribution-NonCommercial-NoDerivs (CC BY-NC-ND) license.

Contents

About the Editors vii

Preface ix

Heinz Langhals
How the Concept of Solvent Polarity Investigated with Solvatochromic Probes Helps Studying Intermolecular Interactions
Reprinted from: *Liquids* 2023, 3, 31, doi:10.3390/liquids4020015 1

Nikolay O. Mchedlov-Petrossyan, Vladimir S. Farafonov and Alexander V. Lebed
Solvatochromic and Acid–Base Molecular Probes in Surfactant Micelles: Comparison of Molecular Dynamics Simulation with the Experiment
Reprinted from: *Liquids* 2023, 3, 21, doi:10.3390/liquids3030021 32

Stefan Spange
Polarity of Organic Solvent/Water Mixtures Measured with Reichardt's B30 and Related Solvatochromic Probes—A Critical Review
Reprinted from: *Liquids* 2024, 4, 10, doi:10.3390/liquids4010010 89

Omar A. El Seoud, Shirley Possidonio and Naved I. Malek
Solvatochromism in Solvent Mixtures: A Practical Solution for a Complex Problem
Reprinted from: *Liquids* 2024, 4, 3, doi:10.3390/liquids4010003 129

Jia Lin Lee, Gun Hean Chong, Masaki Ota, Haixin Guo and Richard Lee Smith, Jr.
Solvent Replacement Strategies for Processing Pharmaceuticals and Bio-Related Compounds—A Review
Reprinted from: *Liquids* 2024, 4, 18, doi:10.3390/liquids4020018 152

Franco Cataldo
Conventional and Green Rubber Plasticizers Classified through Nile Red [E(NR)] and Reichardt's Polarity Scale [$E_T(30)$]
Reprinted from: *Liquids* 2024, 4, 15, doi:10.3390/liquids4020015 182

Stuart J. Brown, Andrew J. Christofferson, Calum J. Drummond, Qi Han and Tamar L. Greaves
Exploring Solvation Properties of Protic Ionic Liquids by Employing Solvatochromic Dyes and Molecular Dynamics Simulation Analysis
Reprinted from: *Liquids* 2024, 4, 14, doi:10.3390/liquids4010014 199

Javier Catalán and Henning Hopf
The Photophysics of Diphenyl Polyenes Analyzed by Their Solvatochromism
Reprinted from: *Liquids* 2024, 4, 13, doi:10.3390/liquids4010013 216

Daniela Babusca, Andrei Vleoanga and Dana Ortansa Dorohoi
Solvatochromic and Computational Study of Some Cycloimmonium Ylids
Reprinted from: *Liquids* 2024, 4, 9, doi:10.3390/liquids4010009 226

Cynthia M. Dupureur
Use of DFT Calculations as a Tool for Designing New Solvatochromic Probes for Biological Applications
Reprinted from: *Liquids* 2024, 4, 7, doi:10.3390/liquids4010007 246

W. Earle Waghorne
Solvent Polarity/Polarizability Parameters: A Study of Catalan's SPP^N, Using Computationally Derived Molecular Properties, and Comparison with π^* and $E_T(30)$
Reprinted from: *Liquids* **2024**, *4*, 8, doi:10.3390/liquids4010008 .... **261**

Manish Kumar, Abhishek Kumar and Siddharth Pandey
Effective Recognition of Lithium Salt in (Choline Chloride: Glycerol) Deep Eutectic Solvent by Reichardt's Betaine Dye 33
Reprinted from: *Liquids* **2023**, *3*, 24, doi:10.3390/liquids3040024 .... **269**

Elaheh Rahimpour and Abolghasem Jouyban
Prediction of Paracetamol Solubility in Binary Solvents Using Reichardt's Polarity Parameter Combined Model
Reprinted from: *Liquids* **2023**, *3*, 32, doi:10.3390/liquids3040032 .... **278**

William E. Acree, Jr. and Andrew S. I. D. Lang
Reichardt's Dye-Based Solvent Polarity and Abraham Solvent Parameters: Examining Correlations and Predictive Modeling
Reprinted from: *Liquids* **2023**, *3*, 20, doi:10.3390/liquids3030020 .... **288**

Pedro P. Madeira, Luisa A. Ferreira, Vladimir N. Uversky and Boris Y. Zaslavsky
Polarity of Aqueous Solutions
Reprinted from: *Liquids* **2024**, *4*, 5, doi:10.3390/liquids4010005 .... **299**

About the Editors

William E. Acree, Jr.

William E. Acree, Jr., received his bachelor's degree (1975), master's degree (1977), and doctorate degree (1981) in chemistry from the University of Missouri at Rolla. He taught at Kent State University from September 1982 to August 1988 before moving to the University of North Texas, where he is currently a professor in the Chemistry Department. His research interests include spectrofluormetric probe studies, thermodynamic properties of nonelectrolyte solutions, solid-liquid equilibrium in organic nonelectrolyte solutions, and the development of linear free energy relationships to describe mathematically solute transfer processes of chemical importance. To date, his research has resulted in the publication of more than 1,050 peer-reviewed research articles, one research monograph titled "Thermodynamic Properties of Nonelectrolyte Solutions", six volumes in the IUPAC-NIST Solubility Data Series, one *Liquid*'s Special Issue Reprint, as well as several encyclopedia articles, book chapters, and educational articles in the *Journal of Chemical Education*.

Franco Cataldo

Franco Cataldo is a habilitated ordinary professor of general and inorganic chemistry. He has taught at the University of Rome, Tor Vergata, and Tuscia University. He has also worked as a researcher in the petrochemical, rubber, oil, and fat industries, where he is currently a scientific consultant. He has published about 500 peer-reviewed research articles and is the author of 52 patents. He is the editor-in-chief of the international journal *Fullerenes, Nanotubes, and Carbon Nanostructures,* as well as the editor-in-chief of Springer's book series titled *"Carbon Materials Chemistry and Physics"*. His main research interests include fullerene chemistry, carbon materials, and nanostructures, as well as physical organic chemistry topics, including solvatochromic probes, solvents, and plasticizer properties. Other fields of his research activity cover polymer chemistry, radiation chemistry, and industrial chemistry.

Preface

This Special Issue honors Prof. Dr. Christian Reichardt, who is a pioneer and leading expert in the area of solvatochromic probe studies. His most notable research in the area involves the development of the ET(30) and normalized ETN solvent polarity scales based on the UV-visible absorption spectrum of the zwitterionic 2,6-diphenyl-4-(2,4,6-triphenyl-1-pyridinium)phenolate dye molecule (commonly referred to as Reichardt's dye). ET(30) values have been used to examine preferential solvation in binary solvent mixtures, to measure trace water concentrations in organic mono-solvents, to probe the surface polarities of chemically functionalized silica particles, and to examine the interfacial regions of aqueous micelles formed by cationic surfactants.

Specific topics covered in this Special Issue include five review articles as well as ten original research papers and communications that report the synthesis of new solvatochromic probe molecules, measurement of new spectroscopic data, and development of new computational methods. Each article is authored by leading experts in their respective fields.

The review articles contained in this Special Issue describe the following: (a) the behavior of solvatochromic and molecular acid-base probes dissolved in molecularly organized aqueous-surfactant solvent media; (b) the polarity of aqueous-organic cosolvent mixtures as measured with Reichardt's B30 and related solvatochromic probe molecules; (c) the strategies employed by industrial chemicals in identifying replacement solvents for the processing of active pharmaceutical ingredients (APIs) and related compounds; and (d) the application of solvatochromic probes to quantify solute-solvent interactions in water, organic mono-solvents, and binary mixtures. Each review article cites between 60 and 245 published papers. Other topics covered in this Special Issue include the classification of both conventional and green rubber plasticizers using the Nile Red and Reichardt's ET(30) polarity scales; determination of the solvation properties of protic ionic liquids through spectroscopic probe studies and molecular dynamic simulations; examination of the lithium-cation induced changes in the physicochemical properties of deep eutectic solvents using solvatochromic probe molecules; critical comparisons between the different published polarity and hydrogen-bonding scales; synthesis and photophysical properties of new classes of solvatochromic probe molecules; use of DFT computations as a tool for designing new solvatochromic probes for biological applications; and utilization of solvent polarity and hydrogen-bonding scale values to predict the solubility of crystalline nonelectrolyte solutes in organic mono-solvents and binary solvent mixtures. The large range of topics covered illustrates that solvatochromic probe studies have been and still continue to be a very important, active research area.

William E. Acree, Jr. and Franco Cataldo
Editors

Review

How the Concept of Solvent Polarity Investigated with Solvatochromic Probes Helps Studying Intermolecular Interactions

Heinz Langhals

Department of Chemistry, LMU University of Munich, Butenandtstr. 13, 81377 Munich, Germany; langhals@lrz.uni-muenchen.de

Abstract: Intermolecular interactions form the basis of the properties of solvents, such as their polarity, and are of central importance for chemistry; such interactions are widely discussed. Solvent effects were reported on the basis of various polarity probes with the $E_T(30)$ polarity scale of Dimroth and Reichardt being of special interest because of its sensitivity, precise measurability and other advantages, and has been used for the investigation of solvent interactions. A two-parameter equation for the concentration dependence of medium effects has been developed, providing insights into structural changes in liquid phases. Moving from condensed gases to binary solvent mixtures, where the property of one solvent can be continuously transformed to the other, it was shown how the polarity of a solvent can be composed from the effect of polar functional groups and other structural elements that form the matrix. Thermochromism was discussed as well as the effect of very long-range interactions. Practical applications were demonstrated.

Keywords: solvent polarity scales; $E_T(30)$ scale; condensed phases; binary mixtures; solvent structure

1. Introduction

Most chemical syntheses were carried out in the liquid phases [1] where the average distances between molecules are small [2]. Noncovalent intermolecular interactions [3] are important in such condensed phases; they dominate the properties of chemical materials, and will subsequently be analyzed below. Adjusting the pressure of gases is a useful method for setting intermolecular distances and allows the systematic study of such effects.

2. General Intermolecular Interactions

The impact of a molecular chemical material such as the pressure p of gases depends on the numbers of particles in a volume, more conveniently indicated by the molar concentration c. Starting from highly diluted gases with essentially independent particles, the impact p is expected to be proportional to the concentration c giving the differential Equation (1).

$$dp = const_1 \cdot dc \qquad (1)$$

As the concentration increases, the impact is changed more and more due to interactions of the particles, and the changes are expected to increase with the concentration of particles. Inverse proportionality implies a basic approach leading to the differential Equation (2), where the constant E stands for the energetic effect of c on p.

$$dp = E/c \cdot dc \qquad (2)$$

Multiplying Equation (2) by c and dividing by the scaling factor c^* for the discrimination between the two extrema of Equations (1) and (2) and adjusting the dimensions yields Equation (3), where $c = 0$ can be included ($E/c^* = const_2$).

$$c/c^* \cdot dp = const_2 \cdot dc \tag{3}$$

The differential Equation (4) results from the addition of Equations (1) and (3) and the subsequent separation [4] of the variables p and c.

$$dp = \frac{const_2}{\left(\frac{c}{c^*} + 1\right)} dc \tag{4}$$

Finally, Equation (5) is obtained by the integration of Equation (4) with the integration constant p_o; the latter is zero for diluted gases because the pressure p vanishes for infinite dilution of gases.

$$p = E \cdot \ln\left(\frac{c}{c^*} + 1\right) + p_o \tag{5}$$

Equation (5) ends up in the ideal gas equation for diluted gases by a Taylor series expansion and truncation after the linear term forming Equation (6), where $c \ll c^*$ and the concentration c is the number of moles n over the volume V.

$$p \approx \frac{E}{(c + c^*)} \cdot c \approx \frac{E}{c^*} \cdot c = \frac{E}{c^*} \frac{n}{V} \tag{6}$$

Thus, the ideal gas equation (7) is obtained with $R \cdot T = E/c^*$.

$$p \cdot V \approx n \cdot \frac{E}{c^*} = n \cdot R \cdot T \tag{7}$$

Equation (5) is an exact description [5] of the concentration dependence of the pressure of gases such as is shown for the precisely measured isotherms by Michels and Michels [6]. A typical linear correlation (performance of target to actual comparison) according to Equation (5) is shown in Figure 1 (a high accuracy of the linear correlation is achieved where deviations from experimental data are within 10 to about 20 ppm).

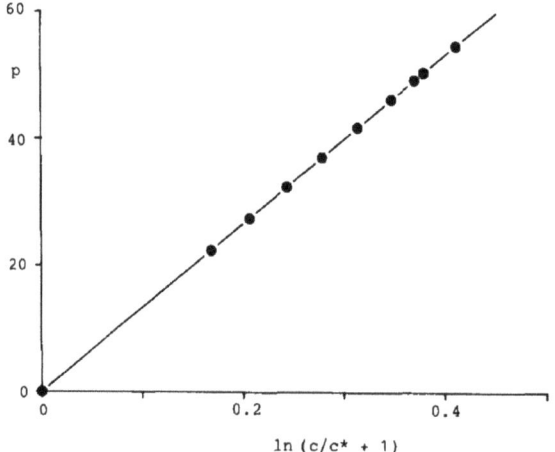

Figure 1. Typical linear correlation between p and $\ln(c/c^* + 1)$ according to Equation (5): CO_2 at 75.260 °C, $c < 3.1$ mol·L^{-1}; $E = 133.4$ at, $c^* = 4.5$ mol·L^{-1}, $r = 0.999\,998$, $n = 10$; deviations in the correlation from experimental data of 10 until 20 ppm.

When the CO_2 concentration c exceeds a threshold value c_k(I-II) of 3.1 mol·L^{-1}, the dependence on the pressure in the low concentration region I changes abruptly to reach the concentration region II (see Figure 2). Equation (5) is also valid for region II, but with different parameters E and c^*, and p_o is different from 0.

$$p = p_o \cdot e^{E \cdot c} + p^* \tag{8}$$

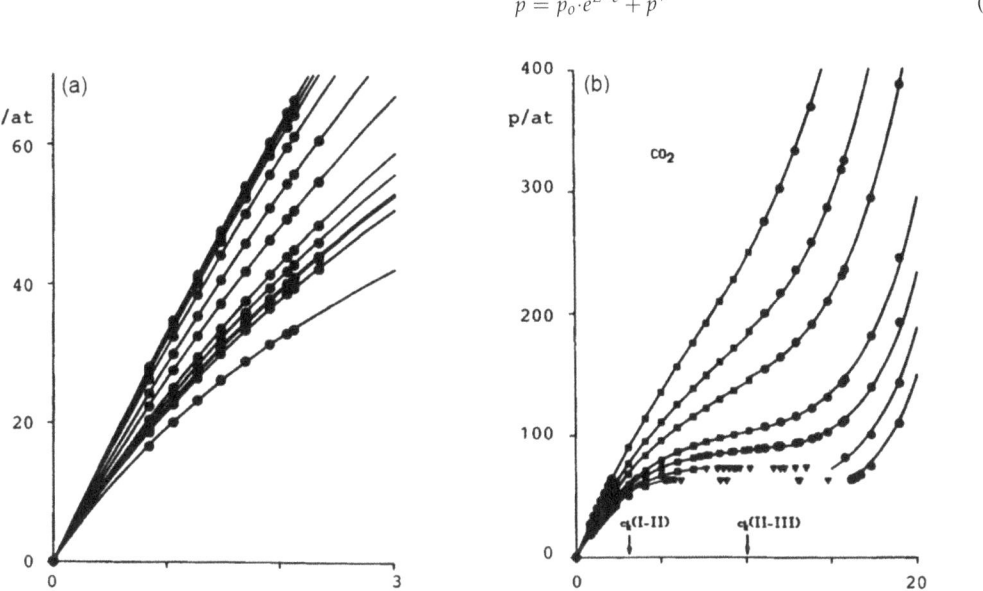

Figure 2. Diagram of state of CO_2: Discontinuous measurements were taken from Refs. [6–8], and curves were calculated [5] by means of Equations (5) and (8) and the application of the method of least squares to fit the experimental data. Measurements at very high pressure were included, but not shown in the right-hand diagram for graphical clarity. Temperatures (rounded): 25, 32, 40, 50, 75, 100, and 140 °C (see Ref. [9]) c_k(I-II) $\cong$ 3.1 mol·L^{-1} (concentration for the change from solvent structure I to II), c_k(II-III) $\cong$ 10 mol·L^{-1} (concentration for the change from solvent structure II to III). (**a**) Enlarged region I ($c < 3.1$ mol·L^{-1}). (**b**) Complete region I, II, and III; filled circles for region I and III, squares for region II, triangles for the biphasic region.

At even higher pressure, when c exceeds c_k(II-III) of 10 mol·L^{-1} the region III is reached. There, an exponential increase in p with concentration c was found and can be described by Equation (8) where p^* means a base pressure.

This more generally observed behavior of gases was interpreted in terms of structural changes in the medium; such effects are also considered to be fundamentally important for non-covalent interactions in solvent effects and will be more differentiated. The concentration c_k(I-II) for the transition from region I to region II of about 3.1 mol·L^{-1} corresponds to an edge length of 8.1 Å of a cube for one molecule, with mean intermolecular distance of 4.5 Å because of the statistical molecular movement [2]. This means about twice the molecular length between the nuclei of oxygen of 2.3 Å [10]. The concentration c_k(II-III) for the transition from region II to region III of about 10 mol·L^{-1} corresponds cube lengths of 5.5 Å and a mean intermolecular distance of 3.0 Å, not far away from the molecular dimensions and the intermolecular distance of 2.76 Å between the oxygen atoms in solid CO_2. There is a minimal restriction of the molecular motion in the region I, with a comparably abrupt change to permanent molecular contacts in region II. Finally, when c_k(II-III) is exceeded, the molecules are essentially densely packed, with further increases in concentration and density, respectively, is restricted by the prevailing effects of molecular compressibility, causing an exponential increase in pressure. The formation of the liquid phase is observed in region II. This may be due to limited miscibility of densely packed molecular aggregates with loosely packed assemblies. The surface tension between such structures is responsible for the formation of two phases, becomes smaller with an increasing temperature, and with lowering the range of two phases, and finally vanishes at the critical point [11] at 31.0 °C for carbon dioxide.

Further information can be obtained from the temperature dependence of the parameters of Equation (5). The values of E and c^* in region I increase monotonically and smoothly with temperature T (see Figure 3), while in region II a slight increase is observed at lower temperatures, rising rapidly between 70 and 120 °C and increasing again slightly when this range is exceeded (this behavior is clearly observed, even being appreciable above the critical point of 31.0 °C). The influence of the temperature is interpreted in terms of region I essentially being dominated by isolated molecules, whereas region II has a complex dynamic arrangement of isolated molecules and aggregates. As a result, two phases formed at low temperatures due to their limited miscibility which at higher temperatures cause the rapid change in parameters shown in the inset of Figure 3, for which thermally induced dissociation of such aggregates is held responsible; the latter resembles the melting of clusters. Such assemblies are important for solvent effects and will be further discussed below.

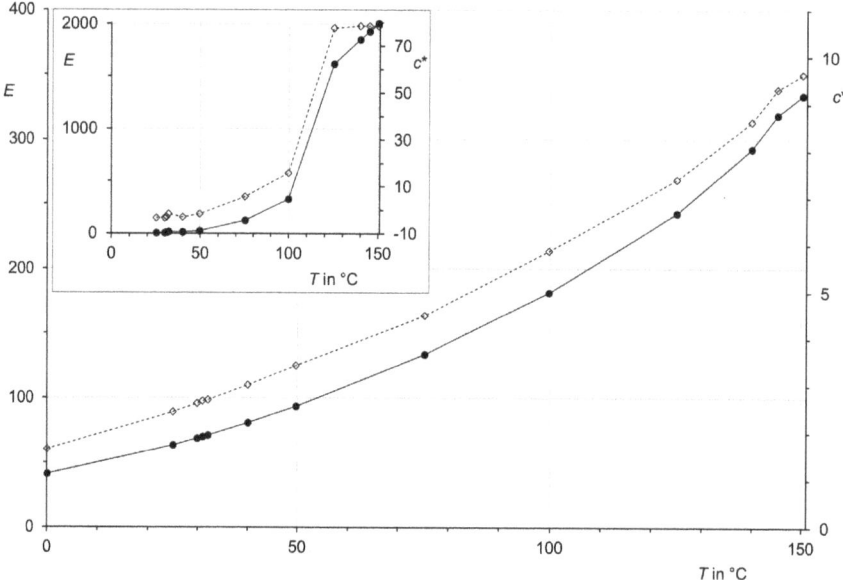

Figure 3. Temperature dependence (T) of the parameters E (solid curve and filled circles) and c^* (dashed curves and open diamonds) of Equation (5) for CO_2 and region I. Inset: parameters for region II.

3. General Solvent Effects

Condensed phases are widely used as media in chemistry and technology. Liquid phases are not only obviously used as solvents for an easier and more efficient handling of dissolved solids and to disperse them until the molecular level, but are also capable of interacting with substrates and imparting new properties through solvation. For example, the handling of the ethyl anion is very difficult in the gas phase because of the strongly exothermic tendency to lose one electron to form the ethyl radical [12,13]; on the other hand, a complexed and solvated ethyl anion such as the ethyl Grignard reagent [14] in ether is fully stable and applied in routine preparative chemistry. Moreover, an appreciable influence of solvents on the rate of chemical reactions was found, documented early on by Berthelot [15] in esterification reactions. More extended investigations [16] were made by Menschutkin, who studied the quarternization [17,18] of triethylamine [19]; he not only found a considerable influence of the solvent on the reaction rate, but also noted strong effects due to the ions involved. The polarity of a medium seemed to be of central importance. Menschutkin's system was not ideal for the general study of solvent effects on chemical reaction rates because the effects involve both the electrophilic substrate for nucleophilic substitution reactions and the nucleophile itself; the separation of individual effects becomes difficult. Clausius Mossotti [20] and Kirkwood [21,22] took a physical approach to solvent effects using the relative dielectric permittivity (dielectric constant ε_r) as the fundamental physical property of a medium in which the solute occupies a spherical or more ellipsoidal volume in the homogenous solvent. The local electric field of the solute causes a polarization of the medium with pronounced effects of charged particles. Debye [23,24] described two extrema of polarization, the orientation polarization where the positions of atomic nuclei are altered, and the shift polarization when charge separation is induced by displacement of electrons only. Orientation polarization dominates in media consisting of sufficiently large dipoles and can be determined using the static permittivity or at low frequencies compared with the correlation time of molecular orientation; the Clausius–Mossotti function (Figure 4) was derived (solute occupies a spherical volume) to describe the effect proportional to $(\varepsilon_r - 1)/(\varepsilon_r + 2)$. Higher frequencies as in the optical range are useful for determining shift polarization because the comparably heavy atomic nuclei can no more follow an external field; the index of refraction is a measure of this because it is quadratic with permittivity at such high frequencies [25]. However, the index of refraction is a complex number where the imaginary part concerns the optical absorption of the material. In general, solvents are colorless and thus transparent in the visible range, with absorption bands only in the UV and NIR causing anomalous dispersion there. As a consequence, the center of the visible range is far away from such interference and particular suitable for obtaining the real part of the index of refraction. The sodium D emission line at 589 nm was a versatile light source for the determination of the refractive index and proved to be a good choice because it is located approximately in the middle of the visible range, far from the absorption bands mentioned above. The polarization polarity is calculated from the index of refraction n at the wavelengths of the D line by the Lorentz–Lorenz function [25] $(n^2 - 1)/(n^2 + 2)$.

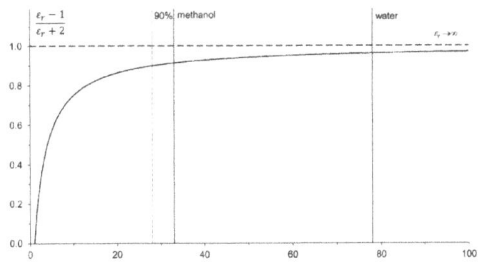
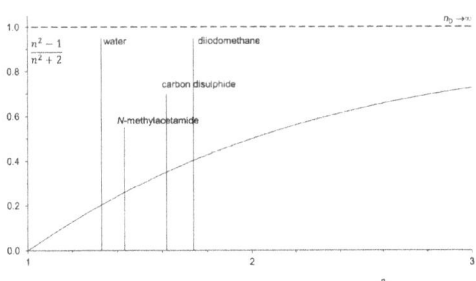

Figure 4. (**Left**) Kirkwood function (vertical axis) as a function of the relative dielectric permittivity ε_r. The dashed vertical line corresponds to 90% of the maximal value of the Kirkwood function [22]. The ε_r values of methanol (33) and water (78) are indicated by vertical lines. (**Right**) The index of refraction n_D as a measure of the polarizability of solvents. The region of n_D of solvents essential for preparative chemistry extends between the values for water (1.33) and diiodomethane (1.74). The values of the highly polarizable carbon disulfide (1.62) as versatile solvent and N-methylacetamide (1.43) with high ε_r are marked by vertical lines.

According to the Debye function, solvent polarity by orientation (named dipolarity for short) and by displacement polarization (named polarizability for short) were discussed individually. The use of the relative dielectric permittivity (dielectric constant ε_r; see Figure 4, left) to describe the dipolarity of solvents [26] is not consistent with general chemical experience because of the rapid saturation at which 90% of the effect is achieved for $\varepsilon_r = 28$ (this is even worse in models for ellipsoidal occupancies). Moreover, the appreciable and chemically important difference in chemical solvent polarity between methanol and water is barely reflected in the model, and solvents such as N-methylformamide ($\varepsilon_r = 182$) and N-methylacetamide ($\varepsilon_r = 191$) would be incorrectly estimated to be even appreciably more polar than water ($\varepsilon_r = 78$). The Lorentz–Lorenz [27] function [28] in Figure 4, right, indicates a comparably narrow range of n_D values relevant for solvents most commonly used in preparative chemistry, between the lower limit given by polar water ($n_D = 1.33$) and the highly polarizable carbon disulfide ($n_D = 1.62$). The upper limit is given by the rarely used solvent diiodomethane ($n_D = 1.74$). Again, water is estimated to be less polar than N-methylacetamide ($n_D = 1.43$). Linear combinations between functions of ε_r and n_D did not yield substantial improvement [29–33]. Finally, a theory of dipole swarms was developed [34] but is limited to minor polar solvents. The mentioned discrepancies may be caused by the fact that a macroscopic volume property of a medium is characterization by permittivity, while chemical reactions and solvation are determined by microscopic, molecular effects.

1   **2**

$$lg\ (k/k_{80E}) = m \cdot Y \qquad (9)$$

Significant progress was made by using the solvolysis of tert-butyl chloride, compound **1**, as a molecular polarity probe by Grunwald, Winstein, and Jones [35–37]. The logarithm of the rate constant k of the unimolecular ionization (S_N1 reaction) of an alkyl halide over the rate constant k_{80E} of the reference reaction of solvolysis of tert-butylchloride, compound **1**, in 80% ethanol/water (80E) was set to the product of m as the reaction-characterizing parameter and Y as the solvent-characterizing parameter in Equation (9). This linear free energy relationship (LFE) [38] is equivalent to Hammett's equation [39] for substituent effects. By definition, m becomes one for the solvolysis of tert-butylchloride (**1**) and Y becomes zero for the reference 80% ethanol/water (80E). The Y values for solvent polarity correspond well to chemical experience, but the polarity scale is not universal being limited to sufficiently strong ionizing solvents such as alcohols and mixtures of alcohols and water. Substrates more complex than **1** can lead to deviations [40].

$$Z = 28{,}591\ [\text{nm} \cdot \text{kcal} \cdot \text{mol}^{-1}]/\lambda_{max} \qquad (10)$$

Kosower introduced [41] the more universal solvatochromic polarity probe **2** in which ionization to generate charge was replaced by the reverse process, the loss of charge by an optically induced charge transfer (see Figure 5). Solvation affects the polar ground state of salt **2** and lowers its energy, while the electrically neutral electronically excited state is only slightly affected by solvent effects, due to the lost charge. The molar energy of electronic excitation of the maximum (λ_{max}) of the CT band of **2** is obtained by Equation (10) and was defined as the Z polarity scale; the dimension is not given in SI units because of the more convenient kcal/mol at the period of publications and is also retained here for polarity scales to avoid confusion (the energy has to be multiplied by 4.184 kJ·kcal^{-1} to obtain SI units).

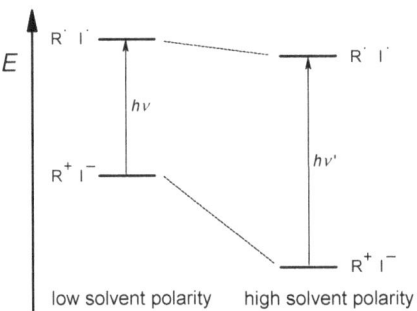

Figure 5. Schematic representation of the solvent effect on the energetic position (E) of dye **2** in the electronic ground and excited states: The energy difference is related to the wavelengths λ of the absorbed light in the CT transition $\Delta E = h\nu = h \cdot c/\lambda$.

The Z scale exhibits a linear correlation with the Y scale for various solvents (see inset of Figure 6). This LFE indicates that the same effect is described. The Z-scale brought about an appreciable progress because it is not limited to strongly ionizing solvents (see open circles in Figure 6), but can be extended to solvents with lower polarity (see filled circles in Figure 6). However, polarity probe **2** is still not optimal because (i) the solubility in low-polarity solvents such as hydrocarbons is very low and impedes experimental work, (ii) the CT band is in the hypsochromic visible region and is shifted to the next absorption band for highly polar solvents so that spectral overlap interferes, (iii) the molar absorptivity of **1** is comparably low and requires appreciable concentrations for UV/Vis measurements with standard equipment, and this addition may alter the polarity of the solvent since it is a polar salt.

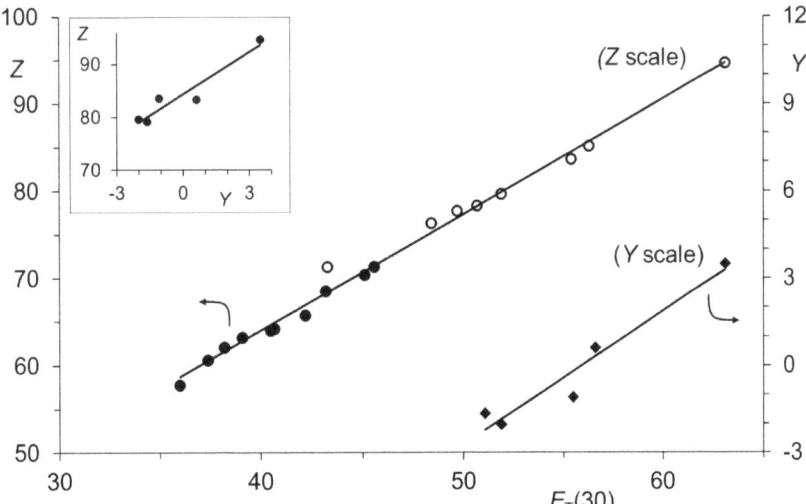

Figure 6. Linear correlation (diamonds) for solvent effects (LFE) between the Y scale (*t*-BuCl **1**) and the $E_T(30)$ scale (compound **3**, right ordinate, correlation number 0.97 with 5 solvents, coefficient of determination 0.94, standard deviation 0.66, slope 0.46, intercept −26; solvents from right to left:

H_2O, $HCONH_2$, CH_3CO_2H, CH_3OH, C_2H_5OH). Linear correlation for the Z scale (compound **2**) and the $E_T(30)$ scale (compound **3**, left ordinate, correlation number 0.995 for 18 solvents, coefficient of determination 0.991, standard deviation 1.0, slope 1.33, intercept 11.0; solvents from right to left: (hydrogen-bonding solvents: hollow circles) H_2O, $HOCH_2CH_2OH$, CH_3OH, C_2H_5OH, 2-C_3H_7OH, n-C_4H_9-1-OH, 1-C_3H_7OH; (non-hydrogen-bonding solvents: Filled circles) CH_3CN, DMSO, t-C_4H_9OH, DMF, $(CH_3)_2CO$, CH_2Cl_2, pyridine, $CHCl_3$, $CH_3OCH_2CH_2OCH_3$, THF, 1,4-dioxane. *Inset:* Linear correlation between the Y scale (t-C_4H_9Cl **1**) and the Z scale (compound **2**, correlation number 0.96 for 5 solvents, coefficient of determination 0.91, standard deviation 2.1, slope 2.6, intercept 84; solvents from right to left: H_2O, $HCONH_2$, CH_3CO_2H, CH_3OH, C_2H_5OH).

4. Pyridinium Phenolate Betaines

Covalent bonding of a negatively charged structural element to the positively charged pyridinium structure in **3** to form a zwitterionic betaine enabled an intramolecular light-induced charge transfer. As a result, substantial improvement was achieved not only by increasing the molar absorptivity, but also by further extending the polarity scales to low-polarity media due to the application of an electrically neutral polarity probe. The dipole moment of a structure with a single bond of an iminium anion to the pyridinium nitrogen atom was still small and resulted in moderate solvatochromism [42] (see further discussion in Refs. [43,44]), whereas with a 4-oxyphenyl anion [45] as the N-substituent, the charges were further separated, inducing strong solvatochromism [46]. Dimroth, Reichardt, and coworkers investigated [47] various N-4-oxyaryl derivatives for solvatochromism and found optimal properties of the betaine 30 (**3**, B30, CAS RN 10081-39-7), which provided a significant advance.

3   **4**

Dye **3** is remarkably strongly solvatochromic, so that solutions are yellow in water, red in methanol, violet in ethanol, blue in 1-butanol, green in acetone, and absorb in 1,4-dioxane in the NIR. The molar energy of light absorption at the maximum is called $E_T(30)$ value of a solvent (transfer energy of betaine No. 30 in the first publication) is calculated analogously to Z in Equation (11) and are meanwhile determined for almost 400 solvents [48].

$$E_T(30) = 28{,}591 \; [\text{nm·kcal·mol}^{-1}]/\lambda_{max} \tag{11}$$

The $E_T(30)$ values correlate linearly with the Y and Z values for various solvents (see Figure 6) and thus describe the same dipolarity effect of solvents. The linear correlation with Z values includes lipophilic solvents (filled circles) and polar protic solvents (open circles). The correlation with the Z values is of special importance for solvent effects because hydrogen bonds, as strong non-covalent interactions, are unimportant for the ion pair **2** (HI's remarkably low boiling point of −35.4 °C, in spite of its high molecular weight, can be seen as an indicator of the absence of hydrogen bonds to I (see also the related discussion in ref. [49])). As a consequence, hydrogen bonds for **3** are estimated to be insignificant. Apparently, the five peripheral phenyl groups in **3** shield the betaine to such an extent that specific solvent effects play a minor role, but still allow the exceptionally strong solvatochromism. Dye **3** is a highly sensitive optical probe for other solvent effects as will be discussed later.

The *tert*-butyl derivative **4** (No. 26 in the first publication [47]) also exhibits very pronounced solvatochromism ($E_T(26)$ values); however, the phenolate oxygen atom does not appear to be as efficiently shielded as in **3**. A good linear correlation is obtained between $E_T(30)$ and $E_T(26)$ values (performance of target to actual comparison) for lipophilic solvents, shown as filled circles and the solid linear regression line in Figure 7. For hydrogen-bonding solvents (open circles and the dashed linear regression line), deviations from this line are observed forming a second linear regression. Obviously, the effect of hydrogen bonding causes a much stronger influence of solvents on the solvatochromic chromophore. The scattering of points within this second correlation is significantly larger than within the first, indicating a stronger contribution of specific solvent effects; such effects were already referred in the first reference to **4** [47].

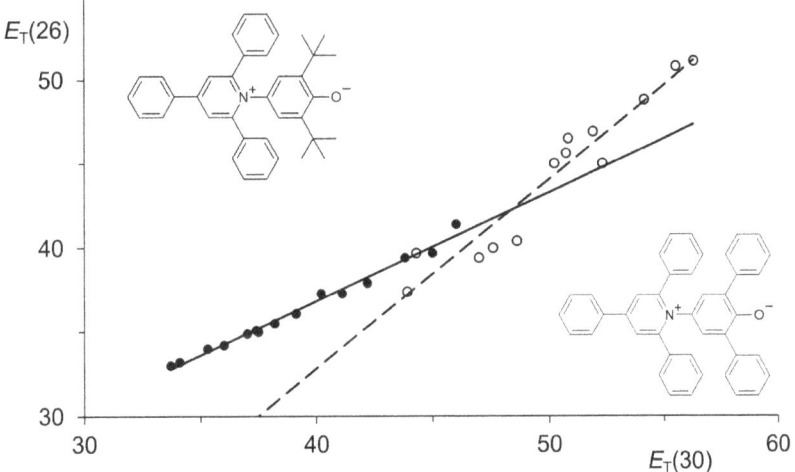

Figure 7. Linear correlation between the $E_T(30)$ (dye **3**) and $E_T(26)$ (dye **4**) values for pure solvents from Ref. [47] the standard deviations are indicated. Filled circles and solid line: non-hydrogen-bonding solvents; slope: 0.64; intercept: 11.5 correlation number 0.993 for 15 solvents, standard deviation 0.30; coefficient of determination 0.987. Open circles and dashed line: Hydrogen-bonding solvents; slope: 1.1; intercept: −12.1; correlation number 0.960 for 13 solvents; standard deviation 1.31; coefficient of determination 0.924. Solvents with increasing $E_T(30)$ values: $C_6H_5CH_3$, C_6H_6, $C_6H_5OC_6H_5$, 1,4-dioxane, 2,6-lutidine, $(CH_2)_4O$, C_6H_5Cl, $CH_3OCH_2CH_2OCH_3$, $CHCl_3$, C_5H_5N, CH_2Cl_2, CH_3COCH_3, $HCON((CH_3)_2$, $(CH_3)_3COH$, $C_6H_5NH_2$, CH_3SOCH_3, CH_3CN, $(CH_3)_2CHCH_2CH_2OH$, 2,6-$(CH_3)_2C_6H_3OH$, $CH_3CHOHCH_3$, $CH_3CH_2CH_2CH_2OH$, $CH_3CH_2CH_2OH$, $C_6H_5CH_2OH$, CH_3CH_2OH, $HOCH_2CH_2OCH_3$, $HCONHCH_3$, CH_3OH, $HOCH_2CH_2OH$.

The solvatochromism of a derivative of **4** without peripheral phenyl groups was studied by Sander and co-workers [50] and showed a linear correlation with E_T^N and $E_T(30)$ values, respectively, for both hydrogen-bonding and non-hydrogen-bonding solvents. This indicates the dominance of dipolarity concerning solvatochromism such as for **3**. The measurements of UV/Vis absorption spectra at low temperature in matrices of solid noble gases at 3 K and slightly higher temperatures such as 25 K gave spectra similar to those expected in the gas phase. The targeted addition of a small amount of water allowed the investigation of its influence and the formation of hydrogen bonds at low temperatures.

5   **6**

The limitations in the application of **3** could be overcome by slightly adapted derivatives. Dye **3** could not applied for acidic solutions such as acetic acid because of the protonation of the basic phenolate anion; Kessler and Wolfbeis [51] developed dye **5** in which the electron depletion by the chlorine atoms lowers the basicity sufficiently to study acidic media, while the solvatochromic behavior is similar to **3** so that extrapolations by linear correlation are possible. Wolfbeis named the polarity values obtained with **5** $E_T(33)$ in continuation of the initial publication by Dimroth, Reichardt, and co-workers. The measurements of highly polar hydrogen bond-forming solvents and solution of salts becomes problematic with **3** because of its low solubility and tendency to aggregate [52]. Reichardt developed dye **6** with the more hydrophilic structure elements of pyridinyl [53] in the periphery of **3**; the *m*-positions of the phenolic aryl substituents in **6** appear to cause minimal interference with solvatochromism, presumably because of the non-conjugation of the phenolate oxygen atom. As a result, the $E_T(30)$ values of many polar and electrolyte-containing solvents can be estimated by extrapolation using **6**.

7

The measurement of the $E_T(30)$ values of media with very low dipolarity such as aliphatic hydrocarbons or tetramethylsilane is restricted due to the minimal solubility of the dye. The solubility-increasing effect of *tert*-butyl groups in lipophilic media [54] was applied to dye **7**, whereby a substitution in the periphery allowed an increase in solubility in such media without significantly affecting solvatochromism so that the corresponding $E_T(30)$ values could be obtained by extrapolation by means of a linear free energy relationship [55].

Finally, the experimental $E_T(30)$ values were scaled for easier comparability, first according to Reichardt as a relative polarity measure (*RPM*) [56] where the value for *n*-hexane was set to 0.000 and $RPM = [E_T(30)_{n\text{-hexane}} - E_T(30)_{\text{solvent}}]/E_T(30)_{n\text{-hexane}}$. Next, the scaled E_T^N values were defined [57] where the value for tetramethylsilane was set to 0 and for water set to 1, such that $E_T^N = [E_T(30)_{solvent} - E_T(30)_{tetrametylsilane}]/[E_T(30)_{water} - E_T(30)_{tetramethylsilane}]$. The experimental $E_T(30)$ values were retained here, although scaling is attractive, because all E_T^N values depend on only two references and their accuracy; if their values might be altered, such as by more precise measurements, the entire set of values would change. Moreover, the value for tetramethylsilane is less reliable because it was obtained by extrapolation, and the value for water is at the limit of direct measurements because of the low solubility of **3**. On the other hand, each experimental $E_T(30)$ value has obtained its individual accuracy, which is determined only by the particular experimental procedure, and is considered to be more reliable.

The molar energy of excitation of dye **3** in various solvents as $E_T(30)$ values was taken from the maximum of the solvatochromic band since an efficient and precise measurement is possible. On the other hand, the band can be split [58] into more than a single Gaussian [59] band, and the point of gravity may be of greater physical significance than the maximum because it represents the thermal energy of chromophores. However, accurate determination is more difficult and requires a clear cut between the most bathochromic band and other partially overlapping bands of higher electronic energy, leading to uncertainties. As a consequence, the molar energy of excitation is retained for the following discussions, while trying to avoid over-interpretation.

5. Polarizability: Further Solvent Effects

The concept of estimating solvent polarity on the basis of solvatochromism was developed by Brooker [60] and co-workers based on the study of a large number of merocyanines in early work. He found essentially two branches of solvent effects [61] that appear to be orthogonal to each other, causing blue and red shifts in the absorption spectra as the polarity of the solvent is increased.

The most pronounced blue shift was observed for **8** (RN 3210-95-5) where the polarity scale based on the molar energy of electronic excitation was named χ_B, while a pronounced red shift was found for **9** (RN 2913-22-6), giving the polarity scale χ_R. A linear correlation between the χ_B scale and the Z scale was found for several pure solvents, and there is also a linear correlation between χ_B scale and the $E_T(30)$ scale indicating that the same effect is described, essentially the dipolarity of solvents. Further study of **8** and **9** is hampered by their difficult synthesis; there is only one paper by Brooker (and a patent) referred to for **8** by the *Chem. Abstr.* and six papers for **9**. As a consequence, χ_B is replaced here by a more accessible and similar $E_T(30)$ scale and further discussion is focused to χ_R. The solvent effect in the χ_R scale correlates better [62] with a function of the refractive index of a medium than with dielectric permittivity. As a result, χ_R is mainly better attributed to the polarizability of a medium than its dipolarity. This was further confirmed by finding the missing link [63] between the two polarity scales: χ_B (or, respectively, the $E_T(30)$ scale) and χ_R.

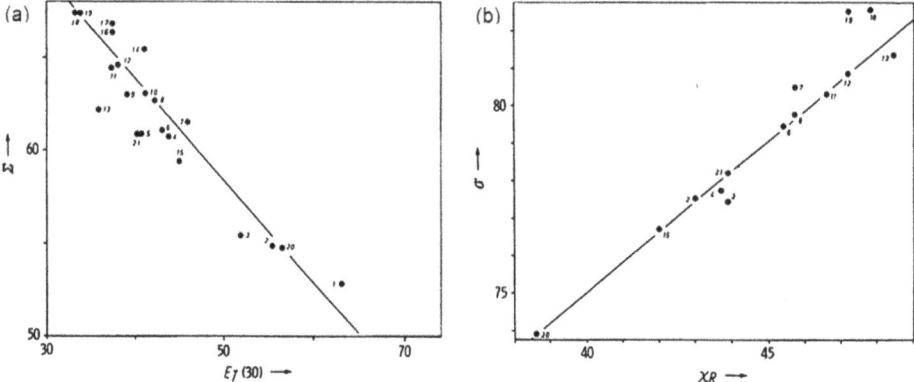

10

The 4-aminophthalimide **10** is solvatochromic in fluorescence and is similar to N-methyl-4-aminophthalimide, the fluorescent dye for Zelinski's [64] universal polarity scale S.

The molar energy of fluorescence of **10** was defined as the Σ scale for solvent polarity and correlates linearly with the $E_T(30)$ scale for pure solvents (see Figure 8a). The light absorption of **10** is also solvatochromic and was defined as the σ scale for solvent polarity and correlates linearly with the χ_R scale (see Figure 8b). As a consequence, all solvent polarity scales can be interrelated.

Figure 8. Linear correlation of polarity scales for the pure solvents water (1), methanol (2), ethanol (3), dimethylformamide (DMF, 4), dimethylphthalate (5), butyronitrile (6), acetonitrile (7), acetone (8), chloroform (9), diethylphthalate (10), tetrahydrofurane (THF, 11), ethylacetate (12), 1,4-dioxane (13), dichloromethane (14), dimethylsulfoxide (DMSO, 15), bromobenzene (16), chlorobenzene (17), m-xylene (18), toluene (19), formamide (20), and pyridine (21). (**a**) Molar energy of the fluorescence of **10** (Σ scale) versus the $E_T(30)$ scale; (**b**) molar energy of the optical excitation of **10** (σ scale) versus the χ_R scale. The larger scattering of points compared to previous plots is attributed to some contributions of specific solvent interactions.

The correlation in Figure 8b between σ and χ_R is interpreted in terms of the dynamics of light absorption of **10** with a low dipole moment electronic ground state and a much larger dipole moment in the electronically excited state (see Figure 9). The energy of the electronic ground state is only slightly affected by polar solvents. Light-induced electronic excitation increases the dipole moment of **10**; however, solvation by re-orientation of the solvent molecules cannot follow this rapid, essentially vertical process. As a result, the polarizability of the solvent remains important and characterizes the solvent effect on this transition; the same effect is important for χ_R and explains the linear correlation between these two polarity scales shown in Figure 8b. The fluorescent lifetime of the electronically excited state of **10** in the order of nanoseconds is long enough for relaxation due to the re-orientation of the solvent molecules, and this process is energetically strongly influenced by the dipolarity of the solvent; consequently, a linear correlation of the molar energy of fluorescence ($\sum$ scale) with the $E_T(30)$ scale can be explained because both scales are affected by the same process of solvation. In summary, for the $E_T(30)$ scale, solvation by dipolar re-orientation of the solvents is crucial, while the χ_R scale describes the polarizability of solvents.

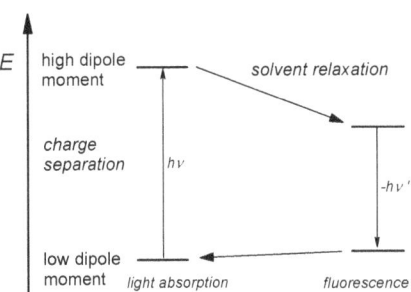

Figure 9. Schematic representation of the solvent effect on the energetic position (E) of dye **10** in the electronic ground and excited state ($\Delta E = h\nu = h \cdot c/\lambda$). The electronic excitation of **10** causes a charge separation, with subsequent solvent relaxation affecting the wavelength of fluorescence.

Strong solvatochromic effects of other similar merocyanines were reported [65].

The high dipolarity of water (solvent 1) and aliphatic alcohols is manifested in Figure 8a and corresponds to chemical experience, while the polarizability of these solvents shown in Figure 8b is only in the middle range and is clearly exceeded by solvents such as formamide (20) and DMSO (15).

6. Various Solvent Effects: Multi-Parameter Equations

Two orthogonal solvent properties, the dipolarity and polarizability, have been described above and are the microscopic, molecular counterpart to Debye's equation [23,24] of the linearly independent macroscopic orientation and shift polarization. This may stimulate the search [66] for further types of microscopic solvent effects that are linear independent from each other and thus should describe orthogonal molecular solvent properties. The hydrogen-bonding ability of solvents has already been discussed here with dye **4**. Other solvent properties such as their Lewis acidity and basicity have been the subject of polarity scales such as Gutmann's [67,68] donor and acceptor numbers, hydrogen donor and acceptor numbers, and more [69]. Different and specialized solvent scales are appropriate for different problems. Consequently, the characterization of solvents with a universal parameter set would be attractive. Various approaches were described in the literature, such as those by Kamlet, Taft and Abboud [70], Catalán et al. [71,72], Koppel and Palm [73], and Vitha et al. [74], two examples of which are detailed here.

$$XYZ = XYZ_0 + s\pi^* + a\alpha + b\beta \tag{12}$$

Kamlet, Taft, and Abboud developed the solvatochromic comparison method as multi-parameter fit in which solvent-dependent properties or processes denoted XYZ in Equation (12), such as the $E_T(30)$ scale, depend on solvents for a linear set of parameter products; XYZ_0 is the intercept of such correlations, and s, a, and b are parameters to be individual to the solvent-dependent system. The parameters π^*, α, and β describe characteristic properties of the respective solvent and are tabulated.

The multi-parameter Equation (12) was applied to the $E_T(30)$ and Z polarity scales where the optimal parameters s, a, and b for a linear correlation were calculated by the application of the least squares method (see Figure 10). Acceptable linear correlations were found; the interpretation of the parameter *a* as a hydrogen-bond-donating property remains problematic, in particular for the Z scale. The value is comparably high, although no hydrogen bonds to I$^-$ of **2** are expected.

$$A = A_0 + b\text{SA} + c\text{SB} + d\text{SP} + e\text{SdP} \tag{13}$$

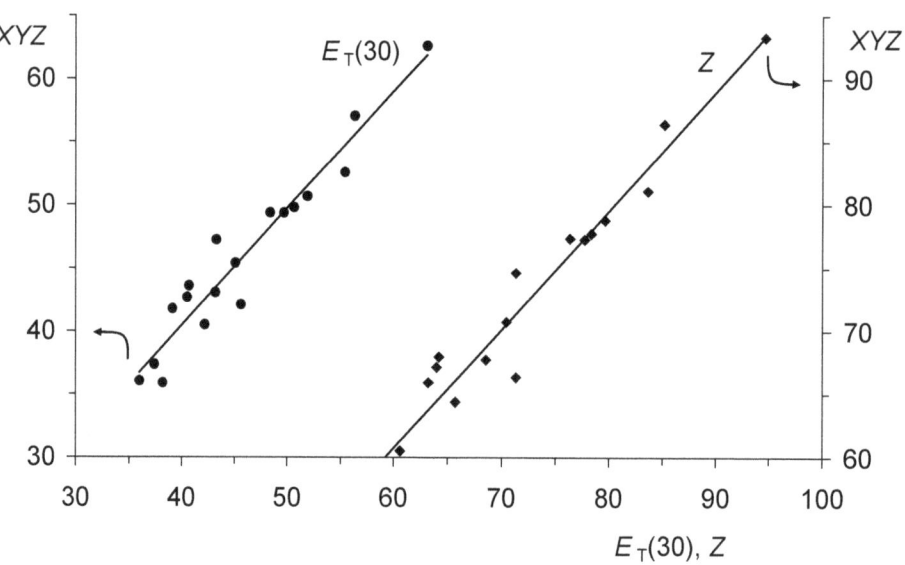

Figure 10. Linear correlations between the $E_T(30)$ and the Z scale, respectively, as the abscissa and the adjusted XYZ scale of Kamlet and Taft as ordinate. Slope 0.927 for $E_T(30)$ has a correlation number of 0.96 (18 measurements; circles), a standard deviation of 2.0, and a coefficient of determination of 0.93, where the parameters are $s = 16.9$, $a = 15.6$, $b = 4.5$, and $XYZ_0 = 25.1$. Slope 0.943 for Z has a correlation number of 0.97 (18 measurements; diamonds), a standard deviation of 2.4, and a measure of determination of 0.94, where the parameters are $s = 20.8$, $a = 21.2$, $b = 7.0$, and $XYZ_0 = 44.6$.

Catalán followed a similar multi-parameter approach with Equation (13), where A describes a solvent-dependent process; *b*, *c*, *d*, and *e* are the process-dependent parameters and A_0 is the intercept of a linear correlation, while the tabulated parameters SdP (dipolarity), SP (polarizability), SA (acidity), and SB (basicity) characterize the medium.

The $E_T(30)$ and Z scale were approached by the multi-parameter Equation (13), where the scale-dependent parameter *b*, *c*, *d*, and *e* were determined by the least squares method for 18 solvents (see Figure 11). The correlations were slightly better than in Figure 10; however, an additional parameter was required.

Several improvements to the multi-parameter concept have been developed [75], such as those by Spange [75] and co-workers. Hunter and co-workers [76,77] investigated competing hydrogen-bonding systems as polarity probes. Progress in correlations has been made for individual descriptions of groups of selected solvents; a universal and precise description of solvent effects is still lacking [78]. This may be caused by the interference of residual specific solvent effects with the polarity probes used. Furthermore, the terms of the applied multi-parameter fits might be not fully linearly independent so that cross terms become significant; this may become increasingly important for stronger anisotropic, more ellipsoidal-like solvent molecules where stronger solvent patterning is induced (this might result in liquid crystals) (see also below). As a result, empirical solvent polarity scales and their extension to multi-parameter equations are very useful for practical applications; however, an over-interpretation should be avoided because of their limitations. The size of solvent shells may be of further influence and will be the subject of the next chapter.

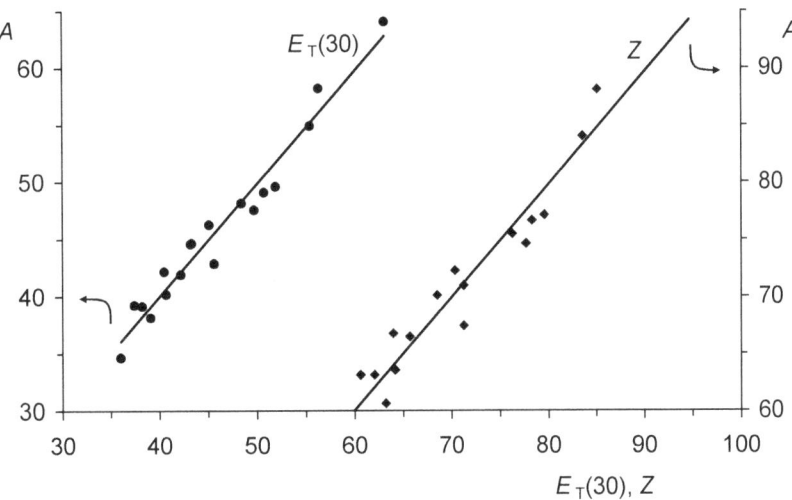

Figure 11. Linear correlations between the $E_T(30)$ and the Z scale, respectively, as the abscissa and the fitted Catalán's A scale as the ordinate. Slope 0.985 for $E_T(30)$ with a correlation number of 0.98 (18 measurements, circles), a standard deviation of 1.6 and a coefficient of determination of 0.96, with parameters are $b = 21.4$, $c = 4.27$, $d = 5.05$, $e = 12.7$, and $A_0 = 25.1$. Slope 0.984 for Z with a correlation number of 0.98 (18 measurements. diamonds), a standard deviation of 2.1 and a measure of determination of 0.96 where the parameters are $b = 28.6$, $c = 7.40$, $d = 5.08$, $e = 16.6$, and $A_0 = 44.6$.

7. Solvent Shell

Solvated molecules are surrounded by directly molecular interacting solvent molecules; the extension of such interactions into the solvent is of special interest for solvent effects. This topic was investigated using the interaction of strong dipoles with various polar solvent molecules. Betaine B30 (**3**), the solvatochromic probe of the $E_T(30)$ scale, exhibits a comparably large dipole moment of about 12 D [79–82] with a negative charge on the oxygen atom extending into the pyridinium ring with a positive charge. The interaction of the solvent molecules with this dipole moment causes the solvatochromism of **3**. The large and extended electric dipole of **3** generates an electric field that extends into the solvent and can be used as a probe to estimate the extension of a significant solvent shell. For such an assay, two chromophores of **3** were combined [58] *anti*-colinearly to form the dyad **11**.

The *anti*-colinearly dipoles in **11** compensate for large distances because no net dipole moment remains so that solvent effects become weak, while at small distances, the effects are still individual and strong so that the solvent effects of each chromophore remain unaffected by the attached second chromophore. The molar energy of excitation of **11** is calculated analogously to **3** and denoted $E_T(30\text{dyad})$.

The solvent effect on the UV/Vis spectra of **3** and **11** is very similar, as is indicated by the linear correlation between $E_T(30)$ and $E_T(30\text{dyad})$ for various solvents (even benzyl alcohol, although an exception, in many cases, is acceptably included) (see Figure 12). The slope of such a correlation can be taken as a measure of the compensation of the dipoles in **11** where a slope of zero should be obtained for complete compensation and a slope of 1 for the absence of any compensation. The slope found experimentally is close to unity, indicating that there is no compensation of the *anti*-collinear dipoles; consequently, the significant solvent shell must be very thin, being not much more than one molecular layer of solvent thick. A possible artifact due to residual conjugation in **11** between the two chromophores could be excluded by using an aliphatic cage in **12** as an isolator; however, **12** gave the same result as **11** [58].

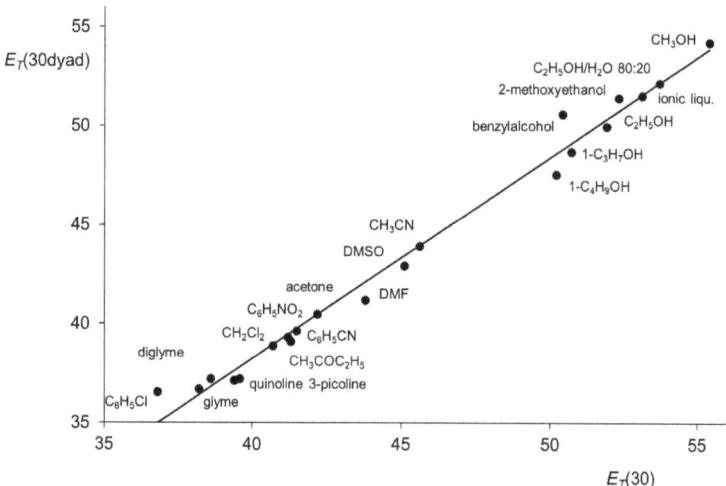

13

Finally, the very rigid diamantane was used as a spacer in **13**. $E_T(30\text{diam})$ as the molar excitation energy of **13** was calculated analogously to **3**.

Figure 12. Linear correlation between the $E_T(30)$ values of betaine **3** and $E_T(30\text{dyad})$ values of dyad **11** for various solvents [58] (ionic liqu.: 1-butyl-3-methylimidazolium-tetrafluoroborate). Slope 1.02, intercept −2.43, correlation number 0.993 for 20 points, standard deviation 0.7, coefficient of determination 0.987.

The linear correlation between $E_T(30)$ and $E_T(30\text{diam})$ for various pure solvents is shown in Figure 13; the slope of this correlation of close to unity indicates identical solvent effects in the $E_T(30)$ and $E_T(30\text{diam})$, thus ruling out significant compensation of the dipoles in **13** with respect to the solvent shell.

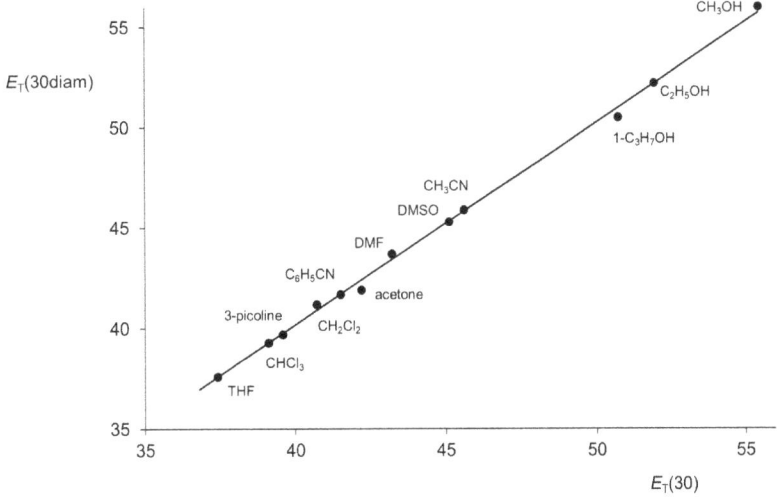

Figure 13. Linear correlation between $E_T(30)$ values of betaine **3** and the $E_T(30\text{diam})$ values of dyad **13** for various solvents [58]. Slope 1.008, intercept −1.33, correlation number 0.9989 for 12 points, standard deviation 0.3, coefficient of determination 0.998.

As a consequence, the solvatochromism of the dyads (**11**, **12** and **13**) based on phenolatebetaine **3** provides no evidence for compensation effects due to *anti*-colinear dipoles and indicates a very thin solvent shell, not much more than one molecular layer thick. The surprisingly thin solvent shell finds its counterpart in the thin surface of liquids such as the phase boundary of water and the gas phase, where investigations [83] indicate an essentially monomolecular active layer, with the hydrogen atoms directed toward the interior of the liquid phase, while the oxygen atoms were found to be in contact with the gas phase. The extraordinarily thin active solvent shell may explain some unaccounted solvent effects such as problems with the relative permittivity (dielectric constant) as a measure of solvent polarity due to some appreciable exceeding of water (see above) although water exhibits a higher polarity according to the chemical experience: The relative permittivity is a three-dimensional property of the volume of a liquid as it is for the refractive index, whereas solvent effects involve two-dimensional molecular surface effects. These can be similar; however, they need not to be identical.

8. Liquid Mixtures

Liquid mixtures are of particular interest for chemists and physicist because the properties of one component can be continuously transformed into the other by changing the composition. Polar molecules in a solvent are surrounded by solvent molecules in diluted solutions and move free of forces in the solvent because the outer sphere corresponds to the solvent. Energy and properties are independent from the positions of the polar molecules such as molecules in the gas phase at low pressure. As a consequence, the same formalism described in Section 2 for gases can be applied to polar additives to solvents: The effect p of the polar component at low concentration is expected to be proportional to its concentration c ($dp = const_1 \cdot dc$) (see also Equation (1)). However, at higher concentrations, this component is expected to interact with itself reducing the effect ($dp = E/c \cdot dc$), where the constant E is a measure of the energetic interaction (see also Equation (2)). The next steps according to Section 2, combination and integration, result in Equation (5). Neither the nature of interaction of the polar component with the solvent nor the interaction with itself was specified. As a consequence, Equation (5) is to be experimentally verified for its application in binary liquid mixtures; p was set equal to $E_T(30)$ so that Equation (14) is obtained where c^* is the concentration of the polar component at which the interaction of the polar component becomes important and $E_T(30)_0$ is the $E_T(30)$ value of the pure component of lower polarity.

$$E_T(30) = E \cdot \ln\left(\frac{c}{c^*} + 1\right) + E_T(30)_o \qquad (14)$$

Further approaches [84,85] to the polarity of binary mixtures were reported in the literature.

The validity of Equation (14) for real systems was tested (performance of target to actual comparison) with the mixture N-tert-butylformamide/benzene for the $E_T(30)$ polarity scale; a *tert*-butyl group was attached to the formamide to increase the solubility [86] in the aromatic hydrocarbon benzene and to be able to set a broad polarity range. A strongly curved relationship was obtained between the $E_T(30)$ values and the concentration c of the more polar N-tert-butylformamide as is shown in the inset of Figure 14. The entire range of concentration c, spanning three decades, could be linearized by Equation (14) (see Figure 14). The strong curvature in the inset is a consequence of the comparably low c^* value of 0.0167 mol·L^{-1}.

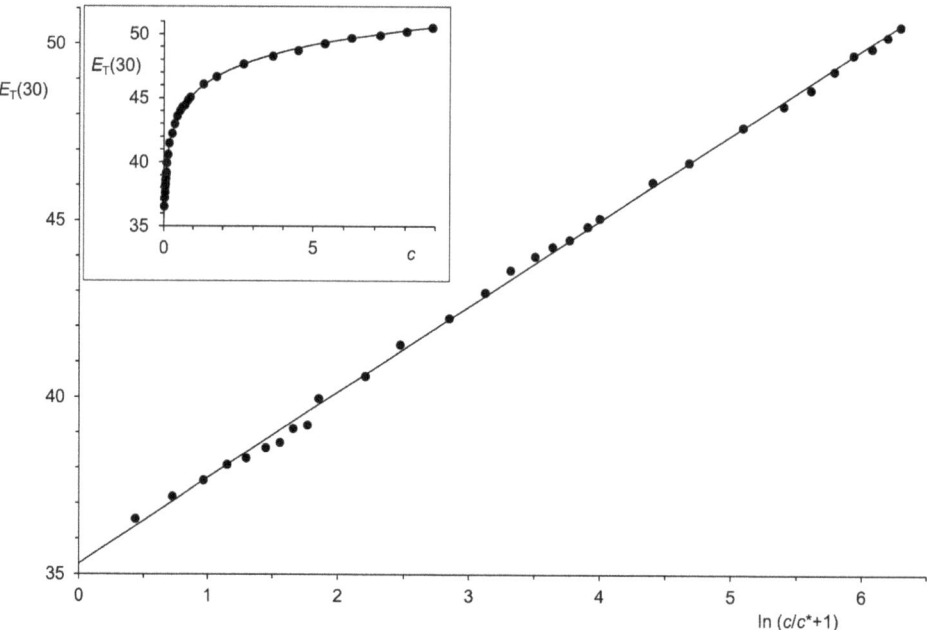

Figure 14. Linear correlation between the $E_T(30)$ values of the mixture N-*tert*-butylformamide/benzene [87] and ln (c/c^* +1) where c is the molar concentration of the former and $c^* = 0.0167$ mol·L^{-1}; slope E = 2.43 kcal·mol^{-1}, intercept $E_T(30)_0$ = 35.3 kcal·mol^{-1}, correlation number r = 0.9993, coefficient of determination r^2 = 0.999, standard deviation σ = 0.17 kcal·mol^{-1}. Inset: $E_T(30)$ values of the mixture N-*tert*-butylformamide/benzene as a function of the concentration c in mol·L^{-1} of the former: Measured $E_T(30)$ values (filled circles) and the calculated with Equation (14) (solid curve).

The description of real binary liquid systems using Equation (5) is not limited to the $E_T(30)$ polarity scale, but is more universal [87], as shown by its application to common polarity scales in Figure 15. The validity of Equation (5) has been demonstrated for more than 80 binary mixtures. On the other hand, Equation (5) and Equation (14), respectively, can be used to investigate special properties of liquid mixtures. This will be shown in the next chapter on hydrogen-bonding solvents.

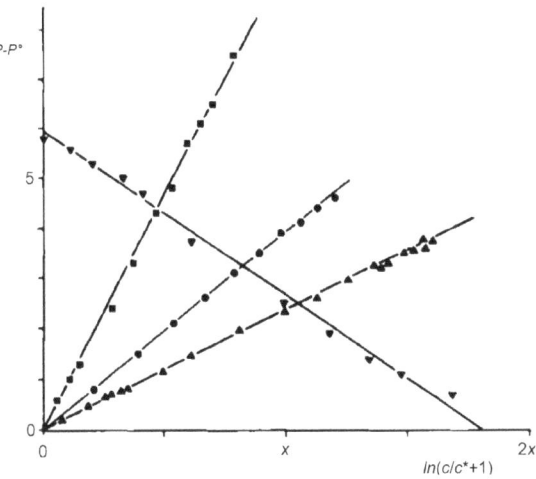

Figure 15. Linear relationship between P and $\ln(c/c^* + 1)$ for various polarity scales according to Equation (5). Diamonds: $P = E_T(30)$ (methanol/acetone); $x = 2$. Circles: $P = Y$ (water/methanol); $x = 1$. Squares: $P = Z$ (methanol/acetone); $x = 2$. Triangles: $P = \pi_1^*$ (ethanol/n-heptane); $x = 1$, ordinate: $P - P° + 5.9$.

9. Hydrogen-Bonding Solvents

Particular attention should be paid to hydrogen bond-forming solvents since hydrogen bonds are strong and directed non-covalent molecular interactions. Therefore, patterning of solvents can be performed by such bonds. On the other hand, the lifetime of hydrogen bonds at room temperature is very short and is in the order of picoseconds [88]. However, the bonds are re-formed in a similarly short period of time, so that molecular sticking structures in the liquid are highly dynamic and enable fast exchange processes. Water, as the most prominent strong hydrogen bond donor is studied here because its low molecular weight allows it to reach the high molar concentration of 55.4 mol·L^{-1}. Water is combined with 1,4-dioxane as a moderate and low-polar hydrogen bond acceptor, where complete miscibility is crucial.

The $E_T(30)$ values of water/1,4-dioxane mixtures as a function of the concentration c of water are shown in Figure 16. Their linear correlation between $\ln(c/c^* + 1)$ is obtained for concentrations c below 26 mol·L^{-1} as a critical concentration c_k (Figure 16, left line); this correlation corresponds to Figure 14. At higher concentrations, $c > 26$ mol·L^{-1} ($c > c_k$), a second steeper correlation results (the c^* values are low and similar so that the two lines can be combined in one diagram). This sudden change is already obvious in a simple logarithmic plot because of the low values of c^* in both correlations ($\ln(c/c^* + 1) \approx \ln c$ for $c >> c^*$); see also the inset top left of Figure 16. The change in slope proceeds in a narrow concentration range and is attributed to a change in solvent structure and corresponds to such a condensation-like phase transition as described for carbon dioxide in Section 2. The formation of the hydrogen-bonded structure requires a sufficiently high concentration to overcome the dissociation at lower concentrations. On the other hand, one can extrapolate the correlation on the left side to higher concentrations until the concentration of pure water (dashed line) and obtain an $E_T(30)$ value of 55.9. This value would be expected if water did not form a hydrogen bonding structure, and the difference from the actual value of 63.1 out of 7.2 units indicates the appreciable increase in solvent polarity due to the formation of the hydrogen-bonded structure.

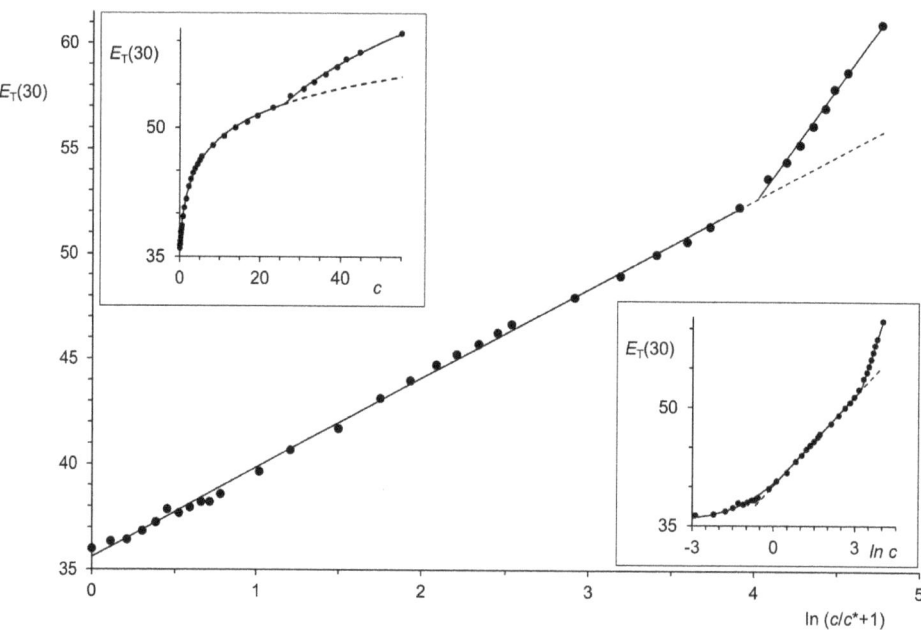

Figure 16. Linear correlation between $\ln(c/c^* + 1)$ and the $E_T(30)$ values of the mixture water/1,4-dioxane [87] ($c^* = 0.48$ mol·L^{-1}; measured values as filled circles; linear correlations as lines) as a function of the concentration c of water. Left correlation: $E = 4.27$, intercept 35.6, standard deviation 0.2, correlation number $r = 0.9992$ for 27 measurements, measure of determination $r^2 = 0.998$, extrapolation with the dashed line. Right correlation: $E = 11.1$, intercept 8.26, standard deviation 0.2, correlation number $r = 0.997$ for 8 measurements, measure of determination $r^2 = 0.993$. Inset top left: $E_T(30)$ values of the water/1,4-dioxane mixture as a function of the water concentration c, curves for calculated $E_T(30)$ values, dashed curve for extrapolation. Inset bottom right: $E_T(30)$ values of water/1,4-dioxane as a function of $\ln c$; The two linear correlations are already obvious in a simple logarithmic plot because of the low values of c^*.

The formation of double linear correlations according to Equation (5) is neither limited to the $E_T(30)$ polarity scale nor to the mixture water/1,4-dioxane, as can be seen in Figure 17 [89]. Remarkably, the critical concentrations c_k for the transition from one linear correlation to the second are independent of the applied polarity scale within the limits of experimental error and are therefore attributed to a property of water (the kink appears at different position in Figure 17 due to the different scales in the coordinate system). It can surmise that water forms clathrate-like [90] structures at sufficiently high concentrations; however, these structures are very dynamic, unlike the crystalline clathrates such as the hydrate of methane.

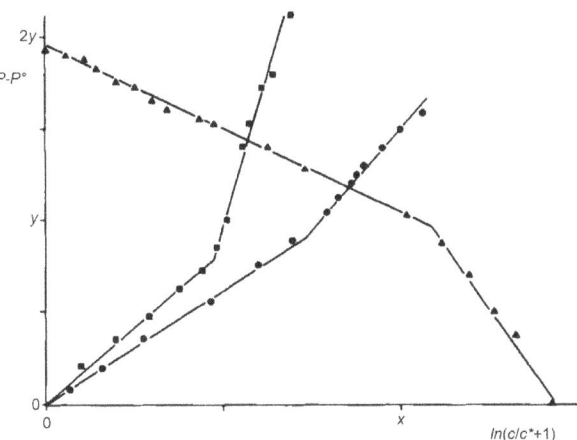

Figure 17. Double linear correlations according to Equation (5) for the water/ethanol mixture and different polarity scales. Filled circles: $P = Y$ ($x = 2, y = 1$); filled squares: $P = E_T(1)$ ($x = 1, y = 4$); filled triangles: $P = \pi_1^* - 7.7$ ($x = 1, y = 4$).

10. Elevation of Polarity

As shown in the last chapter, hydrogen bonds can significantly increase the polarity of the solvent. Consequently, a combination of a moderately polar hydrogen bond donor and a less polar but strong acceptor should increase the solvent polarity.

Ways to increase solvent polarity were tested [91] using the $E_T(30)$ polarity scale and the mixture 1-butanol as the more polar component and hydrogen bond donor and nitromethane as the less polar component and hydrogen bond acceptor. The $E_T(30)$ values of such mixtures as a function of the molar concentration c of 1-butanol pass through a maximum at $c_k = 3.3$ mol·L^{-1} and decrease again to reach the lower $E_T(30)$ value of pure 1-butanol (see Figure 18). The polarity at c_k exceeds the polarity of pure 1-butanol by about 1.6 $E_T(30)$ units. Such an excess of the solvent polarity is interpreted by the formation of a hydrogen-bridged association [91] between 1-butanol as the hydrogen bond donor and nitromethane as the hydrogen bond acceptor, such an associate being more polar than the individual components and means the polar component in Figure 18a bottom right (of course, the molar concentration of this associate is identical with the concentration of 1-butanol because of the 1:1 association). This means that all 1-butanol is captured to form this associate until not enough residual nitromethane is present with increasing concentrations of 1-butanol. Then, a mixture of this associate as the more polar component is formed with 1-butanol so that the polarity of the mixture decreases again because no further associate can be formed. The linear relation between $E_T(30)$ and c_u, the molar concentration of nitromethane, in Figure 18b top right, further supports this interpretation because, according to Equation (5), the polarity increases by the formation of the hydrogen-bonded associate in the medium 1-butanol until all nitromethane is captured in the hydrogen-bonded associate and decreases again to finally reach the lower polarity of pure 1-butanol.

Such mixtures of hydrogen bond donors and acceptors are of particular interest [92] because they offer the possibility to preparing highly polar solvents from components of lower polarity. On the other hand, such mixture exhibits many properties of polar alcohols due to the possibility of rapid intermolecular proton exchange. A completely novel type of polar protic media can be constructed by using chloroform [93] since this solvent exhibits the tendency to form hydrogen bonds to strong hydrogen bond acceptors such as the dipolar aprotic solvents; however, unlike the alcohols, chloroform does not exhibit the tendency to exchange protons under neutral conditions. As a result, comparably polar protic solvents could be constructed without protons exchange. This might be of interest for various reasons.

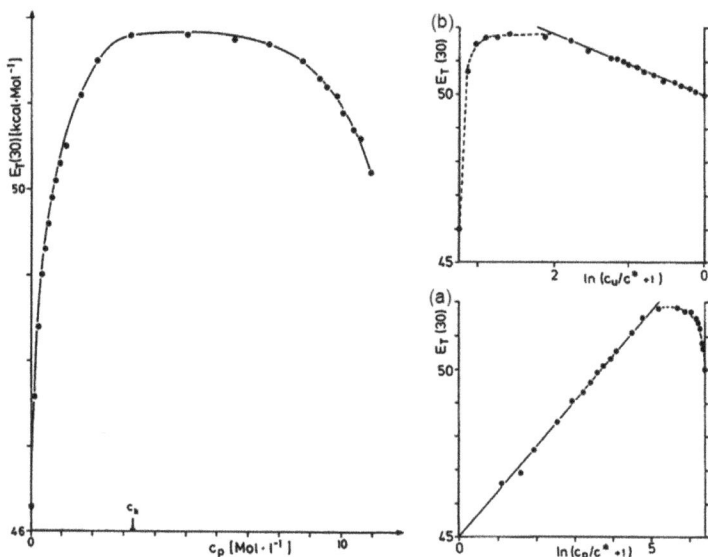

Figure 18. $E_T(30)$ values of mixtures of 1-butanol/nitromethane [91]. (**Left**) $E_T(30)$ values of the mixture nitromethane/1-butanol as a function of the molar concentration c_p of 1-butanol; the maximum is exceeded at $c_k = 3.3$ mol·L^{-1}. (**Bottom right (a)**): Linear correlation between $E_T(30)$ and ln $(c_p/c^* + 1)$ with c_p as the molar concentration of 1-butanol as the solid line and deviations to lower values after passing c_k shown as dashed curve; (**Top right (b)**): Linear correlation between $E_T(30)$ and ln $(c_u/c^* + 1)$ with c_u as the molar concentration of nitromethane as solid line and deviations to lower values after passing c_k shown as dashed curve.

11. Polarity around Miscibility Gaps

Homogeneous binary mixtures of very different polar solvents are possible such as 1,4-dioxane ($E_T(30)$ of 36.0) and water ($E_T(30)$ of 63.1), and one solvent in such mixtures can be used as a solubilizer for applications in complex mixtures. In most of such cases, the solubility of each solvent in the other is limited with the miscibility gap determined by the surface energy (surface tension) between both solvents.

The $E_T(30)$ values of methanol/water mixtures as a function of the molar concentration c_p of water can be described by Equation (5) and form two linear correlations as shown in Figure 19 (squares above) [94]; a change in the solvent structure at high water content forms a kink as shown in Figure 16. The same type of double correlation is found for ethanol/water mixtures (triangles down) with a kink at about the same water concentration as for methanol/water (although at a different position on the abscissa because of the different scaling). The mixture of 1-propanol/water (triangles up) is a further example of this behavior starting with a lower $E_T(30)$ value due to the lower polarity of 1-propanol. A miscibility gap is observed for 1-butanol/water (circles). However, similar linear correlations are found [95] in both the low water content range and high water content range where the expected kink is in the miscibility gap (biphasic region) and can be obtained by extrapolation. The similarity between the mixtures of 1-butanol/water and mixtures of lower alcohols indicate that the two types of solvent structure are miscible for the lower alcohols, but no longer completely for 1-butanol. This behavior, dominated by the solvent structure, seems to be more general.

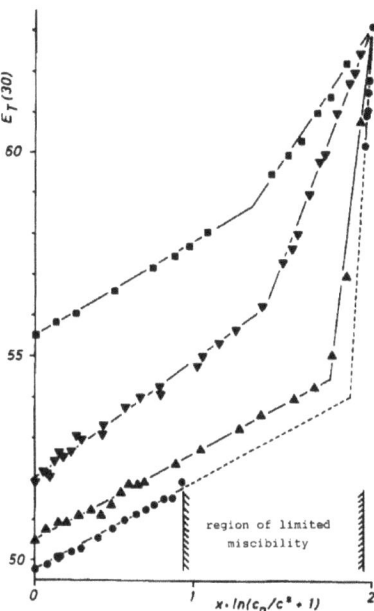

Figure 19. $E_T(30)$ values of binary mixtures between n-alkanols and water as a function of ln $(c_p/c^* + 1)$ according to Equation (5) with c_p as the concentration of the polar water [94]; the ordinate was scaled by x to obtain a uniform value for water. Squares: water/methanol ($x = 0.157$); some values are taken from ref. [47]. Triangles down: water/ethanol ($x = 0.318$); some values are taken from ref. [47]. Triangles up: water/1-propanol ($x = 1.349$). Circles: water/1-butanol ($x = 1.705$). Solid lines result from application of Equation (5) and least squares method; dashed lines are extrapolations into the region of limited miscibility.

12. Polar Functional Groups

The generality of Equation (5) for the polarity of the solvent naturally implies the domination of the polar component in the binary mixture, while the component with lower polarity essentially forms the matrix. The polarity of the more polar component is essentially generated by polar functional groups; as a consequence, the molar concentration of these polar groups should be of fundamental importance.

To verify this concept, the molar concentration of pure alcohols and water was calculated from their density and molecular weight and is identical to the molar concentration of their polar OH group. Plotting the $E_T(30)$ values against their logarithmic concentration in Figure 20 (left) shows a linear relation for 1-propanol (3) and higher homologues [96]. A second steeper line is found for methanol (1) and water (0) indicating the hydrogen-bridging structure of these solvents. Interestingly, the $E_T(30)$ value of ethanol (2) is close to the limit, indicating the special properties of ethanol as a solvent since the hydrogen-bridging structure is just being realized, while slight perturbance such as an increase in temperature or the addition of another component could change the structure to more isolated OH groups with different solvent properties. $E_T(30)$ values for the molar concentrations (ln c_p) of OH groups in mixtures of ethanol with the less polar 1,4-dioxane or n-heptane fit well into the plot for pure 1-alkanoles, as shown in Figure 20 (right). It can be concluded that the polarity of solvents and solvent mixtures, respectively, is dominated by polar functional groups (polar structure elements) and structures or components with lower polarity essentially form the matrix. This concept was further extended and generalized by Spange and co-workers [97].

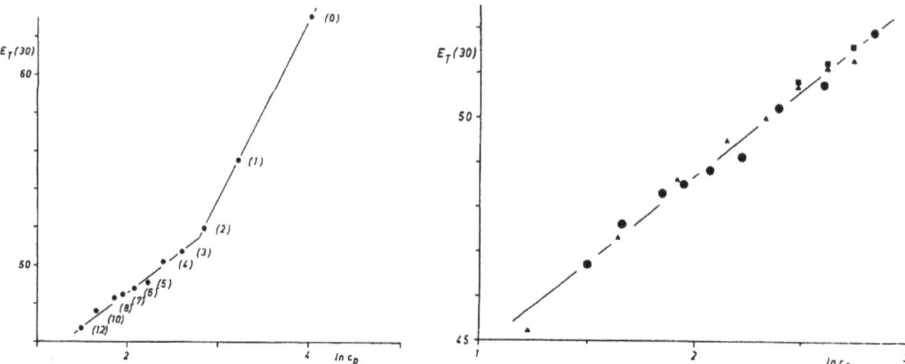

Figure 20. (**Left**) $E_T(30)$ values [56] of homologous linear 1-alkanols as a function of the logarithm of their molar concentration c_p. Their chain lengths n are given in parenthesis such as (0) for water and (1) for methanol. (**Right**) Comparison of $E_T(30)$ values as a function of the logarithm of the molar concentration of OH groups (range of low concentrations of OH groups). Circles: Pure linear 1-alkanols, triangles: Mixtures of ethanol and 1,4-dioxane, squares: Mixtures of ethanol and n-heptane.

The piezochromism of **3**, the pressure-dependent shift in the solvatochromic band, may be a further indicator of the importance of the molar concentration of OH groups for the polarity of the medium. An increase in the pressure [98] from atmospheric pressure to 2000 at shifts the solvatochromic band of **3** in ethanol until $E_T(30) = 52.7$ where the compressibility of ethanol increases the molar concentration from 17.1 to 24.6 mol/L. Such an increase in the concentration of OH groups could be made responsible for the observed increase in the $E_T(30)$ values because the values are within the scope of the second linear correlation in Figure 20.

13. Thermochromism

Most investigations about solvent effects were realized at room temperature or slightly above because of experimental convenience; however, solvent effects are appreciably temperature-dependent. Pronounced effects can be expected for hydrogen-bonding solvents, since such bonds dominate at lower temperatures and become more and more loose at higher temperatures (compare, for example, the high specific heat of water).

The thermochromism of **3** in ethanol [99] is shown in Figure 21, where there is a precise linear relation between the temperature T and the polarity of the solvent as the $E_T(30)$ values. As indicated before, the $E_T(30)$ value of ethanol at room temperature is appreciably lower than of methanol; however, cooling to $-75\,°C$ increases the value so that it overtakes even methanol. The thermal expansion of ethanol [100] may be partially responsible for the effect (see, for example, Figure 20 for the effect of the content of OH groups in alcohols); this expansion is exactly proportional to temperature (compare the use in alcohol thermometers); however, one can estimate that the thermal expansion can only cover about 1/3 of the effect. The residual 2/3 is attributed to the loosening of hydrogen bonds by increasing the temperature. Remarkably, this must also be exactly linearly temperature-dependent because the high precision of the overall linear correlation.

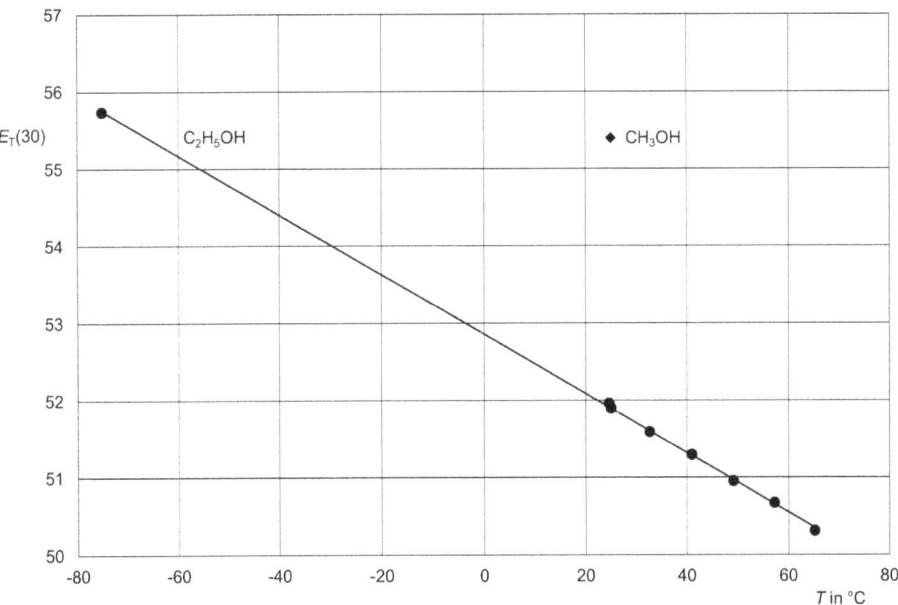

Figure 21. Thermochromism of **3** in ethanol: Filled circles for the measurements (measurements were taken from ref. [99]) and a filled diamond for the polarity of methanol at room temperature for comparison. Linear correlation between T and $E_T(30)$: slope -0.0385, intercept 52.859, correlation number -0.99988 (7 measurements), coefficient of determination 0.9998, standard deviation 0.03.

14. Long-Reaching Noncovalent Interactions

Intermolecular interactions are important for solvent effects where direct molecular contacts appear to dominate; such interactions are strongly damped and become vanishing beyond the direct molecular contact. Longer distances are covered by dipole–dipole interactions such as those according to Förster's mechanism [101,102], where a limit of interactions seems to be reached at more than 10 nm.

However, studies of the concentration dependence of the fluorescence lifetimes τ of strongly fluorescent dyes in Figure 22 indicate [4,103] that there are still interactions even at more than 100 nm because of the large intermolecular distance. Equation (5) was developed without restriction on the nature of interactions. Consequently, the equation can be also used to describe the concentration dependence of τ; see the inset of Figure 22.

The long-reaching interactions were attributed to the effect of evanescent waves of the electronically excited dye molecules. Electronical excitation requires comparably high energy input and is unimportant at room temperature. On the other hand, molecules are vibronically excited even at room temperature. As a consequence, such interactions might be part of solvent effects where an even further reaching can be expected due to the long wavelengths of vibronic energy. This could be the subject of further investigations.

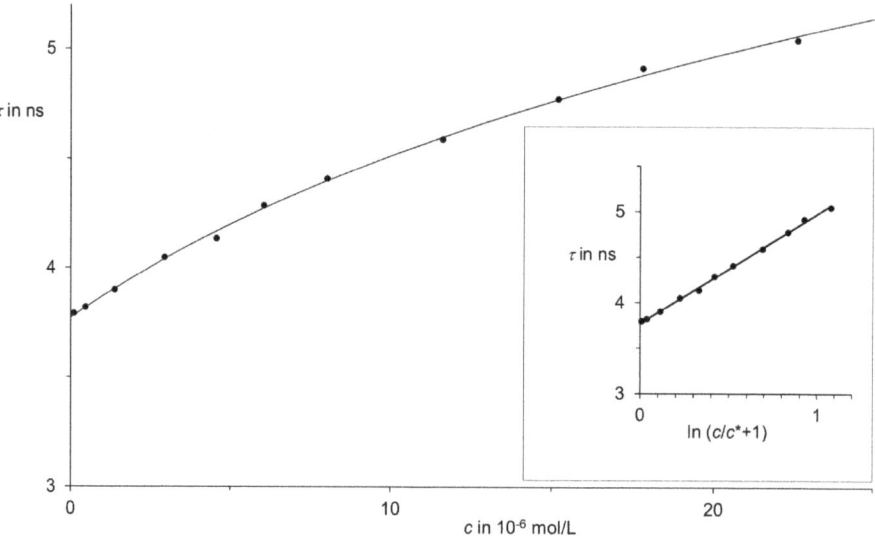

Figure 22. Dependence of the fluorescence lifetime τ on concentration c of highly diluted solutions of the fluorescence dye S-13 (RN 110590-84-6) in chloroform: Measurements as filled circles; curve calculated by means of Equation (5). Inset: linear correlation between τ and $(\ln c/c^* + 1)$ according to Equation (5): $E = 1.199$; $c^* = 1.17 \cdot 10^{-5}$ mol/L; correlation number 0.9992 (11 measurements); coefficient of determination 0.998; standard deviation 0.02.

15. Practical Applications

The visual effect in the solvatochromism of dye 3 facilitates its application in simple color tests. In anhydrous solvents or reagents, residual contents of water can be determined because these induce strong solvatochromic effects. For example, anhydrous *tert*-butylhydroperoxide is required for various applications in preparative chemistry and forms pure blue solutions [104] of 3. A low and unimportant water content shifts the color so that it is no longer pure blue, but acquires a very slight violet component, while a higher water content shifts the color to violet and further to red. A simple color test with 3 allows the rapid identification of solvents, as shown for disinfectants [105] and their components. It can also be used to determine the efficiency of methods for drying solvents, in particular, highly hygroscopic solvents such as ethanol. The measurement of the absorption maximum of dye 3 and application of Equation (5) allows the accurate determination of the composition [106] of binary mixture, in particular, the water content [107] of solvents. The use of solvatochromic fluorescent dyes [108] means an extension because higher diluted solutions can be applied and the optical quality of a sample is of minor importance.

Equation (5) was developed for general intermolecular interactions without specifying their nature or method of detection. As a consequence, a linear correlation according to Equation (5) was obtained for the densities [109] of mixture between various solvents and water such as is shown in Figure 23 (left for water/methanol). Similar linear correlations were obtained for the index of refraction corrected by the density (n_D^{20}/ρ), as shown in Figure 23 (right). Such correlations are useful for various applications such as precise interpolations.

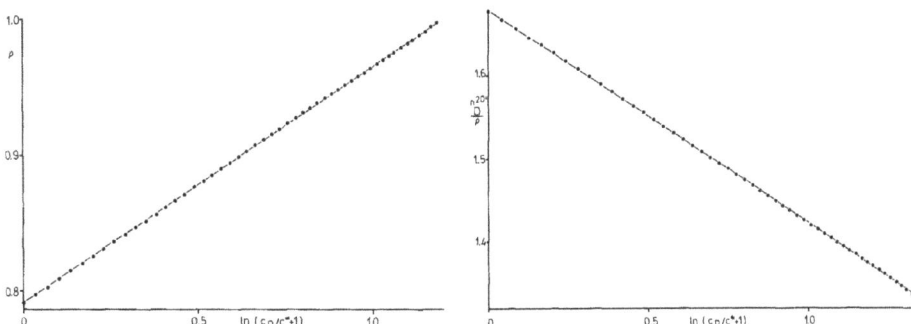

Figure 23. (**Left**) Linear correlation between the density ρ and $\ln(c_p/c^* + 1)$ according to Equation (5); every second measurement is recorded in the graphics for clarity (all measurements were used for calculations); slope 0.174, intercept 0.79093; $c^* = 24.5$, correlation number 0.999947 for 71 measurements; coefficient of determination 0.9999. (**Right**) Linear correlation between the index of refraction over the density n_D^{20}/ρ and $\ln(c_p/c^* + 1)$ according to Equation (5); every second measurement is plotted for clarity; slope -0.257, intercept 1.6797; $c^* = 19.9$, correlation number -0.99998 for 71 measurements; coefficient of determination 0.99996.

16. Conclusions

The chemistry-dominating intermolecular interactions can be efficiently investigated by solvent polarity probes, whereby Dimroth and Reichardt's $E_T(30)$ scale certifies that because of the sensitivity of the optical method and other advantages, this essentially reflects the dipolarity of solvents; the polarizability of solvents is better characterized by Brooker's χ_R scale. Zelinski's universal polarity probe forms a link between both types of polarity scales because the fluorescence-dependent polarity scale S correlates linear with $E_T(30)$ for various solvents, while the absorbance of the dye correlates with Brooker's χ_R scale. Various solvent effects can be measured by means of multi-parameter approaches. Further insight into solvent effects and changes in the solvent structure were obtained by the developed two-parameter equation (5) for concentration-dependent effects. The equation includes condensed phases, binary mixtures where the property of one solvent can be continuously transformed into that of the other by changing the composition of the mixture, and even the effects in pure solvents with polar functional groups: The polar properties of media are dominated by polar groups or polar components, while other elements of structure or components essentially form the matrix. The appreciable thermochromism of **3** leading to a linear correlation between the $E_T(30)$ scale and the temperature (either in °C or K) was reported in relation to the thermal expansion of the solvent, with about 1/3 of the effect attributed to thermal expansion and 2/3 to temperature-induced breaking of hydrogen bonds; both effects are strongly linear with the temperature. Finally, very far-reaching intermolecular interactions were studied using the two-parameter equation (5) indicating intermolecular interactions of electronically excited states even for distances of more than 100 nm. Practical applications were shown for analytics both for the polarity probes and the two-parameter equation (5), the latter of which can be a useful tool for further investigations.

Funding: This research received no external funding.

Data Availability Statement: Not applicable.

Conflicts of Interest: The author declares no conflict of interest.

References

1. Smith, M.B.; March, J. *March's Advanced Organic Chemistry*, 6th ed.; John Wiley & Sons: Hoboken, NJ, USA, 2007; ISBN -13:978-0-471-72091-1.
2. Chandrasekhar, S. Stochastic Problems in Physics and Astronomy. *Rev. Mod. Phys.* **1943**, *15*, 1. [CrossRef]
3. Hunter, C.A. Quantifying Intermolecular Interactions: Guidelines for the Molecular Recognition Toolbox. *Angew. Chem. Int. Ed.* **2004**, *43*, 5310–5324. [CrossRef] [PubMed]
4. Langhals, H.; Schlücker, T. Dependence of the Fluorescent Lifetime τ on the Concentration at High Dilution. *J. Phys. Chem. Lett.* **2022**, *13*, 7568–7573. [CrossRef] [PubMed]
5. Langhals, H. Determination of the Concentration of Gases by Pressure Measurements. *Anal. Lett.* **1987**, *20*, 1595–1610. [CrossRef]
6. Michels, A.; Michels, C. Isotherms of CO_2 between 0° and 150° and pressures from 16 to 250 atm (Amagat densities 18-206). *Proc. Roy. Soc. London Ser. A* **1935**, *153*, 201–214. [CrossRef]
7. Michels, A.; Bijl, A.; Michels, C. Thermodynamic Properties of CO_2 up to 3000 Atmospheres between 25 and 150 °C. *Proc. Royal Soc. London Ser. A Math. Phys. Sci.* **1937**, *160*, 376–384.
8. Michels, A.; Michels, C. Series evaluation of the isotherm data of CO2 between 0 and 150 °C. and up to 3000 atm. *Proc. Royal Soc. London Ser. A Math. Phys. Eng. Sci.* **1937**, *160*, 348–357. [CrossRef]
9. Michels, A.; Michels, C.; Wouters, H. Isotherms of CO_2 between 70 and 3000 atmospheres (Amagat densities between 200 and 600). *Proc. Royal Soc. London. Ser. A Math. Phys. Eng. Sci.* **1935**, *153*, 214–224. [CrossRef]
10. Simon, A.; Peters, K. Single-crystal refinement of the structure of carbon dioxide. *Acta Cryst. B* **1980**, *B36*, 2750–2751. [CrossRef]
11. Michels, A.; Blaisse, B.; Michels, C. The isotherms of CO_2 in the neighbourhood of the critical point and round the coexistence line. *Proc. Royal Soc. London. Ser. A Math. Phys. Eng. Sci.* **1937**, *160*, 358–375. [CrossRef]
12. Grault, S.T.; Squires, R.R. Generation of Alkyl Carbanions in the Gas Phase. *J. Am. Chem. Soc.* **1990**, *112*, 2506–2516. [CrossRef]
13. Tian, Z.; Kass, S.R. Carbanions in the gas phase. *Chem. Rev.* **2013**, *113*, 6986–7010. [CrossRef] [PubMed]
14. Kharasch, M.S.; Reinmuth, O. *Grignard Reactions of Nonmetallic Substances*; Constable & Company Ltd.: London, UK, 1954; OCLC-No. 545586.
15. Berthelot, M.; Péan de Saint Gilles, L. *Recherches Sur Les Affinités. De la Formation Et de la Decomposition Des Éthers*; Mallet-Bachelier: Paris, France, 1862; pp. 385–422.
16. Menschutkin, N. Über die Geschwindigkeit der Esterbildung. *Z. Phys. Chem.* **1887**, *1*, 611–630. [CrossRef]
17. Menschutkin, N. Über die Affinitätskoeffizienten der Alkylhaloide. *Z. Phys. Chem.* **1890**, *5*, 589–600. [CrossRef]
18. Menschutkin, N. Beiträge zur Kenntnis der Affinitätskoeffizienten der Alkylhaloide und der organischen Amine. Zweiter Teil. *Z. Phys. Chem.* **1890**, *6*, 41–57. [CrossRef]
19. Menschutkin, N. Zur Frage über den Einfluss chemisch indifferenter Lösungsmittel auf die Reaktionsgeschwindigkeiten. *Z. Phys. Chem.* **1900**, *34*, 157–167. [CrossRef]
20. Feynman, R.P.; Leighton, R.B.; Sands, M. *The Feynman Lectures on Physics: Definitive Edition*; Addison-Wesley: New York, NY, USA, 2005; Volume 2, Chapter 11-5.
21. Kirkwood, J.G. The Dielectric Polarization of Polar Liquids. *J. Chem. Phys.* **1939**, *7*, 911–919. [CrossRef]
22. Langhals, H. Heterocyclic structures applied as efficient molecular probes for the investigation of chemically important interactions in the liquid phase. *Chem. Heterocycl. Compd.* **2017**, *53*, 2–10. [CrossRef]
23. Debye, P. Der Rotationszustand von Molekülen in Flüssigkeiten. *Physik. Zeits.* **1935**, *36*, 100–101.
24. Eucken, A.; Wicke, E. Grundriss der physikalischen Chemie. In *Akademische Verlagsgesellschaft*, 10th ed.; Geest & Portig: Leipzig, Germany, 1958.
25. Kragh, H. The Lorenz-Lorentz Formula: Origin and Early History. *Substantia* **2018**, *2*, 7–18. [CrossRef]
26. Birge, R.T. A New Table of Values of the General Physical Constants (as of August, 1941). *Rev. Mod. Phys.* **1941**, *13*, 233–239. [CrossRef]
27. Lorenz, L. Ueber die Refractionsconstante. *Ann. Der Phys.* **1880**, *247*, 70–103. [CrossRef]
28. Lorentz, H.A. *Collected Papers*; Martinus Nijhoff: The Hague, The Netherlands, 1936; Volume 2, pp. 1–119.
29. Nee, T.-W.; Zwanzig, R. Theory of Dielectric Relaxation in Polar Liquids. *J. Chem. Phys.* **1970**, *52*, 6353–6363. [CrossRef]
30. Heckmann, A.; Lambert, C.; Goebel, M.; Wortmann, R. Synthese und photophysikalische Eigenschaften einer neutralen organischen gemischtvalenten Verbindung. *Angew. Chem.* **2004**, *116*, 5976, *Angew. Chem. Int. Ed.* **2004**, *43*, 5851. [CrossRef]
31. Amthor, S.; Lambert, C.; Dümmler, S.; Fischer, I.; Schelter, J. Excited Mixed-Valence States of Symmetrical Donor−Acceptor−Donor π Systems. *J. Phys. Chem. A* **2006**, *110*, 5204–5214. [CrossRef] [PubMed]
32. del Castillo, L.F.; Dávalos-Orozco, L.A.; García-Colín, L.S. Ultrafast dielectric relaxation response of polar liquids. *J. Chem. Phys.* **1997**, *106*, 2348–2354. [CrossRef]
33. Domínguez, M.; Caroli Rezende, M. Towards a unified view of the solvatochromism of phenolate betaine dyes. *J. Phys. Org. Chem.* **2010**, *23*, 156–170. [CrossRef]
34. Liptay, W.; Becker, J.; Wehning, D.; Lang, W.; Burkhard, O. The Determination of Molecular Quantities from Measurements on Macroscopic Systems.II. The Determination of Electric Dipole Moments. *Z. Naturforsch. A* **1982**, *37A*, 1396–1408. [CrossRef]
35. Winstein, S.; Grunwald, E.; Jones, H.W. The Correlation of Solvolysis Rates and the Classification of Solvolysis Reactions into Mechanistic Categories. *J. Am. Chem. Soc.* **1951**, *73*, 2700–2707. [CrossRef]

46. Fainberg, A.H.; Winstein, S. Correlation of Solvolysis Rates. III. *t*-Butyl Chloride in a Wide Range of Solvent Mixtures. *J. Am. Chem. Soc.* **1956**, *78*, 2770–2777. [CrossRef]
47. Winstein, S.; Fainberg, A.H. Correlation of Solvolysis Rates. IV.1 Solvent Effects on Enthalpy and Entropy of Activation for Solvolysis of t-Butyl Chloride. *J. Am. Chem. Soc.* **1957**, *79*, 5937–5950. [CrossRef]
48. Chapman, N.B.; Shorter, J. (Eds.) *Correlation Analysis in Chemistry—Recent Advances*; Plenum Press: New York, NY, USA; London, UK, 1978. [CrossRef]
49. Hammett, L.P. *Physical Organic Chemistry*, 2nd ed.; McGraw-Hill: New York, NY, USA, 1970; ISBN -10:0070259054.
50. Wilputte-Steinert, L.; Fierens, P.J.C. Etude cinétique des réactions de solvolyse IV. Discussion Générale. *Bull Soc. Chim. Belges* **1955**, *64*, 308–332. [CrossRef]
51. Kosower, E.M. The Effect of Solvent on Spectra. I. A New Empirical Measure of Solvent Polarity: Z-Values. *J. Am. Chem. Soc.* **1958**, *80*, 3253–3260. [CrossRef]
52. Dimroth, K.; Arnoldy, G.; von Eicken, S.; Schiffle, G. Beziehungen zwischen Farbe, Konstitution, Lösungsmittel und chemischer Reaktionsfähigkeit. Untersuchungen über N^+, N^--Betaine der Pyridinreihe. *Justus Liebigs Ann. Chem.* **1957**, *604*, 221–251. [CrossRef]
53. Dimroth, K. Über den Einfluß des Lösungsmittels auf die Farbe organischer Verbindungen. *Symp. Über Farbenchem. Angew. Chem.* **1960**, *72*, 782–784. [CrossRef]
54. John, W. Die Anwendung der Solvatochromie organischer Farbstoffe zur kolorimetrischen Analyse von Lösungsmittel- und insbesondere von Treibstoffgemischen. *Angew. Chem.* **1947**, *59*, 188–194. [CrossRef]
55. Schneider, W. Strukturchemische Zwischenstufen und ihre kontinuierliche Verschiebung durch Solvatbildung. *Angew. Chem.* **1926**, *39*, 412. [CrossRef]
56. Schneider, W.; Dobling, W.; Cordua, R. Über Mesomerie bei N-Oxyphenyl-pyridinium-Basen. *Ber. dtsch. Chem. Ges.* **1937**, *70B*, 1645–1665, *Chem. Abstr.* **1937**, *31*, 53446. [CrossRef]
57. Dimroth, K.; Reichardt, C.; Siepmann, T.; Bohlmann, F. Über Pyridinium-N-Phenol-Betaine und ihre Verwendung zur Charakterisierung der Polarität von Lösungsmitteln. *Justus Liebigs Ann. Chem.* **1962**, *661*, 1–37. [CrossRef]
58. Available online: https://www.uni-marburg.de/de/fb15/arbeitsgruppen/ag-reichardt/et30-werte-prof-reichardt (accessed on 23 November 2023).
59. Taft, R.W.; Kamlet, M.J. The solvatochromic comparison method. 2. The. alpha.-scale of solvent hydrogen-bond donor (HBD) acidities. *J. Am. Chem. Soc.* **1976**, *98*, 2886–2894. [CrossRef]
60. Plenert, A.C.; Mendez-Vega, E.; Sander, W. Micro- vs Macrosolvation in Reichardt's Dyes. *J. Am. Chem. Soc.* **2021**, *143*, 13156–13166. [CrossRef] [PubMed]
61. Kessler, M.A.; Wolfbeis, O.S. ET(33), a solvatochromic polarity and micellar probe for neutral aqueous solutions. *Chem. Phys. Lipids* **1989**, *50*, 51–56. [CrossRef]
62. Langhals, H. The Polarity of Solutions of Electrolytes. *Tetrahedron* **1987**, *43*, 1771–1774. [CrossRef]
63. Reichardt, C.; Che, D.; Heckenkemper, G.; Schäfer, G. Syntheses and UV/Vis-Spectroscopic Properties of Hydrophilic 2-, 3-, and 4-Pyridyl-Substituted Solvatochromic and Halochromic Pyridinium N-Phenolate Betaine Dyes as New Empirical Solvent Polarity Indicators. *Eur. J. Org. Chem.* **2001**, *12*, 2343–2361. [CrossRef]
64. Langhals, H. Increasing the solubility of aromatic compounds. *Ger. Offen.* DE 3016764 (April 30, **1980**). *Chem. Abstr.* **1982**, *96*, P70417x.
65. Reichardt, C.; Harbusch-Görnert, E. Über Pyridinium-N-phenolat-Betaine und ihre Verwendung zur Charakterisierung der Polarität von Lösungsmitteln, X. Erweiterung, Korrektur und Neudefinition der ET-Lösungsmittelpolaritätsskala mit Hilfe eines lipophilen penta-*tert*-butyl-substituierten Pyridinium-*N*-phenolat-Betainfarbstoffes. *Liebigs Ann. Chem.* **1983**, *1983*, 721–743. [CrossRef]
66. Reichardt, C. Empirische Parameter der Lösungsmittelpolarität als lineare „Freie Enthalpie"-Beziehungen. *Angew. Chem.* **1979**, *91*, 119–131, *Angew. Chem. Int. Ed.* **1979**, *18*, 98–110. [CrossRef]
67. Reichardt, C. Solvatochromic Dyes as Solvent Polarity Indicators. *Chem. Rev.* **1994**, *94*, 2319–2358. [CrossRef]
68. Langhals, H.; Braun, P.; Dietl, C.; Mayer, P. How many molecular layers of polar solvent molecules control chemistry? The concept of compensating dipoles. *Chem. Eur. J.* **2013**, *19*, 13511–13521. [CrossRef]
69. Langhals, H. The rapid identification of organic colorants by UV/Vis-spectroscopy. *Anal. Bioanal. Chem.* **2002**, *374*, 573–578. [CrossRef]
70. Brooker, L.G.S.; Keyes, G.H.; Heseltine, D.W. Color and constitution. XI. Anhydronium bases of *p*-hydroxystyryl dyes as solvent polarity indicators. *J. Am. Chem. Soc.* **1951**, *73*, 5350. [CrossRef]
71. Brooker, L.G.S.; Craig, A.C.; Heseltine, D.W.; Jenkins, P.W.; Lincoln, L.L. Color and constitution. XIII. Merocyanines as solvent property indicators. *J. Am. Chem. Soc.* **1965**, *87*, 2443–2450. [CrossRef]
72. Bekárek, V.; Bekárek, V., Jr. Effect of medium on electronic spectra of 4-dimethylamino-ω-nitrostyrenes. Effective Kirkwood-Onsager functions of medium. *Coll. Czech. Chem. Commun.* **1987**, *52*, 287–298. [CrossRef]
73. Langhals, H. Untersuchung des Lösungsmitteleinflusses auf Absorption und Emission bei Fluoreszenzfarbstoffen. *Z. Phys. Chem.* **1981**, *127*, 45–53. [CrossRef]
74. Zmyreva, I.A.; Zelinsky, V.V.; Kolobkov, V.P.; Krasnickaja, N.D. Universal Scale fort the Action of Solvents Upon Electron Spectra of Organic Compounds. *Dokl. Akad. Nauk SSSR* **1959**, *129*, 1089–1092.

65. Kulinich, A.V.; Mikitenko, E.K.; Ishchenko, A.A. Scope of negative solvatochromism and solvatofluorochromism of merocyanines. *Phys. Chem. Chem. Phys.* **2016**, *18*, 3444–3453. [CrossRef] [PubMed]
66. Griffiths, T.R.; Pugh, D.C. Correlations Among Solvent Polarity Scales, Dielectric Constant and Dipole Moment, and Means to Reliable Predictions of Polarity Scale Values from Current Data. *Coord. Chem. Revs.* **1979**, *29*, 129–211. [CrossRef]
67. Mayer, U.; Gutmann, V.; Gerger, W. The acceptor number—A quantitative empirical parameter for the electrophilic properties of solvents. *Monatsh. Chem.* **1975**, *106*, 1235–1257. [CrossRef]
68. Gutmann, V. Solvent effects on the reactivities of organometallic compounds. *Coord. Chem. Rev.* **1976**, *18*, 225–255. [CrossRef]
69. Katritzky, A.R.; Fara, D.C.; Yang, H.; Tamm, K.; Tamm, T.; Karelson, M. Quantitative Measures of Solvent Polarity. *Chem. Rev.* **2004**, *104*, 175–198. [CrossRef]
70. Kamlet, M.J.; Abboud, J.L.; Taft, R.W. The solvatochromic comparison method. 6. The.pi.* scale of solvent polarities. *J. Am. Chem. Soc.* **1977**, *99*, 6027–6038. [CrossRef]
71. Catalán, J. Toward a Generalized Treatment of the Solvent Effect Based on Four Empirical Scales: Dipolarity (SdP, a New Scale), Polarizability (SP), Acidity (SA), and Basicity (SB) of the Medium. *J. Phys. Chem. B* **2009**, *113*, 5951–5960. [CrossRef] [PubMed]
72. Catalán, J.; Hopf, H. Empirical Treatment of the Inductive and Dispersive Components of Solute−Solvent Interactions: The Solvent Polarizability (SP) Scale. *Eur. J. Org. Chem.* **2004**, *2004*, 4694–4702. [CrossRef]
73. Koppel, A.; Palm, V.A. *Influence of the Solvent on Organic Reactivity*; Chapman, N.B., Shorter, J., Eds.; Advances in Linear Free Energy Relationships; Plenum Press: London, UK, 1972; pp. 203–280.
74. Weckwerth, J.D.; Vitha, M.F.; Carr, P.W. The development and determination of chemically distinct solute parameters for use in linear solvation energy relationships. *Fluid Phase Equilibria* **2001**, *183–184*, 143–157. [CrossRef]
75. Spange, S.; Weiß, N. Empirical Hydrogen Bonding Donor (HBD) Parameters of Organic Solvents Using Solvatochromic Probes—A Critical Evaluation. *ChemPhysChem* **2023**, *24*, e202200780. [CrossRef] [PubMed]
76. Cook, J.L.; Hunter, C.A.; Low, C.M.R.; Perez-Velasco, A.; Vinter, J.G. Solvent Effects on Hydrogen Bonding. *Angew. Chem. Int. Ed.* **2007**, *46*, 3706–3709. [CrossRef] [PubMed]
77. Cabot, R.; Hunter, C.A. Molecular probes of solvation phenomena. *Chem. Soc. Rev.* **2012**, *41*, 3485–3492. [CrossRef] [PubMed]
78. Acree, W.E., Jr.; Lang, A.S.I.D. Reichardt's Dye-Based Solvent Polarity and Abraham Solvent Parameters: Examining Correlations and Predictive Modeling. *Liquids* **2023**, *3*, 303–313. [CrossRef]
79. Schweig, A.; Reichardt, C. π-Elektronen-Dipolmoment eines Pyridinium-N-phenol-betains. *Z. Naturforsch.* **1966**, *21a*, 1373–1376. [CrossRef]
80. Liptay, W.; Schlosser, H.J.; Dumbacher, B.; Hünig, S. Die Beeinflussung der optischen Absorption von Molekülen durch ein elektrisches Feld. *Z. Naturforsch.* **1968**, *23a*, 1613–1625. [CrossRef]
81. Beard, M.C.; Turner, G.M.; Schmuttenmaer, C.A. Measurement of Electromagnetic Radiation Emitted during Rapid Intramolecular Electron Transfer. *J. Am. Chem. Soc.* **2000**, *122*, 11541–11542. [CrossRef]
82. Beard, M.C.; Turner, G.M.; Schmuttenmaer, C.A. Measuring Intramolecular Charge Transfer via Coherent Generation of THz Radiation. *J. Phys. Chem. A* **2002**, *106*, 878–883. [CrossRef]
83. Stiopkin, I.V.; Weeraman, C.; Pieniazek, P.A.; Shalhout, F.Y.; Skinner, J.L.; Benderskii, A. Hydrogen bonding at the water surface revealed by isotopic dilution spectroscopy. *Nature* **2011**, *474*, 192–195. [CrossRef] [PubMed]
84. Hodges, A.M.; Kilpatrick, N.W.; McTigue, P.; Pereram, J.M. The solvation potential at the interface between water and methanol + water mixtures. *J. Electroanal. Chem.* **1986**, *215*, 63–82. [CrossRef]
85. Bosch, E.; Rosés, M. Relationships between E_T Polarity and Composition in Binary Solvent Mixtures. *J. Chem. Soc. Faraday Trans.* **1992**, *88*, 3541–3546. [CrossRef]
86. Langhals, H. *Primary Methods of Generating Solar Power by Using the Targeted Modification of Fluorescent Systems.* Translation of: *Prinzipielle Wege für Die Gewinnung von Solarenergie Über Gezielte Modifizierung Fluoreszierender Systeme*; Habilitationsschrift; Albert-Ludwigs-Universität Freiburg: Freiburg, Germany, 1981. [CrossRef]
87. Langhals, H. Polarität binärer Flüssigkeitsgemische. *Angew. Chem.* **1982**, *94*, 739–749, *Angew. Chem. Int. Ed. Engl.* **1982**, *21*, 724–733. [CrossRef]
88. Keutsch, F.N.; Saykally, R.J. Water clusters: Untangling the mysteries of the liquid, one molecule at a time. *Proc. Nat. Acad. Sci. USA* **2001**, *98*, 10533–10540. [CrossRef]
89. Langhals, H. Die quantitative Beschreibung der Lösungsmittelpolarität binärer Gemische unter Berücksichtigung verschiedener Polaritätsskalen. *Chem. Ber.* **1981**, *114*, 2907–2913. [CrossRef]
90. Makogon, Y.F. *Hydrates of Hydrocarbons*; Penn Well Publishing Company: Tulsa, OK, USA, 1997; ISBN 0-87814-718-7.
91. Langhals, H. Ungewöhnliches Polaritätsverhalten binärer Flüssigkeitsgemische. *Nouv. Journ. Chim.* **1981**, *5*, 511–514.
92. Koppel, I.; Koppel, J. ET parameters of binary mixtures of alcohols with DMSO and acetonitrile. Synergetic solvent effect of high intensity. *Org. React.* **1983**, *20*, 523–546, *Chem. Abstr.* **1984**, *101*, 110180.
93. Testoni, F.M.; Ribeiro, E.A.; Giusti, L.A.; Machado, V.G. Merocyanine solvatochromic dyes in the study of synergistic effects in mixtures of chloroform with hydrogen-bond accepting solvents. *Spectrochim. Acta Part A* **2009**, *71*, 1704–1711. [CrossRef] [PubMed]
94. Langhals, H. Polarität von Flüssigkeitsgemischen mit begrenzt mischbaren Komponenten. *Z. Phys. Chem.* **1987**, *268*, 91–96. [CrossRef]
95. Langhals, H. Polarity of Liquid Mixtures with Components of Limited Miscibility. *Tetrahedron Lett.* **1986**, *27*, 339–342. [CrossRef]

96. Langhals, H. Die Beschreibung der Polarität von Alkoholen als Funktion ihres molaren Gehalts an OH-Gruppen. *Nouv. Journ. Chim.* **1982**, *6*, 265–267, English translation: The description of the polarity of alcohols as a function of their molar content of OH groups.
97. Spange, S.; Weiß, N.; Mayerhöfer, T.G. The Global Polarity of Alcoholic Solvents and Water—Importance of the Collectively Acting Factors Density, Refractive Index and Hydrogen Bonding Forces. *Chem. Open* **2022**, *11*, e202200140. [CrossRef] [PubMed]
98. Tamura, K.; Yoshiaki Ogo, Y.; Imoto, T. Effect of pressure on the solvent polarity parameter: ET value. *Chem. Lett.* **1973**, *2*, 625–628. [CrossRef]
99. Dimroth, K.; Reichardt, C.; Schweig, A. Über die Thermochromie von Pyridinium-N-Phenol-Betainen. *Liebigs Ann. Chem.* **1963**, *669*, 95–105. [CrossRef]
100. Forsythe, W.E. *Smithsonian Physical Tables*, 9th ed.; Smithsonian Institution Press: Washington, DC, USA, 1969; ISBN-10 0874740150; ISBN 13 978-0874740158.
101. Förster, T. Energiewanderung und Fluoreszenz. *Naturwiss* **1946**, *33*, 166–175, *Chem. Abstr.* **1947**, *41*, 36668. [CrossRef]
102. Valeur, B.; Berberan-Santos, M. Excitation Energy Transfer. In *Molecular Fluorescence: Principles and Applications*, 2nd ed.; Wiley-VCH: Weinheim, Germany, 2012. [CrossRef]
103. Langhals, H.; Schlücker, T. Reply to "Comment on 'Dependence of the Fluorescent Lifetime τ on the Concentration at High Dilution'": Extended Interpretation. *J. Phys. Chem. Lett.* **2023**, *14*, 1457–1459. [CrossRef]
104. Langhals, H.; Fritz, E.; Mergelsberg, I. Die Trocknung von *tert*-Butylhydroperoxid nach einem einfachen, erstmals gefahrlosen Verfahren. *Chem. Ber.* **1980**, *113*, 3662–3665. [CrossRef]
105. Langhals, H. The Quality Control of Alcoholic Components of Disinfectants by a Simple Colour Test. *S. Afr. J. Chem.* **2020**, *73*, 81–83. [CrossRef]
106. Langhals, H. Ein neues, unkompliziertes Verfahren zur Bestimmung der Zusammensetzung binärer Flüssigkeitsgemische. *Zeitschr. Analyt. Chem.* **1981**, *308*, 441–444. [CrossRef]
107. Langhals, H. Ein neues, unkompliziert auszuführendes Verfahren zur Bestimmung kleiner Konzentrationen an Wasser in organischen Lösungsmitteln. *Zeitschr. Analyt. Chem.* **1981**, *305*, 26–28. [CrossRef]
108. Langhals, H. Bestimmung der Zusammensetzung binärer Flüssigkeitsgemische mit Hilfe von Fluoreszenzmessungen. *Zeitschr. Analyt. Chem.* **1982**, *310*, 427–428. [CrossRef]
109. Langhals, H. Der Zusammenhang zwischen dem Brechungsindex und der Zusammensetzung binärer Flüssigkeitsgemische. *Z. Phys. Chem.* **1985**, *266*, 775–780, English translation: The relationship between the refractive index and the composition of binary liquid mixtures. [CrossRef]

Disclaimer/Publisher's Note: The statements, opinions and data contained in all publications are solely those of the individual author(s) and contributor(s) and not of MDPI and/or the editor(s). MDPI and/or the editor(s) disclaim responsibility for any injury to people or property resulting from any ideas, methods, instructions or products referred to in the content.

Review

Solvatochromic and Acid–Base Molecular Probes in Surfactant Micelles: Comparison of Molecular Dynamics Simulation with the Experiment

Nikolay O. Mchedlov-Petrossyan *, Vladimir S. Farafonov and Alexander V. Lebed

Department of Physical Chemistry, V. N. Karazin Kharkiv National University, 61022 Kharkiv, Ukraine; farafonov@karazin.ua (V.S.F.); alebed@karazin.ua (A.V.L.)
* Correspondence: nikolay.mchedlov@gmail.com

Abstract: This article summarizes a series of seventeen publications by the authors devoted to molecular dynamics modeling of various indicator dyes (molecular probes) enclosed in surfactant micelles. These dyes serve as generally recognized tools for studying various types of organized solutions, among which surfactant micelles in water are the simplest and most explored. The modeling procedure involves altogether 50 to 95 surfactant molecules, 16 to 28 thousand water molecules, and a single dye molecule. The presentation of the simulation results was preceded by a brief review of the state of experimental studies. This article consists of three parts. First, despite numerous literature data devoted to modeling the micelles itself, we decided to revisit this issue. The structure and hydration of the surface of micelles of surfactants, first of all of sodium *n*-dodecylsulfate, SDS, and cetyltrimethylammonium bromide, CTAB, were studied. The values of the electrical potential, Ψ, were estimated as functions of the ionic strength and distance from the surface. The decrease in the Ψ value with distance is gradual. Attempts to consider both DS$^-$ and CTA$^+$ micelles in water without counterions result in a decay into two smaller aggregates. Obviously, the hydrophobic interaction (association) of the hydrocarbon tails balances the repulsion of the charged headgroups of these small "bare" micelles. The second part is devoted to the study of seven pyridinium *N*-phenolates, known as Reichardt's dyes, in ionic micelles. These most powerful solvatochromic indicators are now used for examining various colloidal systems. The localization and orientation of both zwitterionic and (colorless) cationic forms are generally consistent with intuitive ideas about the hydrophobicity of substituents. Hydration has been quantitatively described for both the dye molecule as a whole and the oxygen atom. A number of markers, including the visible absorption spectra of Reichardt's dyes, enable assuming a better hydration of the micellar surface of SDS than that of CTAB. However, our data show that it is more correct to speak about the more pronounced hydrogen-bonding ability of water molecules in anionic micelles than about better hydration of the SDS micelles as compared to CTAB ones. Finally, a set of acid–base indicators firmly fixed in the micellar pseudophase were studied by molecular dynamics. They are instruments for estimating electrostatic potentials of micelles and related aggregates as $\Psi = 2.303RTF^{-1}\left(pK_a^i - pK_a^{app}\right)$, where pK_a^i and pK_a^{app} are indices of so-called intrinsic and apparent dissociation constants. In this case, in addition to the location, orientation, and hydration, the differences between values of pK_a^{app} and indices of the dissociation constants in water were estimated. Only a semi-quantitative agreement with the experimental data was obtained. However, the differences between pK_a^{app} of a given indicator in two micellar solutions do much better agree with the experimental data. Accordingly, the experimental Ψ values of ionic micelles, as determined using the pK_a^{app} in nonionic micelles as pK_a^i, are reproduced with reasonable accuracy for the corresponding indicator. However, following the experimental data, a scatter of the Ψ values obtained with different indicators for given micelles is observed. This problem may be the subject of further research.

Keywords: surfactant micelles; molecular dynamics modeling; solvatochromic Reichardt's dyes; hydration; acid–base indicators; apparent dissociation constants; electrical surface potential

Citation: Mchedlov-Petrossyan, N.O.; Farafonov, V.S.; Lebed, A.V. Solvatochromic and Acid–Base Molecular Probes in Surfactant Micelles: Comparison of Molecular Dynamics Simulation with the Experiment. *Liquids* **2023**, *3*, 314–370. https://doi.org/10.3390/liquids3030021

Academic Editors: William E. Acree, Jr., Franco Cataldo and Enrico Bodo

Received: 8 May 2023
Revised: 22 July 2023
Accepted: 2 August 2023
Published: 16 August 2023

Copyright: © 2023 by the authors. Licensee MDPI, Basel, Switzerland. This article is an open access article distributed under the terms and conditions of the Creative Commons Attribution (CC BY) license (https:// creativecommons.org/licenses/by/ 4.0/).

1. Introduction

Organized solutions belong to the most explored liquid media for performing numerous chemical processes [1]. The origin of the term "organized solutions" can be found in [2,3]. Among these systems, surfactant micellar solutions are the oldest [4–6] and well-defined [7–12]. Among the most popular tools for studying the properties of surfactant micelles in water are indicator dyes of various types [10,13–23]. They are utilized as acid–base [10,13–20], solvatochromic [1,19,21], and fluorescent [22–26] molecular probes.

These compounds have long been used in sensor devices [27–31], for examining polyelectrolyte solutions [32,33], gelatin films [34], nanodiamond and fullerenol dispersions [35,36], calixarene [37,38] and cucurbituril solutions [39], surfactant monolayers on water/air interfaces [40–42], monolayers on solid/liquid interfaces [43], multilayers of different types [44–46], and even water/air interfaces without introducing additional surfactants [47–49].

With the help of such tools, the electrical surface potential, polarity, hydration, effective permittivity, and viscosity of the micellar pseudophases and related systems can be examined. The problems associated with the molecular probes are a matter of detailed consideration and discussion, and the corresponding publications are huge in number. Some of them are regarded in the above-cited articles, reviews, and books.

In addition to this abundant, but often quite contradictory information, we decided to investigate the "micelle + molecular probe" systems, first "micelle + indicator dyes" using molecular dynamics (MD) simulations. Despite numerous studies devoted to MD modeling of surfactant micelles in water, the systems containing molecular probes were almost overlooked. Some works were devoted to dye molecules in phospholipid bilayers [50,51]. An example of such a study of micelles is the MD investigation of the location of a dye sulforhodamine B in the micellar pseudophase [52].

Within the course of our study, we used first sodium n-dodecylsulfate, SDS, and n-hexadecyl- (or cetyl-) trimethylammonium bromide, CTAB, micelles in water as the most well-defined surfactant systems. In addition, six other surfactants were occasionally used. In total, seventeen molecular probes were examined mainly in SDS and CTAB micellar solutions. The results of our research group have been reported in seventeen articles from 2016 to the present [53–69]. We thought it is expedient to summarize the data obtained and the regularities found. In particular, a comparison of the theoretically derived models with the experimental data is of a special interest.

Since theoretical methods for studying solvation began to be considered as an integral part of the knowledge of solutions, it was important to apply modern MD modeling approaches not only to the structure of micelles, but also to the placement and hydration of molecular probes in the surfactant micelles.

Several theoretical models were proposed in order to qualitatively describe the structure of surfactant micelles. At the same time, molecular dynamics simulation may serve as the most physically grounded model to validate the other ones because it explicitly takes into account most of the interactions present in the solutions discussed. Such simulations go back several decades and revealed the features, which are hardly accessible otherwise.

The works devoted to MD simulations of surfactants and surfactant micelles are numerous. Here, we mention only some of them as an example [70–74], as well as a new review article [75], and a MD modeling of surfactant monolayers on the water/air interface [76]. There are some studies of surfactant micelles [77,78] or bilayers [79] based on other theoretical methods, including those analyzing the systems with incorporated lauric acid [77,79].

Within the course of our studies, the following ionic surfactants attracted our attention: sodium n-dodecylsulfate, SDS; sodium n-hexadecylsulfate (or cetylsulfate, SCS); cetyltrimethylammonium bromide, CTAB; n-dodecyltrimethylammonium bromide, DTAB; N-cetylpyridinium chloride, CPC. A zwitterionic surfactant, N-cetyl-N,N-dimethyl-3-ammonio-1-propanesulfonate, CDAPSn, and a nonionic surfactant Triton X-100 (TX-100), were also examined (Scheme 1). In order to compare the results with literature data, N-cetylpyridinium bromide, CPB, and cetylsulfonic acid with a triethanolammonium cation, TEACSn, were also studied (not shown in Scheme 1).

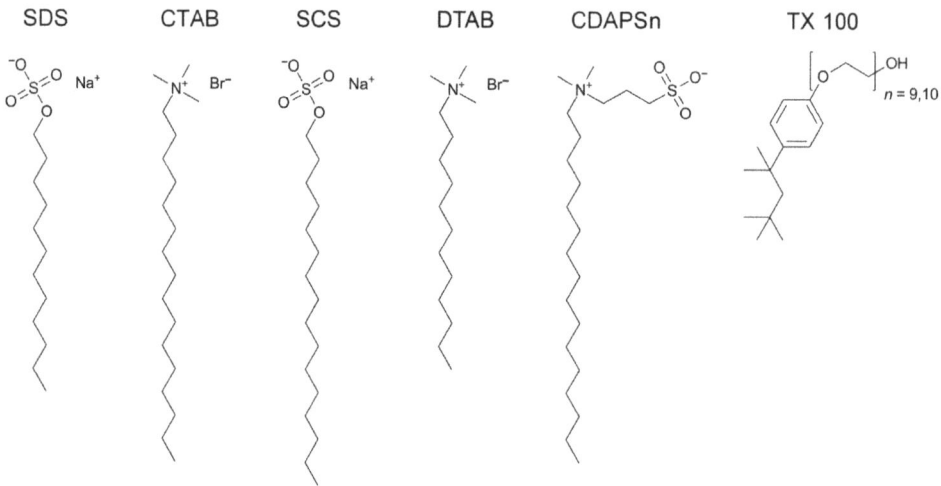

Scheme 1. Molecular structures of surfactants considered in our studies.

The leitmotif of this article is the MD simulation of "micelle + molecular probe" systems and comparison of the obtained results with experiment. We started with the study of solvatochromic dyes, whereas the idea of estimating the surface electric potential of micelles arose in the course of this research. As such, we performed a series of molecular dynamics simulations in order to elucidate the state of a set of dyes enclosed in surfactant micelles. The following characteristics were revealed: localization and spatial orientation of dye molecules with respect to the micelle surface and their degree of hydration. The collected data allowed us to identify the effect of various surfactant and dye parameters on these characteristics:

- Charge of the surfactant headgroup;
- Length of the surfactant hydrocarbon tail;
- Micellar size.
- Substituents in the dye molecule;
- Protonation state of the dye;

2. Simulation Setup

The potential models for all the dyes were manufactured in the framework of the widely used and well-validated OPLS-AA (Optimized Potentials for Liquids Simulation, All-Atom) force field, following the methodology recommended therein. The atomic charges were calculated anew in a two-stage process. Firstly, the geometry of the molecule was optimized in a quantum chemical calculation using Hartree–Fock method and 6–31G(d) basis set. The spatial distribution of the electrostatic potential around the molecule was produced at the same time. Secondly, the distribution was subjected to the RESP (Restrained Electrostatic Potential) algorithm. The RED server was used to carry out the stages [80].

For the surfactants, the OPLS-AA models developed by us were taken [55,56,58,62,65] to ensure the mutual compatibility of all species present in the system. Water was described with the SPC (Simple Point Charges) model.

All simulations were set up uniformly following the same protocol. At first, a surfactant micelle is taken containing the number of monomers corresponding to the range of typical aggregation numbers of the surfactant. It was 60 for SDS, 80 for CTAB and SCS, as well as 50 for DTAB. For CTAB, an aggregation number of 95 was also tried. The micelle is placed in a water-filled cubic box allowing ~2 nm margins around, and the dye of interest in the examined protonation state is placed in the micelle. The resulting number of water molecules is about 16–28 thousands. Three initial configurations are prepared ready to start three production MD runs of the same system, allowing enough sampling to be achieved and the uncertainty of the calculated characteristics to be estimated.

The production runs were carried out at standard conditions (temperature of 298 K and pressure of 1 bar) for 50–130 ns depending on the system. The following parameters and algorithms were used: time step 2 fs for SDS and DTAB or 1.6 fs for longer-chain CTAB and SCS, three-dimensional periodic boundary conditions, Berendsen thermostat and barostat, PME (Particle Mesh Ewald) method for electrostatic interactions, cutoff of van der Waals interactions at 1 nm, LINCS (LINear Constraint Solvent) constraints for all covalent bonds. The beginning of the trajectories was discarded during the analysis as equilibration.

3. Ionic Surfactant Micelles

Although the formation of surfactant micelles in aqueous solutions was proved more than a hundred years ago [5,6], the exact nature of these particles is still a matter of debate.

The ideas about the structure of surfactant micelles found in the literature were generally established quite a long time ago and are periodically refined and improved. It is impossible to consider this issue in detail here. As typical examples of the traditional approach, the model developed by Stigter [81–83] and the article by Larsen and Magid [84] should be mentioned. These and similar works correspond rather to Hartley's model of spherical micelles. At the same time, Menger [85] leans more towards McBain's theory of a lamellar (bilayer, plate-like) surfactant micelle; see also Philippoff's articles [86,87].

Fromherz proposed a hybrid of the two main models, a surfactant-block model, which presumes the contact of water molecules with the hydrocarbon core to some extent ("The entire hydrocarbon chain of all molecules is wettable in time average. The average number of wetted methylenes per chain is given by the wetted alpha-carbon, by the smooth surface of saturating cube configuration—about two methylenes—and by roughness") [88].

The schematic structure of a globular surfactant micelle considered by Grieser and Drummond on the basis of a number of studies by other authors seems to be quite realistic [13]. In addition to the dynamic character of the micelle ⇆ surfactant monomer equilibrium, their model includes two important features. These are as follows. First, "a large hydrocarbon/aqueous solution contact area, i.e., the headgroups do not completely cover the hydrocarbon core". This is in line with the scheme proposed by Israelachvili [89]. Second, "a fluid hydrocarbon core, containing *no* water, in which some chain bending can occur, thereby permitting some of the terminal CH_3 groups to be in contact with the interface." [13]. Today, both statements have been confirmed by MD simulations (see below).

In addition, Gilanyi [90] proposed a concept of fluctuating micelles, based on the small system thermodynamics. Rusanov proposed a detailed consideration of structural, mechanical, thermodynamic, and electrostatic aspects of cylindrical, lamellar, and (first) spherical surfactant micelles in water [91]. Us'yarov [92,93] also considers a spherical model of the SDS micelles even at surfactant concentrations up to 0.9 M.

A key characteristic of all micelles of ionic surfactants is their electrostatic potential, Ψ. A more detailed consideration allows to distinguish between the surface potential, Ψ_0, the Stern layer potential, Ψ_δ, and the electrokinetic potential, ζ. The last is available for direct determination, but corresponds to the share surface (slip plane) located in the diffuse layer of the electrical double layer, EDL. For a charged spherical or cylindrical colloidal particle, the Ψ value can be calculated by an equation proposed by Ohshima, Healy, and White, based on the approximate analytical solution of Poisson–Boltzmann equation [94]. For surfactant spherical micelles, the method was adapted by Hartland et al. [95], to give Equation (1). The left-hand side in this equation is the surface charge density, $\alpha = 1 - \beta$, where β is the degree of counterion binding, s_i is the molecular area of the headgroup.

$$\frac{\alpha}{s_i} = \frac{2\varepsilon\varepsilon_0 kRT}{F}\sinh(Y/2)\left(1 + \frac{2}{kr\cosh^2(Y/4)} + \frac{8\ln[\cosh(Y/4)]}{(kr)^2\sinh^2(Y/2)}\right)^{\frac{1}{2}} \quad (1)$$

Here, $Y = \Psi_\delta F/RT$, κ is the reciprocal Debye length, r is the micellar radius, $\varepsilon_0 = 8.854 \times 10^{-12}$ F m^{-1}, and $\varepsilon = 78.5$ at $T = 298.15$ K. It should be noted, that the calculations require the knowledge of r_i, s_i, and β and therefore are possible only for such well-defined surfactants as CTAB and SDS.

Though the present article is devoted to hydrophilic colloidal systems, an important work concerning the EDL at hydrophobic surfaces [96] should be mentioned, as well as a theoretical approach for the determination of ζ [97].

It is clear that for a better understanding of the nature of surfactant micelles, the α values are of critical importance. These values determine the micelle charge and electrostatic potential, and can be deduced from experimental data. As early as 1951, Philippoff presented α values for nine representative ionic surfactants, although even then there was much more such data; for SDS, $\alpha = 0.183$ [87]. An overview of the literature accumulated over many years [92,95,98] shows a significant spread of the α values for a given surfactant. For instance, let us consider the data for SDS and related anionic surfactants. Sasaki et al. [98] determined $\alpha = 0.27$ by activity measurements of Na$^+$ and DS$^-$ and compared this value with those obtained in thirteen other works by seven different experimental methods: α varies from 0.14 to 0.54, 0.26 on average. Frahm et al. [99] used two methods and determined $\alpha = 0.36$–0.49, while Lebedeva et al. [100] compared values from four different publications, from 0.272 to 0.322, with 0.290 on average.

Still in practice different methods provide dissimilar values. This is because the position of the boundary between bound and free counterions is not an intrinsic characteristic of the micelle but depends on the method used. Hartland et al. [95] noted that micelle mobility methods give α values of approximately 0.3 to 0.4, which is higher than those determined from measurements of sodium ion activity, which give α values of approximately 0.20 to 0.25. Simultaneously, Gilanyi [101] proposed an interpretation of the experimental α values in terms of specific and nonspecific electrostatic interactions between the colloidal particles and small ions.

At the same time, utilization of one and the same method reveals the tendency of α decreasing along with lengthening of the hydrocarbon tails. This is not so pronounced when comparing $C_{10}H_{21}OSO_3Na$ and $C_{12}H_{25}OSO_3Na$ with $C_{14}H_{29}OSO_3Na$, and $C_{16}H_{33}OSO_3Na$ [98], but for seven surfactants, $C_nH_{2n+1}OSO_3Na$, from $n = 8$ and 9 to 13 and 14, α decreases from 0.41–0.42 to 0.19–0.22 (average of values obtained by different methods) [99].

The increase in the concentrations of both surfactant and foreign electrolyte can lead to a decreasing α value. For example, Us'yarov performed a detailed theoretical study of the SDS solutions within a wide concentration range, without and with NaCl additions [92,93]. At SDS concentrations up to 0.9 M and a total Na$^+$ concentration in the bulk aqueous phase from 0.01 to $\approx$0.9 M, the $\alpha = 1 - \beta$ values vary from 0.19 to 0.025–0.04 [92,93].

4. New Molecular Dynamics Modeling Results for Surfactant Micelles

Now let us consider our results concerning the micellar structure. Particularly, the simulations show that water molecules do not penetrate deeply into the hydrocarbon core. In other words, the core is completely dry; see Figure 1A, which is in line with earlier proposed picture [13] and theoretical results [72]. Nevertheless, the continuous thermal motion of surfactant monomers and the character of their packing enable every CH$_2$/CH$_3$ bead of the tail to have occasionally contact to water. This occurs when the bead appears on the micelle surface for some time, which in turn depends on its position in the chain. However, if a micelle adopts a prolate or rod-like shape instead of being close to spherical, then the dry core appears rather thin, as was observed for DTAB micelles. Probably, the most appropriate term for the micellar structure is globular.

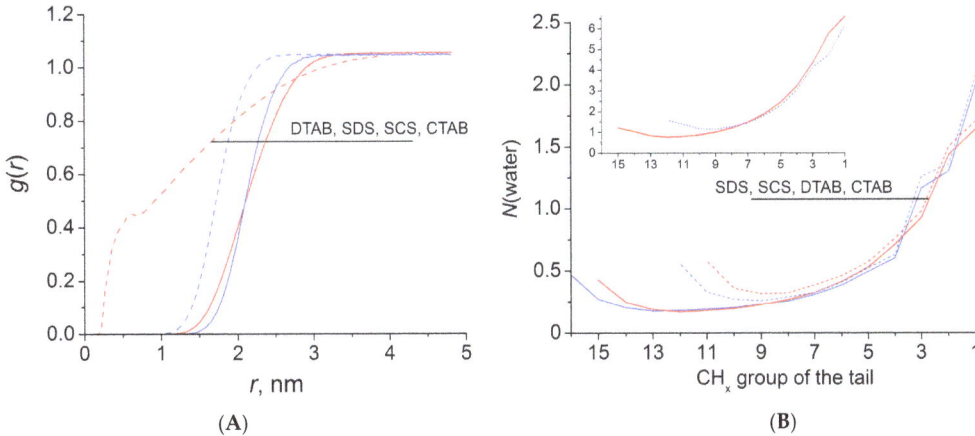

Figure 1. (**A**) Radial distribution functions of water with respect to the micelle center. (**B**) Average number of water molecules within 0.4 nm or 0.5 nm (inset) of the CH$_2$ (CH$_3$) groups of hydrocarbon tail in micelles [54,55]. Here, the red color corresponds to the cationic surfactants, solid lines correspond to the surfactants with a cetyl group.

MD simulation with explicit water allows directly counting the amount of water in micelles using various metrics. On one hand, we estimated the extent of hydration of each CH$_2$/CH$_3$ bead of hydrocarbon tails. Figure 1B shows the average number of water molecules staying within 0.4 nm of each bead in the micelles of SCS and CTAB having a n-C$_{16}$H$_{33}$ hydrocarbon tail, and SDS and DTAB with a n-C$_{12}$H$_{25}$ one. In alkylammonium surfactants, we attributed the first methylene group as belonging to the headgroup, basing on the charge distribution. While the hydration decreases quickly with moving away from the headgroups, it still never reaches zero, and even somewhat increases for the last two beads. The hydration of the first and third beads in the anionic surfactants is ca. 30% higher than in cationic ones, while it is ca. 9% lower for the second and fourth beads. On average, the hydration of the first four beads is somewhat higher in the anionic surfactants. However, this does not mean that the surface layer contains more water because the beads may share water molecules. Instead, if hydration is calculated using a radius of 0.5 nm then CTAB ions have some preference. Hence, we conclude the DS$^-$ and CTA$^+$ ions in micelles are hydrated to a similar extent.

On the other hand, we estimated the complete amount of water within the surface layer of SDS, SCS, DTAB, and CTAB micelles. As such, we used a simplified geometric model where micelle and its hydrocarbon core were represented as concentric triaxial ellipsoids; see Figure 2. Axes of the latter ellipsoid were chosen to make its volume and moments of inertia equal the values computed by MD simulation. The axes of the former ellipsoid were 2 × δ nm longer to reflect the δ-thick surface layer around the core. Then, the water molecules within δ nm of the core were enumerated and a time-average number was computed using the MD trajectories. Having this input data, we found the hydration H of the complete micelle as volume fraction of water in the Stern layer via Equation (2).

$$H(\delta) = \frac{V_{water}}{V_{Stern\,layer}} = \frac{N_{water} \cdot V_{1mol}}{V_{micelle} - V_{core}} \quad (2)$$

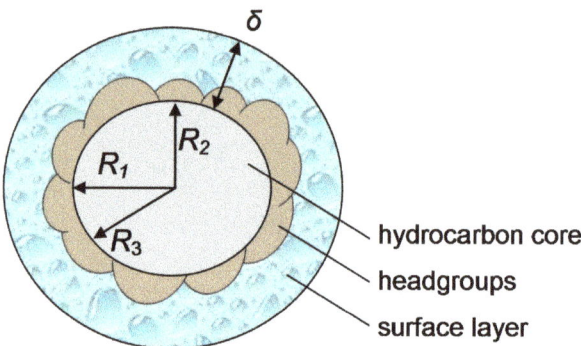

Figure 2. The geometric model used for estimating the hydration of micelles.

Here, δ is the Stern layer thickness; $V_{Stern\ layer}$, V_{water}, $V_{micelle}$, V_{core} are volumes of the surface layer, water, micelle ellipsoid, and the hydrocarbon-core ellipsoid, respectively; N_{water} is the number of water molecules in the surface layer; and V_{1mol} is the volume of a single water molecule (0.030 nm^3).

The results indicate that, despite the difference in the headgroup charge and micelle size, the $H(\delta)$ parameter appears almost equal for all four surfactants considered (the spread is no more than 0.05 units). This agrees well with the previously shown similarity of hydration of individual CH$_2$ beads of the surfactant ions.

As mentioned before, the degree of counterions binding, β (and the related degree of dissociation, α), is an essential quantity determining the behavior of an ionic micelle in solution. MD simulation allows examining the complete distribution of counterions instead of a single-point value. The calculated β values for some of the inspected surfactants are collected in Table 1 for two values of the boundary position. The lower boundary is found from the distribution of counterions around headgroups, and corresponds to the first minimum in the radial distribution function of Na$^+$ (Br$^-$) with respect to S (N) atoms. The upper one is somewhat arbitrarily chosen as 0.9 nm. The comparison indicates that considering the most tightly bound counterions only (the lower boundary) leads to underestimation of β, so the boundary must be chosen farther from the micelle. According to our simulations of SDS, potential models that provide higher counterion binding within the lower boundary also lead to deformation of structure of large micelles [55]. The upper boundary provides β values, which fit in the experimental range, except of CTAB where small underestimation of ca. 0.05 remains.

Table 1. Computed and experimental degrees of counterions binding of micelles, β.

Surfactant	MD	Experiment
SDS	0.23–0.67	0.45–0.86
CPC	0.37–0.65	0.45–0.67
CTAB	0.59–0.69	0.74–0.84

Note. Experimental β values for SDS, CTAB, and CPC were taken from various publications [11,102,103] and [104–107], respectively.

In order to elucidate the role of counterions, we applied an unusual approach. By means of MD simulations one can easily study systems, which cannot be prepared in the laboratory. We performed simulations of "bare" micelles, composed of DS$^-$ or CTA$^+$ ions only, without any other ions present in the aqueous solution. Usually, counterions are considered as a necessary factor for the formation of a micelle, which compensates the electrostatic repulsion between like-charged surfactant ions. In the simulations, the DS$^-$ micelle of 60 monomers expelled a small aggregate and reduced to 43 monomers. The CTA$^+$ micelle of 80 monomers was split into two comparable micelles; see Figure 3. This result indicates that the hydrophobic interaction itself is able to keep the aggregate together without additional stabilization up to some size.

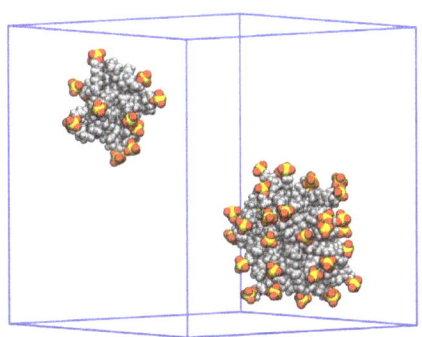
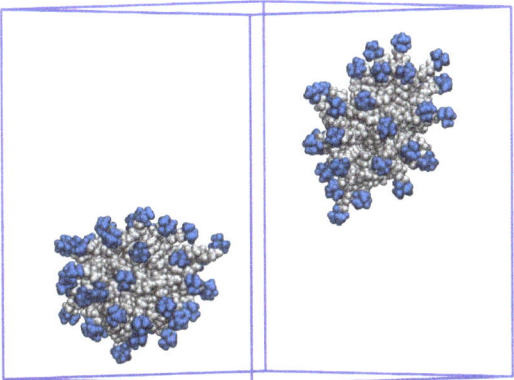

Figure 3. The results of 20 ns simulation of DS$^-$ (**left**) and CTA$^+$ (**right**) micelles without counterions. The small micelles differ only in the aggregation number.

It is well proven and documented that in mixed nonionic + ionic surfactant micelles at low content of the second component, the fraction of the counterions is small or even zero [14,108–110]. In such cases, the hydrophilic portions of the nonionic component ensured micelle stabilization. In any case, the diffuse layer of counterions exists around the micelle. In the case of polyelectrolytes, according to Manning's theory, there is a critical degree of counterions condensation [111]. It should be noted that some early theories do not assume an adsorption layer of counterions in the monolayers of ionic surfactants at the water/air and water/hydrocarbon interfaces [112].

The formation of pre-micellar, or sub-colloidal, species (DS$^-$)$_2$ occurs rather without than with the participation of the sodium ions [113,114], though at SDS and NaClO$_4$ concentrations of 10^{-4} M, the (DS$^-$)$_2$Na$^+$ particles were fixed by the electrospray spectrometry [115].

Though the hydration of ionic groups certainly takes place in the case of counterion-free dodecylsulfate aggregates, the MD prediction of formation of such aggregates can be considered as an illustration of the efficiency of the hydrophobic interaction.

A related issue is the profile of the electrostatic potential around the micelle. This problem was extensively studied previously, but simplified geometries (sphere, cylinder) were mostly used because they enable analytical solutions. We performed the calculation of $\Psi(r)$ considering micelles of natural shapes, which were produced by MD simulation. The suitable method of calculation was the numerical solution of the Poisson–Boltzmann equation. The discrete character of charge distribution and complex shape of micelle were taken into account, as well as the finite size of counterions. Solvent and micelle were treated as dielectric continuums, as is usually done in such calculations. The profiles $\Psi(r)$ predict the expected fade of Ψ with distance and ionic strength, Figures 4 and 5. They were then interpreted in terms of the commonly used characteristics: surface potential Ψ_0 and Stern layer potential, Ψ_δ. For SDS and CTAB micelles, the values $\Psi_0 = -100$ mV and +(149–151) mV, $\Psi_\delta = -75$ mV and +(70–78) mV were found (at $I = 0.05$ M), which stays in line with the estimations of other methods. The Stern layer boundary δ was chosen 0.32 nm for SDS and 0.45 nm for CTAB according to the distribution of counterions around the micelle surface, more details may be found in our previous work [66].

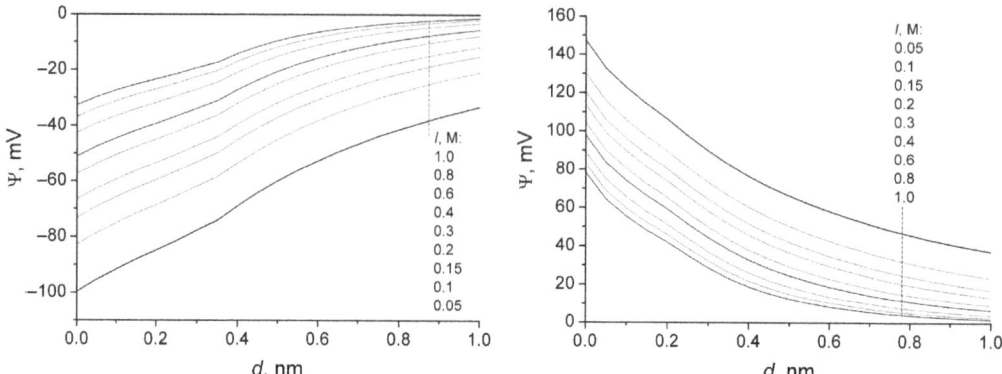

Figure 4. Profiles of electrostatic potential fade with distance for SDS (**left**) and CTAB (**right**) micelles at different ionic strengths. Reprinted from [66] with permission.

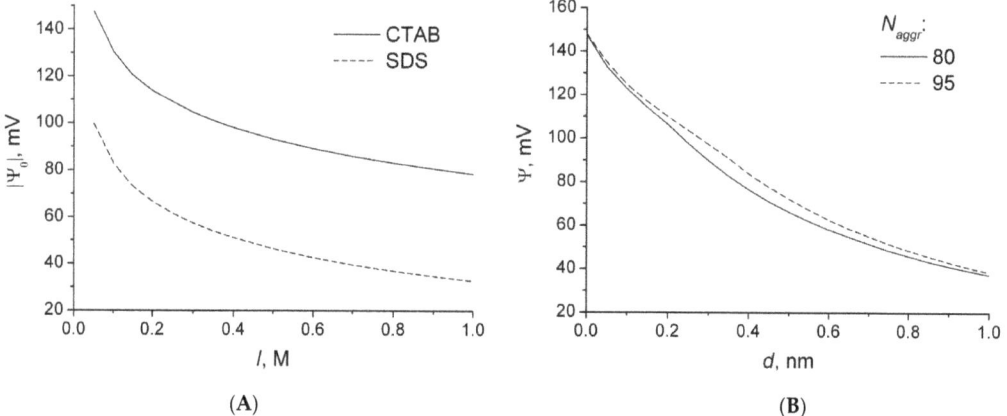

Figure 5. (**A**): Dependence of surface electrostatic potential on the ionic strength for SDS and CTAB micelles; (**B**): The profile of the electrostatic potential fade with distance for CTAB micelles of different size at $I = 0.05$ M. Reprinted from [66] with permission.

Note that Figures 4 and 5B do not show a sharp change in the potential near the surface.

The method chosen by us allows us to calculate the surface electrostatic potential of "bare" micelles without counterions. We used the micelles consisting of 41–43 monomers produced as a result of MD simulations presented before. The Ψ_0 values are −420 mV and +380 mV. Importantly, these numbers are of lesser magnitude than found for the original size micelles, where Ψ_0 is −535 mV and +630 mV. Evidently, a decrease in the aggregation number led to a reduction in the surface charge density and headgroup—headgroup repulsion up to the magnitudes, which can be compensated by hydrophobic interactions. On the other hand, calculations of the electrostatic potential of the dodecylsulfate micelle with "empty", i.e., counterions-free Stern layer using Equation (1) lead to a value of −(193–199) mV at $\alpha = 1$ [14,95]. The screening of the surface charge by the diffuse layer was accounted for in this case, contrary to the above mentioned MD simulations. Therefore, the influence of the diffuse layer corresponds to −340 mV.

From the computational perspective, we identified the importance of bonded parameters for potential models of surfactants. It was previously identified for higher hydrocarbons and lipids and led to derivation of improved potential models, but it was not applied for common surfactants. Usually, the non-bonded parameters (point charges, Lennard–Jones potential σ and ε) are in focus during the parameterization. However, for SDS, the parameters of dihedral (torsion) angles governing the rotation around C–C bonds strongly affect the micelle properties, as well. Too rigid tails (i.e., with high potential barriers for the rotation) superfluously tend to adopt a parallel arrangement. In small micelles (<100 monomers), this affects only the degree of counterions binding because the headgroups become more grouped. However, large micelles may artificially adopt a bilayer-like shape instead of a rod-like one. Whether this occurs depends also on the non-bonded parameters. The problem was solved by making the tails more flexible, which was achieved by employing the dihedral parameters proposed for description of higher hydrocarbons and lipids.

A crucial feature of micelles is their ability to bind (adsorb) molecules and ions present in solutions. The properties of the molecule often undergo a change after adsorption, and experimental registration of this change is the basis of the method of investigating micelle properties by means of molecular probes. Still, this method heavily relies on the a priori information about the state of the molecular probe, which is needed to interpret the registered changes. This is a severe problem because the state itself cannot be observed directly and must be deduced from some other experiments. By using properly prepared and mutually consistent potential models it is possible to recapitulate the adsorption process in MD simulation and examine the state of the molecular probe including its localization, local environment, and diffusion coefficient. For example, we simulated the adsorption of $N(C_3H_7)_4^+$ ions in SDS micelle, and observed a significant reduction in its diffusion coefficient as result [53]. Still, even for these oppositely charged species adsorption may demand a quite long simulation time; therefore, we concluded it is more practically convenient to have the molecular probe initially located in the micelle at the start of MD. The final result (Figure 6) demonstrates the location of this cation; it is in line with the scheme proposed by Bales et al. [116], who considered the tetra-*n*-propylammonium and other tetraalkylammonium cations in the Stern layer of the micelles of tetraalkylammonium dodecylsulfates.

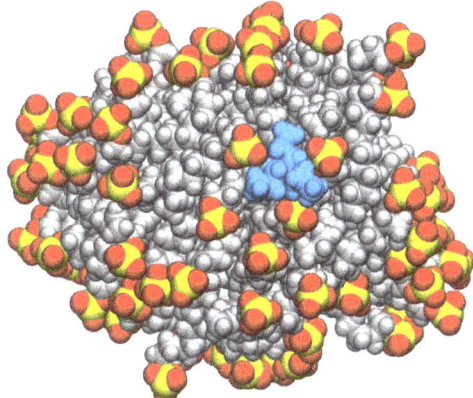

Figure 6. Location of the tetra-*n*-propylammonium cation (colored blue) in a SDS micelle; sodium ions are not shown.

We also paid attention to nonionic micelles with and without an adsorbed dye molecule, since they are used in the indicator method of estimating the surface electrostatic potential [10,13,14].

The micelles of TX-100 with an aggregation number of 110 adapted a curved rod shape in accordance with experimental observations. The diameter of the rods is ~5 nm and their length is ~12 nm (in unbend state), as determined by visual inspection in Visual Molecular Dynamics (VMD) software. The thickness of the hydrophilic shell was approximately 1 nm, while some polyoxyethylene chains extended for 1.5 nm out of it. The detailed structure of the hydrophilic shell is quantified by means of the fraction of O atoms found at a given distance from the hydrocarbon core. The graph is shown in Figure 7; here, a mistake made in this graph in the original paper is corrected. Estimated in such a way, shell thickness equals 1.4 nm, from which the half of O atoms are located within 0.45 nm of the surface [64].

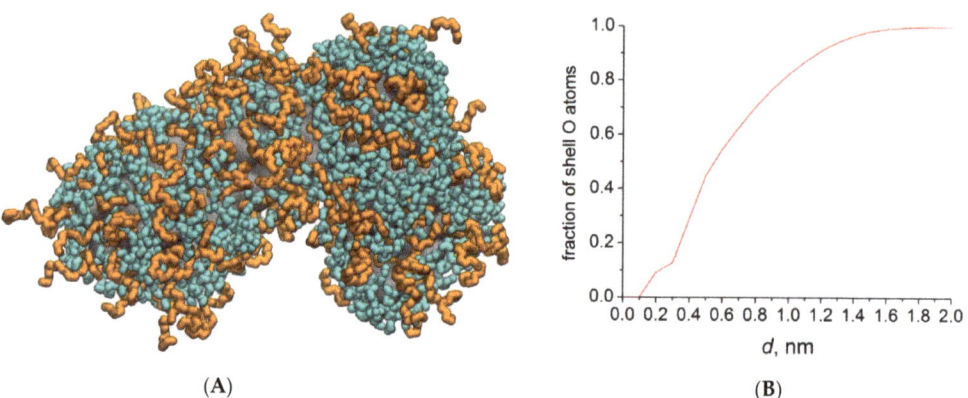

(A)   (B)

Figure 7. (**A**) Rod-like TX-100 micelle. Gray: hydrocarbon core, orange: polyoxyethylene chains, blue: water molecules within 0.7 nm from the core. (**B**) fraction of oxyethylene groups that are within distance d from the core surface.

5. Solvatochromic Indicators in Surfactant Micelles

5.1. Choice of Pyridinium-N-Phenolate Dyes

Solvatochromic dyes have been widely used for studying colloidal systems since long ago [117], and such works are ongoing [1,118–121]. Among other solvatochromic indicators, the pyridinium N-phenolate betaine dyes belong to the most powerful ones [21]. These compounds were often used for studying surfactant micelles and similar nano-sized species [21,117,122,123].

Therefore, we started our research with these dyes in micellar surfactant solutions. The MD simulations were aimed to reveal the locus of the dyes and the solvation/hydration character in the micellar pseudophase. Another reason for this choice is the application of these dyes as acid–base indicators in micellar media. This was first demonstrated for standard Reichardt's dye, a solvatochromic pyridinium N-phenolate dye, i.e., 4-(2,4,6-triphenylpyridinium-1-yl)-2,6-diphenylphenolate (Scheme 2) [19]. Accordingly, not only the colored solvatochromic neutral (zwitterionic) forms (Scheme 3), but also the colorless cations were examined here in micellar environments.

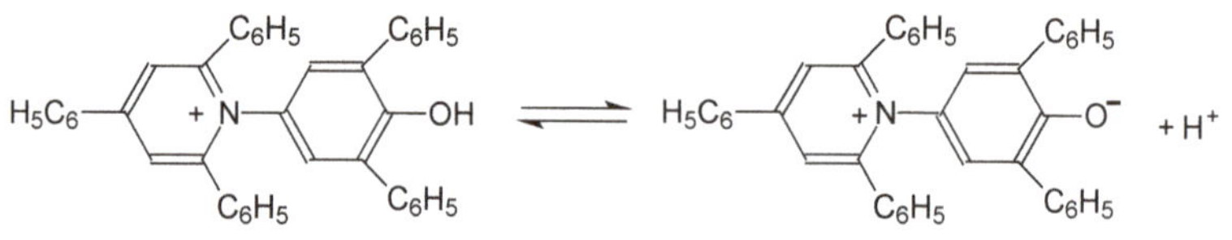

Scheme 2. Acid–base equilibrium of the solvatochromic pyridinium-N-phenolate dye, RD-Ph.

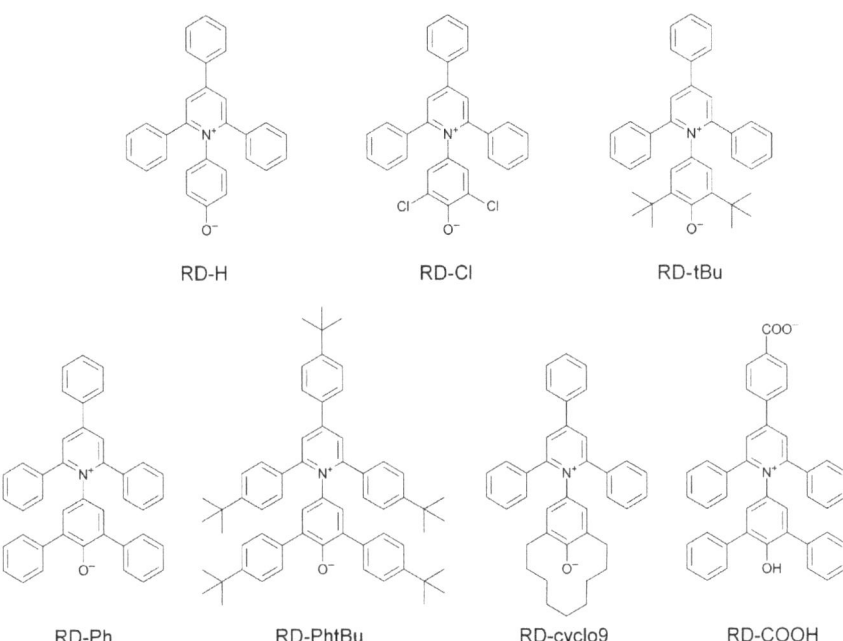

Scheme 3. Molecular formulae of betaine dyes examined in this study.

The main results of our modeling are presented below.

5.2. Localization of the Solvatochromic Betaine Dyes in Surfactant Micelles

Overall, molecules of most of the dyes are located on the surfaces of the micelles staying in contact with both the hydrocarbon core and the solution. Still, the dyes differ by the depth of immersion (penetration) into the micelles. It can be described as the distance between the micelle center and the dye molecule center. As dye center, it is convenient to consider the N atom of the pyridinium ring. The average distances are collected in Table 2 and schematically represented in Figures 8 and 9 as a diagram for convenient comparison. The distances below 1.35 nm (SDS) or 1.55 nm (CTAB) are highlighted gray to indicate the approximate extent of the hydrocarbon core. It is difficult to unambiguously set the border because of the ellipsoidal shape of micelles, so we conventionally defined it here as the distance where water fraction appears at least 5%.

Table 2. Average distance (nm) of the N atom of the betaine dyes to the center of the mass of micelles.

Dye	SDS		CTAB	
	$^+D^-$	^+DH	$^+D^-$	^+DH
RD-H	1.47	1.48	1.73	1.91
RD-Cl	1.52	1.45	1.73	1.94
RD-cyclo9	1.37	1.46	1.67	1.84
RD-tBu	1.31	1.40	1.57	1.89
RD-Ph	1.45	1.41	1.67	1.88
RD-PhtBu	0.96	1.10	1.11	1.39
RD-COOH	1.43	1.44	1.76	1.88
	D^-		D^-	
RD-COOH	1.51		1.81	

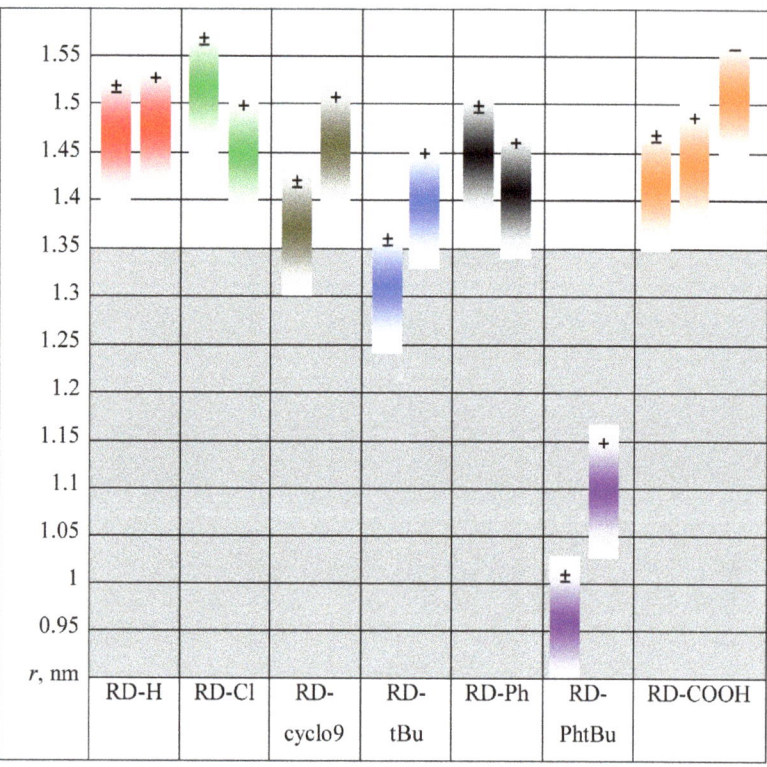

Figure 8. Average distance of the N atom of the betaine dyes to the center of mass of the SDS micelles. The band length corresponds to the range of typical distances.

Less convenient but a more informative characteristic is the distribution function of this distance, $p(r)$, which was computed over MD trajectories and served to find the time-average value. An example of such graphs is shown in Section 5.5 below.

Three observations stem immediately from the diagrams:

- In CTAB micelles, protonation pushes the dye molecule considerably towards the bulk solution by 0.2–0.3 nm, while in SDS micelles the effect is smaller and depends on the dye;
- For zwitterionic forms, the immersion correlates with the steric demands of the substituents. Compact dyes RD-H and RD-Cl are in average located closer to the bulk solution than bulky dyes RD-cyclo9, RD-tBu, RD-Ph, and RD-COOH, with rare exceptions. This becomes most evident in the extreme case of RD-PhtBu, which is deeply buried in the hydrocarbon cores of both micelles.
- In SDS micelles, the dye RD-PhtBu is located ca. 0.35–0.4 nm deeper in the micelle as compared with other dyes. In CTAB micelles, this difference is ca. 0.5–0.55 nm, which is somewhat larger.

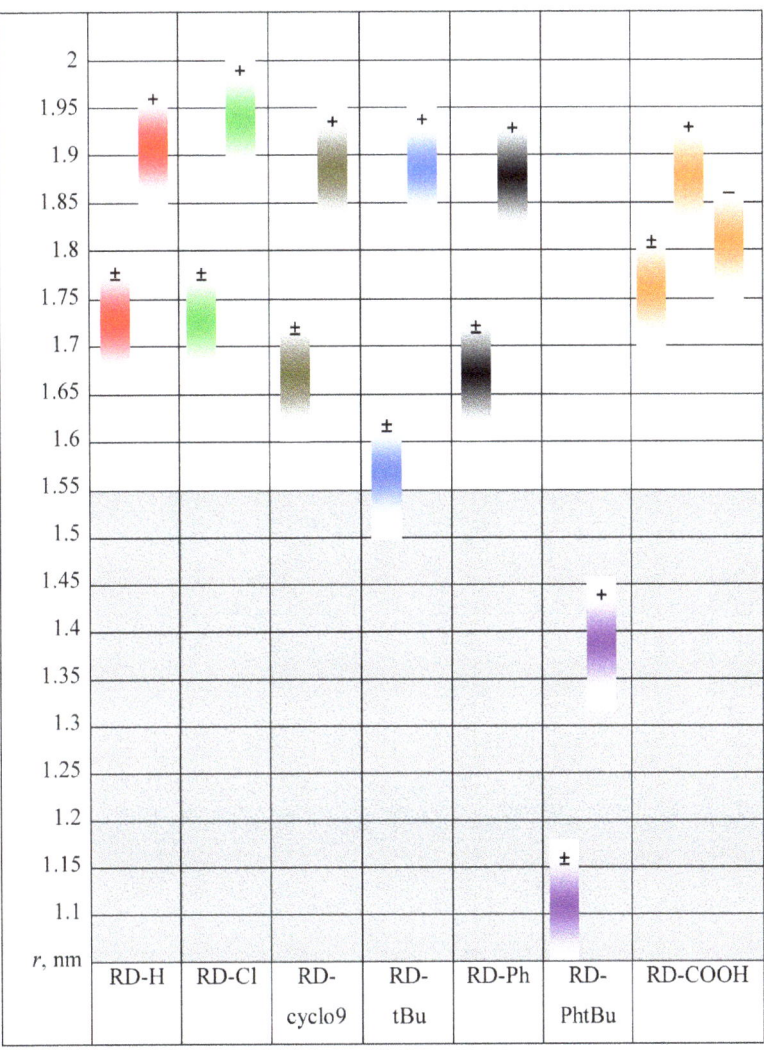

Figure 9. Average distance of the N atom of the betaine dyes to the center of mass of the CTAB micelles. The band length corresponds to the range of typical distances.

5.3. Orientation of the Solvatochromic Betaine Dyes in Surfactant Micelles

An important aspect of the dye localization is the orientation, or placement, of the dye molecule with respect to the micelle surface. Typically, the molecule is placed either roughly parallel to the surface or perpendicular to it (Figure 10). Sometimes both orientations are sampled by turns. The particular placement depends mainly on the substituents, while other factors show a smaller effect. The results of our observations are collected in Table 3.

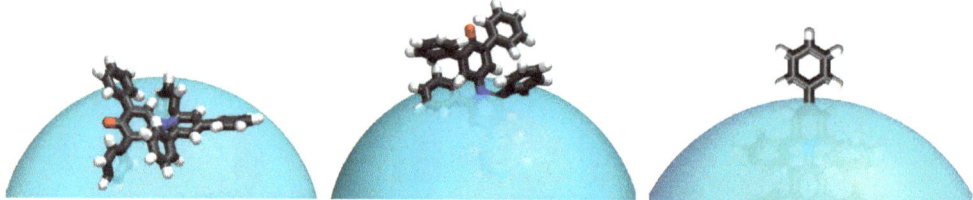

Figure 10. Typical placements of Reichardt's betaine dyes in micelles, shown for the example of the standard dye, RD-Ph. (**Left**) "horizontal" orientation, (**middle**) "vertical" orientation, (**right**) "inverted vertical" orientation.

Table 3. Dominant orientation of Reichardt's dyes molecules in micelles. "H" means horizontal orientation, "V" means the vertical one, "I" means the inverted vertical one. "Sometimes" means a probability of 10–30%; "often" means a probability of >30%.

Dye	SDS		CTAB	
	$^+D^-$	^+DH	$^+D^-$	^+DH
RD-H	H, often V	H	H, sometimes V	H
RD-Cl	H, sometimes V	H	H	H
RD-cyclo9	H, sometimes V	H	H	H
RD-tBu	H, sometimes V	H	H	H
RD-Ph	H, sometimes V	H, sometimes I	H	H, sometimes I
RD-PhtBu	V, often H	I, sometimes H	V, often H	I, sometimes H
RD-COOH	I, sometimes H	H	H, sometimes I	H
	D^-		D^-	
RD-COOH	H		H	

In more precise terms, we defined the inclination angle of the molecule as ∠(micelle COM, dye N atom, dye O atom) and calculated it for each MD frame. The molecule was said to be in the vertical orientation when θ was less than 40°. The inverted vertical one corresponded to $\theta \leq 130°$, and the horizontal one to $40° < \theta < 130°$. Fraction of time, which the molecule spent in a given orientation, indicates probability of the latter. Still, it was important to perform additional visual examination to check the correct identification of the major orientation: simple geometric criteria were insufficient for characterizing such soft and nonspherical particles as micelle + dye aggregates. It was especially important in the case of RD-PhtBu having very deep immersion into micelles. More elaborate analysis was made using distribution functions of θ, $p(\theta)$, the example of such functions is shown below in Section 5.5.

The horizontal orientation is observed as the most often. The dyes RD-Cl, RD-cyclo9, and RD-tBu behave quite similarly, while each of the other dyes has some unique features. The inverted vertical orientation is sometimes adopted by protonated standard dye RD-Ph and is dominant for protonated RD-PhtBu. The neutral RD-PhtBu mostly stays vertical, and neutral RD-COOH in SDS stays in inverted vertical orientation.

We see that the same dye in the same state shows rather similar behavior in both anionic and cationic micelles, which indicates a minor influence of the headgroup charge on this characteristic. This is a quite unexpected result due to the opposite signs of the charge–charge interactions with the micelles.

The substituents also affect the orientation weakly: the zwitterions of RD-Cl, RD-cyclo9, RD-tBu, and RD-Ph behave similarly. Only the extreme cases of the absence of R^3 substituents (RD-H) or with very bulky substituents $R^1 = R^2 = R^3$ ones (RD-PhtBu) express some deviations. The dye RD-COOH is also an individual case because in the zwitterionic form the negative charge is located on the COO^- group instead of O^-.

The most influential factor appears to be the protonation state of the betaine dyes, which determines the overall charge of the molecule. The protonated forms of all betaine dyes (except of the bulkiest RD-PhtBu) are coherent in having a horizontal orientation in micelles of both surfactants.

To be more precise, for zwitterionic forms, the horizontal orientation is usually skewed towards having the phenolate part somewhat pushed towards the bulk solution. Oppositely, for cationic forms, this orientation implies the molecule being accurately parallel to the surface, or sometimes having the pyridinium part advanced towards the solution.

It is important to compare our data with the information on the locus of the colored zwitterionic form of these dyes in micelles obtained by ^{1}H NMR spectroscopy. Zachariasse et al. [122] studied the dye RD-Ph in CTAB micelles, with a surfactant: dye ratio of 20:1. It was demonstrated that the dye exerts the largest influence on the methyl groups of $N(CH_3)_3^+$ and the hydrogen atoms of the α and β methylene groups. Hence, the dye is located in the micellar surface region. Tada et al. [124] have found that the dichloro derivative RD-Cl is rather deeply penetrated into the cationic micelles, though our MD simulations enable concluding that the locus of the RD-Ph dye is deeper (Figure 9).

For the dye RD-H in cationic micelles, Plieninger and Baumgärtel [125] deduced from the ^{1}H NMR data the location of the anionic center of the molecule among the $N(CH_3)_3^+$ groups, and exposition of the cationic pyridinium part to water. A reverse picture was proposed for the same dye in the SDS micelles. Here, the pyridinium center of the molecule is situated in the plane of the $O-SO_3^-$ groups, whereas the phenolate moiety is "immersed in water layers" [125]. Our MD simulations did not confirm such a pronounced penetration of this betaine dye towards the aqueous phase. Additionally, the change in the direction of the zwitterion was observed only for three dyes: RD-Ph (protonated form), RD-PhtBu, and RD-COOH. This is the above-mentioned inverted vertical state.

In Figures 11 and 12, the most probable states of Reichardt's dyes in zwitterionic and cationic forms in SDS and CTAB micelles, elicited from the observation of the MD data, are presented. The orientation of the dyes with the phenolic part placed inside the micelle occurs due to either their extremely pronounced hydrophobicity (RD-PhtBu) or the good solvation of the COOH and COO^- groups (RD-COOH).

The special case of the acid–base equilibria of the last dye is presented in Scheme 4.

In this case, the colored solvatochromic species is the anion, because of the much stronger basic properties of the phenolate as compared with the carboxylate.

All the above information concerning the location and orientation of different dye species in SDS and CTAB micelles should be understood as an average picture of a huge number of snapshots. It reflects the most probable states of the dyes.

5.4. Hydration of the Solvatochromic Betaine Dyes in Surfactant Micelles

Another essential question that we examined is the local environment, or microenvironment, of the dye molecules. In the surface layer of ionic surfactant micelles, water, hydrocarbon, and electrolyte contact the dye simultaneously, and their proportion affects its visible absorption spectrum. We employed the following approach to elucidate this point. At each time point, the atoms located within 0.4 nm from the dye molecule were enumerated and then divided into three categories: belonging to the micelle hydrocarbon core, belonging to the surfactant headgroups, or belonging to the water molecules. The time-average number of atoms of each category was then computed and converted to the molar fraction of the corresponding component (hydrocarbon, headgroups, or water) in the dye microenvironment. The values for the discussed dyes are presented in Figure 13. The protonated forms of all the dyes (except RD-PhtBu) have a similar hydration in both SDS and CTAB micelles, regardless of the substituent and surface charge. Only RD-PhtBu is considerably less hydrated, which is natural for its deep localization in the micelles.

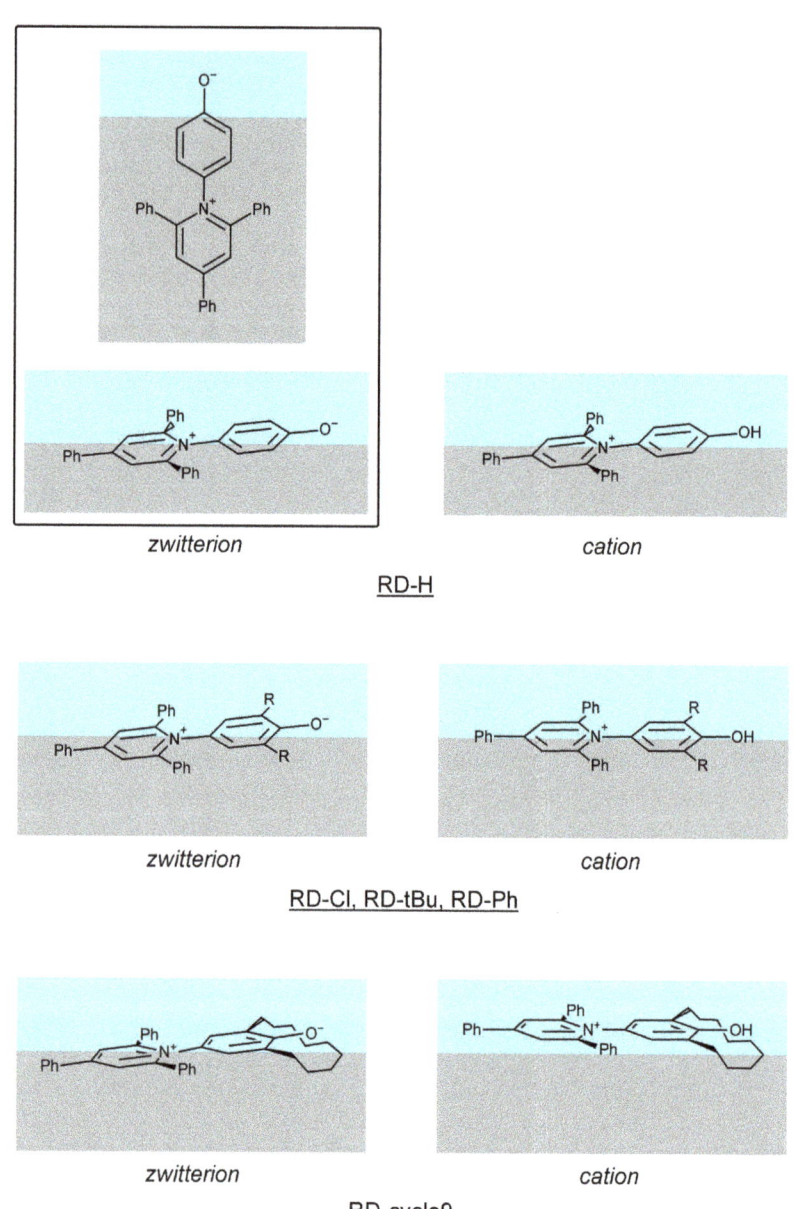

Figure 11. Most probable orientations of the dyes RD-H, RD-Cl, RD-tBu, RD-Ph, RD-cyclo9 in ionic surfactant micelles as deduced from MD simulations.

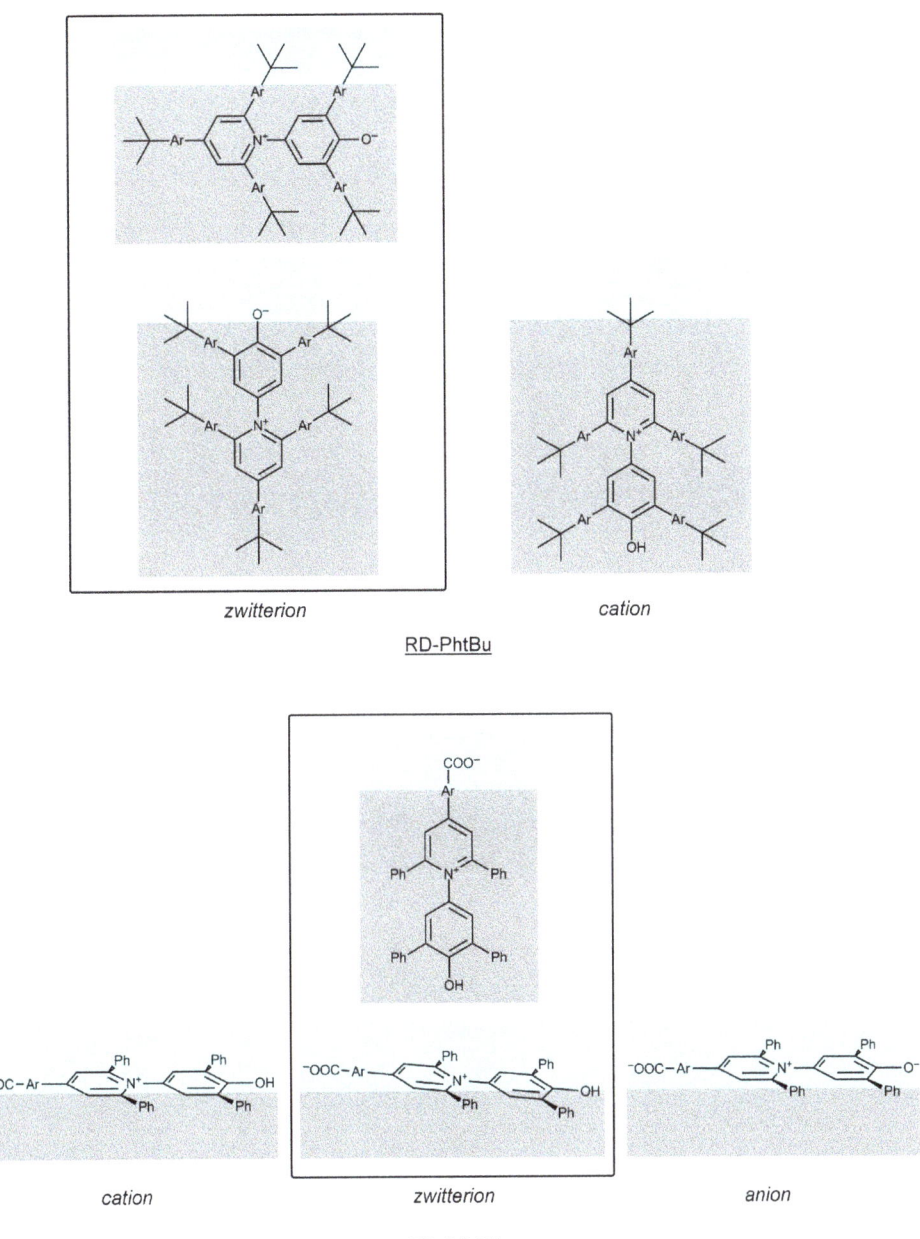

Figure 12. Most probable orientations of the dyes RD-PhtBu and RD-COOH in ionic surfactant micelles as deduced from MD simulations. Adapted from [61].

Cation, colorless Zwitterion, colorless Anion, colored

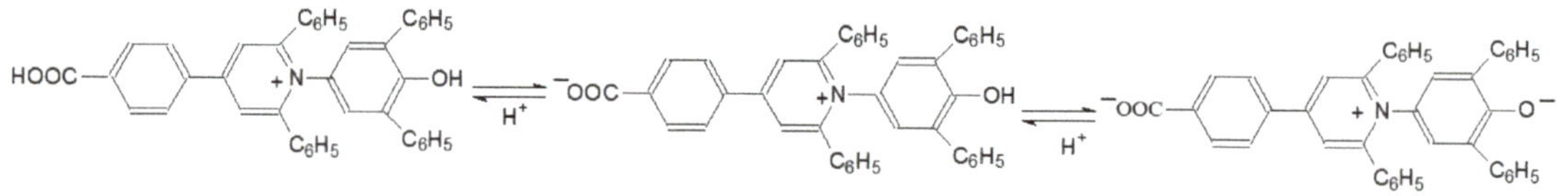

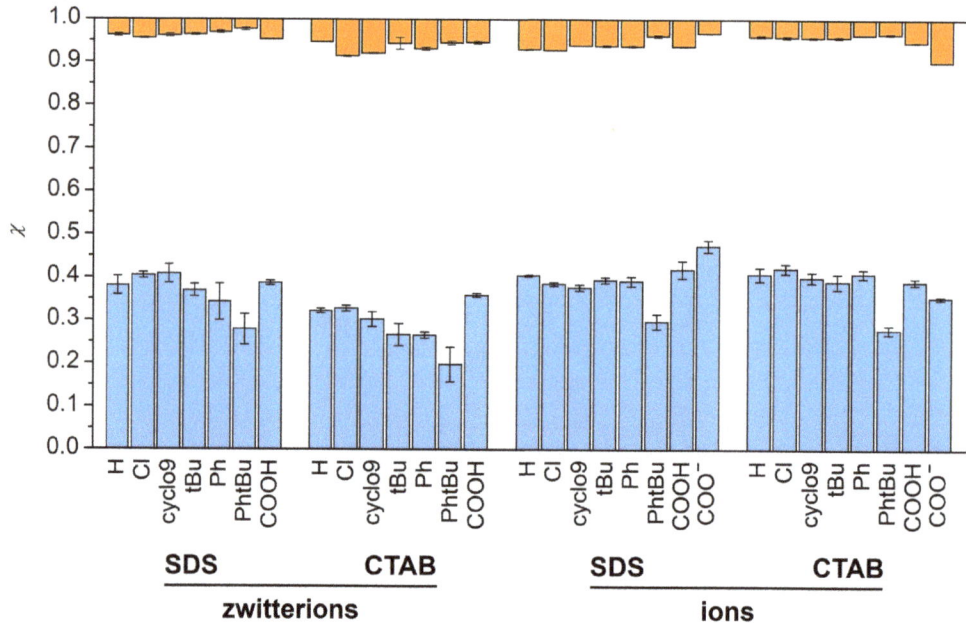

Scheme 4. Protolytic equilibria of the carboxyl-containing dye RD-COOH. For this dye, only the anion is colored and hence solvatochromic.

Figure 13. Molar fraction of water atoms (blue columns below) and headgroup atoms (orange columns above) around molecules of different dyes in various conditions. For the dye RD-COOH, zwitterion is colorless and contains OH and COO^- groups. In the ions section, COOH denotes RD-COOH cation and COO^- denotes its anion.

- For zwitterions, the influence of substituents is somewhat stronger.
- Zwitterions in CTAB micelles are ~20% less hydrated than in SDS ones (the average difference of χ(water) is ~0.08 out of ~0.38).
- The molar fraction of headgroups around the dye molecules is short (<0.1). Zwitterions in CTAB and cations in SDS have ca. 2-fold more headgroups around than in the other cases.

Overall, if RD-PhtBu is not considered, the degree of hydration varies in a rather narrow interval. This can be explained by the large size of the betaine dye molecules, which facilitates contact with water, and their quite similar localization.

To get additional insight, similar calculations were performed for the individual phenolic O atom, Figure 14. As will be discussed below, its microenvironment is of a particular importance.

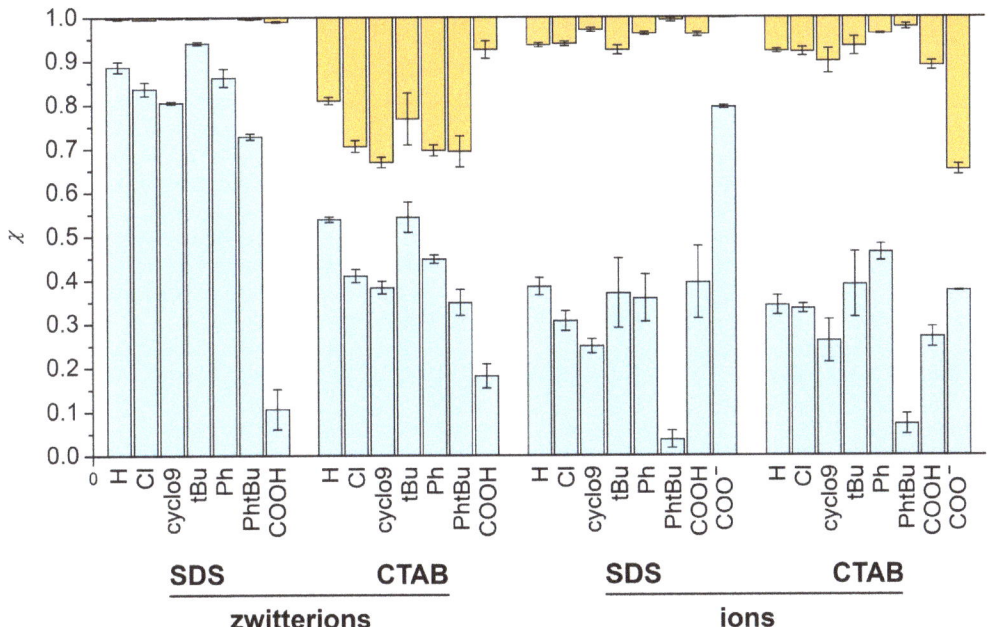

Figure 14. Molar fraction of water atoms (blue columns below) and headgroup atoms (orange columns above) around phenolic O atom of different dyes in various conditions. For the dye RD-COOH, zwitterion is colorless and contains OH and COO$^-$ groups. In the ions section, COOH denotes RD-COOH cation and COO$^-$ denotes its anion.

Here, the substituents pronouncedly affect χ(water) in micelles of both kinds. This is natural because of their proximity to the phenolic O atom.

- The hydration of the OH group in protonated forms depends only weakly on the surface charge, as it was observed for the whole molecule.
- Oppositely, the O$^-$ atom of zwitterions is much better hydrated in SDS than in CTAB micelles: the χ(water) difference is ca. 0.4.
- While PhtBu shows the weakest hydration of the OH group among the cations (it is almost isolated from water), RD-COOH is at this place among zwitterions. This occurs due to the competition for water with the COO$^-$ group in the latter case: the molecule is oriented in such a way that COO$^-$ is hydrated well, while OH is mostly immersed into the micellar hydrocarbon core.
- Surprisingly, the effect of substituents does not correlate with their steric demand. Under the same conditions, RD-tBu and RD-Ph have the highest χ(water) together with RD-H despite of the bulky R^3 groups. Instead, RD-Cl and RD-cyclo9 have lower χ(water), succeeded by RD-PhtBu.
- The O$^-$ atom of zwitterions strongly attracts cationic headgroups and repulse anionic ones. Instead, the OH group of cations interacts similarly with both kinds of headgroups.
- The hydration efficiency of O$^-$ atom of the zwitterions $^+$D$^-$ in SDS micelles decreases in the following sequence: RD-tBu > RD-H > RD-Ph > RD-Cl >, etc. In turn, in CTAB micelles, an interaction of the negative oxygen with the positively charged alkylammonium groups takes place. The difference between the hydration of cations $^+$DH in SDS and CTAB micelles is not significant. The special cases of the RD-PhtBu and the anion COOH (RD-COO$^-$) in CTAB micelles are quite understandable.

5.5. Influence of Other Factors on the Aforementioned Characteristics

The influence of the length of the hydrocarbon chain of a surfactant on the state of the dye molecule in the micelles was studied separately. For this purpose, simulations of the RD-Ph zwitterion in DTAB and SCS micelles were carried out. In the latter case, the temperature was 50 °C, since at room temperature SCS does not form micelles. To ensure the possibility of a correct comparison with SCS, the simulation of RD-Ph in SDS micelles was repeated at this temperature. It was found that the indicator dye is localized deeper in DTAB micelles than in SDS micelles with the same hydrocarbon chain.

Compared with CTAB micelles, in DTAB micelles the pyridinium part of the indicator molecule is immersed deeper due to the fact that the probability of the orientation shown in Figure 10b (Figure 15B) is higher. On the contrary, in the micelles of SCS and SDS (at 50 °C), the molecule is localized and oriented almost similarly.

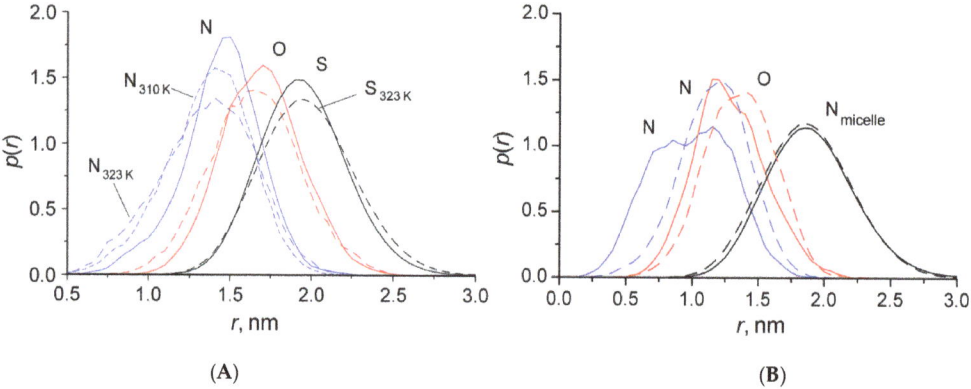

Figure 15. Distance distribution functions between micelle COM and N, O atoms of RD-Ph and N (S) atoms of surfactants. (**A**) SDS at 298 K (solid), 310 K (short dash), 323 K (long dash). (**B**) DTAB and CTAB. The curves for CTAB are dashed and shifted left by 0.5 nm.

Thus, the position of the molecule of standard Reichardt's dye in surfactant micelles during the transition from a hydrocarbon chain of one length to another may change differently, depending on the character of the headgroups.

With regard to the microenvironment, in surfactant micelles with a longer hydrocarbon chain, the hydration of both the molecule as a whole and its O atom somewhat decreases, while the fraction of hydrocarbon around it increases [57].

In addition to anionic and cationic surfactants, RD-Ph was studied in micelles of the zwitterionic surfactant CDAPSn. It was found that the location and microenvironment of the dye in these micelles are generally similar to those in the micelles of CTAB. The dye molecule is located 0.1–0.2 nm (zwitterionic form) or 0.3–0.4 nm (cationic form) closer to the surface than in CTAB micelles. It is important to note that the molecule is almost not in contact with the negatively charged SO_3^- fragment, i.e., it is actually surrounded by positively charged $[CH_2N(CH_3)_2CH_2]^+$ fragments. This makes its microenvironment in this zwitterionic surfactant close to that in the cationic surfactant CTAB.

Finally, using the example of SDS micelles, the effect of temperature on the state of the solubilized dye in the zwitterionic form was studied. It can be seen that, as the temperature increases from 298 K to 310 K and 323 K, the dye gradually immerses into the micelle, although the difference is small (~0.1 nm) and is observed primarily for the pyridinium fragment of the molecule (Figure 15A). The orientation of the dye remains generally unchanged (Figure 16).

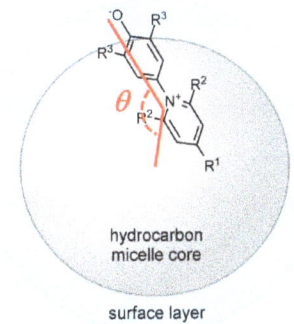 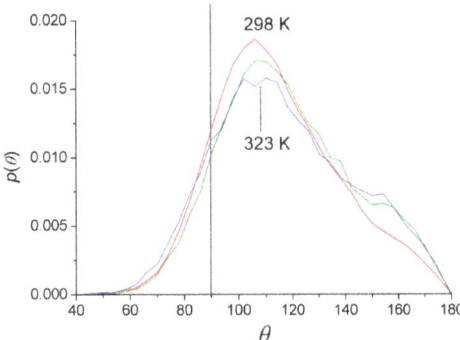

Figure 16. Definition of inclination angle θ (**left**) and its distribution functions $p(\theta)$ (**right**) for RD-Ph in SDS micelles at 298 K, 313 K, 323 K. Adapted from [57,61].

The effect of temperature on the microenvironment of the molecule is also insignificant, but in the microenvironment of the O atom the water content gradually decreases from 10.30 ± 0.04 at 298 K to 10.28 ± 0.14 and 9.97 ± 0.09 atoms at 310 K and 323 K, respectively. This can be caused by the thermal expansion of water, because the microenvironment of a constant volume accommodates fewer molecules [57].

5.6. Hydrogen Bonding of the Solvatochromic Betaine Dyes in Surfactant Micelles

Formation of hydrogen bonds through the phenolic O atom was shown to affect the spectrum of Reichardt's dyes. MD simulation allows calculating the average number of hydrogen bonds formed between the chosen species, n_{hb}; therefore, we examined this point in more detail for standard Reichardt's dye RD-Ph. We used the following criteria to identify an H-bond: (a) donor–acceptor distance no longer than 0.35 nm, and (b) angle ∠ (H atom; donor; acceptor) no more than 30°. As the acceptor, the phenolic O atom was chosen, the donors were water molecules. The results for micellar solutions, as well as for pure water solutions, are listed in Table 4 (data for selected entire organic molecular solvents and an ionic liquid bmim$^+$PF$_6^-$ are given for comparison). The values of E_T^N parameter of these solutions are shown alongside for the Discussion below.

This parameter is defined as follows:

$$E_T^N = \frac{E_T(30) - 30.7}{32.4} \qquad (3)$$

The numbers 30.7 and 32.4 are chosen so that the E_T^N values for water and tetramethylsilane are 1.00 and 0.00, respectively. Reichardt's $E_T(30)$ parameter is:

$$E_T(30) = hc\tilde{\nu}N_A = 28591 \times \frac{1}{\lambda} (\text{kcal/mol}) \qquad (4)$$

Here, h is the Planck constant, c is the speed of light, $\tilde{\nu}$ is the wavenumber, cm^{-1}, N_A is the Avogadro number, and λ is the wavelength, nm, of the charge-transfer absorption band of the dye. The indicator dye denoted here as RD-Ph was numbered "30" in the original paper, which described a series of similar compounds [126]. This dye is now known as standard Reichardt's solvatochromic dye.

Table 4. Calculated number of hydrogen bonds formed by the O atom of the standard dye RD-Ph with water and the experimentally measured E_T^N parameter of micelles and organic solutions (*in italic*).

Surfactant	t, °C	n_{hb}	E_T^N [a]
No (pure water)	25	2.17	1.000
No (pure water)	50	2.08	0.969 [b]
SDS	25	1.85	0.828
SDS	50	1.78	0.810
SCS	50	1.74	0.783
Methanol	25	—	*0.762*
DTAB	25	1.16	0.716 [c]
CTAB	25	1.12	0.687 [d,e]
bmim$^+$PF$_6^-$	25	—	*0.667*
CDAPSn	25	1.01	0.657 [d]
Ethanol	25	—	*0.654*
Acetonitrile	25	—	*0.460*
Dimethyl sulfoxide	25	—	*0.444*

Note. [a] [14], c(surfactant) = 0.01 M; [b] [127]; [c] [128]; or 0.724 [19]; [d] c(surfactant) = 0.001 M, [14]; [e] 0.705 [19].

An interesting observation follows from Table 4: the E_T^N (n_{hb}) dependence is monotonic (Figure 17). The formation of extra hydrogen bonds with surrounding water proportionally increases the measured polarity parameter. The points in Figure 16 form three separate groups: alkylammonium surfactants (left), alkylsulfate surfactants (middle), and water (right). Within each group, the dependence is close to linear with almost the same slope: 0.39 for the first two groups and 0.34 for the third one.

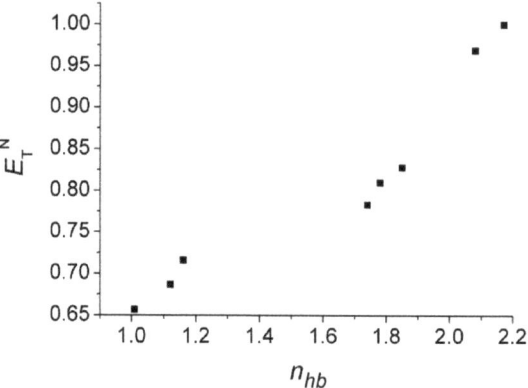

Figure 17. E_T^N parameter of a solution vs. the number of hydrogen bonds between the RD-Ph O atom and water.

This observation has a clear theoretical justification: in the total excitation energy of a Reichardt's dye molecule a term can be distinguished that corresponds to the difference in the hydrogen-bonding energy between the electronic ground and excited states. It is natural to assume that the number of H-bonds in both states is the same, because according to the Franck–Condon principle, the system has no time to rearrange during the excitation–relaxation act. Therefore, this term is proportional to the number of H-bonds in the ground state (Equation (5)). This explains the linearity of the dependence E_T^N vs. n_{hb} of micellar solutions.

$$E_T(30) = E_1 - E_0 = \Delta E_{hb} + \Delta E_{non\text{-}hb} = (e_{hb\,1} - e_{hb\,0}) n_{hb} + \Delta E_{non\text{-}hb} \qquad (5)$$

where E_0 and E_1 are the ground and excited state energies of the RD-Ph molecule, ΔE_{hb} and $\Delta E_{non\text{-}hb}$ are contributions of hydrogen bonding and of other interactions to $E_T(30)$,

$e_{hb\ 0}$ and $e_{hb\ 1}$ are the average energies of a single RD-Ph—water H-bond in the ground and excited states.

Beginning from the publication by Zachariasse et al. [122], there is a number of papers that give evidence for higher values of the E_T^N parameter of SDS and other anionic surfactants as compared with that of cationic ones [14,19]. For example, in our laboratory, the values for 16 different anionic surfactant systems, at different concentrations, ionic strength, and temperature, including microemulsions, are determined to be within the range of 0.779–0.835 [14]. For nine different cationic surfactant systems, E_T^N was 0.623 to 0.708, whereas for seven nonionic surfactants, it was 0.645–0.702 [14]. On the first glance, this allows concluding that the surface of the anionic surfactants is better hydrated than that of the cationic ones.

Moreover, it would not be an exaggeration to say that at least a dozen of publications give evidence for a more hydrated or more polar state of different molecular probes in SDS micelles as compared with those in the case of CTAB. Such information is provided by visible spectra of nitroxides [129], fluorescence of 3,3-dimethyl-2-phenyl-3H-indole [130], fluorescence of rose Bengal B [131], absorption and pK_a of this dye [132], and position of the visible absorption maximum of the anion of 4-n-heptadecyl-7-hydroxy coumarin [133]. The formation of the nonpolar lactone of 5-(N-octadecanoyl)aminofluorescein is less expressed in the SDS micellar aqueous solution as compared with cationic and nonionic surfactants micellar systems [134].

MD simulations demonstrate similar hydration of surfactant ions in anionic and cationic micelles if the tail length is the same (Figure 1). In contrast, it should be noted that the Krafft temperature for SCS is approximately 30 °C higher than that of CTAB. This allows assuming that the Stern layer of the SCS is less hydrated than in the case of CTAB. At the same time, the E_T^N value for SCS, 0.783, even after taking into account the temperature influence (see Table 4), can be estimated at 25 °C as 0.801 and is substantially higher than the value $E_T^N = 0.687$ for CTAB. Additionally, the E_T^N value for SDS, 0.828, is substantially higher than that for DTAB, 0.716–0.724. The corresponding values for CTAC and CPC are close to that for CTAB, and the same is true for DTAB and DTAC [14,19]. At the same time, Figure 1 shows similar hydration of the Stern layers of DTAB and SDS.

Therefore, the different behavior of numerous molecular probes in SDS and CTAB is caused by a more pronounced hydration of foreign molecules involved in the Stern layer of SDS, despite comparable water content on the surfaces of the two micelles. Hence, a hypothesis about the more pronounced hydrogen bond ability of water molecules in anionic micelles can be deduced. In any case, the resulting effect manifests itself in the form of an (apparently) more polar surface of SDS micelles.

Here, it is appropriate to recall a new paper, that allows to shed additional light upon the origin of the λ_{max} and, accordingly, E_T^N values [135]. The authors consider the problem of one of the most common solvatochromic indicators, the aforementioned Reichardt's dye, in terms of micro- and macrosolvation. In this work, based on the study of IR spectra in an argon matrix with water additives, it was shown that the direct formation of a hydrogen bond between a water molecule and the oxygen atom of the indicator dye does not itself introduce fundamental changes in the dye, but enhances the effect of the entire solvent continuum. So, even if the true polarity of the micellar surface of CTAB and SDS is similar, more numerous hydrogen bonds in the last case cause an E_T^N increase.

In some cases, the location character of the solvatochromic or solvatofluoric dyes is more specific. For example, Pal et al. report that some fluorescent 3H-indoles are able to recognize two different sites in SDS micelles, while only one type of locus is observed in the case of CTAB micelles [136]. Another very popular fluorescent molecular probe is pyrene [137]. Thiosemicarbazide and thiosemicarbazide pyrene derivatives are also used for examining CTAB and SDS micelles [137]. The location of pyrene in surfactant micelles will be considered in Section 9.

6. Acid–Base Indicators as Useful Tools for Studying Micelles

6.1. Acid–Base Dissociation in Surfactant Micelles

The application of acid–base indicators began from the papers by Hartley [138] and Hartley and Roe [139]. In reviews by El Seoud [140] and Jana and Pal [141], data for a large number of molecular probes, mainly indicators, in surfactant micelles and related systems were accumulated. The key characteristic is the so-called apparent dissociation constant, K_a^{app}:

$$pK_a^{app} = pH + \log\frac{[HB]}{[B]} \quad (6)$$

Here, pH refers to the aqueous phase, the values in brackets are equilibrium concentrations of the acidic and basic forms of the molecular probe; charges are omitted for simplicity. They may or may not be fully bonded by the micellar pseudophase, with the former being preferred. The articles by Mukerjee and Banerjee [142], Fromherz [143,144], and Funasaki [145–147] enabled substantiating the relation between the electrical surface potential of micelles, Ψ, and the pK_a^{app} value of an indicator (or other molecular probe) completely bound by the micellar pseudophase. This equation therefore may be named as the Hartley–Mukerjee–Fromhertz–Funasaki, or HMFF, equation:

$$pK_a^{app} = pK_a^w + \log\frac{\gamma_B}{\gamma_{HB}} - \frac{\Psi F}{RT \ln 10} = pK_a^i - \frac{\Psi F}{RT \ln 10} \quad (7)$$

Here, K_a^w is the dissociation constant in water, γ is the activity coefficient of transfer from water to the micellar pseudophase, and K_a^i is the so-called intrinsic dissociation constant. Hence, the Ψ value can be determined as given below:

$$\Psi = \frac{RT \ln 10}{F}\left(pK_a^i - pK_a^{app}\right) \quad (8)$$

From these positions, one may consider data for a variety of compounds: nitrophenols [148], hydroxyanthraquinones [149], substituted diarylamines [150], fluorescein dyes [17,18,132,134], rhodamines [31,151–155], azo dyes [156]; acridines [157], and many other indicators [13,19,97,133,140–147,158–166].

Sulfonephthaleins (Scheme 5) belong to the most popular acid–base indicators, and it is not surprising that they were widely used in examining surfactant micelles [142,145–147,164,166–176].

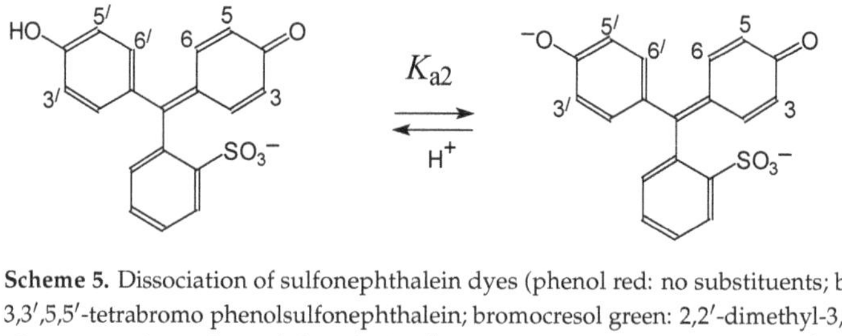

Scheme 5. Dissociation of sulfonephthalein dyes (phenol red: no substituents; bromophenol blue: 3,3′,5,5′-tetrabromo phenolsulfonephthalein; bromocresol green: 2,2′-dimethyl-3,3′,5,5′-tetrabromo-; bromocresol purple: 3,3′-dimethyl-5,5′-dibromo-; bromothymol blue: 2,2′-dimethyl-5,5′-di-*iso*-propyl-3,3′-dibromo-; *o*-cresol red: 3,3′-dimethyl-; *m*-cresol purple: 2,2′-dimethyl-; thymol blue: 2,2′-dimethyl-5,5′-di-*iso*-propyl phenolsulfonephthalein).

Some pK_a^{app} values of these dyes, which were used for the Ψ determination, will be analyzed below.

Coumarin dyes are of particular interest due to their frequent use in determining the electrical potential of interfaces. The indicator properties of the relatively hydrophilic mother compound were demonstrated as early as 1970–1973 [177,178]. In order to ensure the fixation of the reporter molecule to lipid and surfactant assemblies, long hydrophobic hydrocarbon chains were tailored to the coumarin dye [25,95,97,143,144,179–184]. The hydrophobic coumarins (Scheme 6) were basic compounds for developing the quantitative interpretation of the pK_a^{app} values in micelles [95,97,133,144,160,161,184].

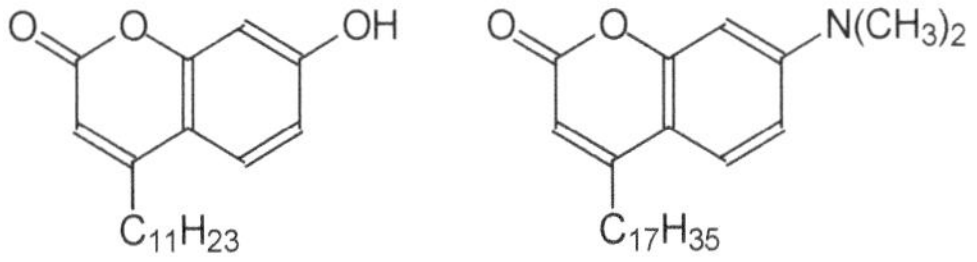

Scheme 6. Molecular structures of two hydrophobic coumarins.

Petrov and Möbius used the hydrophobic coumarin for studying mono- and multilayers [40,41,43–45], and Yamaguchi et al. applied it to pH spectrometry both in surfactant monolayers on water [42] and on pure water interface [48,49].

As mentioned above, the standard solvatochromic Reichardt's dye was applied as an acid–base indicator for micelles [19]. The large size of the dye raises questions about the credibility of the information about the micelle itself, because the obtained data rather reflect the properties of the dye [185]. Nevertheless, the Ψ values obtained via Equation (8) [13,14,20] did not differ significantly from those determined using other indicators. The pK_a^{app} values of different Reichardt's dyes were determined also in surfactant micelles [186] and microemulsions [187]. These and other results are referred to in a review paper by Machado et al. [21].

Several rarely used molecular probes should also be mentioned. For example, Zakharova et al. [188] determined the pK_a^{app} values of a NH-acid, 4-nitroanilide of bis(chloromethyl) phosphinic acid (Scheme 7), in CTAB and SDS micellar solutions:

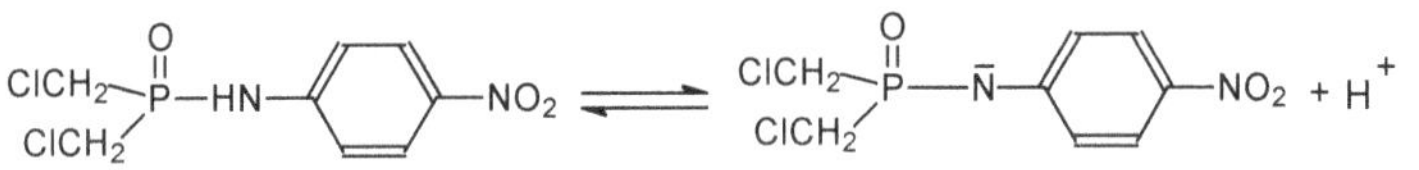

Scheme 7. Dissociation of the 4-nitroanilide of the *bis*(chloromethyl) phosphinic acid.

Even earlier, a set of derivatives of phosphonic acid were examined in CTAB, CTAC, and SDS solutions [189]. In both studies, the dependence of the pK_a^{app} values on the surfactant concentrations was obtained in order to estimate the pK_a^{app}s under conditions of complete binding to micelles, which is difficult to reach experimentally.

Khaula et al. [162] used the following two indicators (Scheme 8): in solutions of CTAB, SDS, and a nonionic surfactant Tween 80.

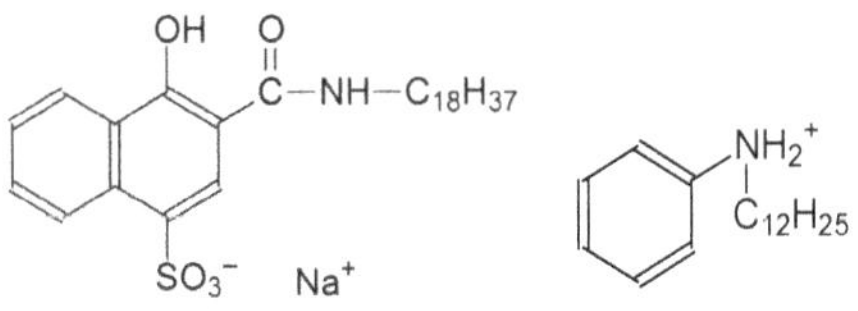

Scheme 8. Hydrophobic indicators used for studying micellar solutions [162].

In addition to such small molecules, large-sized ones are also used in studying colloidal and biocolloidal systems. For instance, relatively recently, Clear et al. [190] applied polymethine dyes (Scheme 9) as optical liposome pH sensors:

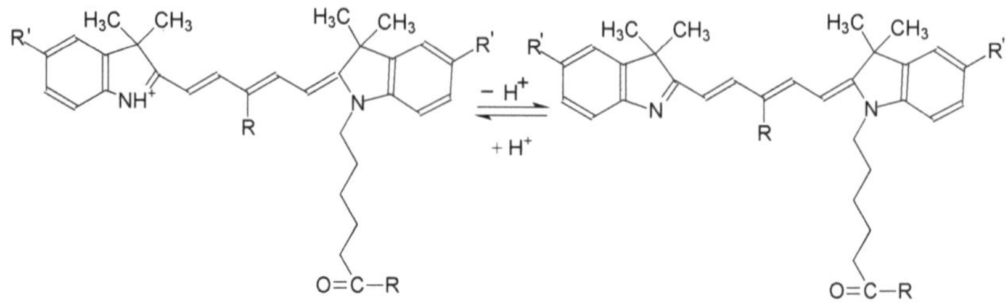

Scheme 9. Acid–base equilibrium of polymethine dyes proposed for pH sensing [190].

Of special interest is the experimental and theoretical study of the acidic dissociation of the so-called GFP (or green fluorescent protein) fluorophore (Scheme 10) [191,192]:

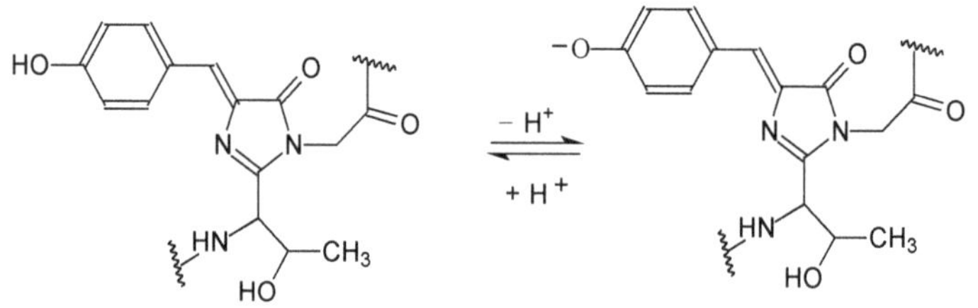

Scheme 10. Acid–base equilibrium of Green fluorescent protein fluorophore (GFP) [191,192].

In addition to spectrophotometric and fluorimetric methods, potentiometric determination of the pK_a^{app} values of molecular probes, including carboxylic and phosphororganic acids, in surfactants were described in detail [193–201].

Maeda performed a detailed thermodynamic analysis of potentiometric acid–base titrations in micellar systems [202,203]. Popović-Nikolić et al. [204,205] used the potentiometric method in studying some biologically active compounds in water in the presence of surfactants.

The dependence of the ^{13}C NMR shifts on pH was used for determination of the pK_a^{app} value of tetradecanoic acid in micelles of a sugar-derived surfactant [206]. ESR probes were also used as acid–base indicators in different colloidal systems [207–212].

6.2. Binding of Indicators by Surfactant Micelles

In any case, the problem of the completeness of binding by micelles is of key significance. A potentiometric study of eleven carboxylic acids with concentrations of 5×10^{-4} M in 0.01 M Triton X-100 solutions allowed Chirico et al. [196] to conclude that, judging by their pK_a^{app}s, acids from acetic to valeric were not bound by the micelles. Even pelargonic and undecanoic acids are included into the micelles almost completely only on increase in the TX-100 concentration. The results obtained by Boichenko et al. with propanoic, butanoic, pentanoic, and hexanoic acids in Brij 35 and SDS solutions [199] and by Eltsov and Barsova with undecanoic, tetradecanoic, and hexadecanoic acids in solutions of CTAB, SDS, Brij 35, and CDAPS [197] confirm these observations.

The binding of indicator dyes can be described by so-called binding constants, K_b; see Equation (9).

$$K_b = \frac{[i_m]}{[i_w](c_{surf} - cmc)} \quad (9)$$

Here, the equilibrium concentrations of the dye ion or molecule, i, fixed at micelles (m) or located in the bulk phase (w) are expressed in moles per liter of the whole solution; $c_{surf} - cmc$ corresponds to the micellized surfactant; cmc is the critical micelle concentration. These constants can be determined using the dependence of pK_a^{app} on c_{surf} [188,189]. For the above shown p-nitroanilide of the bis(chloromethyl) phosphinic acid, the binding constants were determined using the dependence of pK_a^{app} on the CTAB concentration [188]. For the anion and molecule in CTAB solutions, K_b = (4.0–4.8) × 10^4 M^{-1} and (300–540) M^{-1}, respectively, depending on the buffer nature. In SDS solvents, the values were much lower, 1 and 64, respectively. These values of binding constants enable estimating the pK_a^{app} values at complete binding [188].

Sarpal et al. [130] estimated the binding constant of the equilibrium forms of 3,3-dimethyl-2-phenyl-3H-indole (Scheme 11), a fluorescence polarity probe, using fluorescence measurements.

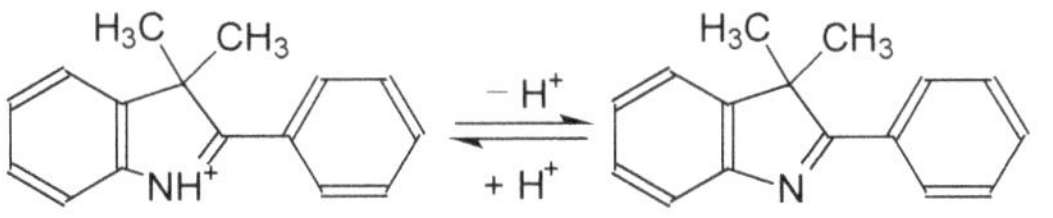

Scheme 11. Acid–base equilibrium of 3,3-dimethyl-2-phenyl-3H-indole.

In SDS micellar solutions, the values K_b = 1.01 × 10^3 M and 12.9 × 10^3 M for the B and HB^+ forms were determined at pH 9.5 and 1.0, respectively. Therefore, the ionic strength is relatively high in the second case. This is in line with the cmc values of SDS determined by the authors in these two cases, (7.0–7.4) × 10^{-3} M and (1.4–1.8) × 10^{-3} M, respectively. Whereas the value in water, pK_a^w, equals to 3.25, the pK_a^{app} value in 0.06 M SDS solution is 4.75. As it follows from the $K_{b,B}$ value, under these conditions the $[B_m]/[B_w]$ ratio is 61, and hence the experimental pK_a^{app} should be considered as a value at complete binding by micelles.

In Table 5, the binding constants obtained using the dependencies of pK_a^{app} vs. c_{surf}, as a rule at 25 °C; the values of the ionic strength, I, are indicated. Here, some typos in the previous paper [14] are corrected.

Table 5. Binding constants, K_b, of indicators and the pK_a^{app} values [±(0.03–0.06)], obtained by extrapolation to complete binding of both equilibrium forms [154,174,213].

Indicator System	Surfactant	I, M	$K_{b,HB}$, M^{-1}	$K_{b,B}$, M^{-1}	pK_a^{app}
Methyl yellow, HB^+/B	Brij 35	0.05	65	5.8 × 10^3	1.12
Rhodamine B, HB^+/$B\pm$	Brij 35	0.05	4.0 × 10^3	5.9 × 10^2	4.08
Bromophenol blue, HB^-/B^{2-}	Brij 35	0.01	1.2 × 10^4	1.25 × 10^3	5.10
Bromophenol blue, HB^-/B^{2-}	Triton X-100	0.01	1.1 × 10^4	1.7 × 10^3	5.00
Bromophenol blue, HB^-/B^{2-}	Triton X-305	0.01	2.4 × 10^3	1.3 × 10^2	4.88
Bromophenol blue, HB^-/B^{2-}	Nonylphenol 12	0.05	2.0 × 10^4	2.7 × 10^3	4.80
Bromophenol blue, HB^-/B^{2-}	Tween 80	0.05	1.3 × 10^4	9.0 × 10^2	5.09
Phenol red, HB^-/B^{2-}	Brij 35	0.01	2.85 × 10^2	32	8.73
Phenol red, HB^-/B^{2-}	CTAB	0.02 (KBr)	3.75 × 10^4	9.06 × 10^4	7.45
Phenol red, HB^-/B^{2-}	CTAB	0.05 (KBr)	1.64 × 10^4	1.55 × 10^4	7.71
Phenol red, HB^-/B^{2-}	CTAB	0.4 (KBr)	2.5 × 10^3	2.6 × 10^2	8.72
Bromothymol blue, HB^-/B^{2-}	SDS	0.2 (NaCl)	1.3 × 10^4	16	9.90

Corresponding values were also determined for indicator dyes in microemulsions [172,173] and phospholipid liposomes [171].

The most popular method for ensuring complete binding of acid–base indicators to micelles is attaching long hydrocarbon chains to them [25,40,41,43–45,95,99,133,134,143, 144,152,153,155,160,162,179–184,190]. Otherwise, the opposite charge of the ionic forms of indicators and inclusion of halogen atoms and nitro groups favors the binding by micelles. Therefore, another issue is the depth of penetration of such modified molecular probes into the pseudophase. This will be considered in the present paper using MD simulations.

Application of the NMR spectroscopy method allows expecting the location of the acid–base indicators as a rule in the Stern layer region [99,122,124,125,160,181,182].

Finally, it is important to know to what extent the pK_a^w values of long-tailed indicators coincide with those of the unmodified compounds [133,134].

The same refers to the pK_a^{app} values, which is easy to check experimentally, contrary to the case of aqueous solutions, where the long-tailed dyes are poorly soluble and can form their own micelles.

Additionally, despite firm fixation to micelles, similar dyes with hydrocarbon tails of unequal length sometimes exhibit different properties. For example, the quenching of fluorescence by N,N-dimethylaniline in CTAB micelles occurs dissimilarly for two eosin derivatives with hydrocarbon length, n = 11 and 15, respectively (Scheme 12) [214].

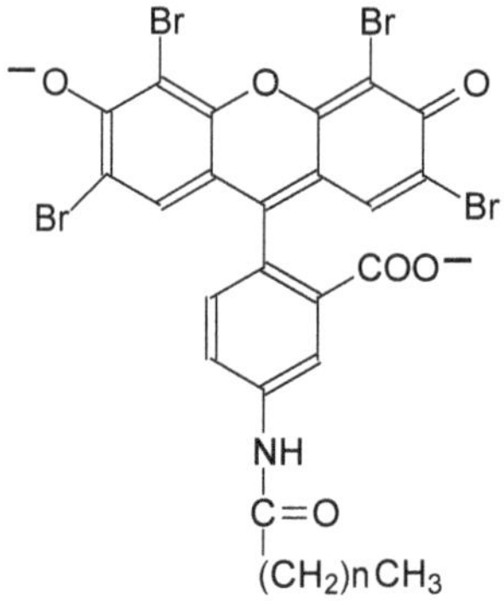

Scheme 12. Molecular structure of hydrophobic derivatives of eosin (the dianionic form).

As the pK_a^w values for long-tailed dyes are difficult to determine in water, the comparison was made in aqueous 1-butanol (Table 6).

Table 6. Indices of the dissociation constants of esters of fluorescein dyes in aqueous 1-butanol and CTAC micelles.

Dye	n-Butanol—Water, 82: 18 by Mass [215]		CTAC Micelles, 4.0 M KCl [18]	
	pK_{a0}	pK_{a1}	pK_{a0}^{app}	pK_{a1}^{app}
Ethylfluorescein	2.68 ± 0.02	8.44 ± 0.07	1.86 ± 0.02	6.59 ± 0.03
n-Decylfluorescein	2.53 ± 0.02	8.56 ± 0.03	2.13 ± 0.01	6.61 ± 0.07
n-Hexadecylfluorescein	—	—	1.58 ± 0.04	7.06 ± 0.03
Ethyleosin	—	3.71 ± 0.05	—	1.11 ± 0.03
n-Decyleosin	—	3.86 ± 0.04	—	1.18 ± 0.05

Additionally, the data in CTAC micelles at high ionic strength of the bulk phase shed some light upon the role of the length of the hydrophobic radical on the acid–base properties of the firmly bound indicator.

Bissell et al. created earlier a very interesting complicated construction [216]. These authors synthesized a number of fluorescent photoinduced electron-transfer sensors with targeting/anchoring modules called by the authors as "molecular versions of submarine periscopes" for mapping membrane-bounded protons. The pK_a^{app}s in micellar solutions of CTAC, SDS, and Triton X-100 were determined; the complete binding is observed at proper hydrophobicity of the anchoring tail group [216].

On the other hand, an approach was developed for estimating the Ψ values in the diffuse layer beyond the micellar surface with the help of indicators—derivatives of the benzoic acid [15,16].

In addition to examining surfactant micelles, the pK_a^{app} values of indicator dyes were determined in phospholipid liposomes [19,25,133,143,171,178–183,190,207], polyelectrolyte solutions [32,33], gelatin films [34], solid surfaces [35,36,217], mono- and multilayers of surfactants [40–45] and polymers [46,152], and on pure water surface [47,48,218–223], including air bubbles in water [49].

7. Experimental Determination of the Surface Electrical Potential of Micelles Using Acid–Base Indicators

Now it is important to consider different methods used for the Ψ determination (Equation (7)), first, for evaluation of the pK_a^i in ionic micelles. This question has been systematized by Grieser and Drummond [13] and afterwards discussed in several articles from this group [10,14,224].

The first approach is based on equating the pK_a^i to pK_a^w [139,145]. However, it was revealed that as a rule, the pK_a^{app} value in nonionic surfactant micelles substantially differs from the value in water [142,146,225]. The above assumption is reasonable only if the surface is well hydrated, and there are some other reasons for using such an approach [35,36].

The second approach consists in screening the surface charge of micelles by high concentrations of an electrolyte in the bulk (aqueous) phase [18,169,170,226]. This allows consider the pK_a^{app} value in the same ionic micelles at high ionic strength as the pK_a^i. However, there is some evidence for incomplete charge screening even at several moles per liter of electrolytes, and too high concentrations also strongly influence the viscosity and other properties of the aqueous phase [10,14,18,156,157,163,164]. Contrary to the micellar solutions of CTAB and nonionic surfactants, in the case of SDS the precipitation at high NaCl concentrations may occur [149]. High ionic strength often leads to polymorphic transformations of the micelles. Additionally, the interpretation of the pH measurements using glass electrodes in cells with liquid junction in such salt solutions becomes less obvious.

The third approach uses the pK_a value of an indicator in water–organic solvents for estimating the pK_a^i value. In this case, two steps are to be made: (i) the composition of a water–organic solvent "s" should be selected, and (ii) the pK_a^i value should be equated to the difference ($pK_a - \log {}^w\gamma_{H^+}^s$) [19,133,134,144,147,156,157,163,164,227,228]. The choice of the appropriate composition of the binary solvent is made using the absorption maximum of the dye. This value must coincide with that in the micelles, and the relative permittivity of the mixed solvent is equated to that of the micellar pseudophase. However, it should be remembered that both λ_{max} values and pK_as in isodielectric solvents sometimes differ substantially. Additionally, the ${}^w\gamma_{H^+}^s$ values used in the cited papers differ from those estimated by the commonly accepted tetraphenylborate assumption [10,14,224].

The fourth approach proposes to use the pK_a^{app} value in micelles of nonionic surfactants as the pK_a^i in ionic micelles [10,13,14,62,95,144,146,149,229]. This method is probably the most often used. The idea of equating the pK_a^i to the pK_a^{app} of the same indicator located at a noncharged surface was also implemented at studying mono- and multilayers [40,41,43–45].

The fifth approach presumes utilization of two indicators, which are cationic and nonionic acids, HB$^+$ and HA, respectively [99,144,162,183]. The main idea consists in the assumption ${}^w\gamma_{HB^+}^s = {}^w\gamma_{A^-}^s$ for the two indicators. This is based on comparing the dependence of the pK_as of a cationic and neutral hydrophobic coumarins [144]. Our experiments with the n-decylfluorescein that is a dual cationic and neutral acid two-step indicator demonstrated that the ${}^w\gamma_{H_2R^+}^s = {}^w\gamma_{R^-}^s$ assumption is not universal [14].

However, the determination of Ψ values both by different approaches and by different indicators, as a rule, leads to different results for the same ionic micelle.

In Tables 7–11, some examples of the Ψ scatter are presented. Here, the pK_a^{app} values in micelles of nonionic surfactants were used as the pK_a^i in ionic micelles, Equation (8). The data refer to the complete binding by all kinds of micelles under consideration, as a rule, at 25 °C or room temperature. The ionic strength is indicated if the exact value is available in the cited papers. At low ionic strength, its variation has a particularly noticeable effect on the Ψ value.

Table 7. Value of Ψ of the CPC micelles at an ionic strength of 0.05 M [14].

Indicator	$\Delta pK_a^{app} = pK_a^{app} - pK_a^w$		Ψ, mV
	In CPC Micelles, 0.05 M Cl$^-$	In Nonionic Micelles	
Bromophenol blue, −/=	−2.16	0.61	+164
Bromocresol green, −/=	−1.68	1.03	+160
Bromocresol purple, −/=	−1.32	0.91	+132
Bromothymol blue, −/=	−0.94	1.64	+152
n-Decyleosin, 0/−	−1.9	0.71	+154
n-Decylfluorescein, 0/−	−1.39	0.69	+123
n-Decylfluorescein, +/0	−2.15	−0.80	+80
Reichardt's dye, +/±	−1.55	0.46	+118
N,N'-di-n-octadecylrhodamine,+/±	−0.76	0.97	+102

Table 8. Value of Ψ of the CTAB micelles at low ionic strengths [10,148,160,197].

Indicator	pK_a^{app}		Ψ, mV	Ionic Strength, M
	In CTAB Micelles	In Brij-35 Micelles		
Bromophenol blue, −/=	2.26	5.10	+168	0.011
Bromothymol blue, −/=	6.59	9.19	+153	0.011
2-Nitro-4-n-nonylphenol, 0/−	5.86	8.52	+157	<0.01
4-Octadecylnaphthoic acid, 0/−	4.20	6.60	+142	0.014
N,N'-di-n-octadecylrhodamine, +/±	2.24	4.12	+111	0.019
Tetradecanoic acid, 0/−	5.34	6.36	+60	<0.01

Table 9. Value of Ψ of the SDS micelles at an ionic strength of 0.05 M [10,14,62].

Indicator	ΔpK_a^{app}		Ψ, mV
	In SDS Micelles, 0.05 M Na$^+$	In Nonionic Micelles	
2,6-Dinitro-4-n-dodecylphenol, 0/−	1.51	−0.09	−95
n-Decyleosin, 0/−	2.63	0.71	−113
n-Decylfluorescein, 0/−	2.65	0.69	−116
n-Decylfluorescein, +/0	2.23	−0.80	−179
Reichardt's dye, +/±	2.06	0.46	−94
N,N'-di-n-octadecylrhodamine, +/±	1.97	0.97	−59
Rhodamine B, +/±	2.10	1.0	−65
Hexamethoxy red, +/0	2.14	−0.90	−179
Neutral red, +/0	≈2.3	≈−0.8	−183
Methyl yellow, +/0	1.56	−2.13	−218

Table 10. Value of Ψ of the SDS micelles at low ionic strengths [10,148,156,160,197].

Indicator	pK_a^{app}		Ψ, mV	Ionic Strength, M
	In SDS Micelles	In Brij-35 Micelles		
2-Nitro-4-n-nonylphenol, 0/−	10.05	8.52	−90	
4-Octadecylnaphthoic acid, 0/−	8.10	6.60	−88	
Tetradecanoic acid, 0/−	8.45	6.36	−123	
N,N'-di-n-octadecylrhodamine, +/±	5.52	4.12	−83	0.02

Table 10. Cont.

Indicator	pK_a^{app}		Ψ, mV	Ionic Strength, M
	In SDS Micelles	In Brij-35 Micelles		
Rhodamine B, +/±	5.70	4.08	−96	0.015
Hexamethoxy red, +/0	5.89	2.10	−224	0.01
Neutral red, +/0	9.17	5.64	−208	
Methyl yellow, +/0	5.28	1.12	−245	0.01
Acridinium, +/0	7.01	3.69	−196	

Table 11. Value of Ψ of surfactant—1-pentanol—benzene microemulsions; 1.3 vol.% pseudophase; ionic strength 0.05 M [14,173,187].

Indicator	Ψ, mV	
	CPC-Based	SDS-Based
Bromophenol blue, −/=	+179	—
Bromocresol green, −/=	+200	—
Bromocresol purple, −/=	+165	—
Bromothymol blue, −/=	+152	—
n-Decylfluorescein, +/0	+71	−150
n-Decylfluorescein, 0/−	+101	−90
Reichardt's dye, +/±	+65	−86
N,N'-di-n-octadecylrhodamine, +/±	+98	−47

The ΔpK_a^{app} values of sulfonephthaleins in cationic micelles gradually increase from bromophenol blue to thymol blue [167,168,175]. For example, Politi and Fendler [167] reported the values ΔpK_a^{app} = −1.25 (bromocresol green); −1.05 (bromophenol red); −0.9 (bromothymol blue); and −0.2 (thymol blue) for a 0.008 M CTAB solution, ionic strength 0.01 M NaCl. Kulichenko et al. [175] reported the values ΔpK_a^{app} = −1.89 (bromophenol blue); −1.51 (bromocresol green); −1.21 (bromocresol purple); −0.70 (bromothymol blue); −0.85 (phenol red); −0.54 (m-cresol purple); −0.74 (cresol red); and −0.20 (thymol blue) for a 0.06 M tridecylpyridinium bromide solution. Rosendorfová and Čermáková studied three sulfonephthaleins in micellar solutions of a cationic surfactant Septones, 1-(ethoxycarbonyl)-pentadecyl-trimethylammonium bromide (Scheme 13) [168]:

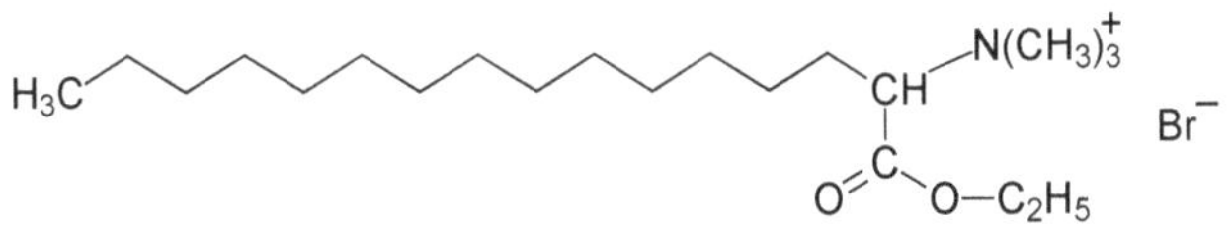

Scheme 13. Molecular structure of the cationic surfactant Septones.

At ionic strength 0.2 M, the values ΔpK_a^{app} = −0.80 (bromophenol blue); −0.45 (bromocresol green); and +0.12 (phenol red) were determined [168]. Our study for eight sulfonephthaleins in CPC micelles at ionic strength 0.05 M demonstrated the increase in ΔpK_a^{app} from −2.16 for bromophenol blue to −0.37 for thymol blue [230]. In CTAB at KBr + HBr or KBr + buffer concentrations of 0.005 M, the ΔpK_a^{app}s of four sulfonephthaleins change from −2.11 to −0.43 for bromophenol blue to thymol blue, respectively [231]. Importantly, the same sequence of the ΔpK_as of the standard series of sulfonephthaleins is observed in binary water–acetone solvents [232] and in other mixtures of non-hydrogen bond donor solvents (acetonitrile, dimethyl sulfoxide) [213]. In the last named systems, the ΔpK_a values are substantially positive. It should be recalled here that, while studying acid–base reactions in CTAB micelles, Minch et al. [233] noted that "intramicellar water has a lower tendency to hydrogen bond with organic molecules than "ordinary" water". Though the ΔpK_a^{app}s in cationic micelles are negative owing to the positive Ψ values, the differentiating influence of this kind of the organized solutions is the same as in the above solvents. The more pronounced the delocalization of the negative charge in the R^{2-} anion is, the lower is the ΔpK_a value. However, the ΔpK_a^{app} values in nonionic micelles obey other regularities. Here, the total hydrophobicity of the compounds is of key significance. Obviously, the use of the pK_a^{app} values of these indicators in nonionic micelles will lead to different Ψ values for the same ionic micelles as calculated according to Equation (8).

The scatter of the Ψ values in Tables 7 and 8 reaches 100 mV.

Even more substantial it is in Tables 9 and 10.

The data gathered in Tables 7–11 may be supplemented by the results reported by Pal and Yana, who determined the pK_a^{app}s of seven hydroxyanthraquinones (Scheme 14) in micelles of cationic, anionic, and nonionic surfactants at an ionic strength of 0.3 M, 25 °C [149].

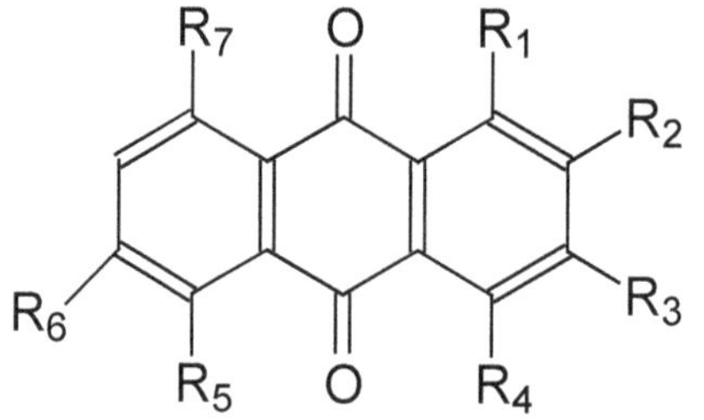

Scheme 14. Molecular structure of hydroxyanthraquinones used as acid–base indicators in micellar solutions. I: R_1 = OH; II: R_1 = R_2 = OH; III: R_1 = R_4 = OH; IV: R_1 = R_7; V: R_1 = R_2 = R_7; VI: R_1 = R_2 = R_5 = R_7; VII: R_1 = R_2 = OH, R_3 = SO_3Na (the nonspecified substituents: R = H).

The authors compare the electronic absorption spectra and fluorescence in water, organic solvents, and micellar media, and demonstrate the increase and decrease in pK_a^{app} values along with increasing concentration of NaCl in CTAB and SDS micellar solutions, respectively. This study presents an abundant material demonstrating the role of the dye location in the interfacial region. The Ψ values, as determined by Equation (8) using the pK_a^{app} values in Triton X-100 micelles as pK_a^i of ionic micelles, vary in a wide range, and in some cases are positive even in SDS micelles [149].

Fernandez and Fromherz [144] used the long-tailed indicators 4-undecyl-7-hydroxycoumarin and 4-heptadecyl-7-(dimethylamino)coumarin. At CTAB and SDS concentrations of 0.024 M, the Ψ values at low bulk ionic strength were determined as +148 and -137 mV, respectively. The pK_a^i values of the indicators, 8.85 and 1.25, were determined in nonionic micelles of Triton X-100 [144]. The indicator 4-heptadecyl-7-hydroxycoumarin was used by Hartland et al. [95] for determination of the Ψ value of the SDS micelles at different NaCl concentrations; the pK_a^i = 9.10 of the indicator in $C_{12}E_8$ micelles was used for calculations. At 0.02 M SDS, the Ψ values are -141; -125; -110 mV at 0.007; 0.02; and 0.065 M NaCl, respectively [95]. Simultaneously, the Ψ values for micelles of series surfactants were determined with this indicator [133].

Concluding, the Ψ value of the interfacial region of a given micelle at a fixed ionic strength can vary within a wide range, as obtained with different acid–base indicators.

8. Theoretical Determination of the Surface Electrical Potential of Micelles

8.1. Formulation of the Problem

For well-defined micellar systems, different methods were used for theoretical calculation of the Ψ values. As an example, Equation (1) in Section 3 can be mentioned. Lukanov and Firoozabadi [78] developed more detailed approaches in this direction. For the SDS micelles, the calculated electrostatic potential values at 0.1 nm from the charged surface [78] agree quite well with the experimental Ψ values, determined with acid–base indicators at various NaCl concentrations [95,114,229]. Us'yarov calculated the whole set of parameters, including α and Ψ [92,93], within a wide range of SDS and NaCl concentrations [92,93]. Some other papers were devoted to the theoretical estimations of the pK_a^{app} values in micelles and different biocolloidal systems [77,79,192]. However, the most popular method so far is the experiment with acid–base indicators.

Below we present the main results obtained by our group. Our approach was based on the utilization of the acid–base indicators. Namely, the MD simulations were used for evaluation of the ΔpK_a^{app} of indicators, and the difference between the ΔpK_a^{app}s in nonionic and ionic micelles allows the estimation of the Ψ values of the latter. This study was accompanied by consideration of the localization, orientation, and hydration of the equilibrium forms of the acid–base indicators following the methodology described above for the betaine solvatochromic dyes.

The first step in this direction was devoted to the indicator 4-[(E)-([1,1'-biphenyl]-4-yl)diazenyl]-2-nitrophenol, which was first used for this purpose in a study published by Hartley and Roe as early as 1940 [139]. In this case, triethanolammonium cetylsulfonate, TEACSn, and N-cetylpyridinium bromide, CPB, (Scheme 15) were used in the modeling [59] because the same surfactants were used by the cited authors [139]. Sodium cetylsulfonate, SCSn, was also involved in our simulations.

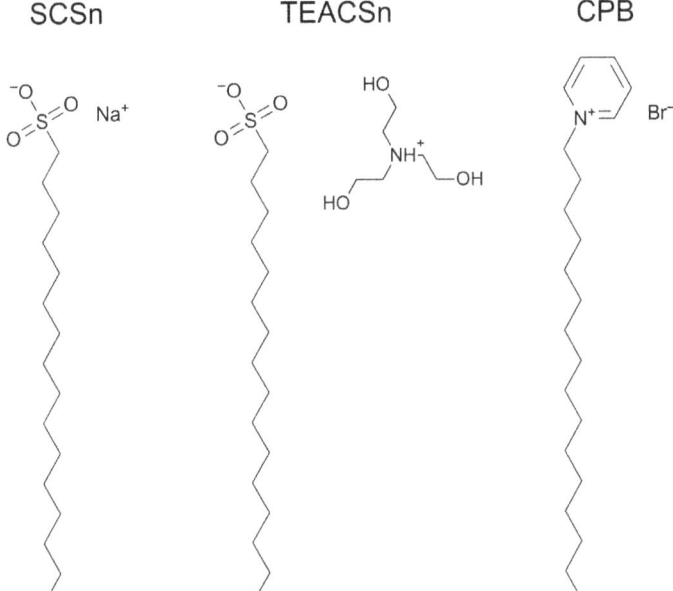

Scheme 15. Molecular structure of the surfactants used in modeling with the Hartley indicator.

The acid–base dissociation of this and other indicators selected for modeling is presented in Scheme 16. In addition to the Hartley indicator, the 4-n-heptadecyl-7-hydroxycoumarin used by many authors for the Ψ determination (see above), was also involved in our theoretical study. Other three indicators were used in our experimental determinations of the electrostatic potential. They are as follows: 2,6-dinitro-4-n-dodecylphenol [62], n-decylfluorescein [14,173,213], and N,N'-di-n-octadecylrhodamine [14,152,153]. The results are published in a set of publications [59,62,64,65,67–69].

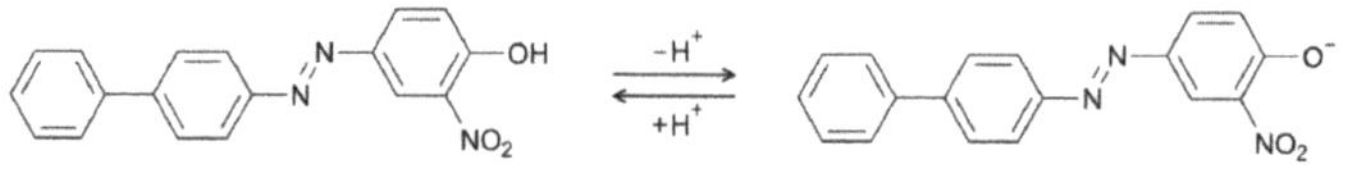

Scheme 16. Dissociation of the acid–base indicators used in molecular dynamics modeling. HD: Hartley indicator; DDP: 2,6-dinitro-4-*n*-dodecylphenol; HHC: 4-*n*-heptadecyl-7-hydroxycoumarin; DF: *n*-decylfluorescein; DR: N,N'-di-*n*-octadecylrhodamine.

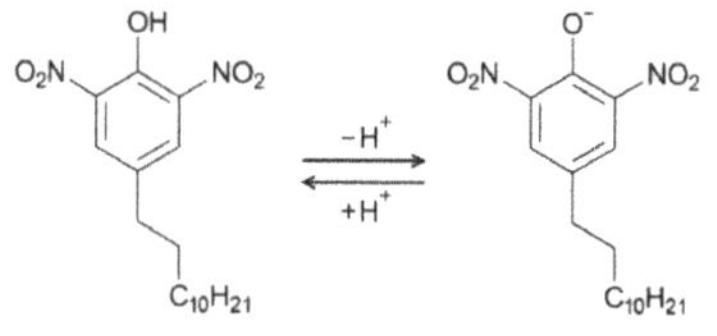

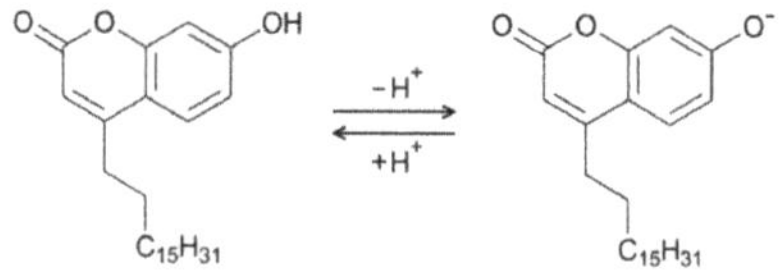

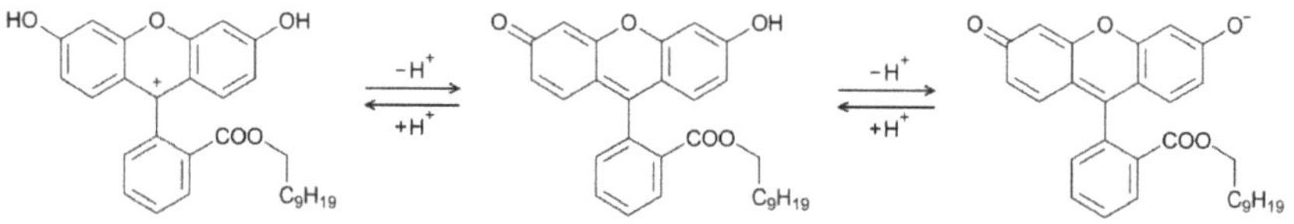

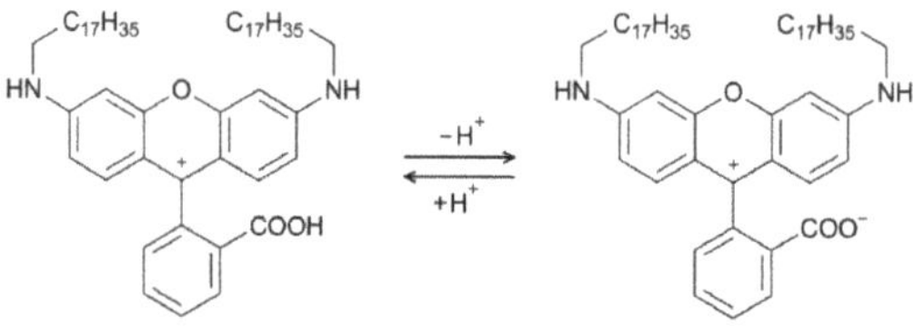

8.2. Location of Acid–Base Indicators in Micelles

These phenomena were considered in detail for the dyes HD [59], DDP [62,64], and coumarin and xanthenes dyes [67].

The chromophoric moieties of the molecules are located either on the surface of the micelles between the hydrocarbon core and solution, or within the hydrocarbon core having only hydroxyl groups and adjacent atoms contacting with water. Both locations may be typical for the same molecule and alternate. Consequently, the dyes differ by the depth of immersion into micelles. Following the betaine dyes, we quantified it as the distance between the micelle center and a chosen atom of the dye molecule (the N atom bound to the nitrophenol moiety of HD, the C atom of the α-CH$_2$ group of the hydrocarbon tail of DDP and HHC, and the C atom carrying the positive charge in carbocations of DF and DR). The average distances are collected in Table 12 and schematically represented in Figures 18 and 19. As before, the distances below 1.35 nm (SDS) or 1.55 nm (CTAB, CPB) are highlighted gray to indicate the approximate extent of the hydrocarbon core, conventionally defined as the distance where the water fraction appears to be at least 5%. The data were obtained from the distribution function of the distance $p(r)$, computed over MD trajectories.

Table 12. Average distance (nm) of acid–base indicator molecules to the center of mass of micelles.

Dye	SDS		CTAB	
	Acidic Form	Basic Form	Acidic Form	Basic Form
HD	1.62 [a]	1.93 [a]	1.37 [b]	1.56 [b]
DDP	1.23	1.16	1.48	1.30
HHC	1.29	1.42	1.48	1.50
DR	1.37	1.47	1.64	1.60
DF	1.47	1.35	1.90	1.65
		anion		anion
DF		1.44		1.61

[a] Data for SCSn at 323 K; [b] data for CPB at 308 K.

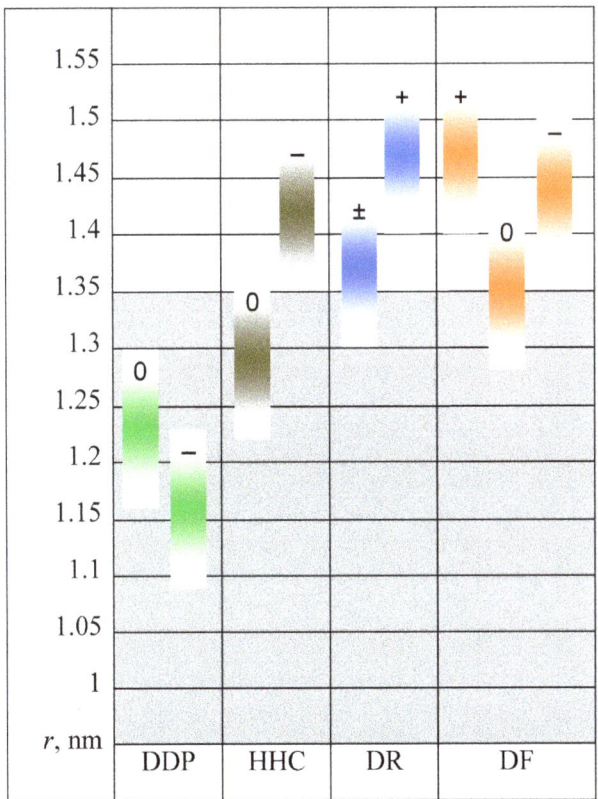

Figure 18. Average distance of the acid–base indicator molecules to the center of mass of SDS micelle. The band length corresponds to the range of typical distances.

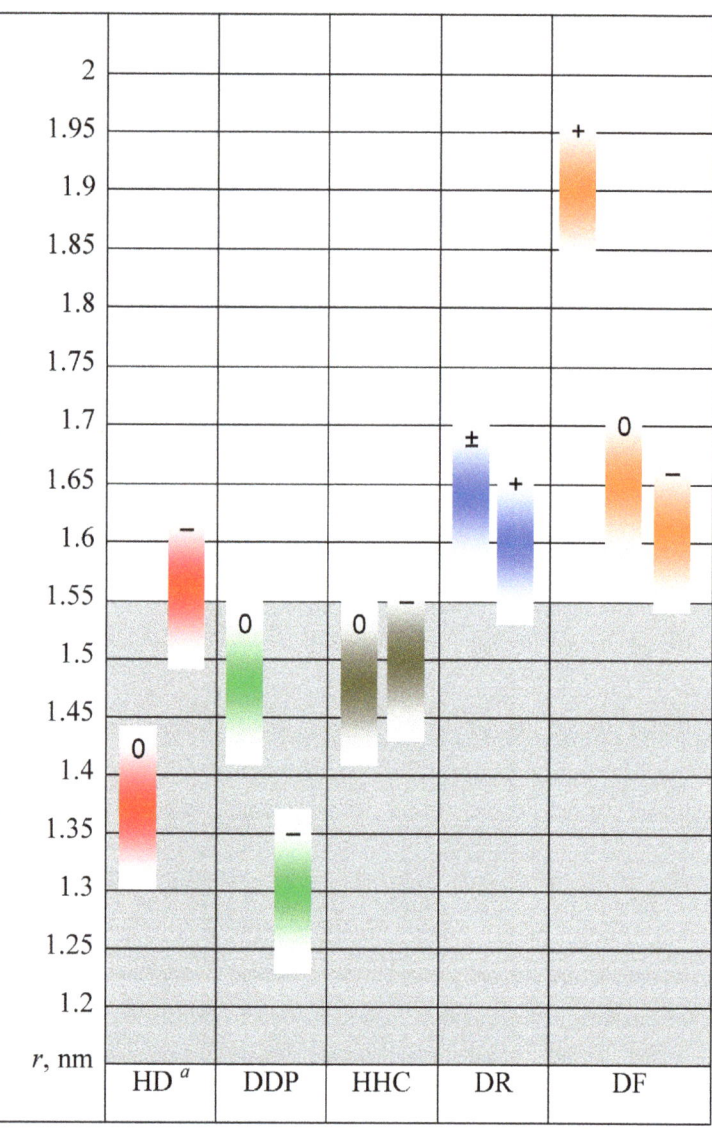

Figure 19. Average distance of the acid–base indicator molecules to the center of mass of CTAB micelles. The band length corresponds to the range of typical distances. [a] For HD, data for CPB are given.

Hence, the distance for the acidic forms in SDS micelles is 1.27–1.47 nm, while for the basic forms it is 1.16–1.47 nm. In CTAB micelles, the distances are 1.48–1.90 and 1.30–1.65 nm, respectively.

In SDS micelles, introducing charge to the molecule (by protonation or deprotonation) considerably pushes it towards the bulk solution, the only exception being DDP. In the cationic ones, the effect depends on the dye: for HD, DF cation, and HHC (a little) the trend is the same, while for the DR cation and the anions of DDP and DF acquiring charge immerses the molecule in average deeper into micelles. The exceptional advancement of the DF cation in CTAB micelles is accompanied with reorientation of the molecule, as shown below. Similar exceptionally far location is observed for HD anion in anionic SCS (not shown in Figure 18) that is explained by electrostatic repulsion and absence of a long hydrocarbon radical that would hold the molecule in micelle.

8.3. Orientation of Acid–Base Indicator Dyes in Surfactant Micelles

Despite structural differences from Reichardt's betaine dyes (in particular, the presence of a long hydrocarbon tail), molecules of the considered acid–base indicators show similar placements: they are either roughly parallel or perpendicular to the micelle surface (Figure 20). Often both orientations are sampled by turns. The summary of our observations from MD trajectories is collected in Table 13. For analysis, the approach used for RD was employed (see Section 5.3 for details). The inclination angle of the molecule θ was defined for each dye. The fraction of time a molecule had $\theta < 40°$ was assumed to be the probability of "vertical" orientation, and the rest of the time corresponded to the "horizontal" one. The inverted vertical orientation was not observed for these dyes, in contrast to Reichardt's betaines. For HD, the results of calculations were corrected according to visual examination of the MD trajectories.

Figure 20. Typical placements of acid–base indicator dyes in micelles shown using DDP as an example. Left: vertical orientation, right: horizontal orientation.

Table 13. Dominant orientations of acid–base indicator molecules in micelles. "V" means the vertical orientation, "H" means the horizontal one. "Sometimes" means a probability of 10–30%, "often" means a probability of >30%, "and" means a probability of approximately 50%.

Dye	SDS		CTAB	
	Acidic Form	Basic Form	Acidic Form	Basic Form
HD [a]	V and H	H, sometimes V	V, sometimes H	V
DDP	V and H	V	H, often V	V
HHC	V	V	V	V
DR	V and H	H	V	H, sometimes V
DF	H	V	H	V
	anion		anion	
DF		V		V

[a] For HD, the data are for SCSn and CPB instead of SDS and CTAB, respectively.

In Figures 21 and 22, the most frequently observed state of the indicators in SDS and CTAB micelles is shown in accordance with the MD data. It summarizes the information of Table 13.

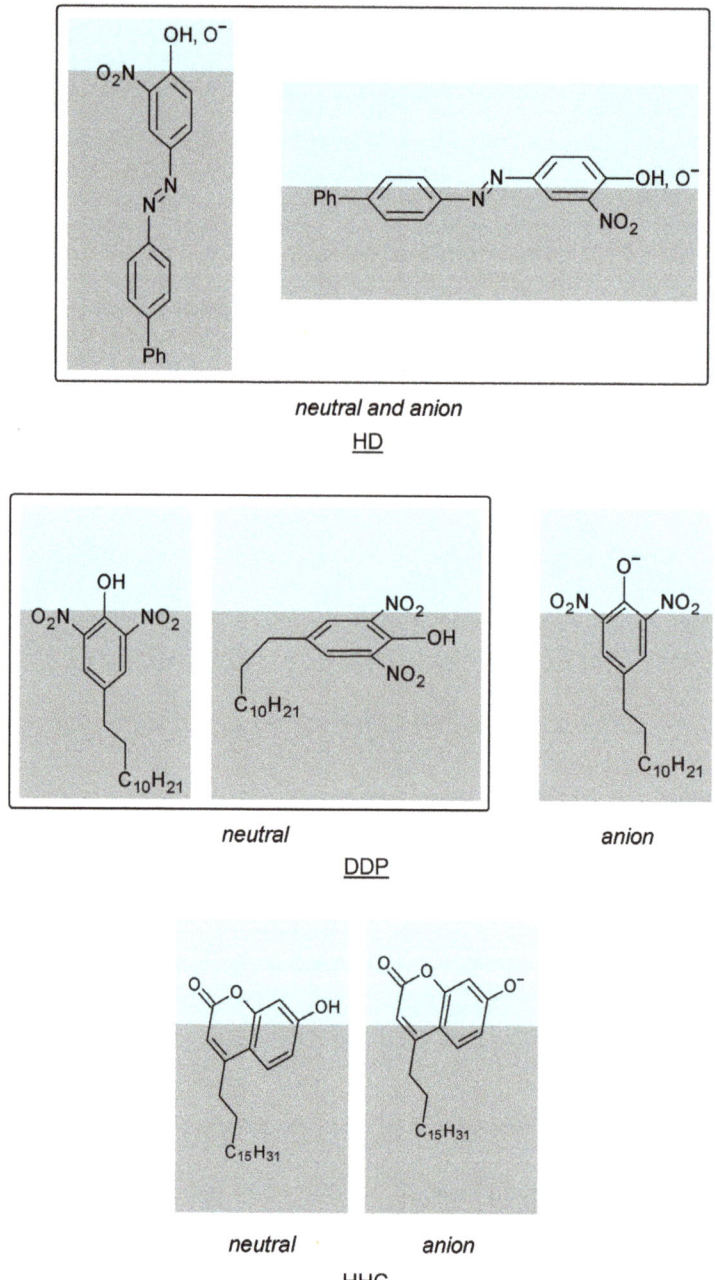

Figure 21. Most probable orientations of the dyes HD, DDP, and HHC in ionic surfactant micelles as deduced from the MD simulations.

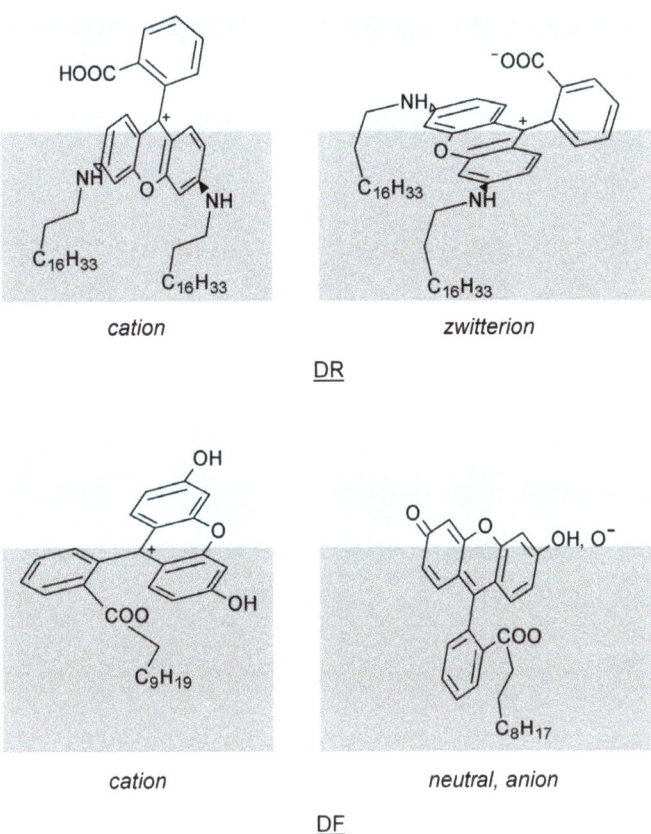

Figure 22. Most probable orientations of the dyes DR and DF in ionic surfactant micelles as deduced from the MD simulations. Adapted from [67].

The presence of a long hydrocarbon tail in a molecule (or even two of them in the case of DR) does not fix the position of its chromophoric moiety and does not prevent its rotation within a micelle. The particular placement of the molecule is difficult to predict on the basis of its molecular structure. Still, the common point is that the ionizing group (both protonated or deprotonated) continuously preserves contact with water molecules in the bulk solution.

8.4. Hydration of Acid–Base Indicator Dyes in Surfactant Micelles

Following the procedure used above for describing the local environment of betaine dyes (see Section 5.4), we examined it for acid–base indicators. The molar fractions of water and headgroups are depicted in Figure 23 for entire molecules and in Figure 24 for O atoms of ionizing groups.

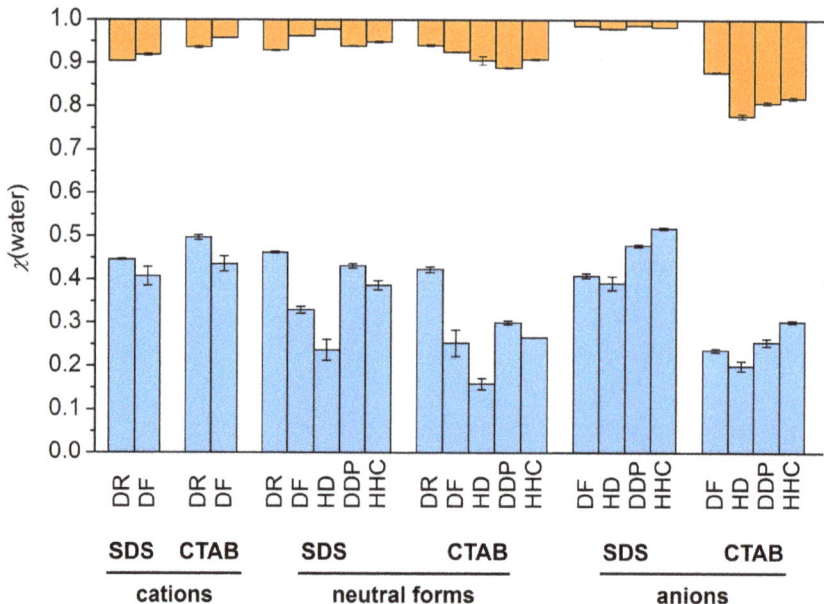

Figure 23. Molar fraction of water atoms (blue columns below) and headgroup atoms (orange columns above) around molecules of different acid–base dyes. For HD, the data for SCSn and CPB are given.

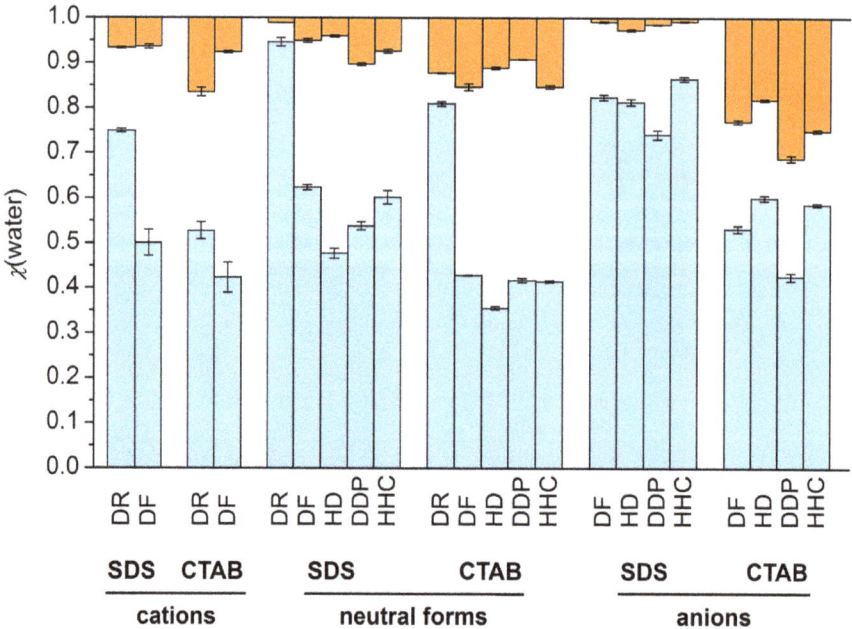

Figure 24. Molar fraction of water atoms (blue columns below) and headgroup atoms (orange columns above) around O atom(s) of the ionizing group of different acid–base dyes. For HD, the data for SCSn and CPB are given.

The most evident trend is higher hydration of anionic forms of the dyes in SDS as compared with CTAB, the difference approaches 0.2. The same is true for neutral forms, too, albeit to a smaller extent. The main reason for the effect is the close contact of anions with cationic headgroups of CTAB and their repulsion from the anionic ones of SDS.

The microenvironment of the ionizing groups quantitatively differs from that of the whole molecules. Still, the trend stated above is true, as well: transferring neutral and anionic forms from SDS to CTAB reduces the amount of water molecules around the groups. In all cases, deprotonation improves hydration of the group or, in few cases, keeps it constant.

8.5. Effect of Other Factors on the above Considered Characteristics

Following the original experiment, we investigated the Hartley's dye in two kinds of anionic micelles, which differ in counterions, namely, SCSn and TEACSn [59]. Both the localization and orientation of the molecule were found to be almost equal; however, hydration of the dye in the latter micelles was 1.4-fold less. The reason was that TEA$^+$ ions considerably screened the dye and its hydroxyl group from water. An important consequence of this observation is that the two characteristics (placement and hydration) are not directly and unambiguously connected.

With the example of DDP we studied the role of the length of the hydrocarbon chain, of substituents, and of surfactant head groups on the molecule location. Firstly, an analogue of DDP with a shorter hydrocarbon tail, namely, 4-*n*-pentyl-2,6-dinitrophenol, was examined in SDS [59]. Both forms are situated 0.1–0.2 nm closer to the water phase than DDP, which proves that the effect is present albeit it is rather small.

Secondly, the impact of substituents was studied with the example of 2-nitro-4-*n*-nonylphenol in CTAB. The molecule has a single nitro group and a one bead shorter tail than DDP. Interestingly, this change affected neutral and anionic forms differently: while the former shifted deeper by ca. 0.15 nm, the latter advanced towards water by ca. 0.07 nm. As a result, the molecular form is hydrated 15% weaker than that of DDP, while the anionic form has 30% more water around than DDP anion.

Lastly, DDP was simulated in zwitterionic micelles of CDAPSn [62]. The neutral form was found to be immersed 0.1–0.2 nm deeper into the hydrocarbon core than in CTAB, while location of the anionic form was very similar. Orientation was similar, as well. Still, the bulkier headgroups of CDAPSn made the dye 10–15% less hydrated because of expulsion of water from Stern layer. Importantly, despite the zwitterionic headgroup having both positively and negatively charged moieties, $[CH_2N(CH_3)_2CH_2]^+$ and SO_3^-, the indicator contacted almost exclusively with the former ones. This observation explained the highly positive value of Ψ determined for CDAPSn micelles with DDP.

8.6. Estimation of the ΔpK_a^{app} and Ψ Values

The opportunities provided by the MD simulation extend much beyond revealing geometric characteristics of "micelle + indicator" aggregates. One of the features of the method is evaluation of free energy changes that accompany both physical and chemical processes. We utilized this feature to computationally reproduce the described above experimental method of determination of Ψ.

The in silico method is based on computing the free energy change in deprotonation of an indicator molecules, ΔG_{deprot}, in water and in micelles. A well-suitable technique for this task is so-called alchemical transformation [65]. The difference of these ΔG_{deprot} values is proportional to the pK_a shift of the indicator on transfer from water to micelles, as is illustrated by Figure 25 and Equation (10).

$$pK_a^w = \frac{\Delta G_a^w}{RT \ln 10}; \quad pK_a^{app} = \frac{\Delta G_a^{app}}{RT \ln 10}; \quad \Delta pK_a^{app} = \frac{\Delta G_a^{app} - \Delta G_a^w}{RT \ln 10}$$

$$\Delta G_a^{app} - \Delta G_a^w = \Delta G_{deprot}^{app} - \Delta G_{deprot}^w \qquad (10)$$

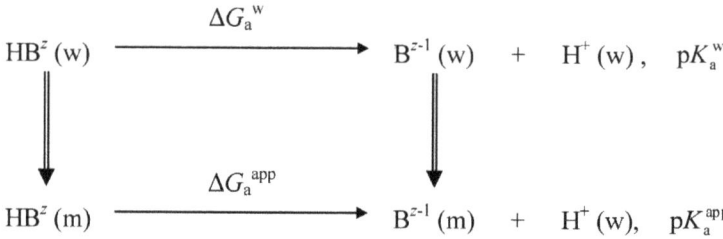

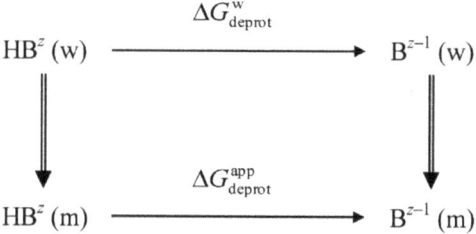

Figure 25. Thermodynamic cycle for computing pK_a shifts. Double arrows denote transfer between media. ΔG_a is the free energy change corresponding to HB^z dissociation. Adapted with permission from [65]. Copyright 2020 American Chemical Society.

This equation is possible because both pK_a^w and pK_a^{app} contain proton activity in bulk solution.

Estimated in this way ΔpK_a^{app} values appeared considerably more negative than expected (in average, by 1 unit in SDS and 3 units in CTAB), but overall a pronounced linear proportionality was found for each individual indicator. Furthermore, it was similar for different indicators (Figure 26A), and the order of increase in ΔpK_a^{app} values was mostly reproduced. We attributed the distortion to inaccurate reproducing indicator—water interactions by the used potential models [65]. In SDS micelles, the indicators are well-hydrated; hence transfer to water causes limited dehydration. Its contribution to $\Delta pK_{a\,calc}^{app}$ is limited, as well. Conversely, when an indicator is transferred from water to CTAB micelles, it loses a large fraction of interacting water molecules around, and the energy effect of dehydration is high. If the energy effect is reproduced inaccurately, the distortion is bigger in the case of CTAB where the effect itself is higher. Other sources of errors in computed ΔG_{deprot} values were also discussed, but their role was identified as minor.

From the computational perspective we pointed out the importance of incorporating corrections to so-called finite-size effects, which appear when ΔG between states of different charge is computed in an MD cell of a finite size. In our simulation setup, their magnitude was found to be 0.2–0.7 units (if converted to $\Delta pK_{a\,calc}^{app}$) for SDS and CTAB, and even ≥ 1 units for CDAPSn and TX-100 [234].

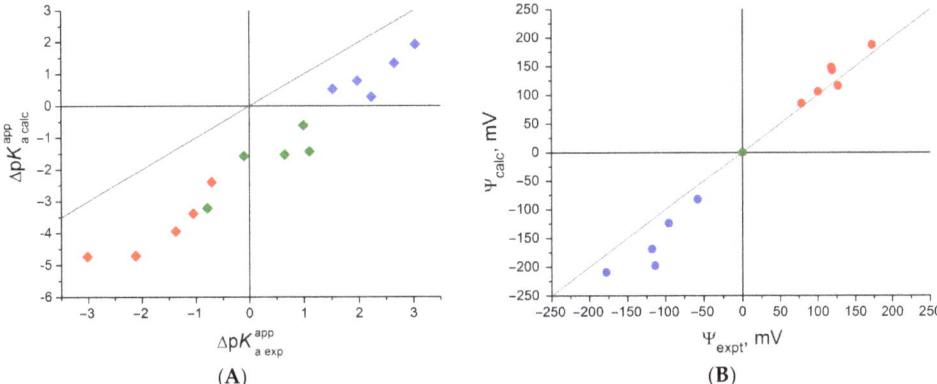

Figure 26. Calculated vs. experimental values of pK_a shifts (**A**) and Ψ (**B**). Blue, red, and green points are values for SDS, CTAB, TX-100 micelles, respectively; gray line is $y = x$.

Following the experimental method of estimating Ψ, the calculations of pK_a shifts were done in ionic (SDS, CTAB) and nonionic (TX-100) micelles, and their difference gives Ψ:

$$\Psi_{calc} = \frac{RT \ln 10}{F} \left(\Delta p K_{a\,calc}^{app}(\text{nonionic}) - \Delta p K_{a\,calc}^{app}(\text{ionic}) \right) \quad (11)$$

It also may be found directly from ΔG_{deprot} values, if the water phase is replaced with nonionic micelles in the thermodynamic cycle shown in Figure 25 [68].

$$\Psi_{calc} = \frac{1}{F} \left(\Delta G_{deprot}(\text{nonionic}) - \Delta G_{deprot}(\text{ionic}) \right) \quad (12)$$

Surprisingly, despite the large distortion of the input $\Delta p K_{a\,calc}^{app}$ values, the obtained Ψ_{calc} values were in fine agreement with experimental results [68,69] (Table 14 and Figure 26B). For CTAB, the distortion was often of the order of magnitude of the experimental uncertainty (± 5 mV), while for SDS the coincidence was reasonable (± 30 mV), except for two cases where only the order of magnitude was preserved. We think this stems naturally from our explanation of $\Delta p K_a^{app}$ distortion: there, the hydration energy error was high upon transfer from water to CTAB (and TX-100) where the indicator is poorly hydrated; while it was small upon transfer to SDS. Here, the transfer is done from TX-100 to CTAB having a similarly low hydration of the indicator; and between TX-100 and SDS where it differs strongly. This matches fine accuracy of Ψ_{calc} values determined in the former case and some distortion in the latter case.

Table 14. Experimental and calculated values of surface electrostatic potential of micelles. The values correspond to the ionic strength of 0.05 M. Adapted with permission from [68]. Copyright 2023 American Chemical Society.

Indicator	SDS		CTAB	
	Ψ_{calc}, mV	Ψ_{exp}, mV	Ψ_{calc}, mV	Ψ_{exp}, mV
DDP	−123	−96	189	172
HHC	−198	−114	115	127
DR	−82	−58	105	100
DF (cation)	−209	−178	85	79
DF (neutral)	−169	−118	142	119

The results of this in silico computation of surface electrostatic potential by means of indicator dyes may be compared with those presented above computed for entire micelles.

While the two methods differ by the basic principle, treatment of water, usage of probe, they still share atomic models of surfactant micelles. Importantly, the obtained values lie in the same interval (70–200 mV by magnitude). Ψ_{calc} estimated with probes are close to Ψ_0 from Figure 4: average Ψ_{calc}(SDS) = −156 mV, Ψ_{calc}(CTAB) = +127, while Ψ_0 was found −100 mV and +(149–151) mV, respectively. Taking into account the ca. 35% overestimation of Ψ_{calc}(SDS) with respect to experimental values, the correspondence is notable, which proves the consistency of our calculations. Furthermore, this shows that the indicator method provides rather surface potential Ψ_0 than Stern layer potential Ψ_δ, albeit probe molecules are proven to be located in the latter layer.

An essential point is that the computations reproduce the experimental variation in Ψ values of the same micelles as determined by different indicators. On one hand, this confirms the validity of the computations, on the other hand, this indicates that the spread is an intrinsic problem of the discussed method of estimating Ψ. Using the MD trajectories we investigated the problem deeper and deduced two origins of the spread. At first, pK_a^{app} in nonionic micelles of TX-100 is predicted to often mismatch the pK_a^i in ionic micelles because of different hydration of a dye molecule in these micelles. Second, each micelle + indicator aggregate has an individual structure, which also should affect the average value of electrostatic potential sensed by the indicator. Still, for the studied set of indicators, we did not find a correlation between these characteristics and reported Ψ values [69].

9. Some Other Reporter Molecules

Apart from indicator dyes, another instrument used to provide information about hydration of micelles are spin probes. We investigated a single probe, namely, methyl-5-doxylstearate (Scheme 17) that was previously examined in SDS and DTAB micelles [235,236].

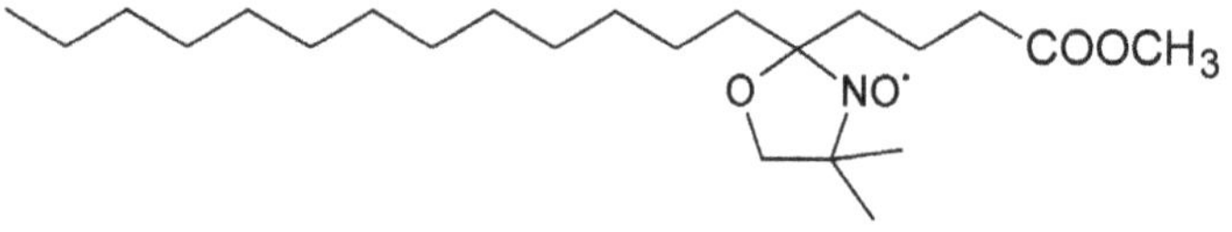

Scheme 17. Molecular structure of methyl-5-doxylstearate.

The EPR measurements were originally interpreted as the molar fraction of water in the Stern layer, which was found to be 0.68 in the anionic micelles and 0.44 in the cationic ones. This stays in line with the information from previously discussed dyes, which in general shows higher hydration of the former micelles. We hypothesized that the values are relevant rather to the microenvironment of the spin label alone than for the entire Stern layer. Our simulations of these systems showed that hydration of the O atom carrying the unpaired electron in these micelles is 0.64 and 0.39 if compared with pure water solution, and number of O˙—water hydrogen bonds reduces in a similar proportion. At the same time, hydration of the whole molecule is almost the same in both micelles [63].

Fluorescent dyes are another important group of molecular probes [22–31,130,131, 134,149,237–241]. An example of a fluorescent molecular probe that is not a dye in the traditional sense is the above mentioned pyrene [137]. Its solubility in aqueous micellar solutions is almost five orders of magnitude higher than in water [242]. The specific features of pyrene emission were used to determine the cmc values of surfactants, micellar polarity, and to investigate the extent of water penetration in micellar systems [243,244], to study the partition of n-pentanol [95], n-hexanol, and n-heptanol [245] between bulk water and the SDS micelles. MD simulations of the state of pyrene in SDS and CTAB micelles were studied in [246,247], respectively. The study revealed that the pyrene molecule is located in the last-named micelles in a more polar region than in SDS micelles [137,246]. The reason is the interaction of the electron-rich pyrene ring with the quaternary ammonium groups [137,247]. As a result, the pyrene molecule is located not only in the interior cavity, but also in the palisade layer of the CTAB micelles [247]. By contrast, in the SDS micelles, the pyrene molecule is located in the hydrophobic core region only [246].

The acid–base indicators considered in the present paper also belong to fluorophores, e.g., coumarins [23,25], rhodamines [24,31,51,52,151–155], and fluorescein and its derivatives [24,29,30,131,134,240]. Additionally, fluorescence is used also for determination of the pK_a^{app} and Ψ values not only for coumarins (see Section 6.1), but also for fluorescein dyes [248]. Therefore, the results of MD modeling of these compounds in micellar media presented above can be useful for a better understanding of the systems where they are used as fluorescent molecular probes, too.

At last, a recent application of our approach for understanding the behavior of substrates in calixarene-based catalytic systems should be mentioned [249].

10. Conclusions

10.1. Molecular Dynamis Modeling of Micelles

Molecular dynamics modeling of SDS, CTAB, and some other surfactant micelles in water makes it possible to reveal their structure, hydration, and counterion binding. The electrical potential, Ψ, as a function of the distance from the micellar surface and bulk ionic strength may be computed and interpreted in terms of the surface potential Ψ_0 and the Stern layer potential Ψ_δ. Numerical methods of solving the Poisson–Boltzmann equation enable using atomic models of micelles instead of idealized spherical and cylindrical models.

Consideration of the surfactant ions DS^- and CTA^+ in water without counterions results in appearance of two small micelles in both cases. Hence, the hydrophobic association of the hydrocarbon tails balances the repulsion of the charged headgroups of these small "bare" micelles. The experimental verification is impossible in this case.

10.2. Location, Orientation, and Hydration of Seven Reichardt's Solvatochromic Dyes in Micelles

In CTAB micelles, the dye zwitterions, $^+D^-$, penetrate deeper than the cations. This is not the case in SDS micelles. The dye RD-PhtBu in both forms is situated in the hydrocarbon region. The same is found but much less pronounced for the zwitterions of RD-tBu and RD-cyclo9 in SDS micelles and that of RD-tBu in CTAB.

For the zwitterions, $^+D^-$, of the dyes RD-H, RD-Cl, RD-cyclo9, and RD-tBu, the horizontal orientation is most probable, and the vertical one is sometimes found in SDS micelles. For the $^+$DH cations, the horizontal orientation predominates in both kinds of micelles. The inverted orientation, with the phenol part directed to the micellar center, is probable for the cation of the most hydrophobic dye, RD-PhtBu, and for the RD-COOH dye in both forms bearing the OH group.

For the standard solvatochromic dye, 4-(2,4,6-triphenylpyridinium-1-yl)-2,6-diphenylphenolate, RD-Ph, the horizontal orientation is more typical. The orientation, intermediate between inverted and horizontal, can be sometimes expected only for the colorless cation.

The obtained data do not contradict intuitive expectations based on the concepts of hydrophobicity and hydrophilicity of molecules, ions, and their individual fragments. For example, the dye RD-PhtBu, bearing five additional $C(CH_3)_3$ groups, is always most deeply immersed into the micelles of both types and hence less hydrated, though some contacts with water molecules nevertheless take place owing to the dynamic character of the surfactant micelles.

This is not the case for solvation of the oxygen atom of the O^- and OH groups. Indeed, the hydration of the phenolate oxygen of the $^+D^-$ forms is much more pronounced in SDS than in CTAB micelles. This is more than a convincing confirmation of the higher value of the E_T^N value in SDS micelles than in the CTAB ones.

10.3. Hydration of Surfactant Micellar Surfaces: SDS vs. CTAB

MD modeling does not demonstrate a serious difference in the hydration of micellar surfaces of anionic surfactants compared to cationic ones at the same tail length. The consideration of the Krafft temperature allows expecting even a better hydration in the case of CTAB micelles. Therefore, the pronounced exceeding of the E_T^N values for SDS and other anionic surfactant micelles as compared with those for CTAB is caused by better hydration of the betaine dye. The same can be said about the data obtained with other molecular probes. Hence, it is more correct to speak about the more pronounced hydrogen-binding ability of water molecules in anionic micelles than about better hydration of the anionic micelles themselves.

10.4. Acid–Base Indicators in Surfactant Micelles: Molecular Dynamics Modeling

Location, orientation, and hydration of several popular acid–base indicators, used in experiments of Ψ determination, were studied using molecular dynamics simulation. They are 4-[(E)-([1,1'-biphenyl]-4-yl)diazenyl]-2-nitrophenol, a dye proposed in the pioneering work by Hartley and Roe as early as 1940 [139], 4-*n*-heptadecyl-7-hydroxycoumarin, a standard indicator used for this purpose, and three other dyes.

The simulation results generally consistent with the intuitive assumption about the location of acid–base indicators. The dyes are located in such a way that the ionizing groups remain in contact with water and on the surface, while the chromophore part can either lie at the interface between the hydrocarbon core and water, or be immersed in the core. Two positions can be observed for the same molecule as alternating. The immersion depth nontrivially depends on the state of protonation.

The vertical orientation is typical for both forms of 2,6-dinitro-4-*n*-dodecylphenol and 4-*n*-heptadecyl-7-hydroxycoumarin, neutral and anionic forms of *n*-decylfluorescein, cationic form of N,N'-di-*n*-octadecylrhodamine, and for neutral form of the Hartley dye; this is true for micelles of both charges. The horizontal orientation is typical for anionic forms of the Hartley dye and *n*-decylfluorescein, for the neutral form of N,N'-di-*n*-octadecylrhodamine; and it is also observed for some other cases.

Regarding the hydration of acid–base dyes, the important factor is the protolytic form and micelle charge. The neutral (including zwitterionic) forms are slightly more hydrated in anionic micelles, while the anionic forms are much more hydrated. This is observed both for entire dye molecule and for its ionizing group; however, the particular values are dissimilar.

10.5. Determination of the Electrical Surface Potential of Micelles Using Acid–Base Indicators

Utilization of techniques for computing free energy changes allows finding ΔpK_a^{app} values for transferring a dye molecule from water to micelles. The results appear strongly distorted. However, the correlation between experimental and computed values is pronounced and linear. The distortion is explained by the poor reproduction of the hydration energy by used potential models.

Nevertheless, if ΔpK_a^{app} is computed for both ionic and nonionic micelles, then Ψ can be found from the difference between the values in nonionic and ionic micelles. It may be called in silico implementation of the "wet" method. The calculated values agree well with the experiment: there is almost a quantitative match for CTAB and 35% (in average) overestimation for SDS. The reason is the partial cancellation of error of input ΔpK_a^{app} values.

The in silico calculation of Ψ is accurate enough to reproduce the variation in Ψ values of the same micelles as determined by different indicators. Hence, this variation is an intrinsic problem of the indicator method of estimating Ψ. We did not find a correlation between reported Ψ values and such characteristics of micelle + indicator aggregates as depth of immersion of the indicator in micelle and its hydration.

Still, the set of dyes currently under consideration is limited, and only TX-100 was studied as nonionic surfactant and used in the assumption that pK_a^i (in ionic micelles) = pK_a^{app} (in nonionic micelles). Therefore, there is room for further developing this issue.

Author Contributions: Conceptualization, N.O.M.-P., V.S.F. and A.V.L.; methodology, N.O.M.-P., V.S.F. and A.V.L.; software, V.S.F. and A.V.L.; validation, N.O.M.-P., V.S.F. and A.V.L.; formal analysis, V.S.F.; investigation, N.O.M.-P., V.S.F. and A.V.L.; resources, V.S.F. and A.V.L.; data curation, N.O.M.-P. and V.S.F., writing—original draft preparation, N.O.M.-P. and V.S.F., writing—review and editing, N.O.M.-P. and V.S.F.; visualization, V.S.F., supervision, N.O.M.-P. and A.V.L.; project administration, N.O.M.-P., funding acquisition, N.O.M.-P., V.S.F. and A.V.L. All authors have read and agreed to the published version of the manuscript.

Funding: This study was partially supported by the Ministry of Education and Science of Ukraine via grant number 0122U001485.

Data Availability Statement: The data is available upon request.

Acknowledgments: The authors express their gratitude to Christian Reichardt (Philipps University, Marburg, Germany) for detailed discussion of the manuscript, as well as for many years of collaboration in the field of physical chemistry of betaine dyes. The authors are also grateful to Dmitry Nerukh (Aston University, Birmingham, UK) for the provided computational facilities.

Conflicts of Interest: The authors declare no conflict of interest.

References

1. Reichardt, C. Solvation effects in organic chemistry: A short historical overview. *J. Org. Chem.* **2022**, *87*, 1616–1629. [CrossRef] [PubMed]
2. Shinoda, K. The significance and characteristics of organized solutions. *J. Phys. Chem.* **1985**, *89*, 2429–2431. [CrossRef]
3. Hartley, G.S. Organised structures in soap solutions. *Nature* **1949**, *163*, 767–768. [CrossRef]
4. Vincent, B. McBain and the centenary of the micelle. *Adv. Colloid Int. Sci.* **2014**, *203*, 51–54. [CrossRef]
5. McBain, J.W. Mobility of highly charged micelles. *Trans. Faraday Soc.* **1913**, *9*, 99–101. [CrossRef]
6. Reychler, A. Beiträge zur Kenntnis der Seifen. *Z. Chem. Ind. Koll.* **1913**, *12*, 277–283. [CrossRef]
7. Romsted, L.S. (Ed.) *Surfactants Science and Technology. Retrospects and Prospects*; CRC Press: Boca Raton, FL, USA; Taylor & Francis Group: Abingdon, UK, 2014; 583p. [CrossRef]
8. Kralchevsky, P.A.; Danov, K.D.; Denkov, N.D. Chemical physics of colloid systems and interfaces. In *Handbook of Surface and Colloid Chemistry*; Birdi, K.S., Ed.; CRC Press: Boca Raton, FL, USA, 2009; Chapter 7; pp. 197–377.
9. Reichardt, C.; Welton, T. *Solvents and Solvent Effects in Organic Chemistry*, 4th ed.; Wiley-VCH: Weinheim, Germany, 2011; 718p.
10. Mchedlov-Petrossyan, N.O.; Vodolazkaya, N.A.; Kamneva, N.N. Acid-base equilibrium in aqueous micellar solutions of surfactants. In *Micelles: Structural Biochemistry, Formation and Functions & Usage*; Bradburn, D., Bittinger, J., Eds.; Nova Science Publishers: New York, NY, USA, 2013; pp. 1–71.
11. Moroi, Y. *Micelles: Theoretical and Applied Aspects*; Plenum Press: New York, NY, USA, 2013; 252p.
12. Zueva, O.S.; Rukhlov, V.S.; Zuev, Y.F. Morphology of Ionic Micelles as Studied by Numerical Solution of the Poisson Equation. *ACS Omega* **2022**, *7*, 6174–6183. [CrossRef]
13. Grieser, F.; Drummond, C.J. The physicochemical properties of self-assembled surfactant aggregates as determined by some molecular spectroscopic probe techniques. *J. Phys. Chem.* **1988**, *92*, 5580–5593. [CrossRef]
14. Mchedlov-Petrossyan, N.O. Protolytic equilibrium in lyophilic nanosized dispersions: Differentiating influence of the pseudophase and salt effects. *Pure Appl. Chem.* **2008**, *80*, 1459–1510. [CrossRef]
15. Lacerda, C.D.; Andrade, M.F.C.; Pessoa, P.S.; Prado, F.M.; Pires, P.A.R.; Pinatto-Botelho, M.F.; Wodtke, F.; Dos Santos, A.A.; Dias, L.G.; Lima, F.S.; et al. Experimental mapping of a pH gradient from a positively charged micellar interface to bulk solution. *Colloids Surf. A* **2021**, *611*, 125770. [CrossRef]
16. Souza, T.P.; Zanette, D.; Kawanami, A.E.; de Rezende, L.; Ishiki, H.M.; do Amaral, A.T.; Chaimovich, H.; Agostinho-Neto, A.; Cuccovia, I.M. pH at the micellar interface: Synthesis of pH probes derived from salicylic acid, acid–base dissociation in sodium dodecyl sulfate micelles, and Poisson–Boltzmann simulation. *J. Colloid Interface Sci.* **2006**, *297*, 292–302. [CrossRef] [PubMed]
17. De Freitas, C.F.; Estevão, B.M.; Pellosi, D.S.; Scarminio, I.S.; Caetano, W.; Hioka, N.; Batistella, V.R. Chemical equilibria of Eosin Y and its synthetic ester derivatives in non-ionic and ionic micellar environments. *J. Mol. Liq.* **2020**, *327*, 114794. [CrossRef]
18. Mchedlov-Petrossyan, N.O.; Vodolazkaya, N.A. Protolytic Equilibria in Organized Solutions: Ionization and Tautomerism of Fluorescein Dyes and Related Indicators in Cetyltrimethylammonium Chloride Micellar Solutions at High Ionic Strength of the Bulk Phase. *Liq.* **2021**, *1*, 1–24. [CrossRef]
19. Drummond, C.J.; Grieser, F.; Healy, T.W. A single spectroscopic probe for the determination of both the interfacial solvent properties and electrostatic surface potential of model lipid membranes. *Faraday Discuss.* **1986**, *81*, 95–106. [CrossRef]
20. Healy, T.W.; Drummond, C.J.; Grieser, F.; Murray, B.S. Electrostatic surface potential and critical micelle concentration relationship for ionic micelles. *Langmuir* **1990**, *6*, 506–508. [CrossRef]
21. Machado, V.G.; Stock, R.I.; Reichardt, C. Pyridinium N-Phenolate Betaine Dyes. *Chem. Rev.* **2014**, *114*, 10429–10475. [CrossRef]
22. Waggoner, A.S. Dye indicators of membrane potential. *Ann. Rev. Biophys. Bioeng.* **1979**, *8*, 47–68. [CrossRef]

23. Pal, R.; Petri, W.A.; Ben-Yashar, V.; Wagner, R.R.; Barenholz, Y. Characterization of the fluorophore 4-heptadecyl-7-hydroxycoumarin: A probe for the head-group region of lipid bilayers and biological membranes. *Biochemistry* **1985**, *24*, 573–581. [CrossRef]
24. Memon, N.; Balouch, A.; Hinze, W.L. Fluorescence in Organized Assemblies. In *Encyclopedia of Analyt Chem*; Meyers, R.A., Ed.; Wiley: New York, NY, USA, 2008; pp. 1–94. [CrossRef]
25. Krämer, R. Interaction of membrane surface charges with the reconstituted ADP/ATP-carrier from mitochondria. *Biochim. Biophys. Acta* **1983**, *735*, 145–159. [CrossRef]
26. Kachel, K.; Asuncion-Panzalan, E.; London, E. The location of fluorescence probes with charged groups in model membranes. *Biochim. Biophys. Acta* **1998**, *1374*, 63–76. [CrossRef]
27. Janata, J. Do optical sensors really measure pH? *Analyt. Chem.* **1987**, *59*, 1351–1356. [CrossRef]
28. Lobnik, A.; Oehme, I.; Murkovic, I.; Wolfbeis, O.S. pH optical sensors based on sol–gels: Chemical doping versus covalent immobilization. *Analyt. Chim. Acta* **1998**, *367*, 159–165. [CrossRef]
29. Weidgans, B.M.; Krause, C.; Klimant, I.; Wolfbeis, O.S. Fluorescent pH sensors with negligible sensitivity to ionic strength. *Analyst* **2004**, *129*, 645–650. [CrossRef] [PubMed]
30. Schröder, C.R.; Weidgans, B.M.; Klimant, I. pH Fluorosensors for Use in Marine Systems. *Analyst* **2005**, *130*, 907–916. [CrossRef] [PubMed]
31. Li, D.; Zhang, M.; Zhou, L.; Li, Z.; Xu, Q.; Pu, X.; Sun, Y.; Zhang, Y. Monitoring the pH value of an aqueous micellar solution in real-time using a fiber optofluidic laser. *J. Light. Technol.* **2023**, *41*, 362–366. [CrossRef]
32. Neumann, M.G.; Pastre, I.A.; Chinelatto, A.M.; El Seoud, O.A. Effects of the structure of anionic polyelectrolytes on surface potentials of their aggregates in water. *Colloid Polym. Sci.* **1996**, *274*, 475–481. [CrossRef]
33. Kharchenko, A.Y.; Moskaeva, O.G.; Klochaniuk, O.R.; Marfunin, M.O.; Mchedlov-Petrossyan, N.O. Effect of poly (sodium 4-styrenesulfonate) on the ionization constants of acid-base indicator dyes in aqueous solutions. *Colloids Surf. A* **2017**, *527*, 132–144. [CrossRef]
34. Nikitina, N.A.; Reshetnyak, E.A.; Svetlova, N.V.; Mchedlov-Petrossyan, N.O. Protolytic properties of dyes embedded in gelatin films. *J. Brazil. Chem. Soc.* **2011**, *22*, 855–864. [CrossRef]
35. Mchedlov-Petrossyan, N.O.; Kamneva, N.N.; Marynin, A.I.; Kryshtal, A.P.; Ōsawa, E. Colloidal properties and behaviors of 3nm primary particles of detonation nanodiamond in aqueous media. *Phys. Chem. Chem. Phys.* **2015**, *17*, 16186–16203. [CrossRef]
36. Kharchenko, A.Y.; Marfunin, M.O.; Semenov, K.N.; Charykov, N.A.; Mchedlov-Petrossyan, N.O. *Fullerenol Aqueous Solutions as Media for Acid-Base Reactions: Neutral Red as Molecular Probe, XV International Congress 'YoungChem 2017', Lublin, Poland: Abstract Book*; Chemical Scientific Society 'Flogiston'; Warsaw University of Technology: Warsaw, Poland, 2017; p. 92.
37. Cheipesh, T.A.; Zagorulko, E.S.; Mchedlov-Petrossyan, N.O.; Rodik, R.V.; Kalchenko, V.I. The Difference between the aggregates of a short-tailed and a long-tailed cationic calix[4]arene in water as detected using fluorescein dyes. *J. Mol. Liq.* **2014**, *193*, 232–238. [CrossRef]
38. Cheipesh, T.A.; Mchedlov–Petrossyan, N.O.; Bogdanova, L.N.; Kharchenko, D.V.; Roshal, A.D.; Vodolazkaya, N.A.; Taranets, Y.V.; Shekhovtsov, S.V.; Rodik, R.V.; Kalchenko, V.I. Aggregates of cationic calix[4]arenes in aqueous solution as media for governing protolytic equilibrium, fluorescence, and kinetics. *J. Mol. Liq.* **2022**, *366*, 119940. [CrossRef]
39. Chandra, F.; Kumar, P.; Koner, A.L. Encapsulation and modulation of protolytic equilibrium of β-carboline-based norharmane drug by cucurbit[7]uril and micellar environments for enhanced cellular uptake. *Colloids Surf. B* **2018**, *171*, 530–537. [CrossRef] [PubMed]
40. Petrov, J.G.; Möbius, D. Fluorometric titration of 4-heptadecyl-7-hydroxycoumarin in neutral monolayers at the air/water interface. *Langmuir* **1989**, *5*, 523–528. [CrossRef]
41. Petrov, J.G.; Möbius, D. Determination of the electrostatic potential of positively charged monolayers at the air/water interface by means of fluorometric titration of 4-heptadecyl-7-hydroxycoumarin. *Langmuir* **1990**, *6*, 746–751. [CrossRef]
42. Yamaguchi, S.; Bhattacharyya, K.; Tahara, T. Acid-base equilibrium at an aqueous interface: pH spectrometry by heterodyne-detected electronic sum frequency generation. *J. Phys. Chem. C* **2011**, *115*, 4168–4173. [CrossRef]
43. Petrov, J.G.; Möbius, D. Strong influence of the solid substrate on the acid-base equilibrium of 4-heptadecyl-7-hydroxy coumarin in monolayers at solid/liquid interfaces. *Langmuir* **1991**, *7*, 1495–1497. [CrossRef]
44. Petrov, J.G.; Möbius, D. Dependence of pK of 4-heptadecyl-7-hydroxycoumarin at a neutral multilayer/water interface on the multilayer thickness. *Langmuir* **1993**, *9*, 756–759. [CrossRef]
45. Petrov, J.G.; Möbius, D. Interfacial acid-base equilibrium and electrostatic potentials of model Langmuir–Blodgett membranes in contact with phosphate buffer. *Colloids Surf. A* **2000**, *171*, 207–215. [CrossRef]
46. Mchedlov-Petrossyan, N.O.; Bezkrovnaya, O.N.; Vodolazkaya, N.A. Polymeric Langmuir–Blodgett films functionalized by pH-sensitive dyes. *Chem. Phys. Technol. Surf.* **2020**, *11*, 72–99. (In Russian) [CrossRef]
47. Wang, H.; Zhao, X.; Eisenthal, K.B. Effects of monolayer density and bulk ionic strength on acid-base equilibria at the air/water interface. *J. Phys. Chem. B* **2000**, *104*, 8855–8861. [CrossRef]
48. Yamaguchi, S.; Kundu, A.; Sen, P.; Tahara, T. Communication: Quantitative estimate of the water surface pH using heterodyne-detected electronic sum frequency generation. *J. Chem. Phys.* **2012**, *137*, 151101. [CrossRef] [PubMed]
49. Mchedlov-Petrossyan, N.O.; Kharchenko, A.Y.; Marfunin, M.O.; Klochaniuk, O.R. Nano-sized bubbles in solution of hydrophobic dyes and the properties of the water/air interface. *J. Mol. Liq.* **2019**, *275*, 384–393. [CrossRef]

50. Loura, L.M.S.; Ramalho, J.P.P. Location and dynamics of acyl chain NRD-labeled phosphatidylcholine (NRD-PC) in DPPC bilayers. A molecular dynamics and time-resolved fluorescence anisotropy study. *Biochim. Biophys. Acta* **2007**, *1768*, 467–478. [CrossRef]
51. Kyrychenko, A. A molecular dynamics model of rhodamine-labeled phospholipid incorporated into a lipid bilayer. *Chem. Phys. Lett.* **2010**, *485*, 95–99. [CrossRef]
52. Polat, B.E.; Lin, S.; Mendenhall, J.D.; VanVeller, B.; Langer, R.; Blankschtein, D. Experimental and molecular dynamics investigation into the amphiphilic nature of sulforhodamine B. *J. Phys. Chem. B* **2011**, *115*, 1394–1402. [CrossRef]
53. Farafonov, V.S.; Lebed, A.V. Investigating localization of a tetrapropylammonium ion in a micellar solution of sodium dodecylsulfate by means of molecular dynamics simulation. *Kharkov Univer. Bull. Chem. Ser.* **2016**, *26*, 73–79. [CrossRef]
54. Farafonov, V.S.; Lebed, A.V. Molecular dynamics simulation study of cetylpyridinum chloride and cetyltrimethylammonium bromide micelles. *Kharkov Univer. Bull. Chem. Ser.* **2016**, *27*, 25–30. [CrossRef]
55. Farafonov, V.S.; Lebed, A.V. Developing and validating a set of all-atom potential models for sodium dodecyl sulfate. *J. Chem. Theory Comput.* **2017**, *13*, 2742–2750. [CrossRef]
56. Farafonov, V.S.; Lebed, A.V.; Mchedlov-Petrossyan, N.O. Character of localization and microenvironment of the solvatochromic Reichardt's betaine dye in SDS and CTAB micelles: MD simulation study. *Langmuir* **2017**, *33*, 8342–8352. [CrossRef]
57. Farafonov, V.S.; Lebed, A.V.; Mchedlov-Petrossyan, N.O. Solvatochromic Reichardt's dye in micelles of sodium cetyl sulfate: Md modeling of location character and hydration. *Kharkov Univer. Bull. Chem. Ser.* **2017**, *28*, 5–11. [CrossRef]
58. Farafonov, V.S.; Lebed, A.V.; Mchedlov-Petrossyan, N.O. Solvatochromic betaine dyes of different hydrophobicity in ionic surfactant micelles: Molecular dynamics modeling of location character. *Colloids Surf. A* **2018**, *538*, 583–592. [CrossRef]
59. Mchedlov-Petrossyan, N.O.; Farafonov, V.S.; Lebed, A.V. Examining surfactant micelles via acid-base indicators: Revisiting the pioneering Hartley–Roe 1940 study by molecular dynamics modeling. *J. Mol. Liq.* **2018**, *264*, 683–690. [CrossRef]
60. Farafonov, V.S.; Lebed, A.V.; Mchedlov-Petrossyan, N.O. Examining solvatochromic Reichardt's dye in cationic micelles of different size via molecular dynamics. *Vopr. Khimii Khimicheskoi Tekhnologii* **2018**, *5*, 62–68.
61. Farafonov, V.S.; Lebed, A.V.; Mchedlov-Petrossyan, N.O. An MD simulation study of Reichardt's betaines in surfactant micelles: Unlike orientation and solvation of cationic, zwitterionic, and anionic dye species within the pseudophase. *Kharkov Univer. Bull. Chem. Ser.* **2018**, *30*, 27–35. [CrossRef]
62. Mchedlov-Petrossyan, N.O.; Farafonov, V.S.; Cheipesh, T.A.; Shekhovtsov, S.V.; Nerukh, D.A.; Lebed, A.V. In search of an optimal acid-base indicator for examining surfactant micelles: Spectrophotometric studies and molecular dynamics simulations. *Colloids Surf. A* **2019**, *565*, 97–107. [CrossRef]
63. Farafonov, V.S.; Lebed, A.V. Nitroxyl spin probe in ionic micelles: A molecular dynamics study. *Kharkov Univer. Bull. Chem. Ser.* **2020**, *34*, 57–64. [CrossRef]
64. Farafonov, V.S.; Lebed, A.V.; Khimenko, N.L.; Mchedlov-Petrossyan, N.O. Molecular dynamics study of an acid-base indicator dye in triton X-100 non-ionic micelles. *Vopr. Khimii Khimicheskoi Tekhnologii* **2020**, *1*, 97–103. [CrossRef]
65. Farafonov, V.S.; Lebed, A.V.; Mchedlov-Petrossyan, N.O. Computing pKa shifts using traditional molecular dynamics: Example of acid-base indicator dyes in organized solutions. *J. Chem. Theory Comput.* **2020**, *16*, 5852–5865. [CrossRef]
66. Farafonov, V.S.; Lebed, A.V.; Mchedlov-Petrossyan, N.O. Continuum electrostatics investigation of ionic micelles using atomistic models. *Ukrainian Chem. J.* **2021**, *87*, 55–69. [CrossRef]
67. Farafonov, V.S.; Lebed, A.V.; Mchedlov-Petrossyan, N.O. Localization of hydrophobized coumarin and xanthene acid–base indicators in micelles. *Theor. Exper. Chem.* **2022**, *58*, 181–189. [CrossRef]
68. Farafonov, V.S.; Lebed, A.V.; Nerukh, D.A.; Mchedlov-Petrossyan, N.O. Estimation of Nanoparticle's Surface Electrostatic Potential in Solution Using Acid–Base Molecular Probes I: In Silico Implementation for Surfactant Micelles. *J. Phys. Chem. B* **2022**, *126*, 1022–1030. [CrossRef] [PubMed]
69. Farafonov, V.S.; Lebed, A.V.; Nerukh, D.A.; Mchedlov-Petrossyan, N.O. Estimation of nanoparticle's surface electrostatic potential in solution using acid-base molecular probes II: Insight from atomistic simulations of micelles. *J. Phys. Chem. B* **2022**, *126*, 1031–1038. [CrossRef] [PubMed]
70. Bruce, C.D.; Berkowitz, M.L.; Perera, L.; Forbes, M.D.E. Molecular dynamic simulations of dodecylsulfate micelle in water: Micellar structural characteristics and counterion distribution. *J. Phys. Chem. B* **2002**, *106*, 3788–3793. [CrossRef]
71. Bruce, C.D.; Senapati, S.; Berkowitz, M.L.; Perera, L.; Forbes, M.D.E. Molecular dynamic simulations of dodecylsulfate micelle in water: The behavior of water. *J. Phys. Chem. B* **2002**, *106*, 10902–10907. [CrossRef]
72. Roussed, G.; Michaux, C.; Perpete, E.A. Multiscale molecular dynamics simulations of sodium dodecyl sulfate micelles: From coarse-grained to all-atom resolution. *J. Mol. Model* **2014**, *20*, 2469. [CrossRef]
73. Verma, R.; Mishra, A.; Mitchell-Koch, K.R. Molecular modeling of cetylpyridinium bromide, a cationic surfactant, in solutions and micelle. *J. Chem. Theory Comput.* **2015**, *11*, 5415–5425. [CrossRef] [PubMed]
74. Belyaeva, E.A.; Vanin, A.A.; Anufrikov, Y.A.; Smirnova, N.A. Molecular-dynamic simulation of aliphatic alcohols distribution between the micelle of 3-methyl-1-dodecylimidazolium bromide and their aqueous surrounding. *Colloids Surf. A* **2016**, *508*, 93–100. [CrossRef]
75. Taddese, T.; Anderson, R.L.; Bray, D.J.; Warren, P.B. Recent advances in particle-based simulation of surfactants. *Current Opin. Coll. Int. Sci.* **2020**, *48*, 137–148. [CrossRef]
76. Pantoja-Romero, W.S.; Estrada-Lopez, E.D.; Picciani, P.H.S.; Oliveira, O.N.; Lachter, E.R.; Pimentel, A.S. Efficient molecular packing of glycerol monoestearate in Langmuir monolayer at the air-water interface. *Colloids Surf. A* **2016**, *508*, 85–92. [CrossRef]

77. Morrow, B.H.; Koenig, P.H.; Shen, J.K. Constant pH simulations of pH responsive polymers. *J. Chem. Phys.* **2012**, *137*, 194902. [CrossRef]
78. Lukanov, B.; Firoozabadi, A. Specific ion effects on the self-assembly of ionic surfactants: A molecular thermodynamic theory of micellization with dispersion forces. *Langmuir* **2014**, *30*, 6373–6383. [CrossRef] [PubMed]
79. Morrow, B.H.; Wang, Y.; Wallace, J.A.; Koenig, P.H.; Shen, J.K. Simulating pH titration of a single surfactant in ionic and nonionic surfactant micelles. *J. Phys. Chem. B* **2011**, *115*, 14980–14990. [CrossRef] [PubMed]
80. Vanquelef, E.; Simon, S.; Marquant, G.; Garcia, E.; Klimerak, G.; Delepine, J.C.; Cieplak, P.; Dupradeau, F.-Y. R.E.D. Server: A web service for deriving RESP and ESP charges and building force field libraries for new molecules and molecular fragments. *Nucleic Acids Res.* **2011**, *39*, W511–W517. [CrossRef]
81. Stigter, D. Micelle Formation by Ionic Surfactants. II. Specificity of head groups, micelle structure. *J. Phys. Chem.* **1974**, *78*, 2480–2485. [CrossRef]
82. Stigter, D. Micelle formation by ionic surfactants. III. Model of Stern layer, ion distribution, and potential fluctuations. *J. Phys. Chem.* **1975**, *79*, 1008–1014. [CrossRef]
83. Stigter, D. Micelle formation by ionic surfactants. IV. Electrostatic and hydrophobic free energy from Stern-Gouy ionic double layer. *J. Phys. Chem.* **1975**, *79*, 1015–1022. [CrossRef]
84. Larsen, J.W.; Magid, L.J. Calorimetric and counterion binding studies of the interactions between micelles and ions. Observation of lyotropic series. *J. Am. Chem. Soc.* **1974**, *96*, 5774–5782. [CrossRef]
85. Menger, F.M. On the structure of micelles. *Acc. Chem. Res.* **1979**, *12*, 111–117. [CrossRef]
86. Philippoff, W. Mechanische Eigenschaften von Seifenlösungen in ihrer Beziehung zur Struktur. *Kolloid-Zeitschrift* **1941**, *96*, 255–261. [CrossRef]
87. Philippoff, W. The micelle and swollen micelle on soap micelles. *Discuss. Faraday Soc.* **1951**, *11*, 96–107. [CrossRef]
88. Fromherz, P. Micelle structure: A surfactant-block model. *Chem. Phys. Lett.* **1981**, *77*, 460–466. [CrossRef]
89. Israelachvili, J.N. *Intermolecular and Surface Forces*, 3rd ed.; Academic Press: New York, NY, USA, 2011; 674p.
90. Gilanyi, T. Fluctuating micelles: A theory of surfactant aggregation 2. Ionic surfactants. *Colloids Surf. A* **1995**, *104*, 119–126. [CrossRef]
91. Rusanov, A.I. *Micellization in Surfactant Solutions*; Harwood Academic Publishes: Reading, UK, 1997.
92. Us'yarov, O.G. The electrical double layer of micelles in ionic surfactant solutions in the presence of background electrolyte: 1. Diluted micellar solutions of sodium dodecyl sulfate. *Colloid J.* **2007**, *69*, 95–102. [CrossRef]
93. Us'yarov, O.G. The electrical double layer of micelles in ionic surfactant solutions in the presence of background electrolyte: 1 Moderately concentrated micellar solutions of sodium dodecyl sulfate. *Colloid J.* **2007**, *69*, 103–110. [CrossRef]
94. Ohshima, H.; Healy, T.W.; White, L.R. Accurate analytic expressions for the surface charge density/surface potential relationship and double-layer potential distribution for a spherical colloidal particle. *J. Colloid Int. Sci.* **1980**, *90*, 17–26. [CrossRef]
95. Hartland, G.V.; Grieser, F.; White, L.R. Surface potential measurements in pentanol-sodium dodecyl sulphate micelles. *J. Chem. Soc. Faraday Trans. 1* **1987**, *83*, 591–613. [CrossRef]
96. Maduar, S.R.; Belyaev, A.V.; Lobaskin, V.; Vinogradova, O.I. Electrodynamics of near hydrophobic surfaces. *Phys. Rev. Lett.* **2015**, *114*, 118301. [CrossRef] [PubMed]
97. Predota, M.; Machesky, M.L.; Wesolowski, D.J. Molecular origins of the zeta potential. *Langmuir* **2016**, *32*, 10189–10198. [CrossRef]
98. Sasaki, T.; Hattori, M.; Sasaki, J.; Nukina, K. Studies of Aqueous Sodium Dodecyl Sulfate Solutions by Activity Measurements. *Bull. Chem. Soc. Jpn.* **1975**, *48*, 1397–1403. [CrossRef]
99. Frahm, J.; Diekman, S.; Haase, A. Electrostatic properties of ionic micelles in aqueous solutions. *Ber. Bunsenges. Phys. Chem.* **1980**, *84*, 566–571. [CrossRef]
100. Lebedeva, N.V.; Shahine, A.; Bales, B.L. Aggregation number-based degrees of counterion dissociation in sodium *n*-alkyl sulfate micelles. *J. Phys. Chem. B* **2005**, *109*, 19806–19816. [CrossRef]
101. Gilanyi, T. On the counterion dissociation of colloid electrolytes. *J. Colloid Int. Sci.* **1988**, *125*, 641–648. [CrossRef]
102. Kralchevsky, P.A.; Ananthapadmanabhan, K.P. Determination of the aggregation number and charge of ionic surfactant micelles from the stepwise thinning of foam films. *Adv. Colloid Interface Sci.* **2012**, *183–184*, 55–67. [CrossRef]
103. Loginova, L.P.; Boichenko, A.P.; Galat, M.N.; Nguen, K.N.K.; Kamneva, N.N.; Varchenko, V.V. Micelle formation characteristics of sodium dodecylsulphate and cetylpyridinium chloride in the presence of aliphatic alcohols and carboxylic acids. *Kharkov Univer. Bull. Chem. Series* **2010**, *18*, 47–55. (In Russian)
104. Koya, P.A.; Ahmad, T.; Ismail, W.K. Conductometric studies on micellization of cationic surfactants in the presence of glycine. *J. Solut. Chem.* **2015**, *44*, 100–111. [CrossRef]
105. Rafati, A.A.; Gharibi, H.; Iloukhani, H. Micellization of cetylpyridinium chloride using conductometric technique. *Phys. Chem. Liq.* **2001**, *39*, 521–532. [CrossRef]
106. Di Michele, A.; Brinchi, L.; Di Profio, P.; Germani, R.; Savelli, G.; Onori, G. Head group size, temperature and counterion specificity on cationic micelles. *J. Colloid Int. Sci.* **2011**, *358*, 160–166. [CrossRef]
107. Ribeiro, A.C.F.; Lobo, V.M.M.; Valente, A.J.M.; Azevedo, E.F.G.; da Miguel, M.G.; Burrows, H.D. Transport properties of alkyltrimethylammonium bromide surfactants in aqueous solutions. *Colloid Polym. Sci.* **2004**, *283*, 277–283. [CrossRef]
108. Treiner, C.; Amar, A.; Fromon, M. Counter ion condensation on mixed anionic/nonionic surfactant micelles: Bjerrum's limiting condition. *J. Colloid Int. Sci.* **1989**, *128*, 416–421. [CrossRef]

109. Treiner, C.; Mannebach, M.H. Counter ion condensation on mixed cationic/nonionic micellar systems: Bjerrum's electrostatic condition. *Colloid Polym. Sci.* **1990**, *268*, 88–95. [CrossRef]
110. Treiner, C.; Fromon, M.; Mannebach, M.H. Evidences for Noncondensed Counterions in the Nonionic-Rich Composition Domain of Mixed Anionic/Nonionic Micelles. *Langmuir* **1989**, *5*, 283–286. [CrossRef]
111. Manning, G.S. Limiting laws and counterion condensation in polyelectrolyte solutions. III. An analysis based on the Mayer ionic solution theory. *J. Chem. Phys.* **1969**, *51*, 3249–3252. [CrossRef]
112. Davies, J.T.; Rideal, E.K. *Interfacial Phenomena*; Academic Press: New York, NY, USA; London, UK, 1961.
113. Podchasskaya, E.S.; Us'yarov, O.G. The effect of a background electrolyte on the premicellar association and average activity of sodium dodecyl sulfate ions. *Colloid J.* **2005**, *67*, 177–183. [CrossRef]
114. Reshetnyak, E.A.; Chernysheva, O.S.; Mchedlov-Petrosyan, N.O. Premicellar aggregation in water–salt solutions of sodium alkyl sulfonates and dodecyl sulfate. *Colloid J.* **2016**, *78*, 647–651. [CrossRef]
115. Jakubowska, A. Interactions of univalent counterions with headgroups of monomers and dimers of an anionic surfactant. *Langmuir* **2015**, *31*, 3293–3300. [CrossRef] [PubMed]
116. Bales, B.L.; Tiguida, K.; Zana, R. Effect of the nature of the counterion on the properties of anionic surfactants. 2. Aggregation number-based micelle ionization degree for micelles of tetraalkylammonium dodecylsulfates. *J. Phys. Chem. B* **2004**, *108*, 14948–14955. [CrossRef]
117. Reichardt, C. Pyridinium N-phenolate betaine dyes as empirical indicators of solvent polarity: Some new findings. *Pure Appl. Chem.* **2004**, *76*, 1903–1919. [CrossRef]
118. Rezende, M.C.; Mascayano, C.; Briones, L.; Aliaga, C. Sensing different micellar microenvironments with solvatochromic dyes of variable lipophilicity. *Dyes Pigment.* **2011**, *90*, 219–224. [CrossRef]
119. Kedia, N.; Sarkar, A.; Purkayastha, P.; Bagchi, S. An electronic spectroscopic study of micellisation of surfactants and solvation of homomicelles formed by cationic or anionic surfactants using a solvatochromic electron donor acceptor dye. *Spectrochim. Acta A* **2014**, *131*, 398–406. [CrossRef]
120. Xie, X.; Bakker, E. Determination of effective stability constants of ion-carrier complex in ion selective nanospheres with charged solvatochromic dyes. *Analyt. Chem.* **2015**, *87*, 11587–11591. [CrossRef]
121. Adamoczky, A.; Nagy, T.; Fehér, P.P.; Pardi-Tóth, V.; Kuki, Á.; Nagy, L.; Zsuga, M.; Kéki, S. Isocyanonaphthol Derivatives: Excited-State Proton Transfer and Solvatochromic Properties. *Int. J. Mol. Sci.* **2022**, *23*, 7250. [CrossRef]
122. Zachariasse, K.; Van Phuc, N.; Kozankiewicz, B. Investigation of micelles, microemulsions, and phospholipid bilayers with the pyridinium-N-phenolbetaine $E_T(30)$, a polarity probe for aqueous interfaces. *J. Phys. Chem.* **1981**, *85*, 2676–2683. [CrossRef]
123. Reichardt, C. Solvatochromic Dyes as Solvent Polarity Indicators. *Chem. Rev.* **1994**, *94*, 2319–2358. [CrossRef]
124. Tada, E.B.; Novaki, L.P.; El Seoud, O.A. Solvatochromism in cationic micellar solutions: Effects of the molecular structures of the solvatochromic probe and the surfactant headgroups. *Langmuir* **2001**, *17*, 652–658. [CrossRef]
125. Plieninger, P.; Baumgärtel, H. Eine ^{1}H NMR-spektroskopische Untersuchung zur Einlagerung von Pyridinium-N-Phenoxidbetainen in Micellen. *Justus Liebigs Ann. Chem.* **1983**, *1983*, 860–875. [CrossRef]
126. Dimroth, K.; Reichardt, C.; Siepmann, T.; Bohlmann, F. Über Pyridinium-N-phenol-betaine und ihre Verwendung zur Characterisierung der Polarität von Lösungsmitteln. *Justus Liebigs Ann. Chem.* **1963**, *661*, 1–37. [CrossRef]
127. Liotta, C.L.; Hallett, J.P.; Pollet, P.; Eckert, C.A. Reactions in nearcritical water. In *Organic Reactions in Water: Principles, Strategies and Applications*; Lindstrom, U.M., Ed.; Blackwell Publishing: Oxford, UK, 2007.
128. Novaki, L.P.; El Seoud, O.A. Microscopic polarities of interfacial regions of aqueous cationic micelles: Effects of structures of the solvatochromic probe and the surfactant. *Langmuir* **2000**, *16*, 35–41. [CrossRef]
129. Ramachandran, C.; Pyter, R.A.; Mukerjee, P. Microenvironmental effects on transition energies and effective polarities of nitroxides solubilized in micelles of different charge types and the effect of electrolytes on the visible spectra of nitroxides in aqueous solutions. *J. Phys. Chem.* **1982**, *86*, 3198–3205. [CrossRef]
130. Sarpal, R.S.; Belletête, M.; Durocher, G. Fluorescence probing and proton-transfer equilibrium reactions in water, SDS, and CTAB using 3,3-dDimethyl-2-pheny1-3H-indole. *J. Phys. Chem.* **1993**, *97*, 5007–5013. [CrossRef]
131. Reed, W.; Politi, M.J.; Fendler, J.H. Rotational diffusion of rose bengal in aqueous micelles: Evidence for extensive exposure of the hydrocarbon chains. *J. Am. Chem. Soc.* **1981**, *103*, 4591–4593. [CrossRef]
132. Mchedlov-Petrossyan, N.O.; Rubtsov, M.I.; Lukatskaya, L.L. Acid-base equilibrium of Bengal Rose B in micellar solutions of anionic surfactants. *Russ. J. Gen. Chem.* **2000**, *70*, 1177–1183.
133. Drummond, C.J.; Grieser, F. Absorption spectra and acid-base dissociation of the 4-alkyl derivatives of 7-hydroxycoumarin in self-assembled surfactant solution: Comments on their use as electrostatic surface potential probes. *Photochem. Photobiol.* **1987**, *45*, 19–34. [CrossRef]
134. Kibblewhite, J.; Drummond, C.J.; Grieser, F.; Thistlethwaite, P.J. Lipoidal eosin and fluorescein derivatives as probes of the electrostatic characteristics of self-assembled surfactant/water interfaces. *J. Phys. Chem.* **1989**, *93*, 7464–7473. [CrossRef]
135. Plenert, A.C.; Mendez-Vega, E.; Sander, W. Micro- vs macrosolvation in Reichardt's dyes. *J. Am. Chem. Soc.* **2021**, *143*, 13156–13166. [CrossRef]
136. Sarpal, R.S.; Belletête, M.; Durocher, G. Effect of small chemical variation and functionality on the solubilization behavior and recognition capability of 3H-indoles in SDS and CTAB. *Phys. Chem. Lett.* **1994**, *221*, 1–6. [CrossRef]

137. Maity, A.; Das, S.; Mandal, S.; Gupta, P.; Purkayastha, P. Interaction of semicarbazide and thiosemicarbazide pyrene derivatives with anionic and cationic micelles: Changed character of pyrene due to alteration in charge density induced by the side chains. *RSC Adv.* **2013**, *3*, 12384–12389. [CrossRef]
138. Hartley, G.S. The effect of long-chain salts on indicators: The valence-type of indicators and the protein error. *Trans. Faraday Soc.* **1934**, *30*, 444–450. [CrossRef]
139. Hartley, G.S.; Roe, J.W. Ionic concentrations at interfaces. *Trans. Faraday Soc.* **1940**, *35*, 101–109. [CrossRef]
140. El Seoud, O.A. Effects of organized surfactant assemblies on acid-base equilibria. *Adv. Colloid Int. Sci.* **1989**, *30*, 1–30. [CrossRef]
141. Jana, N.R.; Pal, T. Acid-base equilibria in aqueous micellar medium: A comprehensive review. *J. Surf. Sci. Technol.* **2001**, *17*, 191–212.
142. Mukerjee, P.; Banerjee, K. A study of the surface pH of micelles using solubilized indicator dyes. *J. Phys. Chem.* **1964**, *68*, 3567–3574. [CrossRef]
143. Fromherz, P. A new method for investigation of lipid assemblies with a lipoid pH indicator in monomolecular films. *Biochem. Biophys. Acta* **1973**, *323*, 326–334. [CrossRef] [PubMed]
144. Fernandez, M.S.; Fromherz, P. Lipoid pH indicators as probes of electrical potential and polarity in micelles. *J. Phys. Chem.* **1977**, *81*, 1755–1761. [CrossRef]
145. Funasaki, N. The effect of the solvent property of the surfactant micelle on the dissociation constants of weak electrolytes. *J. Chem. Soc. Jpn. Chem. Ind. Chem.* **1976**, *5*, 722–726. [CrossRef]
146. Funasaki, N. The dissociation constants of acid-base indicators on the micellar surface of dodecyldimethylamine oxide. *J. Colloid. Int. Sci.* **1977**, *60*, 54–59. [CrossRef]
147. Funasaki, N. Micellar effects on the kinetics and equilibrium of chemical reactions in salt solutions. *J. Phys. Chem.* **1979**, *83*, 1998–2003. [CrossRef]
148. Seguchi, K. Effect of surfactants on the visible spectra and acidity of substituted nitriphenols. *Yukagaku* **1979**, *28*, 20–25. [CrossRef]
149. Pal, T.; Yana, N.R. Polarity dependent positional shift of probe in a micellar environment. *Langmuir* **1996**, *12*, 3114–3121. [CrossRef]
150. Guo, Z.-j.; Miyoshi, H.; Komoyoji, T.; Haga, T.; Fujita, T. Quantitative analysis with physicochemical substituent and molecular parameters of uncoupling activity of substituted diarylamines. *Biochim. Biophys. Acta* **1991**, *1059*, 91–98. [CrossRef]
151. Obukhova, E.N.; Mchedlov-Petrossyan, N.O.; Vodolazkaya, N.A.; Patsenker, L.D.; Doroshenko, A.O.; Marynin, A.I.; Krasovitskii, B.M. Absorption, fluorescence, and acid-base equilibria of rhodamines in micellar media of sodium dodecyl sulfate. *Spectrochim. Acta A* **2017**, *170*, 138–144. [CrossRef]
152. Mchedlov-Petrossyan, N.O.; Vodolazkaya, N.A.; Bezkrovnaya, O.N.; Yakubovskaya, A.G.; Tolmachev, A.V.; Grigorovich, A.V. Fluorescent dye N,N'-dioctadecylrhodamine as a new interfacial acid-base indicator. *Spectrochim. Acta. A* **2008**, *69*, 1125–1129. [CrossRef] [PubMed]
153. Mchedlov-Petrossyan, N.O.; Vodolazkaya, N.A.; Yakubovskaya, A.G.; Grigorovich, A.V.; Alekseeva, V.I.; Savvina, L.P. A novel probe for determination of electrical surface potential of surfactant micelles: N,N'-di-n-octadecylrhodamine. *J. Phys. Org. Chem.* **2007**, *20*, 332–344. [CrossRef]
154. Mchedlov-Petrossyan, N.O.; Vodolazkaya, N.A.; Doroshenko, A.O. Ionic equilibria of fluorophores in organized solutions. The influence of micellar microenvironment on protolytic and photophysical properties of rhodamine B. *J. Fluoresc.* **2003**, *13*, 235–248. [CrossRef]
155. Caruana, M.V.; Fava, M.C.; Magri, D.C. A colorimetric and fluorimetric three-input inverted enabled OR logic array by self-assembly of a rhodamine probe in micelles. *Asian J. Org. Chem.* **2015**, *4*, 239–243. [CrossRef]
156. Drummond, C.J.; Grieser, F.; Healy, T.W. Acid-Base Equilibria in Aqueous Micellar Solutions. Part 4.-Azo Indicators. *J. Chem. Soc. Faraday Trans. I* **1989**, *85*, 561–578. [CrossRef]
157. Drummond, C.J.; Grieser, F.; Healy, T.W. Acid-Base Equilibria in Aqueous Micellar Solutions. Part 3.-Azine Derivatives. *J. Chem. Soc. Faraday Trans. I* **1989**, *85*, 551–560. [CrossRef]
158. Dorion, F.; Charbit, G.; Gaboriaud, R. Protonation equilibria shifts in aqueous solutions of dodecyl sulfates. *J. Colloid Int. Sci.* **1984**, *101*, 27–36. [CrossRef]
159. Romsted, L.S. Quantitative treatment of benzimidazole deprotonation equilibria in aqueous micellar solutions of cetyltrimethylammonium ion (CTAX, $X^- = Cl^-$, Br^-, and NO_3^-) surfactants. 1. Variable surfactant concentration. *J. Phys. Chem.* **1985**, *89*, 5107–5113. [CrossRef]
160. Lovelock, B.; Grieser, F.; Healy, T.W. Properties of 4-octadecyloxy-1-naphthoic acid in micellar solutions and in monolayer films absorbed onto silica attenuated total reflectance plates. *J. Phys. Chem.* **1985**, *89*, 501–507. [CrossRef]
161. Drummond, C.J.; Warr, G.G.; Grieser, F.; Ninham, B.W.; Evans, D.F. Surface properties and micellar interfacial microenvironments of n-dodecyl -D-maltoside. *J. Phys. Chem.* **1985**, *89*, 2103–2109. [CrossRef]
162. Khaula, E.V.; Zaitsev, N.K.; Galashin, A.E.; Goldfeld, M.G.; Alfimov, M.V. Surface potential and electrogeneous reactions on the interface between the micellar and aqueous phases. *Zhurn. Fis. Khim.* **1990**, *64*, 2485–2492.
163. Drummond, C.J.; Grieser, F.; Healy, T.W. Acid-base equilibria in aqueous micellar solutions. Part 1.–'Simple' weak acids and bases. *J. Chem. Soc. Faraday Trans. 1* **1989**, *85*, 521–535. [CrossRef]
164. Drummond, C.J.; Grieser, F.; Healy, T.W. Acid-base equilibria in aqueous micellar solutions. Part 2.–Sulphonephthalein Indicators. *J. Chem. Soc. Faraday Trans. 1* **1989**, *85*, 537–550. [CrossRef]

165. Minero, C.; Pelizzetti, E. Quantitative treatments of protonation equilibria shifts in micellar systems. *Adv. Colloid Int. Sci.* **1992**, *37*, 319–334. [CrossRef] [PubMed]
166. Mchedlov-Petrossyan, N.O.; Plichko, A.V.; Shumakher, A.S. Acidity of microheterogeneous systems: Effect of nonionic admixtures on acid-base equilibria of dyes bound to micelles of ionogenic surfactants. *Chem. Phys. Rep.* **1996**, *15*, 1661–1678.
167. Politi, M.J.; Fendler, J.H. Laser pH-jump initiated proton transfer on charged micellar surfaces. *J. Amer. Chem. Soc.* **1984**, *106*, 265–273. [CrossRef]
168. Rosendorfová, J.; Čermáková, L. Spectrometric study of the interaction of some triphenylmethane dyes and 1-carbethoxypentadecyltrimethylammonium bromide. *Talanta* **1980**, *27*, 705–708. [CrossRef]
169. Moller, J.V.; Kragh-Hansen, U. Indicator dyes as probes of electrostatic potential changes on macromolecular surfaces. *Biochemistry* **1975**, *14*, 2317–2323. [CrossRef]
170. Mashimo, T.; Ueda, I.; Shieh, D.D.; Kamaya, H.; Eyring, H. Hydrophilic region of lecithin membranes studied by bromothymol blue and effect of an inhalation anesthetic, enflurane. *Proc. Nat. Acad. Sci. USA* **1979**, *76*, 5114–5118. [CrossRef]
171. Gorbenko, G.P.; Mchedlov-Petrossyan, N.O.; Chernaya, T.A. Ionic equilibria in microheterogeneous systems Protolytic behaviour of indicator dyes in mixed phosphatidylcholine–diphosphatidylglycerol liposomes. *J. Chem. Soc. Faraday Trans.* **1998**, *94*, 2117–2125. [CrossRef]
172. Mchedlov-Petrossyan, N.O.; Isaenko, Y.V.; Tychina, O.N. Dissociation of acid-base indicators in microemulsions based on nonionogeneous surfactants. *Zh. Obsh. Khim. Russ. J. Gen. Chem.* **2000**, *70*, 1963–1971.
173. Mchedlov-Petrossyan, N.O.; Isaenko, Y.V.; Salamanova, N.V.; Alekseeva, V.I.; Savvina, L.P. Ionic equilibria of chromophoric reagents in microemulsions. *J. Anal. Chem.* **2003**, *58*, 1018–1930. [CrossRef]
174. Mchedlov-Petrossyan, N.O.; Timiy, A.V.; Vodolazkaya, N.A. Binding of sulfonephthalein anions to the micelles of an anionic surfactant. *J. Mol. Liq.* **2000**, *87*, 75–84. [CrossRef]
175. Kulichenko, S.A.; Fesenko, S.A.; Fesenko, N.I. Color indicator system for acid-base titration in aqueous micellar solutions of thee cationic surfactant tridecylpyridinium bromide. *J. Analyt. Chem.* **2001**, *56*, 1144–1148. [CrossRef]
176. Kulichenko, S.A.; Fesenko, S.A. Acid-base properties of sufophthalein indicators in water–micellar solutions of sodium dodecylsulfate. *Ukr. Chem. J.* **2022**, *68*, 100–184.
177. Grünhagen, H.H.; Witt, H.T. Umbeliferrone as indicator for pH changes in one turn-over. *Z. Naturforsch.* **1970**, *25b*, 373–386. [CrossRef]
178. Montal, M.; Gitler, C. Surface potential and energy-coupling in bioenergy–conserving membrane-systems. *Bioenergetics* **1973**, *4*, 363–382. [CrossRef] [PubMed]
179. Teissiy, J.; Tocanne, J.F.; Pohl, W.G. 4-Pentadecyl-7-hydroxycoumarin as a probe for the structure of the lipid-water-interface comparative studies with lipid monolayers, black lipid membranes and lipid microvesicles. *Ber. Bunsenges. Phys. Chem.* **1978**, *82*, 875–876. [CrossRef]
180. Fromherz, P.; Arden, W. pH-Modulated pigment antenna in lipid bilayer on photosensitized semiconductor electrode. *J. Am. Chem. Soc.* **1980**, *102*, 6211–6218. [CrossRef]
181. Fromherz, P. Lipid coumarin dye as a probe of interfacial electrical potential in biomembranes. *Methods Enzymol.* **1989**, *171*, 376–387. [CrossRef]
182. Haase, A.; Fromherz, P. Surface potential measurements at lipid membranes with pH-indicators. *Biophys. Struct. Mech.* **1981**, *7*, 299. [CrossRef]
183. Lukac, S. Surface potential at surfactant and phospholipid vesicles as determined by amphiphilic pH indicators. *J. Phys. Chem.* **1983**, *87*, 5045–5050. [CrossRef]
184. Buckingham, S.A.; Garvey, C.J.; Warr, G.G. Effect of head-group size on micellization and phase behavior in quaternary ammonium surfactant systems. *J. Phys. Chem.* **1993**, *97*, 10236–10244. [CrossRef]
185. Fromherz, P. in General Discussion. *Faraday Discuss. Chem. Soc.* **1986**, *81*, 139–140. [CrossRef]
186. Mchedlov–Petrossyan, N.O.; Vodolazkaya, N.A.; Reichardt, C. Unusual findings on studying surfactant solutions: Displacing solvatochromic pyridinium N-phenolate towards outlying areas of rod–like micelles? *Colloids Surf. A* **2002**, *205*, 215–229. [CrossRef]
187. Mchedlov-Petrossyan, N.O.; Isaenko, Y.V.; Goga, S.T. Reichardt betaines as combined solvatochromic and acid-base indicators in microemulsions. *Russ. J. Gen. Chem.* **2004**, *74*, 1741–1747. [CrossRef]
188. Zakharova, L.Y.; Fedorov, S.B.; Kudryavtseva, L.A.; Belskii, V.E.; Ivanov, B.E. Acid-base properties of bis(chloromethyl)phosphinic acid para-nitroanilide in aqueous micellar solutions of surface active agents. *Bull. Acad. Sci. USSR Div. Chem. Sci.* **1990**, *39*, 883–885. [CrossRef]
189. Fedorov, S.B.; Kudryavtseva, L.A.; Belsky, V.E.; Ivanov, B.E. Determination of the dissociation constants of micellar-bound compounds. *Kolloidn. Zhurn.* **1986**, *48*, 199–201.
190. Clear, K.J.; Virga, K.; Gray, L.; Smith, B.D. Using membrane composition to fine-tune pK_a of an optical liposome pH sensor. *J. Mater. Chem. C* **2014**, *4*, 2925–2930. [CrossRef]
191. Slocum, J.D.; First, J.T.; Webb, L.J. Orthogonal electricl field measurements near the green fluorescent protein fluorephore through Stark effect spectroscopy and pK_a shift provide a unique benchmark for electrostatic models. *J. Phys. Chem. B* **2017**, *121*, 6799–6812. [CrossRef]

192. Lin, Y.-C.; Ren, P.; Webb, L.J. AMOEBA force field trajectories improve prediction of accurate pK_a values of the GFP fluorophore: The importance of polarizability and water interactions. *J. Phys. Chem. B* **2022**, *126*, 7806–7817. [CrossRef] [PubMed]
193. Underwood, A.L. Acid–base titrations in aqueous micellar systems. *Anal. Chim. Acta.* **1977**, *93*, 267–273. [CrossRef]
194. Da Silva, F.L.B.; Bogren, D.; Söderman, O.; Åkesson, T.; Jönsson, B. Titration of fatty acids solubilized in cationic, nonionic, and anionic micelles. Theory and experiment. *J. Phys. Chem. B.* **2002**, *106*, 3515–3522. [CrossRef]
195. Söderman, O.; Jönsson, B.; Olofsson, G. Titration of fatty acids solubilized in cationic and anionic micelles. Calorimetry and thermodynamic modeling. *J. Phys. Chem. B.* **2006**, *110*, 3288–3293. [CrossRef] [PubMed]
196. Chirico, G.; Collini, M.; D'Alfonso, L.; Denat, F.; Diaz-Fernandez, Y.A.; Pasotti, L.; Rousselin, Y.; Sok, N.; Pallavicini, P. Micelles as containers for self-assembled nanodevices: A fluorescent sensor for lipophilicity. *ChemPhysChem* **2008**, *9*, 1729–1737. [CrossRef]
197. Eltsov, S.V.; Barsova, Z.V. Ionization of long-chain fatty acids in micellar solutions of surfactants. *Kharkov Univ. Bull. Chem. Ser.* **2008**, *16*, 292–298.
198. Boichenko, A.P.; Markov, V.V.; Cong, H.L.; Matveeva, A.G.; Loginova, L.P. Re-evaluated data of dissociation constants of alendronic, pamidronic and olpadronic acids. *Cent. Eur. J. Chem.* **2009**, *7*, 8–13.
199. Boichenko, A.P.; Dung, L.T.K. Solubilization of aliphatic carboxylic acids (C_3-C_6) by sodium dodecyl sulfate and brij 35 micellar pseudophases. *J. Solut. Chem.* **2011**, *40*, 968–979. [CrossRef]
200. Kamneva, N.M.; Boichenko, A.P.; Ivanov, V.V.; Markov, V.V.; Loginova, L.P. Acid-base properties and complex formation of the alendronate acid in water-ethanol medium and ultramicroheterogeneous micellar medium of Brij 5. *Ukr. Khim. Zh.* **2012**, *78*, 74–78.
201. Meloun, M.; Ferenčíková, Z.; Netolická, L.; Pekárek, T. Thermodynamic dissociation constants of alendronate and ibandronate by regression analysis of potentiometric data. *J. Chem. Eng. Data* **2011**, *56*, 3848–3854. [CrossRef]
202. Maeda, H. A thermodynamic analysis of the hydrogen ion titration of micelles. *J. Coll. Int. Sci.* **2003**, *263*, 277–287. [CrossRef]
203. Maeda, H. Phenomenological approaches in the thermodynamics of mixed micelles with electric charges. *Adv. Coll. Int. Sci.* **2010**, *156*, 70–82. [CrossRef]
204. Popović-Nikolić, M.R.; Popović, G.V.; Agbaba, D.D. The effect of nonionic surfactant Brij 35 on solubility and acid–base equilibria of verapamil. *J. Chem. Eng. Data.* **2017**, *62*, 1776–1781. [CrossRef]
205. Popović-Nikolić, M.R.; Popović, G.V.; Stojilković, K.; Dobrosavljević, M.; Agbaba, D.D. Acid–base equilibria of rupatadine fumarate in aqueous media. *J. Chem. Eng. Data* **2018**, *63*, 3150–3156. [CrossRef]
206. Whiddon, C.R.; Bunton, C.A.; Söderman, O. Titration of fatty acids in sugar-derived (APG) surfactants: A ^{13}C NMR study of the effect of headgroup size, chain length, and concentration on fatty acid pK_a at a nonionic micellar interface. *J. Phys. Chem. B* **2003**, *107*, 1001–1005. [CrossRef]
207. Khramtsov, V.V.; Marsh, D.; Weiner, L.; Reznikov, V.A. The application of pH-sensitive spin labels to studies of surface potential and polarity of phospholipid membranes and proteins. *Biochim. Biophys. Acta* **1992**, *1104*, 317–324. [CrossRef] [PubMed]
208. Smirnov, A.I.; Ruuge, A.; Reznikov, V.A.; Voinov, M.A.; Grigor'ev, I.A. Site-directed electrostatic measurements with a thiol-specific pH-sensitive nitroxide: Differentiating local pK and polarity effects by high-field EPR. *J. Am. Chem. Soc.* **2004**, *126*, 8872–8873. [CrossRef]
209. Molochnikov, L.S.; Kovaleva, E.G.; Golovkina, E.L.; Kirilyuk, I.A.; Grigor'ev, I.A. Spin probe study of acidity of inorganic materials. *Coll. J.* **2007**, *69*, 769–776. [CrossRef]
210. Khlestkin, V.K.; Polienko, J.F.; Voinov, M.A.; Smirnov, A.I.; Chechik, V. Interfacial surface properties of thiol-protected gold nanoparticles: A molecular probe EPR approach. *Langmuir* **2008**, *24*, 609–612. [CrossRef]
211. Voinov, M.A.; Kirilyuk, I.A.; Smirnov, A.I. Spin-labeled pH-sensitive phospholipids for interfacial pK_a determination: Synthesis and characterization in aqueous and micellar solutions. *J. Phys. Chem. B* **2009**, *113*, 3453–3460. [CrossRef]
212. Voinov, M.A.; Rivera-Rivera, I.; Smirnov, A.I. Surface electrostatics of lipid bilayers by EPR of a pH-sensitive spin-labeled lipid. *Biophys. J.* **2013**, *104*, 106–116. [CrossRef]
213. Mchedlov-Petrossyan, N.O. *Differentiation of the Strength of Organic Acids in True and Organized Solutions*; Kharkov National University: Kharkov, Ukraine, 2004.
214. Blatt, E. Ground-state complexation of eosin-labeled fatty acids with dimethylaniline in cetyltrimethylammonium bromide micelles. *J. Phys. Chem.* **1986**, *90*, 874–877. [CrossRef]
215. Mchedlov-Petrossyan, N.O.; Tychina, O.N.; Berezhnaya, T.A.; Alekseeva, V.I.; Savvina, L.P. Ionization and tautomerism of oxyxanthene dyes in aqueous butanol. *Dyes Pigment.* **1999**, *43*, 33–46. [CrossRef]
216. Bissell, R.A.; Bryan, A.J.; de Silva, A.P.; McCoy, C.P. Fluorescent PET (photoinduced electron transfer) sensors with targeting/anchoring modules as molecular versions of submarine periscopes for mapping membrane-bounded protons. *J. Chem. Soc. Chem. Commun.* **1994**, 405–407. [CrossRef]
217. Silina, Y.E.; Kuchmenko, T.A.; Volmer, D.A. Sorption of hydrophilic dyes on anodic aluminium oxide films and application to pH sensing. *Analyst* **2015**, *140*, 771–778. [CrossRef]
218. Rao, Y.; Subir, M.; McArthur, E.A.; Turro, N.J.; Eisenthal, K.B. Organic ions at the air/water interface. *Chem. Phys. Lett.* **2009**, *477*, 241–244. [CrossRef]
219. Bhattacharyya, K.; Sitzmann, E.V.; Eisenthal, K.B. Study of chemical reactions by surface second harmonic generation: p-Nitrophenol at the air–water interface. *J. Chem. Phys.* **1987**, *87*, 1442–1443. [CrossRef]

220. Bhattacharyya, K.; Castro, A.; Sitzmann, E.V.; Eisenthal, K.B. Studies of neutral and charged molecules at the air/water interface by surface second harmonic generation: Hydrophobic and solvation effects. *J. Chem. Phys.* **1988**, *89*, 3376–3377. [CrossRef]
221. Zhao, X.; Subrahmanyan, S.; Eisenthal, K.B. Determination of pK_a at the air/water interface by second harmonic generation. *Chem. Phys. Lett.* **1990**, *171*, 558–562. [CrossRef]
222. Castro, A.; Bhattacharyya, K.; Eisenthal, K.B. Energetics of adsorption of neutral and charged molecules at the air/water interface by second harmonic generation: Hydrophobic and solvation effects. *J. Chem. Phys.* **1991**, *95*, 1310–1315. [CrossRef]
223. Eisenthal, K.B. Liquid interfaces probed by second-harmonic and sum-frequency spectroscopy. *Chem. Rev.* **1996**, *96*, 1343–1360. [CrossRef] [PubMed]
224. Mchedlov-Petrossyan, N.O.; Vodolazkaya, N.A.; Timiy, A.V.; Gluzman, E.M.; Alekseeva, V.I.; Savvina, L.P. Acid-Base and Solvatochromic Indicators in Surfactant Micellar Solutions of Various Types: Is the Common Electrostatic Model Valid? *Kharkov Univ. Bull. Chem. Ser.* **2002**, *9*, 171–208.
225. Funasaki, N. A consideration concerning the surface potential of mixed micelles of sodium dodecyldecaoxyethylene sulfate and dodecyldimethylamine oxide. *J. Colloid Int. Sci.* **1977**, *62*, 189–190. [CrossRef]
226. Mchedlov-Petrossyan, N.O.; Kleshchevnikova, V.N. Influence of the cetyltrimethylammonium chloride micellar pseudophase on the protolytic equilibria of oxyxanthene dyes at high bulk phase ionic strength. *J. Chem. Soc. Faraday Trans.* **1994**, *90*, 629–640. [CrossRef]
227. Drummond, C.J.; Grieser, F.; Healy, T.W. Interfacial properties of a novel group of solvatochromic acid-base indicators in self-assembled surfactant aggregates. *J. Phys. Chem.* **1988**, *92*, 2604–2613. [CrossRef]
228. Hobson, R.A.; Grieser, F.; Healy, T.W. Surface potential measurements in mixed micelle systems. *J. Phys. Chem.* **1994**, *98*, 274–278. [CrossRef]
229. Loginova, L.P.; Samokhina, L.V.; Mchedlov-Petrossyan, N.O.; Alekseeva, V.I.; Savvina, L.P. Modification of the properties of NaDS micellar solutions by adding electrolytes and non-electrolytes: Investigations with decyl eosin as a -probe. *Colloids Surf. A* **2001**, *193*, 207–219. [CrossRef]
230. Mchedlov-Petrossyan, N.O.; Kleshchevnikova, V.N. The influence of a cationic surfactant on the protolytic properties of some triphenylmethane dyes. *Zhurn. Obshch. Khim. Russ. J. Gen. Chem.* **1990**, *60*, 900–911. (In Russian)
231. Mchedlov-Petrossyan, N.O.; Loginova, L.P. Kleshchevnikova. The influence if salts on the ionization of indicators in the Stern layer of cationic micelles. *Zhurn. Fiz. Khim. Russ. J. Phys. Chem.* **1993**, *67*, 1649–1653. (In Russian)
232. Mchedlov-Petrossyan, N.O.; Lyubchenko, I.N. Ionization of sulfophthalein dyes in organic solvents. *Zhurn. Obshch. Khim. Russ. J. Gen. Chem.* **1987**, *57*, 1371–1378. (In Russian)
233. Minch, M.J.; Giaccio, M.; Wolff, R. Effect of cationic micelles on the acidity of carbon acids and phenols. Electronic and ^{1}H nuclear magnetic resonance spectral studies of nitro carbanions in micelles. *J. Am. Chem. Soc.* **1975**, *97*, 3766–3772. [CrossRef]
234. Rocklin, G.J.; Mobley, D.L.; Dill, K.A.; Hünenberger, P.H. Calculating the binding free energies of charged species based on explicit-solvent simulations employing lattice-sum methods: An accurate correction scheme for electrostatic finite-size effects. *J. Chem. Phys.* **2013**, *139*, 184103. [CrossRef] [PubMed]
235. Lebedeva, N.; Bales, B.L. Location of spectroscopic probes in self-aggregating assemblies. I. The case for 5-doxylstearic acid methyl ester serving as a benchmark spectroscopic probe to study micelles. *J. Phys. Chem. B* **2006**, *110*, 9791–9799. [CrossRef]
236. Lebedeva, N.; Zana, R.; Bales, B.L. A reinterpretation of the hydration of micelles of dodecyltrimethylammonium bromide and chloride in aqueous solution. *J. Phys. Chem. B* **2006**, *110*, 9800–9801. [CrossRef] [PubMed]
237. Van Stam, J.; Depaemelaere, S.; De Schryver, F.C. Micelle aggregation numbers–A fluorescence study. *J. Chem. Educ.* **1998**, *75*, 93–98. [CrossRef]
238. De Oliveira, H.P.M.; Gehlen, M.H. Electronic energy transfer between fluorescent dyes with inter- and intramicellar interactions. *Chem. Phys.* **2003**, *290*, 85–91. [CrossRef]
239. Pereira, R.V.; Gehlen, M. Fluorescence of acridinic dyes in anionic surfactant solution. *Spectrochim. Acta A* **2005**, *61*, 2926–2932. [CrossRef]
240. Acharya, S.; Rebery, B. Fluorescence spectrometric study of eosin yellow dye–surfactant interactions. *Arab. J. Chem.* **2009**, *2*, 7–12. [CrossRef]
241. Yarmukhamedov, A.S.; Kurtaliev, E.N.; Jamalova, A.A.; Nizomov, N.; Salakhitdinov, F. Complex formation of some oxazine dyes in aqueous solutions of surfactants. *Uzbek J. Phys.* **2022**, *24*, 206–213. (In Russian)
242. Almgren, M.; Grieser, F.; Thomas, J.K. Dynamic and static aspects of solubilization of neutral arenes in ionic micellar solutions. *J. Amer. Chem. Soc.* **1979**, *101*, 29–291. [CrossRef]
243. Kalyanasundaram, K.; Thomas, J.K. Environmental effects on vibronic band intensities in pyrene monomer fluorescence and their application in studies of micellar systems. *J. Amer. Chem. Soc.* **1977**, *99*, 2039–2044. [CrossRef]
244. Ray, G.B.; Chakraborty, I.; Moulik, S.P. Pyrene absorption can be a convenient method for probing critical micellar concentration (cmc) and indexing micellar polarity. *J. Colloid Int. Sci.* **2006**, *294*, 248–254. [CrossRef]
245. Abuin, E.B.; Lissi, E.A. Partition of n-hexanol in micellar solutions of sodium dodecyl sulfate. *J. Colloid Int. Sci.* **1983**, *95*, 198–203. [CrossRef]
246. Yan, H.; Cui, P.; Liu, C.-B.; Yuan, S.-L. Molecular dynamics simulation of pyrene solubilized in a sodium dodecyl sulfate micelle. *Langmuir* **2012**, *28*, 4931–4938. [CrossRef] [PubMed]

247. Gao, F.; Yan, H.; Yuan, S. Fluorescent probe solubilised in cetyltrimethylammonium bromide micelles by molecular dynamics simulation. *Molec. Simul.* **2013**, *39*, 1042–1051. [CrossRef]
248. Friedrich, K.; Wooley, P. Electrostatic potential of macromolecules measured by pK_a shift of a fluorophore. 1. The 3′ terminus of 16S RNA. *Eur. J. Biochem.* **1988**, *173*, 227–231. [CrossRef] [PubMed]
249. Kharchenko, D.V.; Farafonov, V.S.; Cheipesh, T.A.; Mchedlov-Petrossyan, N.O.; Rodik, R.V.; Kalchenko, V.I. Catalytic properties of calixarene bearing choline groups in the processes of ester hydrolysis. *Theor. Exper. Chem.* **2022**, *58*, 363–371. [CrossRef]

Disclaimer/Publisher's Note: The statements, opinions and data contained in all publications are solely those of the individual author(s) and contributor(s) and not of MDPI and/or the editor(s). MDPI and/or the editor(s) disclaim responsibility for any injury to people or property resulting from any ideas, methods, instructions or products referred to in the content.

Review

Polarity of Organic Solvent/Water Mixtures Measured with Reichardt's B30 and Related Solvatochromic Probes—A Critical Review

Stefan Spange

Department of Polymer Chemistry, Institute of Chemistry, Chemnitz University of Technology, Straße der Nationen 62, 09111 Chemnitz, Germany; stefan.spange@chemie.tu-chemnitz.de

Abstract: The UV/Vis absorption energies (ν_{max}) of different solvatochromic probes measured in co-solvent/water mixtures are re-analyzed as a function of the average molar concentration (N_{av}) of the solvent composition compared to the use of the mole fraction. The empirical $E_T(30)$ parameter of Reichardt's dye **B30** is the focus of the analysis. The Marcus classification of aqueous solvent mixtures is a useful guide for co-solvent selection. Methanol, ethanol, 1,2-ethanediol, 2-propanol, 2-methyl-2-propanol, 2-butoxyethanol, formamide, N-methylformamide (NMF), N,N-dimethylformamide (DMF), N-formylmorpholine (NFM), 1,4-dioxane and DMSO were considered as co-solvents. The $E_T(30)$ values of the binary solvent mixtures are discussed in relation to the physical properties of the co-solvent/water mixtures in terms of quantitative composition, refractive index, thermodynamics of the mixture and the non-uniformity of the mixture. Significant linear dependencies of $E_T(30)$ as a function of N_{av} can be demonstrated for formamide/water, 1,2-ethanediol/water, NMF/water and DMSO/water mixtures over the entire compositional range. These mixtures belong to the group of solvents that do not enhance the water structure according to the Marcus classification. The influence of the solvent microstructure on the non-linearity $E_T(30)$ as a function of N_{av} is particularly clear for alcohol/water mixtures with an enhanced water structure.

Keywords: Reichardts dye; solvatochromism; solvent mixtures; refractive index; solvent composition; average molar concentration

Citation: Spange, S. Polarity of Organic Solvent/Water Mixtures Measured with Reichardt's B30 and Related Solvatochromic Probes—A Critical Review. *Liquids* **2024**, *4*, 191–230. https://doi.org/10.3390/liquids4010010

Academic Editors: William E. Acree, Jr., Franco Cataldo and Enrico Bodo

Received: 1 November 2023
Revised: 8 January 2024
Accepted: 22 January 2024
Published: 17 February 2024

Copyright: © 2024 by the author. Licensee MDPI, Basel, Switzerland. This article is an open access article distributed under the terms and conditions of the Creative Commons Attribution (CC BY) license (https://creativecommons.org/licenses/by/4.0/).

1. Introduction

The development of Reichardt's dye 2,6-diphenyl-4-(2,4,6-triphenyl-1-pyridinium)-phenolate (**B30**) (see Scheme 1) was a milestone in the study of solvent properties [1]. Recall that the original empirical solvent parameter $E_T(30)$ is defined as the molar absorption energy of **B30** expressed in kcal/mol, measured in a given solvent [1]:

$$E_T(30) \text{ (kcal/mol)} = 28{,}591/\lambda_{max} \text{ (nm)} \qquad (1)$$

There are numerous studies in the literature classifying the polarity of pure solvents and solvent mixtures according to their composition, measured with **B30** and other solvatochromic probe molecules [2–35].

Explanatory concepts on solvatochromism can be found in the informative review of [34]. A preliminary summary of the solvent mixtures treated can be found in Table 3 of [35]. In this context, **B30** and related solvatochromic probes have been used to establish so-called hydrogen bond strength (HBD) scales for organic solvents [36–41]. In addition to the HBD classification, there is a definition for the hydrogen-bond-accepting (HBA) strength of solvents [42]. The HBD and HBA classifications reflect the molecular properties of the solvent molecule in relation to hydrogen bonding in terms of the Kamlet–Taft approach and other similar concepts by Catalan and Laurence [36–41]. The interaction of the solvent HBD groups with the phenolate oxygen was considered to be critical in measuring the

HBD strength of solvents and solvent mixtures. However, this approach is only partially justified, as we have recently shown [43]. The concept of determining HBD parameters works quite well for ionic liquids (ILs) and other salts due to the electrostatic interaction between the **B30** phenolate anion and the constituent cation of the IL [41,44–46]. However, due to some contradictions between theory and experimental results, the phenomenon is still under investigation [46].

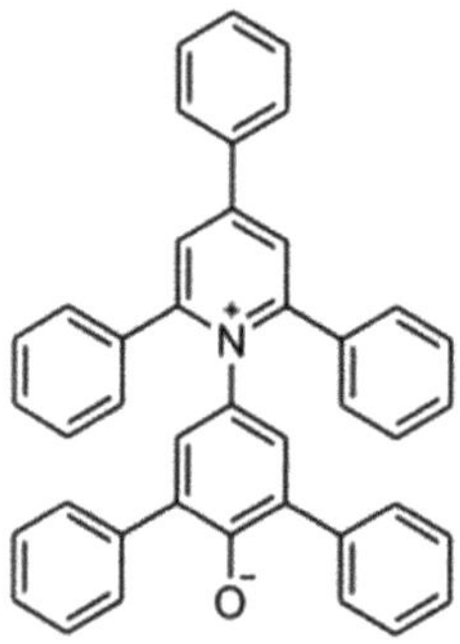

Scheme 1. Reichardt's dye 2,6-diphenyl-4-(2,4,6-triphenyl-1-pyridinium)-phenolate (**B30**).

B30 in particular is routinely used as a polarity indicator for binary solvent mixtures [2,6–9,13–21]. One of the most difficult problems in interpreting the solvatochromism of **B30** in solvent mixtures is the issue of preferential solvation and its influence on the $E_T(30)$ value [10,15–21,24,34]. However, the concept of preferential solvation is interpreted differently in the literature [9–12,24]. The fundamental problem with the definition of preferential solvation was correctly recognized by Ghoneim [24]. Two basic scenarios must be distinguished in this question for **B30**:

i. The solvent mixture (true micelles are a different situation) is inherently inhomogeneous and the solute **B30** is therefore preferentially entrapped by a specific microdomain.

ii. The solute probe such as **B30** preferably forms a specific complex with one of the two solvent components.

A complementary good definition for preferential solvation is given by Morisue and Ueno regarding case ii. "Preferential solvation is a phenomenon, whereby solvent proportion of binary mixed solvent in the vicinity of a solute molecule differentiates from the statistic proportion in bulk" [29]. It is therefore necessary to clearly distinguish whether the probe molecule is specifically solvated by the solvent molecule or is present in a partial volume enriched with a component of the mixture. Scenario i. assumes that the physical structure of the solvent mixture is not affected by the solute **B30**. In the case of scenario ii., the type of probe itself determines the extent to which preferential solvation occurs. Thus, if scenario ii. is true, then different solvatochromic probes should each show different UV/Vis absorption energy dependencies on the quantitative solvent composition for the same mixture. Langhals had already shown in 1981 that different solvatochromic probes used for the same mixture measure the same qualitative dependencies as a function of solvent composition, e.g., for ethanol/water [5]. This crucial finding would rule out scenario ii. But the situation is not so simple.

Despite ambitious work on the subject, the problem of solvatochromism in solvent mixtures has not really been adequately addressed in the literature. It is therefore necessary to explain the chronological development of the concepts for interpreting UV/Vis spectroscopic absorption energy data of probe molecules in solvent mixtures. In the first papers on **B30** [1,2], the $E_T(30)$ values of various binary solvent mixtures, including ethanol/water and 1,4-dioxane/water, were determined. It was shown that $E_T(30)$ depends in a complex way on the quantitative composition of the mixture. Langhals recognized that the solvatochromism of **B30**, the $E_T(30)$ values, can be described empirically as a logarithmic function depending on the concentration of the solvent components [6]. As early as 1982, Langhals also showed that the $E_T(30$ values of the homologous primary n-alcohol series are a linear function of the total molar concentration of the respective alcohol (N) [7]. The core problem was that no theoretical justification for this link had been presented in the past. Perhaps as a result, this very important discovery was not properly understood by many scientists and its significance was not fully appreciated. These seemingly empirical findings [6,7] have an important physical background based on the Lorentz–Lorenz relation [47].

Later in 1986, Haak and Engberts presented a valuable paper on the influence of temperature (T) on the solvatochromic properties of **B30** in aqueous solvent mixtures [8]. It is worth analyzing this study in detail, as the authors have correctly identified the effects of the hydrophobic alcoholic component such as 2-n-butoxyethanol (BE) in water on $E_T(30)$. However, several interpretations need to be re-evaluated in the light of new physical research on specific solvent mixtures, as will be shown in the course of this study.

Since 1982, the general topic of preferential solvation in solvent–water mixtures has been studied in detail by Marcus in numerous papers based on thermodynamics using the Kirkwood–Buff theory for fully miscible aqueous solvent mixtures [48–51]. Even then, Marcus was aware of various discrepancies between thermodynamic results and solvatochromic measurements [12]. He stated: "A single probe, such as the betaine used for the $E_T(30)$ polarity parameter, cannot provide an answer". As early as 1988, Dorsey [9] concluded that **B30** perceived the hydrogen bond network rather than direct hydrogen bonds: "Therefore, it could be that a change in the hydrogen-bonding network of the solution is being sensed by the ET-30 probe in the dilute alcohol concentration as well". This is the thesis that reaches the heart of the matter.

Since 1992, O. Connor and Rosés have independently developed the preferential solvation model [11,15,16]. The preferential solvation model suggests the formation of stoichiometrically defined complexes between the solvatochromic probe (solute) and the two solvents, as well as between solvent molecules. It was assumed that the measured UV/Vis shift was caused by the formation of a complex between the **B30** probe or similar probes and the solvent molecule [1,36,37]. This scenario belongs to case ii. These models assume that the strength of the H-bridge bond to **B30** is linearly correlated with the magnitude of the UV/Vis shift.

In the important paper by Kipkemboi [13], which was not considered further, the solvatochromism of **B30** in 2-methyl-2-propanol/water and 2-amino-2-methylpropane/water mixtures was studied in detail. The authors concluded that preferential solvation cannot be the main reason for the observed effects. Taking into account the refractive index and the partial molar concentration of the components in the qualitative interpretation, both the polarizability and the number of water dipoles have an influence on the solvatochromic shifts of **B30**.

In 2004, however, Bentley took up Langhals' discovery [7] and showed the dependence of $E_T(30)$ on the global polarity of alcohols with respect to N. In addition, the relationships between the $E_T(30)$ values of alcohol/water mixtures were alternatively analyzed as a function of volume or molar fraction of the mixture composition [27]. Bentley concluded that preferential solvation may be overestimated.

An actual preferential solvation could be demonstrated for **B30** in the phenol/acetone and phenol/acetonitrile systems [43,44]. A stoichiometric 1:1 complex of **B30** with phenol can be clearly identified. In phenol/1,2-dichloroethane, depending on the quantitative composition, both effects i. and ii. can be observed simultaneously with different proportions depending on the phenol concentration [43,52–54]. Importantly, these UV/Vis studies have convincingly demonstrated that the effect of specific hydrogen bond formation on $E_T(30)$ is much less than that of the bulk solvent phenol [43]. Recent studies show that preferential solvation can be detected, but the solute/solvent complexes must be unambiguously identified by independent spectroscopic measurements. [55,56].

As mentioned above, the preferential solvation models are based in particular on the assumption that the UV/Vis shift of **B30** and related probes such as 1-ethyl-4-(methoxycarbonyl) pyridinium iodide (**K**), *cis*-dicyano-bis(1,10-phenanthroline) iron II (**Fe**), or Brookers Merocyanine (**BM**), is mainly caused by the formation of specific interactions (hydrogen bonds) between the solvent and the probe. This assumption is a fundamental misunderstanding. This fact can be clearly demonstrated, independently of each other, using three different derivatives of the Reichardt dye family found in the literature [57–59]. C. Reichardt himself ignored the results of the solvatochromism of the thiolate betaine derivative of **B30**, which did not show the desired difference from **B30** when measured in HBD solvents [57]. It was an unpleasant experience for us to discover that the H-bridge bonding patterns at the barbiturate anion substituent of the **B30** derivative caused only a negligible UV/Vis shift compared to the bulk HBD solvents [58]. However, at the time, we did not fully appreciate the implications of this finding for understanding the UV/Vis shift. Unfortunately, we had to abandon the concept of molecular recognition by UV/Vis shift of solvatochromic probes. Recently, the Sander group showed that the [2,6-di-*tert*.-butyl-4-(pyridinium-1-yl)] phenolate forms a defined 1:1 complex with water, leading to only small shifts in the π-π^* transition compared to the influence of the global polarity of the bulk water [59]. Thus, **B30** and other related probes do not fulfil this purported property as an indicator of the HBD strength of the solvent molecule when the bulk solvent is measured [36,37,40,41]. The overall UV/Vis shift of **B30** in pure HBD solvents is mainly due to the effect of the global polarity of the hydrogen bonding network of the solvent and not to direct hydrogen bonding with the dissolved probe [7,9,28,43,60–62]. Sander and co-workers also showed that the stoichiometric **B30**/HBD solvent complex is the true solvatochromic species and not the original **B30**. This result was the missing link in understanding the discrepancies between the different interpretations of the derivatives of Reichardt's dye, as it was known that steric shielding of the phenolate oxygen of the **B30** derivatives leads to a change in the solvatochromic properties [1].

The misinterpretation that the total UV/Vis shift is primarily due to the direct formation of hydrogen bonds at **B30** must be fundamentally corrected, even though many papers have taken this as a defined basis. Accepting this fact will be difficult for many scientists working in this field, as it overturns entrenched patterns of thought. Following Bentley [27], we question the classical preferential solvation approach of special solvatochromic probes for certain alcohol/water and related aqueous binary solvent systems with respect to scenario ii., as reported in [4,19–21,25–27,30–32].

Suppan also concluded that the process of hydrogen bonding between solute and solvent in water can be endergonic, using the preferential solvation index for interpretation [63]. Later, Rezende recognized that the concept of preferential solvation has some weaknesses, and the preferential solvation index was also recommended to overcome some problems in explaining difficult results [64,65].

However, the real scientific problem with the evaluation of UV/Vis absorption energy data of solvatochromic probes in solvent mixtures in the literature is much more serious. Most authors routinely use the mole fraction x of a component of the solvent mixture to define the quantitative composition in physical terms. It has been assumed that a strictly linear dependence of the UV/Vis absorption energy of the dissolved solvatochromic dye on x would indicate ideal mixing behavior [10,14–31]. This thesis must be fundamentally questioned, since only the change in the Gibbs free energy (ΔG) of a solvent mixture can be linearly related to the mole fraction of the components involved [10,66]. The Gibbs free energy is a composite variable [66]. For the UV/Vis absorption energy of a dissolved probe molecule in a solvent mixture, however, the situation is somewhat different. The number of transition moments, i.e., the atoms and molecules that are affected by both the light and solvent in a given volume depends on the average molar concentration ($N_{av,x}$) of the solvent dipoles with respect to x, but not directly on x [47,67,68]. Therefore, the experimentally found curvilinear relationship $E_T(30)$ as a function of x(water) may indicate a preferred solvation [10,14–21], since $N_{av,x}$ is reciprocal to x(water). We assume that the real reason for the curved shape of the function $E_T(30)$ or EP_{HBD} (see later Equation (2)) as a function of x(water) is not necessarily the preferred solvation, but the influence of inhomogeneity due to the difference in mass of the two different solvents. This aspect is particularly relevant for aqueous mixtures due to the low molar mass of water. EP_{HBD} is usually the UV/Vis absorption energy (ν_{max} in cm^{-1}) or in kcal/mol [$E_T(30)$] of the solvatochromic probe such as **B30** measured at λ_{max}, Equations (1) and (2).

$$EP_{HBD} \equiv \nu_{max} \, (\text{cm}^{-1}) \approx aN \, (\text{mol/cm}^3) + b. \qquad (2)$$

N refers to the molar concentration, according to Equation (3), of the solvent dipoles or the polarized solvent molecules according to the Debye–Lorenz, Clausius–Mosotti–Lorenz or Lorentz–Lorenz relation [47]. σ is the physical density and M the molar mass of the pure solvent substance.

$$N \, (\text{mol/cm}^3) = \sigma(\text{g/cm}^3)/M \, (\text{g/mol}). \qquad (3)$$

There are hardly any well-founded studies on the subject of the various quantities of mixture composition, as the mole fraction x seems to have become established as a routine basis for calculation. There are only a few papers that briefly mention the influence of the different composition variables on $E_T(30)$ and qualitatively illustrate it with some examples [9,27,43,50]. Significantly, Marcus already suspected that this topic would raise a number of unanswered questions; he mentioned timidly "the different measures of composition of a binary solvent mixture should be borne in mind" [50].

In addition, it has been empirically found that the EP_{HBD} of pure solvents is linearly correlated with the N (total molar concentration) of the solvent under solvent variation for specific solvent families [7,27,43,61,62]. Furthermore, there is a fundamental relationship between N and the spectroscopic quantities ν_{max} and ε_{max} (Lmol/cm = 10^3 cm^2/mol) as the molar absorption coefficient as shown in Equation (4).

$$N \, (\text{mol/cm}^3) = \nu_{max}(\text{cm}^{-1})/\varepsilon_{max}(\text{cm}^2/\text{mol}),$$
$$\nu_{max} \, (\text{cm}^{-1}) = N \, (\text{mol/cm}^3) \, \varepsilon_{max} \, (\text{cm}^2/\text{mol}). \qquad (4)$$

Equation (4) has been completely overlooked in the past. The relationship is not artificial. The physical relationship between the absorption energy ν_{max}, the molar absorption coefficient ε_{max} and N is theoretically determined through Beer's approximation and the Lorentz–Lorenz relation [67,68]. The fundamental Lorentz–Lorenz relation is given by Equation (5).

$$f\left(n_D^{20}\right) = N \, 4/3\pi \, R_m. \qquad (5)$$

With n_D^{20} the refractive index measured at 589 nm; R_m molar refractivity and $f(n_D^{20}) = [(n_D^{20})^2 - 1]/[(n_D^{20})^2 + 2]$.

It is a matter of identifying the physically correct amount of N in the solvent system [69]. The general factor N in the original Lorentz–Lorenz relation, Equation (5), refers to the molar concentration of the total number of solvent molecules [47]. It has recently been shown that, within homologous series of n-alkane derivatives, the correlation of the refractive index as a function of N results in a negative slope [69], which theoretically does not agree with the original Lorentz–Lorenz relation [47]. Since N is empirically related to ν_{max} by Equation (2), many correlations of ν_{max} with n_D^{20} from the literature are not meaningful. Only when the actual molar concentration of the "chromophore" of the solvent molecule, the C-H bond concentration N_{CH}, is taken into account, is the applicability of the Lorentz–Lorenz relationship for correlation analysis fulfilled. The reason for this is simple, because $N \sim -N_{CH}$. [69]. Therefore, instead of N, the respective concentration of the corresponding functional fractions of the solvent is actually required which is N_{CH} for special solvent families. Accordingly, N should be replaced by N_{CH} when investigating structure–property relationships with respect to refractive index. Then, Equation (6) is obtained:

$$f\left(n_D^{20}\right) = N_{CH}\, 4/3\pi\, R_m. \tag{6}$$

For solvents containing hydroxyl- and/or -CO-NH-groups, the situation is straightforward, as the HBD groups are the dominant dipoles in the solvent volume. Thus, Equation (2) essentially holds when solvent families are treated individually, but is convincingly applicable to HBD solvents [6,27,61]. Indeed, many EP_{HBD} correlate linearly with the physically determined hydroxyl group density, which is proportional to the molar concentration N [Equation (2)], rather than with the acidity in terms of the pKa of the solvent [43]. For non-HBD solvents, linear relationships between EP and N are only found if one stays within the series of a particular solvent family [61].

The reason for the clear result of Equation (2) is that ε_{max} of negative solvatochromic probes, Equation (4), changes inversely linearly with ν_{max} as the solvent is varied [1,70,71]. Equation (2) works only moderately well for positive solvatochromic dyes as the preliminary evaluation of Nile Red shows; see Figure S1a in the supplementary materials; the UV/Vis-spectroscopic data are taken from [72]. In this context, the question is how the molar absorption coefficient ε_{max} of the solvatochromic probe changes systematically linearly with N, since ε_{max} also correlates with the refractive index due to the Kramer–Kronig relation [73]. For positive solvatochromic dyes, ε_{max} remains essentially unchanged within structurally similar solvent series [70,71]. This consideration is in line with older studies by Suppan [74,75]. Since the electromagnetic coupling of the solvent chromophore with the dye is theoretically understood for negative solvatochromic dyes [70,71], only the solvatochromism of such dyes with respect to N is analyzed in this review.

There are several reasons for the motivation of this review and re-evaluation of the $E_T(30)$ parameters of organic co-solvent/water mixtures. Enormous progress has been made in the study of aqueous solvent mixtures, both experimentally and theoretically. Many new insights into their microstructure and dynamics, structure and properties have been gained in recent years for alcohol/water mixtures [76–96] and other co-solvent/water mixtures (see references in the main text). In particular, these new findings on the microstructure of alcohol/water mixtures require a re-evaluation of many older results on the solvatochromism of probes in these mixtures. A crucial argument for testing the solvatochromism of **B30** in aqueous mixtures is that water is not a strongly acidic solvent in the sense of the HBD property, but is one of the most polar solvents due to its exceptionally high molar concentration N and the polarization of the volumetric OH bonds [60]. A very precise distinction must be made between volumetric water and smaller quantities of water as a solute in a mixture [75]. From x(water) < 0.2, the situation is different for aqueous mixtures than in the water-rich range, as water behaves more like a solute than a solvent [75–77].

Another key argument concerns the appropriate use of the various measures of mix composition [50]. Recently, we have shown that $E_T(30)$ is an approximately linear function of the average molar concentration (N_{av}) of ethanol/water and methanol/chloroform mixtures [43]. This is true for certain concentration ranges, then the correlation coefficient for the linear relationship r (regression coefficient) is ~0.99 [43]. It is likely that linear dependencies $E_T(30)$ as a function of N_{av} only arise if the thermal motion of the solvent molecules overcomes the structuring of the solvent mixture. Is the solvatochromic probe measuring an average number of different solvent dipoles as a snapshot in certain compositional ranges? To answer such questions, we need to take a closer look at the dynamics of the solvent mixture [84–86]. Pure alcoholic solvents and alcohol/water mixtures fit into a relationship when the dielectric relaxation time τ_1 and the number of OH dipoles are correlated on the basis of N (see Figure 4 in [86]). Relaxation time of ethanol/water mixtures increases with decreasing number of OH dipoles due to increasing alcohol content. Reminder, the dielectric relaxation time τ_1 is defined as the time it takes 63% of the molecules in the sample to return to disorder [87]. Thus, the degree of ordering of binary alcohol/water mixtures containing two different types of OH-dipoles probably increases with increasing structuring, i.e., concentration of C-C bonds originating from the alcohol molecules.

The following question arises: Can (binary) solvent mixtures can be treated in the same way as pure solvents with regard to the average molar concentration (N_{av}) of relevant solvent dipoles or polarizable solvent molecules? The situation regarding the appropriate measure to use is complicated. To correlate the results of UV/Vis spectroscopy or dielectric spectroscopy, different composition variables, such as the molar and volume fractions of the mixture, are sometimes used alternately [9,27,88,89]. For ternary mixtures or multi-component systems, the determination of the composition in suitable parameters is even more complex. However, the work of F. Martin et al. shows that solvation models can in principle also be used to explain the solvatochromism of probes in ternary mixtures [97,98]. Measuring the physical properties of ternary solvent mixtures in terms of density, refractive index and heat of mixing requires careful and extensive studies. There is not as much data available in this area. Therefore, only binary mixtures will be considered in this review. The fundamental aspect of compositional quantities is covered in the methods chapter of this paper.

2. Methods

The average molar concentration N_{av} is a crucial physical property of all non-homogeneous substances. It must be clearly defined which atoms and molecules are being considered. This study deals with binary solvent mixtures. The N_{av} of any homogeneous binary solvent mixture can be easily calculated from the composition of the two components, their molar masses and the actual physical density of the solvent mixture according to Equation (7) [66].

$$N_{av,Z} = \rho_m / M_{AV,z} = \rho_{m(1,2)} / (Z_1 M_1 + Z_2 M_2) \quad (7)$$

$\rho_{m(1,2)}$ is the actual density (after mixing) of the mixture at given Z.
M_1 and M_2 are the molar masses (g/mol) of solvent 1 and 2, respectively;
$M_{av,z}$ is the average molar mass of the solvent components.
The factors Z_1 and Z_2 are either:
the molar fraction ($Z = x$; $\rightarrow N_{av,x}$),
mass fraction ($Z = w$; $\rightarrow N_{av,w}$), or
volume fraction ($Z = \varphi$; $\rightarrow N_{av,v}$) of solvent 1 and solvent 2 before mixing.

The average molar concentrations $N_{av \cdot z}$ in terms of different $M_{av,z}$ have not yet been fully considered as quantitative composition size in evaluating physical measurands of solvent mixtures. We had underestimated this point in a previous paper [62]. The linearity of a relationship between a measured quantity and a quantitative composition is not necessarily a criterion for physical correctness. It must be emphasized that the decisive quantity is the average molar mass $M_{av,z}$ which can be calculated either by x, w or φ [99]; see Equation (8):

$$M_{av,z} = z_1(M_1 - M_2) + M_2 \quad (8)$$

Therefore, the numerical differences between $N_{av,x}$, $N_{av,w}$ and $N_{av,v}$ are due to the differences in M_1 and M_2 as well as the quantitative ratio of the two solvents, rather than to the density changes, as shown for various alcohol/water mixtures when $N_{av,x}$ is plotted as a function of x(water) (see Figure S1b in the Supplementary Materials).

The problem of average molecular weight is a central one in polymer chemistry. Different physical measurement methods, such as end-group analysis through NMR or acid-based titration, viscosity, osmotic pressure of the polymer solution, light scattering and ultracentrifugation, are used to measure different numerical values of the average molar mass for the same polymer sample [100]. The numerical value of the average molecular weight depends not only on the method of measurement but also on the shape of the molecular weight distribution curve [101]. Note that colligative physical methods measure the number average (M_n) of the polymer sample. This would correspond to the $M_{av,x}$ of solvent mixtures. Non-colligative physical methods (preferably) measure data related to the weight average (M_w). The result of the non-colligative method depends on the nature of the solvent and polymer solute. For example, the refractive index is a non-colligative measurement. It is therefore not surprising that the determination of mixture composition through refractive index measurements is always controversial [102–104].

The non-uniformity of a polymer is defined by the ratio M_w/M_n [100,101]. Following the teachings of polymer chemistry [101], the ratio of $M_{av,w}/M_{av,x} = DI$ has been defined in this work as the dispersion index of a binary solvent mixture. Accordingly, Equation (9) is used in practice as an indicator of the non-uniformity of the solvent mixture. DI is an artificially constructed variable, but the approach is borrowed from polymer chemistry.

$$M_{av,w}/M_{av,x} = DI \qquad (9)$$

For a binary mixture, this approach is straightforward. Figure 1a shows the dependence of DI as a function of x(water) for methanol/water, 2-propanol/water and 2-methyl-2-propanol/water mixtures. $M_{av,x}$ and $M_{av,w}$ are calculated by Equations (10) and (11), respectively.

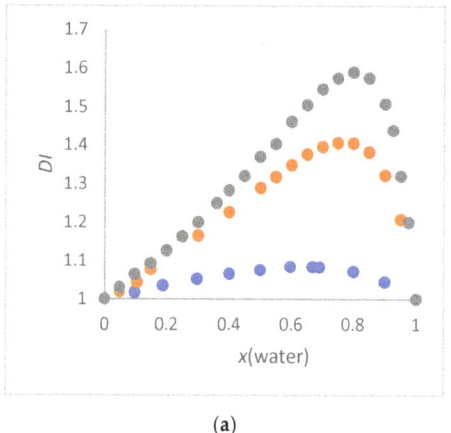
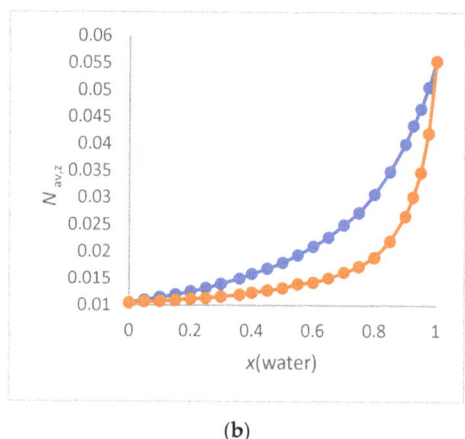

(a) (b)

Figure 1. (a) Dependence of DI as a function of x(water) for methanol/water (blue), 2-propanol/water (orange) and 2-methyl-2-propanol/water (grey); (b) $N_{av,z}$ (in mol/cm^3) of the 2-methyl-2-propanol/water mixture as a function of x(water). $N_{av,x}$ (blue) and $N_{av,w}$ (orange), calculated according to Equations (10) or (11); for data see Table S12a. The connections between the individual points serve to orientate the reader.

The 2-methyl-2-propanol/water mixtures show the greatest inhomogeneity at x(water) = 0.8 (strongest curvature of the graph in Figure 1b), since the quotient $M_{av,w}/M_{av,x}$ as a function of x(water) has its maximum at this position. This x(water) = 0.8 corresponds to $N_{av,x}$ = 0.25 mol/cm^3 and $N_{av,w}$ = 0.15 mol/cm^3, respectively. As would be expected arithmetically, the greater the mass difference, the greater the DI for a given x. The smaller the mass difference, the wider the distribution of DI at DI_{max}. It cannot be overlooked that the position of DI_{max} with respect to x(water) corresponds to both the order of the excess molar volume of water and the excess thermodynamic properties for these alcohol/water mixtures [78–80,83,95]. This is remarkable because the DI only considers the masses and their proportions and does not include any other physical data. It can be assumed that this agreement is rather random for alcohol/water mixtures. Therefore, the suitability of the DI to support the interpretation of $E_T(30)$ as a function of solvent composition in alcohol/water mixtures will be demonstrated in this work.

As explained in the introduction, for the evaluation of UV/Vis spectroscopic absorption data [67,68], the mole fraction (x) is theoretically suitable for the determination of the average molar mass. Therefore, $N_{av,x}$, determined through Equation (10), is preferred in this paper for correlation with $E_T(30)$.

$$N_{av,x} = \rho_m / M_{AV,z} = \rho_{m(1,2)} / (x_1 M_1 + x_2 M_2) \tag{10}$$

The weight fraction w_1 is calculated from the mass fractions m_1 and m_2 of the two components according to $w_1 = m_1/(m_1 + m_2)$. $N_{av,w}$ is obtained from Equation (11).

$$N_{av,w} = \rho_m / M_{AV,w} = \rho_{m(1,2)} / (w_1 M_1 + w_2 M_2) \tag{11}$$

Since $M_{av,w}$ is inherently greater than $M_{av,x}$ [100,101], $N_{av,x}$ is always greater than $N_{av,w}$. For example, Figure 1b shows the relationships between the compositional quantities $N_{av,x}$ and $N_{av,w}$ with x(water) for the binary solvent mixture 2-methyl-2-propanol/water. Figure 1b clearly shows that $N_{av,w}$ reflects the inhomogeneity of the mixture as a function of the quantitative composition more strongly than $N_{av,x}$, since the curve $N_{av,w}$ versus x(water) shows a stronger deviation from linearity than $N_{av,x}$ versus x(water) (see also Figure S1b in the Supplementary Materials section).

The volume fraction can also be used to determine $N_{av,v}$, Equation (12). However, there are still some open questions regarding the physical meaning of this quantity.

$$N_{av,v} = \rho_m / M_{AVv,v} = \rho_{m(1,2)} / (\varphi_1 M_1 + \varphi_2 M_2). \tag{12}$$

This consideration refers to the solvent volume of each solvent component before mixing according to the IUPAC definition of volume fraction: "Volume of a constituent of a mixture divided by the sum of volumes of all constituents prior to mixing" [105]. This definition assumes ideal mixing behavior, which is not the case for most aqueous and non-aqueous solvent mixtures [106]. When two liquids are mixed, neither the total number nor the total mass of molecules changes, but the sum of the volumes may change compared to the volumes before mixing. Therefore, the use of the volume fraction in the determination of $N_{av,v}$ is controversial as to its true physical meaning. The use of $N_{av,v}$ (average molar concentration related to volume fraction) can only serve as an empirical guide.

Because of these well-known problems with volume changes after mixing, the issue is treated thermodynamically in terms of excess molar volume (V_E) by Equation (13) and described semi-empirically by several sophisticated concepts and approaches [66,78,107]. Equation (13) is well established in the textbooks.

$$V_E = (x_1 M_1 + x_2 M_2)/\rho_{m(1,2)} - (x_1 M_1/\rho_{m(1)}) - (x_2 M_2)/\rho_{m(2)}) \tag{13}$$

where $\rho_{m(1)}$ and $\rho_{m(2)}$ are the densities of the pure solvent 1 and 2, respectively. $x1$ and $x2$ are the mole fractions of solvent 1 and solvent 2, respectively. Analyses of V_E as a function of x(solvent 1) and x(solvent 2) can provide valuable information on the partial excess partial molar volumes of solvents 1 and 2 as a function of composition.

If only the molar fraction of the OH groups of a component on N_{av} is considered, i.e., that of the HBD solvent fraction (M_1), then Equation (7) can be modified to Equation (14).

$$N_{av}\,(\text{component1}) = x_1 \cdot \rho_m / M_{AV} = x_1 \cdot \rho_{m(1,2)} / (x_1 M_1 + x_2 M_2) \tag{14}$$

The approach of Equation (14) is useful in determining whether the influence of the proportion of HBD solvents mixed with non-HBD solvents is due to the overall polarity or to the preference of the HBD component. This procedure has been demonstrated for the dependencies of $E_T(30)$ as function of $N_{av,x}$ compared to $N_{av,x}(CH_3OH)$ for methanol/chloroform mixtures [43]. It has been shown that methanol is the dominant solvent according to scenario i of preferential solvation.

Equation (14) can also be used to consider the average number of OH groups ($D_{,av,xDHB}$) of a multifunctional OH component in the mixture, e.g., for dihydric alcohols such as 1,2-

ethanediol [62]. For pure 1,2-ethanediol, then, $2N = D_{HBD}$. See later the treatment of 1,2-ethanediol/water mixtures in relation to $E_T(30)$.

The problem with the average molar concentration is that the sum of the two dipoles is considered, e.g., for methanol and water. This is correct if the sum of the dipoles of the solvent and their effects is proportional to the measured quantity. Recently, we have shown that the total molar concentration N of pure solvents is not suitable to describe the changes in refractive index n_D^{20} as a function of structural variation within homologous series of n-alkane derivatives [69]. Instead, the molar concentration of the C-H bonds (or N-H) is crucial to adequately reflect the theoretically required linear relationship between n_D^{20} and N according to the modified Lorentz–Lorenz Equation (6). Equation (15) is particularly suitable for co-solvent/water mixtures to calculate the average molar concentration of C-H and/or N-H bonds of the co-solvent [71].

$$N_{av,x,CH} = [m\ x(\text{co-solvent})]\ N_{av,x}, \qquad (15)$$

The factor m is the number of C-H and N-H bonds per co-solvent molecule; x is the mole fraction of the respective co-solvent. Since the atomic refraction of the C-H and N-H (amide) bonds are nearly equal [108], no additional correction is necessary for formamide (FA), N-methylformamide (NMF) and N,N-dimethylformamide (DMF). For mixtures of organic solvents, the situation is more complicated because additional chemical bonds contribute to the molar refraction of the individual solvent molecules. This is particularly important for halogenated and aromatic solvents. Therefore, only the co-solvent/water mixtures are straightforward, as water is a weak (negligible) chromophore.

Basically, the general statement of this chapter shows that the absorption energy (EP) of a dissolved dye in a mixture is inversely proportional to the mole fraction due to $M_{av} \sim x(\text{co-solvent}) \sim 1/EP$ according to Equations (4) and (8). These basic relationships are independent of a physical law such as the Lorentz–Lorenz equation.

3. Results

3.1. Selection of the Solvent Mixtures

Because of the huge amount of data, we looked for a common thread to make statements that are as representative as possible and that also reveal fundamental correlations. Marcus distinguishes two groups of aqueous solvent mixtures in which the co-solvent either enhance or does not enhance the water structure. The evaluation is based on the excess partial molar volume or the excess partial molar heat capacity of the water [109,110]. Note that the Marcus classification only applies to the water-rich section of the mixture [$x(\text{water}) > 0.7$, $x_{\text{co-solvent}} < 0.3$] [109,110]. Marcus stated "Some solutes such as ethylene glycol, 1,4-dioxane, acetonitrile, NMF, FA, urea, ethanolamine, and dimethylsulfoxide, many of which hydrogen-bond very strongly with water, do not enhance the water structure" [110]. The selection was made according to this scientifically justified criterion. However, the Marcus evaluation can only be used as a rough guide because some co-solvents can be classified differently depending on whether the excess partial molar volume or the corresponding heat capacity is used. For some co-solvents, such as DMF, acetone, acetonitrile or THF, the classification is borderline [109,110], which shows how difficult the issue is. The binary mixtures acetonitrile/water, acetone/water and THF/water are each unique and will be discussed together in a separate publication. The situation regarding the non-enhancement of the water structure is quite clear for the FA/water, 1,2-ethanediol/water and glycerol/water mixtures [109,110]. Enhancement of the water structure is particularly relevant for the ethanol/water, 2-propanol/water and 2-methyl-2-propanol/water mixtures [109,110]. However, the term "water structure enhancement" sounds mysterious. [78,79]. The problem is that there are qualitatively different microdomains of water in alcohol/water mixtures in terms of structure and size [90–93,111]. Marcus [110] noted that the "Enhancement of the water structure then consists of the changing of some of the dense (water) domains to bulky ones". This phenomenon would inevitably lead to an increase in the average alcohol concentration in the remaining mixed phase compared to the co-existing microdomain water phase or the hypothetical phase resulting from the initial mixing ratio for each composition. Therefore, the overall polarity of the actual ethanol/water mixed phase should be lower than the phase that would result if ethanol and water were statistically completely mixed at a given composition. This should be kept in mind.

The ethanol/water mixture seems to be one of the most difficult solvent mixtures to understand when considering simple systems; see [111] and the references cited. The temperature increase associated with volume shrinkage when ethanol and water are mixed is apparently a thermodynamic anomaly [79]. The strongly negative entropy of the mixing process suggests complex structure formation depending on the composition, as demonstrated through dielectric spectroscopy and a special microscopic technique [86–89,111].

The curves of the solvatochromic parameters as a function of x(water) in [18,19] agree remarkably well with those of the partial excess molar volume as a function of x(water) of methanol/water, ethanol/water, 2-propanol/water and 2-methyl-2-propanol/water [95,96]. Therefore, the physics of alcohol/water mixtures deserves special attention in this study. There has been little discussion of the effect of the microstructure of alcohol/water mixtures on a solvatochromic probe [33].

3.2. Refractive Index of Aqueous Solvent Mixtures

The suitability of Equation (6) in combination with Equation (15) is illustrated for several amide derivative/water, DMSO/water and 1,4-dioxane/water mixtures. These solvent mixtures belong to the class where no enhancement of the water structure is observed [109,110]. References for n_D^{20} data are given in Tables in the Supplementary Materials section. No usable refractive index data could be found in the literature for NMF/water mixtures.

Plotting the refractive index (n_D^{20}) measured at a wavelength of 589 nm as a function of $N_{av,x,CH}$ gives a straight line, as can be seen in Figure 2a and from Equation (16) to Equation (20). The 1,2-ethanediol/water and glycerol/water mixtures, both of which show excellent linearity of n_D^{20} as a function of $N_{av,x,CH}$, are described in Section 3.4.6.

$$n_D^{20} = 1.5\, N_{av,x,CH} + 1.34,$$
$$n = 12\ (FA/water);\ r = 0.9997. \tag{16}$$

$$n_D^{20} = 1.736\, N_{av,x,CH} + 1.334,$$
$$n = 8\ (NFM/water);\ r = 0.9977. \tag{17}$$

$$n_D^{20} = 1.065\, N_{av,x,CH} + 1.34,$$
$$n = 14\ (DMF/water);\ r = 0.988. \tag{18}$$

$$n_D^{20} = 1.5\, N_{av,x,CH} + 1.34,$$
$$n = 18\ (DMSO/water);\ r = 0.9952. \tag{19}$$

$$n_D^{20} = 1.5\, N_{av,x,CH} + 1.34,$$
$$n = 12\ (1,4\text{-dioxane}/water);\ r = 0.9977. \tag{20}$$

The positive slopes $\Delta n_D^{20}/\Delta N_{av,x,CH}$ and the excellent quality of the linear correlations n_D^{20} as a function of $N_{av,x,CH}$ for several co-solvent/water mixtures are a clear proof of the approach of Equations (6) and (15) for solvent mixtures. The quantity $N_{av,x,CH}$ fulfils the theoretical requirements of Beer's approximation and the Lorentz–Lorenz relation [47,67,69]. The $E_T(30)$ parameters of these aqueous solvent mixtures decrease with increasing n_D^{20} (see Figure S2 in Supplementary Materials). These results will be explained at the appropriate place in the following text where the particular mixture is discussed.

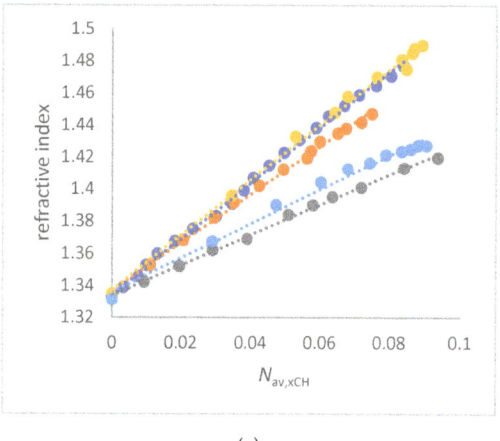
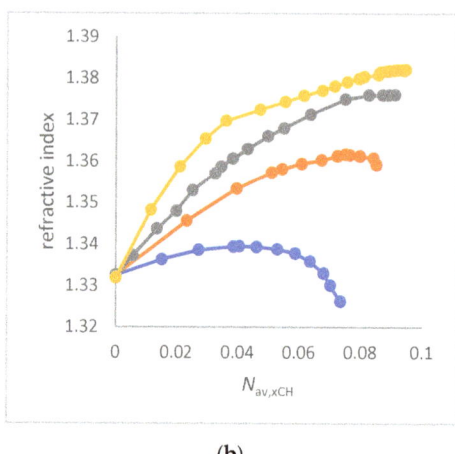

Figure 2. (a) Correlations of refractive index n_D^{20} as a function of $N_{av,x,CH}$ (mol/cm^3) for co-solvents that do not enhance the water structure of co-solvent/water mixtures; (b) Plots of refractive index n_D^{20} as a function of $N_{av,x,CH}$ (mol/cm^3) for co-solvents that enhance the water structure of co-solvent/water mixtures; to (a) FA/water (orange), water/N-formylmorpholine (NFM) (yellow), DMF/water (grey), 1,4-dioxane/water (light blue) and DMSO/water (dark blue); to (b) methanol/water (dark blue), ethanol/water (orange), 2-propanol/water (grey), and 2-methyl-2-propanol/water (yellow). The links between the individual points are a guide for the reader.

The conclusive linear relationships in Figure 2a clearly demonstrate the approach of Equations (6) and (15) when analyzing the refractive index of aqueous solvent mixtures. However, this is only true as long as alcohol/water mixtures are not considered.

Remarkably, the linearity n_D^{20} as a function of $N_{av,x,CH}$ does not apply to alcohol/water mixtures in which the water structure is enhanced [109,110]. In particular, the methanol/water and ethanol/water systems give a maximum curve of n_D^{20} as a function of $N_{av,x,CH}$; see Figure 2b. For the other alcohol/water mixtures, an asymptotic curve is obtained, but with a positive slope along the curve; Figure 2b. In the past there were empirical concepts to get around the non-linearity n_D^{20} as function of composition for methanol/water; i.e., by using the quotient n_D^{20}/density instead n_D^{20} alone [112]. However, the physical background is more complicated and is still under investigation [113–115]. Recent studies have shown that at the mesoscale there are microdomains of water and ethanol/water consisting of different refractive indices [111]. Depending on the balance between segregation and aggregation of these regions [115], the non-linearity of n_D^{20} as a function of composition is due to the coexistence of two different microdomains with different compositions and hence different refractive indices. The ratio of the two domains is a function of the original solvent proportions before mixing. The polarization effects and dipolar dispersion forces relevant to methanol/water mixtures may have an additional influence [60,93,94]. Figure 2b clearly supports the hypothesis of the coexistence of different microdomains of water/alcohol mixtures [90,91,111]. The following preliminary result can be stated: the alcohol/water mixtures that show an enhancement of the water structure according to Marcus do not show a linear dependence n_D^{20} on $N_{av,x,CH}$.

Alcohol/water mixtures will be further discussed in this paper under the aspect of the co-existence of different microdomains.

3.3. Temperature Influence on $E_T(30)$ in Terms of Density Impact

The $E_T(30)$ data of ethanol measured at different temperatures are taken from the original work of Dimroth–Reichardt and Linert to his subject [1,116]. The used data are provided in Supplementary Materials, Table S1. With increasing temperature, $E_T(30)$

decreases due to the decreasing density of the solvent and thus the decreasing number of dipoles per volume, which leads to perfect linear correlations of $E_T(30)$ as a function of $N(T)$; see Equations (21) and (22). The diagram is shown in Figure S3.

$$E_T(30) = 1834\ N\ (T) + 20.9, \qquad\qquad (21)$$
$$n = 8\ (\text{Reichardt}),\ r = 0.9969.$$

$$E_T(30) = 2205\ N\ (T) + 14.3, \qquad\qquad (22)$$
$$n = 7\ (\text{Linert}),\ r = 0.9978$$

The influence of temperature on the $E_T(30)$ value of solutions of **B30** in ethanol and methanol was also investigated by Zhao [117]. The authors hypothesized a de-defined **B30**/methanol complexation with decreasing temperature due to the appearance of an apparent isosbestic point in the UV/Vis spectrum series, in contrast to **B30** in ethanol. This conclusion is not yet clear because the increase in the intensity of the UV/vis absorption band is probably due to volume shrinkage on cooling, for which correction is not included in the reference. It is therefore possible that the isosbestic point is caused by the contribution of two or more different species. The presence of alcohol/**B30** complexes was also suggested by temperature-dependent UV/vis studies performed by El Soud [118]. However, complexation of **B30** with ethanol has not been directly demonstrated through independent spectroscopic measurements. Sanders suggested that the **B30**/HBD solvent complex would be the actual solvatochromic species as derived from theoretical considerations [59]. However, the specific influence of the dye/solvent complex on $E_T(30)$ is much smaller than the volume effect of the global hydrogen bonding network. For these reasons, these few results represent only a snapshot, as much remains to be done to understand the effect of temperature on $E_T(30)$ in terms of density fluctuations associated with structural changes as a function of temperature [118,119]. However, this first inventory shows that the increase in $E_T(30)$ with decreasing temperature is mainly due to an increase in density and thus in N.

3.4. Solvatochromism of B30 in Aqueous Solvent Mixtures

This part of the manuscript is the central concern. It is about correcting many misinterpretations in the literature. Most of the $E_T(30)$ data of the solvent mixtures to be evaluated were taken from [1–4,11–20] and others. Some specific comments on the datasets used are necessary, as several aspects have to be taken into account. It is necessary to check which $E_T(30)$ value corresponds exactly to the given concentration, as mole fractions, weight fractions and volume fractions are used alternatively [1,2,8,11–20].

The densities of the mixtures for each specific composition and temperature are required for evaluation. This was the most difficult problem to solve. Fortunately, the densities of alcohol/water mixtures often correlate significantly with the mole fraction (x) in certain ranges of the composition. Thus, unknown densities for certain compositions can be calculated from correlation equations using accurate data from the literature. References are given in the headings of the figures and tables in the Supplementary Material section.

Fortunately, many of the measured $E_T(30)$ values from the literature are in very good agreement between different authors for series of measurements. We have compared the data of Reichardt [2] and Rosés [18–20] and found that an almost perfect agreement of the measured $E_T(30)$ values as a function of $N_{av,x}$ is found. For an example, see Figure S4a for the ethanol/water mixture. For this task, it was necessary to convert the volume percentages from [1,2] to derive a mole fraction. Despite the very good agreement, a dataset from the same source was generally used for the analysis if sufficient measured values were available. For the FA/water mixture, data from two different references were mixed because the authors' measurements covered different composition ranges [21,120]. The deviations are very small. When staying within one data series, the regression coefficient r approaches one for FA/water. For the NMF/water mixtures, there is no large variation above $x(\text{water}) > 0.2$, see Supplementary Materials of [21].

The high quality of the overall dataset from Rosés should be emphasized. Rosés also used the carboxylate substituted betaine dye of **B30**; the **B30**-COONa to study alcohol/water mixtures due to the low solubility of **B30** in pure water and highly water concentrated mixtures [19]. There is an almost perfect agreement between $E_T(30)$ and $E_T(30\text{-COONa})$ over the whole composition range. This aspect will be taken up again in the discussion section.

The perfect complementarity of the different $E_T(30)$ values for DMSO/water from several references [7,12,14,121,122] should be noted (see Figure S4b). All datasets fit exactly in one relationship (see below). However, there are very small differences [$\Delta E_T(30) \sim 1$ kcal] between the authors' results.

Since the $E_T(30)$ datasets for 1,2-ethanediol/water show some unacceptable differences in the low water concentration range between the data from [12,15], we used only the dataset from [12] which fits well (see Figure S5).

The perfect complementary agreement of the $E_T(30)$ data from [13,19] for the 2-methyl-2-propanol/water mixture at high water concentration is also particularly noteworthy.

An unfortunate and common problem was that many measured UV/Vis data of various solvatochromic dyes were accurately reported neither in the tables nor in the Supplementary Materials [6,10]. Often only the coefficients of the applied solvation models or artificially modified parameters were given instead of the original spectroscopic data.

To support the correlations of $E_T(30)$ as a function of $N_{av,x}$, Kosower's Z-scale was considered appropriate [123–125] because of the linear correlation of Z with the $E_T(30)$ parameter [1,34,35]. However, this proved not to be the case. It is important to clarify the situation of the different Z values for DMSO/water and ethanol/water mixtures in the literature, as only the Z values given by Kosower have been directly determined with **K** [123,124]. The Z values used by Marcus for correlations were calculated by himself indirectly using Brownstein's S values [126] (see note in citation 23 of Marcus' paper) [12]. The same applies to Gowland's Z values, which were also determined indirectly from 4-pyridine-N-oxide via a correlation equation [127]. We are convinced that the main problem is the reproducible measurement of Z values with Kosower's dye, because in [127] it was mentioned that the Z value depends on the concentration of **K** in ethanol/water. Sufficient dilution is necessary or, alternatively, extrapolation to infinite dilution if experimental problems may occur.

To test whether case ii. of preferential solvation is significant, the literature data of other negatively solvatochromic probes such as **B1** [(2,4,6-triphenyl-1-pyridinium)-phenolate] [1], Brooker's Merocyanine (**BM**) [128] or **Fe** [129] were considered, although fewer data points per individual correlation are available. For this purpose, EP of BM or the UV/Vis absorption energy at the peak maximum $\nu_{max}(\textbf{Fe})$ are analyzed as a function of $N_{av,x}$.

3.4.1. 1,2-Ethanediol/Water, Methanol/Water and Ethanol/Water Mixtures

The reason for considering 1,2-ethanediol/water mixtures in comparison to methanol/water and ethanol/water mixtures is as follows. In all three binary solvent mixtures, the enthalpy of mixing is exothermic over the whole composition range [82,83,130]. While 1,2-ethanediol as a co-solvent does not enhance the water structure, methanol and ethanol do [109,110].

As mentioned above, the relationship $E_T(30)$ as function of $x(\text{water})$ resulted in a curved line, regardless whether methanol/water, ethanol/water or 1,2-ethanediol/water mixtures were considered, as seen in Figure 3b. This was discussed in the introduction and is well described in the literature [2,8,10–20]. The greater the difference in molar mass, the more non-uniform the mixture will be. The order of DI_{max} is as follows: ethanediol/water mixtures (green) > ethanol/water mixtures (blue) > methanol/water mixtures (grey) (see Figure 3b). This in turn depends on the $x(\text{water})$ in the mixture. It can be clearly seen that the strongest curvature along a line of $E_T(30)$ as a function of $x(\text{water})$ for each specific co-solvent/water mixture occurs when the *DI* is highest. This is a purely physical effect and has nothing to do with the specific solvation.

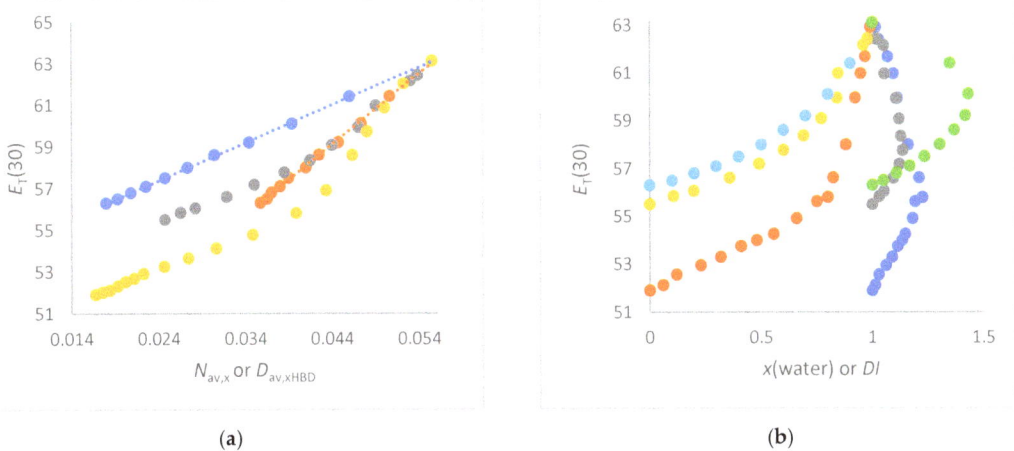

(a) (b)

Figure 3. (a) Comparison of correlations of $E_T(30)$ (kcal/mol) as a function of $N_{av,x}$ (mol/cm^3) for 1,2-ethanediol/water (blue dots) with methanol/water (grey dots) and ethanol/water (yellow dots). The orange dots belong to the correlation of $E_T(30)$ as function of D_{HBD} for 1,2-ethanediol/water; (b) plots of $E_T(30)$ (kcal/mol) as a function of x(water) and *DI*, respectively, for methanol/water (grey and yellow), 1,2-ethanediol/water (light blue and green) and ethanol/water (orange and blue).

The situation is different when $E_T(30)$ is theoretically correctly correlated with $N_{av,x}$ (see Figure 3a). An excellent linear correlation of $E_T(30)$ as a function of $N_{av,x}$ is then obtained for the 1,2-ethanediol/water mixtures (Equations (23) and (24)). This overall result is very significant. This corresponds to the physical finding that the 1,2-ethanediol/water mixtures do not show abrupt structural changes over the entire composition range [77,110,130]. The interpretation of the $E_T(30)$ curve as function of $N_{av,x}$ for the 1,2-ethanediol/water mixtures requires an essential comment, because each 1,2-ethanediol molecule contains two OH groups. Therefore, the number of OH dipoles per 1,2-ethanediol is doubled. For the 1,2-ethanediol/water mixtures, the hydroxyl group concentration is calculated as a function of the number of the total OH dipoles using the D_{HBD} model [62]. The $D_{,av,xHBD}$ quantities are calculated from Equation (14) using the partial OH concentration of the 1,2-ethanediol component in the mixture (see Table S2). The function $E_T(30)$ versus D_{HBD} for 1,2-ethanediol/water mixtures determined according to Equation (14) is the orange dotted line in Figure 3a. This curve is completely congruent with the relationship $E_T(30)$ versus $N_{av,x}$ for methanol/water mixtures in the water-rich range ($N_{av,x} > 0.04$ mol/cm^3) (grey dotted line of Figure 3a). However, it is noteworthy that the correlation $E_T(30)$ versus $N_{av,x}$ of methanol/water mixtures from $N_{av,x} < 0.04$ mol/cm^3 runs parallel to the correlation $E_T(30)$ versus $N_{av,x}$ (dark blue) for 1,2-ethanediol/water mixtures. This agreement illustrates the significant influence of the concentration of OH dipoles on $E_T(30)$. This result also shows the strong influence of the total number of OH groups of binary aqueous mixtures in terms of $D_{,av,xHBD}$ or $N_{av,x}$ on $E_T(30)$ [62]. These clear results completely exclude a preferential solvation of **B30** in methanol/water, ethanol/water and 1,2-ethanediol/water mixtures in the sense of scenario ii. The results for the methanol/water and ethanol/water mixtures do not really correspond to scenario i either. It is always the total number of dipoles per volume that determines the $E_T(30)$ value within certain composition ranges, regardless of structural variations.

A kink can be seen in the correlation line $E_T(30)$ as a function of $N_{av,x}$ for both methanol/water and ethanol/water mixtures in Figure 3a. These noticeable kinks in the graphs of $E_T(30)$ as a function of composition in alcohol/water mixtures have been recognized in several previous studies and attributed to structural changes in the solvent structure [5,6,131,132].

However, the linear correlations of $E_T(30)$ as function of $N_{av,x}$ for each section of the solvent mixture are of excellent quality as shown by Equations (23)–(28).

$$E_T(30) = 181.3\ N_{av,x} + 53.02,$$
$$n = 12\ (1,2\text{-ethanediol/water});\ r = 0.999. \qquad (23)$$

$$E_T(30) = 341.1\ D_{HBD} + 44,$$
$$n = 12\ (1,2\text{-ethanediol/water});\ r = 0.999. \qquad (24)$$

$$E_T(30) = 342\ N_{av,x} + 44.02,$$
$$n = 7\ (\text{methanol/water};\ N_{av,x} > 0.04);\ r = 0.9957. \qquad (25)$$

$$E_T(30) = 162\ N_{av,x} + 51.5$$
$$n = 6\ (\text{methanol/water};\ N_{av,x} < 0.04;\ r = 0.9985. \qquad (26)$$

$$E_T(30) = 500.7\ N_{av,x} + 35.6,$$
$$n = 8\ (\text{ethanol/water};\ N_{av,x} > 0.04;\ r = 0.995. \qquad (27)$$

$$E_T(30) = 158.6\ N_{av,x} + 49.27,$$
$$n = 10\ (\text{ethanol/water};\ N_{av,x} < 0.04);\ r = 0.998. \qquad (28)$$

Various physical data on the properties of methanol-water mixtures indicate a structural variation in the range of $x(\text{water}) = 0.5$ to 0.6; corresponding to $N_{av,x} = 0.035$ and 0.04 mol/cm^3 [86–95].

This wide distribution is also confirmed by the heat of interaction as a function of composition, with the largest measured heat of about -850 kJ/mol in a range from $x(\text{water}) \sim 06$ to 0.75 [81,82]. The refractive index of methanol/water mixtures reaches its maximum at $x(\text{water}) = 0.6$ [112–114]. The highest heat of the exothermic interaction is at $x(\text{water}) = 0.6$ [82,83] ($N_{av,x} = 0.038$ mol/cm^3), which is fully reflected by the DI_{max} of the methanol/water mixtures, which is highest at $x(\text{water}) = 0.6$ (see Figure 1b).

However, the overall situation with these two monohydric alcohol/water mixtures is not entirely clear. For ethanol/water mixtures, the function $E_T(30)$ versus $N_{av,x}$ shows a clear kink at exactly $N_{av,x} = 0.04$ mol/cm^3 corresponding to $x(\text{water}) = 0.8$. The excess molar volume for ethanol/water mixtures is at $x(\text{water}) = 0.6$, but the heat of interaction is highest at $x(\text{water}) = 0.82$ to 0.845 [82,83]. Therefore, the refractive index maximum of ethanol/water mixtures does not correspond to thermodynamics, as is apparently the case for methanol/water mixtures. The different behavior of the composition of methanol/water and ethanol/water mixtures with respect to the refractive index was also noted by Langhals [112]. For the ethanol/water mixtures, the plots $E_T(30)$ as function of $N_{av,x}$ or $x(\text{water})$ are clearly determined through thermodynamics. Exactly at this composition, where the largest heat of interaction is measured, the graphs show a kink in the line indicating the structural change [5,81,82,84,89]. This agreement between the curves in Figure 3a and the thermodynamics or refractive index clearly show the influence of the physical properties of the mixture on $E_T(30)$, as suggested in previous studies [5,6,131,132].

However, there are a number of other aspects to consider. Bentley [27] has shown that the volume fraction correlates better linearly with the static dielectric constant (ε_r) or the $E_T(30)$ values of alcohol/water mixtures than the mole fraction as a composition parameter of alcohol/water mixtures. The volume fraction has also been recommended in a recent publication to explain the $E_T(30)$ as a function of solvent composition more accurately than using the mole fraction [133]. Accordingly, for ethanol/water and methanol/water mixtures, the $N_{av,w}$ and $N_{av,v}$ quantities have been calculated and empirically tested as variables for correlation with $E_T(30)$ [62]. It seems surprising that the $N_{av,w}$ and $N_{av,v}$ quantities give a much better linear relationship with $E_T(30)$ than the use of $N_{av,x}$ when the whole range of composition is considered. The methanol/water and ethanol/water mixtures fit seamlessly into the primary alcohol series when the full dataset $E_T(30)$ of primary alcohols is included; see Equations (29) and (30) and Figure S6a in the Supplementary Materials. The overall correlations with 42 data points are convincing.

$$E_T(30) = 313\, N_{av,v} + 46.7,$$
n = 42 (methanol/water, ethanol/water and primary alcohol); r = 0.994. (29)

$$E_T(30) = 304.8\, N_{av,w} + 46.7,$$
n = 42 (methanol/water, ethanol/water and primary n-alcohol); r = 0.994. (30)

We are therefore in full agreement with the conclusions of [133], that the volume fraction gives better results in terms of linear correlation. For the correlation with $E_T(30)$, however, it makes no qualitative difference whether the mass or the volume fraction is used to determine $N_{av,z}$. Therefore, the motivation for using the volume fraction given in [133] should be reconsidered. Using the mass fraction would give similar results. Whichever alcohol/water mixture is considered, the actual curve $E_T(30)$ versus $N_{av,w}$ or $N_{av,v}$ is not really strictly linear, although a very good regression coefficient for linearity can be calculated. The data points along the relationship lines show a significant pattern like a string of pearls, as can be seen in Figure S6 in the Supplementary Materials. This is an important detail. Thus, the subtleties observed in the correlation of $E_T(30)$ with $N_{av,x}$ do not disappear, but are merely reduced in the plots $E_T(30)$ as a function of either $N_{av,w}$ or $N_{av,v}$. The approximate linearity of $E_T(30)$ as a function of $N_{av,w}$ or $N_{av,v}$ is due to the stronger algorithmic consideration of the inhomogeneity of the solvent components in $N_{av,w}$ or $N_{av,v}$ (see Figure 1b).

These results clearly show that the discussed preferential solvation of **B30** by water is meaningless for methanol/water and ethanol/water mixtures. This is also an indication that the polarization forces and dipolar effects of the molecules in the solvated mixture act collectively on **B30**. In 1963, in the first paper on phenolate betaine dyes, Dimroth and Reichardt also studied the better water soluble **B1** probe in ethanol/water mixtures [1]. For data, see Table S4. There is also a very good correlation of $E_T(1)$ as function of $N_{av,v}$, as can be seen from Equation (31). The correlation of $E_T(1)$ as versus $N_{av,x}$ is equivalent to that of $E_T(30)$ versus $N_{av,x}$.

$$E_T(1) = 216.7\, N_{av,v} + 57.95,$$
n = 10 (**B1** in ethanol/water and water), r = 0.988. (31)

If pure water is omitted from Equation (31), then the correlation quality is significantly improved to r = 0.999. This is also a strong indication that **B1** is preferentially enriched in ethanol/water-rich regions when the mixture is examined. The x_b values of **BM** (x_b is the shift of the UV/Vis peak of **BM** in methanol/water) [128]) correlate very well with $N_{av,x}$; see Equation (32).

$$x_b = 201.2\, N_{av,x} + 57.8,$$
n = 11 (**BM** in methanol/water), r = 0.997. (32)

Consequently, the preferential solvation of **BM** in methanol/water as assumed by Machado [26] or Tanaka [134] is not applicable when $N_{av,x}$ is used instead of x(water) to evaluate solvatochromism. The methanol/water mixtures were also studied by Taha using the **Fe** probe [129]. There is also a linear correlation and no curved curve for $\nu_{max}(\mathbf{Fe})$ as function of $N_{av,x}$, Equation (33).

$$\nu_{max}(\mathbf{Fe})\,[10^3\,\text{cm}^{-1}] = 36.66\, N_{av,x} + 17.32,$$
n = 11 (**Fe** in methanol/water), r = 0.992. (33)

For ethanol/water mixtures, the $\nu_{max}(\mathbf{Fe})$ as function of $N_{av,x}$ shows a similar correlation with excellent quality as previously reported [43]. The correlation of $\nu_{max}(\mathbf{Fe})$ with x(water) in place with $N_{av,x}$ is worse.

These results clearly show that several types of negatively solvatochromic dyes such as **B30**, **B1**, **BM** and **Fe** do not indicate preferential solvation in the methanol/water and ethanol/water mixtures. Thus, the linear correlations of EP parameters as function of $N_{av,x}$ according to Equation (2) are clearly confirmed by other solvatochromic dyes despite the smaller dataset compared to $E_T(30)$. Since the UV/Vis energies of the different solvatochromic probes show the same linear dependencies as a function of $N_{av,x}$, it is quite clear that the solvent structure determines the solvatochromism and not the preferred solvation according to scenario ii. This conclusion is in complete agreement with older results by Langhals [5].

3.4.2. Formamide/Water and other Amide/Water Mixtures

FA/water is the only binary aqueous mixing system considered in this study that fulfils the thermodynamics of ideal mixing [66,110,135,136]. The heat of mixing is endothermic, and the entropy is positive over the whole composition range. The mixing entropy is highest at $x = 0.5$ [135,136].

The best linear correlations (r about 1) of $E_T(30)$ as a function of $N_{av,x}$ over the whole composition range of the solvent mixture were found for FA/water, NMF/water and 1,2-ethanediol/water mixtures (see Figures 3a and 4a).

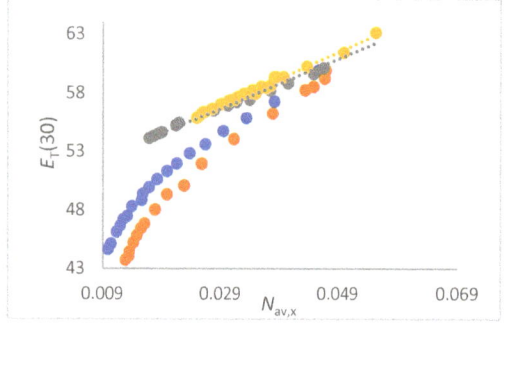
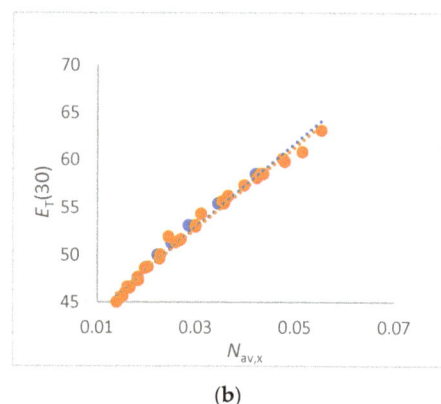

(a)   (b)

Figure 4. (**a**). Correlations of $E_T(30)$ (kcal/mol) as a function of $N_{av,x}$ (mol/cm^3) formamide/water (yellow), NMF/water (grey), N-formylmorpholine/water (blue) and DMF/water (orange) mixtures. For data, see Tables S2–S5; (**b**) correlations of $E_T(30)$ (kcal/mol) as a function of $N_{av,x}$ (mol/cm^3) for DMSO/water mixtures (orange, all data); blue dots are data from [15] (O Connor).

For FA/water mixtures the linear correlations $E_T(30)$ as function of $N_{av,x}$ are of excellent quality; see Figure 4a as well as Equation (34).

$$E_T(30) = 229 \, N_{av,x} + 50.2,$$
$$n = 16 \text{ (FA/water); } r = 0.999. \tag{34}$$

The perfect linearity of $E_T(30)$ as a function of $N_{av,x}$ can be explained by the excellent physical properties of the FA/water mixtures [109,110,135–138]. The water-like structure of FA is due to the fact that water and FA molecules can exchange positions without changing the solvent structure [139]. Only the V_E (Equation (13)) changes a little, as a function of composition [138]. There is no segregation within the FA/water mixtures and the average number of dipoles per volume perfectly determines the $E_T(30)$ at room temperature. A very good linear relationship $E_T(30)$ versus $N_{av,x}$ is also obtained for NMF/water mixtures, see Equation (35).

$$E_T(30) = 212\ N_{av,x} + 50.5,$$
$$n = 17\ (NMF/water);\ r = 0.993. \tag{35}$$

Furthermore, for NFM/water and DMF/water mixtures, there are also excellent linear correlations of $E_T(30)$ as function of $N_{av,x}$ in the section of higher water content range; $x_{co\text{-}solvent} < 0.35$ [31,107,140–142].

The slight kinks in the curves at lower water contents are due to the non-linear change in density as a function of composition [140–142]. The physical data of the NMF/water, DMF/water and NFM/water mixtures are given in Tables S2–S5 in the Supplementary Materials. According to [76], water is considered to be a solute rather than a solvent when $N_{av,x} < 0.035$ mol/cm^3. However, an excellent linear correlation of the refractive index as a function of $N_{av,x,CH}$ is seen for all mixtures (see Figure 2a) over the entire composition range, including the range of low water concentrations.

Marcus also described the aqueous urea solution as a binary solvent mixture system in which no enhancement of the water structure occurs, although pure urea is a solid at room temperature [110]. Accordingly, we analyzed the $E_T(30)$ values of the urea/water and N,N-dimethylpropylene urea/water binary mixtures from the literature [143–145]. There are very good linear correlations of $E_T(30)$ as a function of $N_{av,x}$ for both urea/water and N,N-dimethylpropylene urea/water mixtures with high correlation quality (see Figure S6a in Supplementary Materials). This result shows that solutions of solids in water can also be treated in the same way. If the co-solvent or co-component (urea, N,N-dimethylpropylene urea) can form a three-dimensional hydrogen bond structure with water, then a linear correlation of $E_T(30)$ with $N_{av,x}$ is found.

3.4.3. DMSO/Water Mixture

DMSO/water mixtures represent a physical challenge among binary aqueous solvent systems due to the unclear thermodynamics at higher DMSO contents [146–154]. This was therefore chosen for this fundamental work as an illustrative example. There are a large number of physical studies on these mixtures, so only those relevant to the explanation of solvatochromism in terms of $N_{av,x}$ will be referred to. The following analysis shows where the problems lie. There results a very good linear correlation of $E_T(30)$ as function of $N_{av,x}$ including $E_T(30)$ data from several references, Equation (36) and Figure 4b.

$$E_T(30) = 432\ N_{av,x} + 39.7,$$
$$n = 22\ (DMSO/water)\ r = 0.993. \tag{36}$$

Although the overall correlation $E_T(30)$ with $N_{av,x}$ seems convincing due to the clear linearity, there is a small kink in the linear plot at $N_{av,x} \approx 0.025$ to 0.03 mol/cm^3. The kink becomes more obvious when considering only the data from [15], see Equations (37) and (38).

$$E_T(30) = 414\ N_{av,x} + 40.7,$$
$$n = 9\ (DMSO/water\text{-}rich;\ N_{av,x} > 0.02);\ r = 0.998. \tag{37}$$

$$E_T(30) = 624\ N_{av,x} + 36.2,$$
$$n = 5\ (DMSO/water\ low;\ N_{av,x} < 0.02);\ r = 0.999. \tag{38}$$

This small effect has a significant physical background as the density of the binary solvent mixture changes significantly at this composition [146,148]. However, density measurements for DMSO/water mixtures in the DMSO-rich region are not consistent in the literature [146,148]. In the water-rich range from $N_{av,x} < 0.05541$ mol/cm^3 (pure water) to $N_{av,x} = 0.03$ mol/cm^3, the density of water/DMSO mixtures decreases linearly with increasing water content. The density is almost constant in the range from $N_{avx} = 0.03$ (60% weight DMSO) to 0.014 mol/cm^3 (pure DMSO) (see Table S9). In [148], it was reported that the density even decreases slightly. It should be noted that exactly at this mixture composition $N_{av,x} = 0.028$ mol/cm^3 the plot of $E_T(30)$ as a function of $N_{av,x}$ has a slight, imperceptible kink.

In the literature, there are several investigations on the DMSO/water mixtures using different solvatochromic probes [12,15,129,155–157]. Regardless of the type of solvatochromic probe used, it is clear that at $N_{av,x} \approx 0.03$ mol/cm^3 a slight change in the profile of the parameter values can be observed as a function of the composition. Thus, the physical structural change of the DMSO/water mixtures determines the empirical parameter and not the artificially constructed acid-base properties of the solvent system [155,157]. This result is fully consistent with the prediction in the introduction that no differences should occur in scenario ii. when different probes are used. For reasons of space, the analyses of the Kamlet–Taft (KAT) parameters of DMSO/water mixtures [157] are presented in Figure S9 in the Supplementary Materials. As a consequence of this result, the determination of individual empirical polarity parameters in terms of the KAT or Catalán scale is meaningless for DMSO/water mixtures. Furthermore, a curved function of the $E_T(30)$ value of the solvatochromic probe on x(water) of DMSO/water mixtures is found (see (Figure 5) of [12]). If the x(water) is replaced by $N_{av,x}$, a linear correlation is obtained, as shown in Figure 4b. The correlation of the UV/Vis absorption energy of other probes such as **Fe** [$\nu_{max} 10^{-3}$ cm^{-1} (**Fe**)] [129] as function of $N_{av,x}$ for DMSO/water mixtures clearly shows a linear dependence, see Equation (39).

$$\nu_{max} 10^{-3} \text{ cm}^{-1} \text{ (\textbf{Fe})} = 67.7\, N_{av,x} + 15.8,$$
$$n = 12 \text{ (DMSO/water)},\ r = 0.996. \tag{39}$$

The change in the curve of the solvatochromic parameter after at $N_{av,x}$ about 0.03 mol/cm^3 is clearly due to physical changes in the solvent structure. Furthermore, if the static dielectric constant (ε_r) of DMSO/water mixtures is plotted as a function of $N_{av,x}$, then the kink at $N_{av,x}$ at 0.03 mol/cm^3 becomes also evident (see Figure S7a). The ε_r data are taken from [151]. This property is also shown in the plots of $E_T(30)$ as a function of n_D^{20} (Figure S2). While the correlation of n_D^{20} as a function of $N_{av,CH}$ (Equation (10)) is nearly linear (Figure 2a), the correlation of $E_T(30)$ as function of n_D^{20} shows a slight kink at $N_{av,x} = 0.03$ mol/cm^3.

To return to the DMSO/water mixtures, the concentration of all dipoles (water + DMSO) of the system determines the solvatochromic property and not the preferential solvation. This is a clear result. The only surprising thing is the rather good linearity of the function $E_T(30)$ versus $N_{av,x}$ when many data from the literature are used together. This shows that **B30** is not very sensitive to physical changes in the DMSO/water mixture system at RT. Therefore, the solvatochromic method is not well suited to detecting the physical change in the liquid structure of DMSO/water at different compositions.

What is the reason for the good linearity of $E_T(30)$ as a function of $N_{av,x}$ although major structural changes of the mixture occur at $N_{av,x} = 0.03$ mol/cm^3? The complexity of the water dynamics of DMSO/water mixtures has been thoroughly investigated through ultrafast IR experiments and dielectric spectroscopy [149–151]. These results are very important in partially explaining the results of the correlations in this study. The average lifetime of water-bound DMSO changes (decreases) almost linearly with the mole fraction of water. This result is consistent with $E_T(30)$ increasing almost linearly with water content (see also Figure 5 in [149]). This explains why the barely noticeable kink in the correlation can be neglected, as the water dynamics overcome the local structuring around the dissolved dye. Thus, the lifetime of the water/water component is independent of the water concentration in the high DMSO region $N_{av,x} < 0.03$ mol/cm^3. Obviously, neither water/DMSO nor **B30**/water complexes are relevant for the determination of $E_T(30)$ since the solvent mixture has a high dynamic at 298 K [150,151]. Thus, even if DMSO/water or **B30**/water complexes are present, they cannot be detected using **B30** due to the fast dynamics of the binary solvent system. The situation is similar to other solvatochromic dyes such as **Fe**. Therefore, other physical measurements such as dielectric spectroscopy are more suitable than solvatochromic probe molecules for analyzing the structure of DMSO/water mixtures. The outstanding behavior of the DMSO/water mixtures at higher DMSO contents $N_{av,x} < 0.028$ mol/cm^3 has been the subject of numerous simulation experiments [152–154]. Apparently, the behavior at $N_{av,x} < 0.03$ mol/cm^3 is due to the entropy increase in the system, which is still difficult to understand theoretically [154], since the experimentally determined heat of interaction is exothermic over the whole composition range.

3.4.4. 1,4-Dioxane/Water Mixtures

The 1,4-dioxane/water mixtures were subjected to numerous physical tests [158–169]. The dependence of the UV/Vis-absorption energy maxima of solvatochromic dyes such as **B30**, **Fe**, **M540**, various 7-N,N-diethylaminocoumarins or harmaline as function of dioxane/water composition has been extensively studied in the literature [2,5,70,129,165–169].

The thermodynamics of 1,4-dioxane/water mixtures is characterized by a transition from exothermic to endothermic heat of mixing with increasing 1,4-dioxane content [158]. This is the main difference to the DMSO/water system [147]. The heat of interaction Δ_rH of 1,4-dioxane/water mixtures has its maximum exothermic heat at around x(water) = 0.8 (yellow dot in Figure 5b) corresponding to $N_{av,x} = 0.032$ mol/cm^3 or $N_{av,v} = 0.018$ mol/cm^3. The largest partial molar volume of water in 1,4-dioxane/water mixtures is $x = 0.8$ [167]. Δ_rH is zero at x(water) = 0.52 ($N_{av,x} = 0.02$ mol/cm^3). With this composition, the 1,4-dioxane/water mixture has the highest density and the lowest $-T\Delta S$ value. At x(water) < 0.52, the heat of interaction becomes endothermic. For the evaluation in this paper, the volume fractions of the 1,4-dioxane/water mixtures given in [2] were reconverted to the average molar concentration of the solvent dipoles. Fortunately, there is excellent agreement between the $E_T(30)$ data from four different literature sources, as shown in Table S7. The $E_T(30)$ data from these four different sources fit perfectly into a relationship. To evaluate the influence of the inhomogeneity of the mixture with respect to the composition, we plotted $E_T(30)$ as function of $N_{av,x}$ and $N_{av,v}$ as well as x(water) (Figure 5a,b).

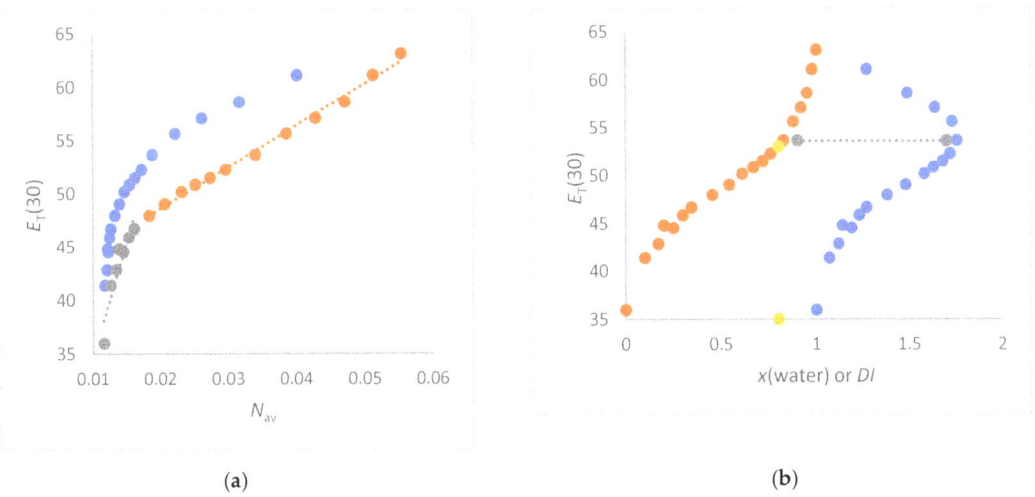

Figure 5. (**a**) Correlations of $E_T(30)$ (kcal/mol) as a function of $N_{av,x}$ (mol/cm^3) (orange dots) and $N_{av,v}$ (blue dots) for 1,4-dioxane/water mixtures at 298 K; (**b**) plots of $E_T(30)$ (kcal/mol) as a function of x(water) for 1,4-dioxane/water mixtures (orange dots) compared with the inhomogeneity (DI) of the system in terms of $M_{av,v}/M_{av,x}$ ratio (blue dots). The yellow dots indicate the composition with the greatest inhomogeneity. The yellow dots indicate the composition with the inflection point and the greatest inhomogeneity. The grey dots show the correspondence between the two curves in terms of maximum inhomogeneity.

The correlation of $E_T(30)$ as a function of $N_{av,x}$ results in two consecutive linear lines with different slopes. The change in the function $E_T(30)$ as function of N_{av}, is at $Nav,x = 0.015$ [x(water) = 0.3] mol/cm^3; see Equations (40) and (41) and Figure 5a.

At this composition (at $E_T(30) \sim 46$ kcal/mol), there is also the strongest curvature in the curve $E_T(30)$ as a function of x(water) in the 1,4-dioxane-rich section (see Figure 5b).

$$E_T(30) = 2997.5\, N_{av,x} + 2.123$$
$$r = 0.944,\ n = 7\ (N_{av,x} < 0.02\ \text{mol/cm}^3,\ 1{,}4\text{-dioxane rich section}) \tag{40}$$

$$E_T(30) = 398{,}8\, N_{av,x} + 40.42$$
$$r = 0.997,\ n = 12\ (N_{av,x} > 0.02\ \text{mol/cm}^3,\ \text{water-rich section}) \tag{41}$$

The correlation of $E_T(30)$ as a function of $N_{av,v}$ (blue dots in Figure 5a) gives an asymptotic curve without linearity of specific sections. This could be explained by the fact that the variable $N_{av,v}$ better reflects the inhomogeneities of the composition.

The 1,4-dioxane/water mixtures are subject to fine structuring over the whole composition range, in which both types of molecules are always involved [161–164]. The volume structure of 1,4-dioxane/water mixtures changes significantly in the range of $N_{av,x} < 0.02$ mol/cm^3. Accordingly, the strongest bend in the graph $E_T(30)$ as function of $N_{av,x}$ corresponds to the composition where the significant change in the volume structure of the 1,4-dioxane/water mixtures takes place. Exactly at $E_T(30) = 47$ kcal/mol ($N_{av,x} = 0.018$ mol/cm^3), the dielectric relaxation time τ_1 passes through a maximum ($\tau_1 \approx 25$ ps) for 1,4-dioxane/water mixtures [162]. The use of $N_{av,X}$(water) according to Equation (14) as the mixture composition parameter gives a similar plot as when $N_{av,x}$ is used (see Figure S8b), indicating that 1,4-dioxane and water are always involved together in the volumetric structure and thus in the dissolution of dissolved **B30**. Thus, 1,4-dioxane does not enhance the water structure in any way, which is in full agreement with the Marcus classification [110].

It is worth analyzing the correlations of $E_T(30)$ as a function of x(water) from the point of view of thermodynamics and the structural change of the 1,4-dioxane/water mixtures as shown in Figure 5b. At x(water) = 0.52 ($N_{av,x} = 0.02$ mol/cm^3), the curve $E_T(30)$ as a function of x(water) shows an inflection point (not marked in Figure 5b). It is precisely at this composition that this binary solvent system behaves in an athermal manner, i.e., ΔrH mixture = 0 [158,159]. The strong curvature $E_T(30)$ as a function of x(water) = 0.8 (marked in yellow) ($N_{av,v} = 0.015$ mol/cm^3) is clearly due to the inherent mass inhomogeneity of the mixture, as shown in the simultaneous plot for the DI (Figure 5b). At this composition ($N_{av,x} = 0.032$ mol/cm^3), mixing has the highest exotherm. This result is consistent with the results from the thermodynamics of methanol/water and ethanol/water mixtures, which are a good indication that x(water) reflects the thermodynamics of the mixture in relation to other quantities more comprehensively than the quantity $N_{av,x}$. Thus, the S-shaped function $E_T(30)$ versus x(water) (Figure 5b, orange dots) is attributed to the change in interaction heat as function of composition. This feature is only partly recognized when $N_{av,x}$ is used as the composition size, as seen in Figure 5a. There is no bend or kink in the plot $E_T(30)$ as function of $N_{av,x}$ for $N_{av,x} \sim 0.032$ mol/cm^3 (largest exothermic heat), but at $N_{av,x} = 0.02$ mol/cm^3 (zero heat).

The linear function of $E_T(30)$ as a function of $N_{av,x}$ in the water-rich region $N_{av,x} > 0.02$ mol/cm^3 is due to the fact that the average concentration of both the water dipoles and 1,4-dioxane molecules determines the $E_T(30)$ value. Both fractions are constantly mixed together and do not segregate [162,163]. The larger $E_T(30)$ in the 1,4-dioxane rich fraction, compared to a hypothetical linear plot of $E_T(30)$ versus $N_{av,x}$, can be easily explained by the results of Buchner: "This indicates a largely microheterogeneous structure for such mixtures, with the presence of water-rich domains of significant size in the dioxane-rich fraction" [163]. Thus, **B30** preferentially measures the water enriched portions of the 1,4-dioxane/water domains within the compositional spectrum. Obviously, the water clusters are solvated by the 1,4-dioxane excess and the **B30** is enriched in the 1,4-dioxane clusters below $N_{av,x} < 0.015$ mol/cm^3.

The refractive index as a function of the composition $N_{av,xCH}$ of the 1,4-dioxane mixture (see Figure 2a, red dots) and Equation (23) give a linear curve. This is consistent with the fact that the static permittivity ε_r of 1,4-dioxane/water mixtures is also a linear function of $N_{av,x}$ including pure 1,4-dioxane (see Figure S7b). For this investigation, the composition data x(water) from [162] were converted to $N_{av,x}$. In contrast, the correlation of $E_T(30)$ as a function of ε_r or n_D^{20} is not linear over the whole composition range, because the values of the pure 1,4-dioxane or the 1,4-dioxane-rich fraction do not fit linearly; see Figure S8a.

As a consequence, the **B30** probe reflects the volumetric structure of 1,4-dioxane/water differently compared to volumetric polarity-related physical measurements such as dielectric spectroscopy or refractive index. In summary, the 1,4-dioxane/water mixtures present a challenge in terms of the formation of solvent structures as a function of the quantitative composition, since different physical methods (UV/vis spectroscopy of **B30**, dielectric spectroscopy, refractive index, calorimetry) register different dependencies of the measurand on the different composition sizes.

Note that only x and $N_{av,x}$ are physically based quantities, referring to thermodynamic and UV/Vis spectroscopic quantities, respectively. Despite this concern, in summary, the complex dependence of $E_T(30)$ on the composition of 1,4-dioxane/water mixtures can be readily interpreted in terms of $N_{av,x}$, $N_{av,v}$ or x(water). This is possible by analyzing the physical properties of this binary solvent system, where thermodynamics, dipole concentration and solvent dynamics play a role. Despite this caveat, it is clear that the specific solvation of **B30** by HBD solvent molecules is not responsible for this UV/Vis shift. The $E_T(30)$ of 1,4-dioxane/water mixtures is mainly determined through the concentration of water dipoles permanently mixed with 1,4-dioxane molecules.

3.4.5. 2-Propanol/Water and 2-methyl-2-propanol/Water Mixtures

The 2-propanol/water and 2-methyl-2-propanol/water mixtures are considered separately because they show a change in the heat of mixing with increasing alcohol content in the sense of a reversal from exothermic to endothermic heat [80,82], similar to the 1,4-dioxane/water mixtures [158]. In particular, the 2-methyl-2-propanol-water mixtures in particular have been the subject of research and speculative interpretations in recent decades [170–179]. A mystical character has been attributed to this particular mixture due to the method-dependent results of the mixture [178].

First, the correlations of $E_T(30)$ as function of $N_{av,x}$ and x(water) are discussed; Figure 6a,b.

The plot of $E_T(30)$ as a function of mole fraction x(water) shows relatively similar curves for all mixtures (see Figure S10 in the Supplementary Materials).

If one compares the curves $E_T(30)$ with the curve of the inhomogeneity (DI) of the solvent mixture, both as a function of x(water), see Figures 3b and 6b, then the same result is obtained for 2-propanol/water and 2-methyl-2-propanol/water, methanol/water, ethanol/water and 1,4-dioxane/water mixtures. The strongest curvature of the plot $E_T(30)$ versus x(water) always occurs immediately after the strongest inhomogeneity. This corresponds "immediately after" to a difference of about 1.5 kcal/mol with respect to $E_T(30)$, which is illustrated by the horizontal lines (grey and green dots) between the two curves in Figure 6b. This scenario can be found in all plots of $E_T(30)$ versus $M_{av,w}/M_{av,x}$, regardless of the type of alcohol/water mixture.

The curves $E_T(30)$ as a function of $N_{av,x}$ for methanol/water, ethanol/water, 2-propanol/water and 2-methyl-2-propanol/water mixtures differ qualitatively for both methanol/water and ethanol/water mixtures compared to both 2-propanol/water and 2-methyl-2-propanol/water mixtures in the low water content range. Therefore, the plots of $E_T(30)$ as a function of $N_{av,x}$ show an inflection points at about 0.031 mol/cm^3 and 0.0273 mol/cm^3 for 2-propanol/water and 2-methyl-2-propanol/water mixtures, respectively.

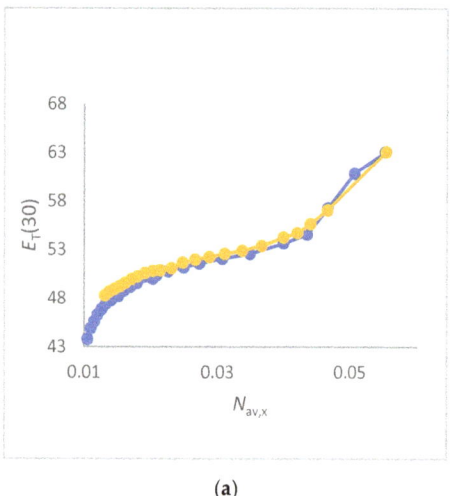 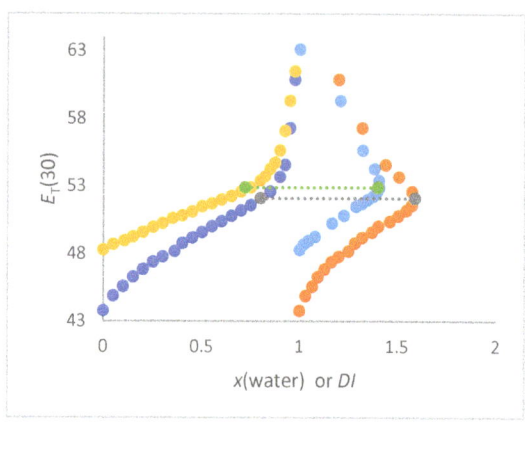

Figure 6. (a) Correlations of $E_T(30)$ (kcal/mol) as a function of $N_{av,x}$ (mol/cm³) for 2-propanol/water (yellow) and 2-methyl-2-propanol/water (blue); (b) plot of $E_T(30)$ (kcal/mol) as a function of x(water) for 2-methyl-2-propanol-water (dark blue and orange) and 2-propanol/water (yellow and light blue) compared with the plot of $E_T(30)$ as a function of the inhomogeneity (DI) of the solvent mixture in terms of $M_{av,w}/M_{av,x}$ (orange and light blue).

At lower water concentrations, the 2-propanol/water mixtures (x(water) < 0.5; $N_{av,x}$ = 0.02 mol/cm³) and 2-methyl-2-propanol mixtures (x(water) < 0.55; N_{avx} = 0.018 mol/cm³) are endothermic in terms of the heat of mixture. At about x = 0.65 to x = 0.5 (water) ($N_{av,x}$ ≈ 0.03, see Figure 6a), both systems behave athermally; i.e., ΔH_{mixing} = 0. Exactly at this composition, the curve $E_T(30)$ as a function of $N_{av,x}$ (see Figure 6a) shows an inflection point. The same result is found for the 1,4-dioxane/water mixtures (see Figure 5b). In the composition range with endothermic heat of mixing, both curves $E_T(30)$ vs. N_{avx} show a higher $E_T(30)$ than would be expected from linearity. In this region, the entropy of mixing is positive and the proportion of water in the mixture determines the $E_T(30)$ proportionally, more than in the water-rich region does. The dielectric relaxation time decreases significantly from low to high water content, i.e., the structure in the water-poor region is more stable in time than in the water-rich region.

The **Fe** complex has also been studied in 2-propanol/water mixtures [129]. Consistent with the correlations of $E_T(30)$ versus $N_{av,x}$, the plot of ν_{max} (**Fe**) as a function of $N_{av,x}$ (see Figure S11) shows a similar pattern to that of Figure 6a. This shows the influence of the physical structure of the solvent as a function of composition.

In addition, there are numerous studies with positive solvatochromic probes such as Nile Red [7], 4-nitroaniline [18,180–182], 4-nitroanisole [18], 4-(1-azetidinyl)-benzonitrile [177] or coumarin 343 and 480 [178] in various alcohol/water mixtures.

While $E_T(30)$ as a function of $N_{av,x}$ for 2-propanol/water and 2-methyl-2-propanol/water mixtures show similar curves, the situation is different for the 2-butoxyethanol/water mixtures.

3.4.6. 2-Butoxyethanol/Water Mixtures

As a final example, the solvatochromism of **B30** in 2-butoxyethanol (BE)/water mixtures is reanalyzed. The $E_T(30)$ data are taken from [8]. BE itself is partially hydrophobic, but it mixes completely with water at room temperature; separation occurs only at higher temperatures [8,183]. Due to their self-structuring properties, BE/water mixtures have been the subject of numerous physical investigations [184–193]. The self-propelled agglomeration of the BE molecules in water has been demonstrated through various scattering methods [184,185]. At $x(BE) > 0.02$ agglomeration begins to occur resulting in an inhomogeneous solvent mixture at the level of about 1 nm [184]. However, other studies have shown that 130 nm aggregates are present [185]. The inhomogeneity of the BE/water mixing system is complicated by the fact that this feature can be observed at different length and time scales [188–191,193].

The mixtures BE/water and 2-methyl-2-propanol/water are often compared for their similarity [193]. We will show that, despite the discussion in the literature, the two solvent mixtures are completely different. The microstructures of both solvent mixtures are very subtle and are strongly influenced by the composition in the water-rich part. However, the heat of mixing is exothermic over almost the whole composition range for BE/water mixtures, but weakly endothermic at high BE concentrations (about 95 wt%) [187,188]. In addition, photo-switchable spiro compounds have been measured in BE/water mixtures [192]. It has been suggested that the solvent structure of BE/water is affected by this type of photo-switching. Therefore, it cannot be excluded that the dissolved probe molecule co-determines the fine structure of BE/water mixtures, complicating the whole situation. Therefore, only the analysis of the $E_T(30)$ values will be discussed here. Unfortunately, the interesting solvatochromic results of El Seoud on this solvent system were not given as original data [22]. Plotting $E_T(30)$ as a function of $N_{av,x}$ for BE/water mixtures gives an asymmetric profile, as shown in Figure 7a (grey dots).

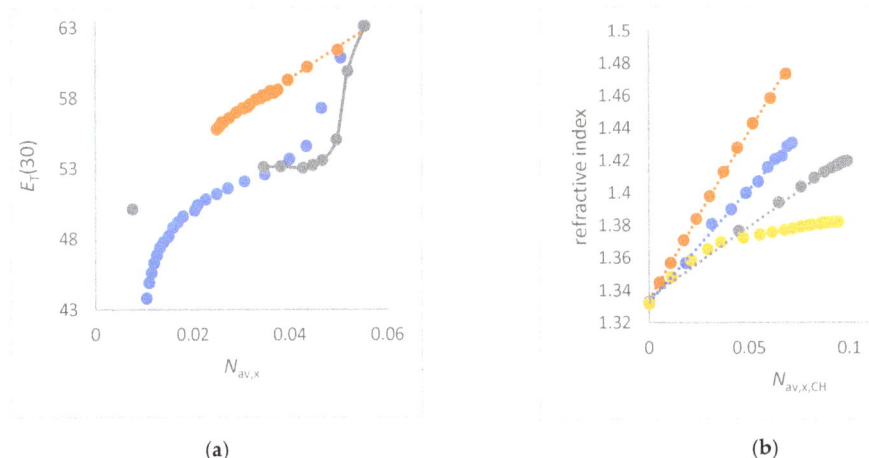

(a) (b)

Figure 7. (a) Relations of $E_T(30)$ (kcal/mol) as a function of $N_{av,x}$ (in mol/cm^3) for BE/water mixtures (grey). The links between the individual grey dots are a guide for the reader. 2-methyl-2-propanol/water mixtures (blue) compared to the FA/water mixtures (orange dots), which have ideal mixing behavior, as a reference; (b) relations of n_D^{20} as a function of $N_{av,x,CH}$ for 2-butoxyethanol/water (grey), 2-methyl-2-propanol/water (yellow), 1,2-ethanediol/water (blue) and glycerol/water mixtures (orange).

The crucial region of low BE content (at $N_{av,x} = 0.0497$ mol/cm^3) deserves special attention, since, at this composition, the agglomeration of BE molecules occurs [$x(BE) \approx 0.02$; or $N_{av,xCH} = 0.017$ mol/cm^3] [184]. B30 dye is apparently absorbed by these BE-rich agglomerates, leading to an abrupt decrease in $E_T(30)$ at $N_{av,x} \approx 0.05$ mol/cm^3, as shown by the grey dotted line in Figure 7a. This is consistent with the fact that $E_T(30)$ decreases abruptly with increasing C-H concentration due to the BE component at $N_{av,CH} = 0.017$ mol/cm^3 (not plotted; for data, see Table S13). There is only a narrow transition.

This result exactly fulfils the preferential solvation scenario i., as explained in the Introduction. The agglomeration of BE is driven by hydrophobic interactions, as also suggested by the analysis of the fluorescence of coumarin and related probe molecules in BE/water mixtures [193,194]. Obviously, the trapping of probes is also determined through the solvent cage property of the BE/water mixtures. This is in contrast to 2-methyl-2-propanol/water mixtures, where strong thermal fluctuations in partial concentrations occur [195,196]. Thus, in 2-methyl-2-propanol/water mixtures there is no true hydrophobic solvation of **B30**, but the partial water structures are changed depending on the composition, as discussed in the previous chapter.

The two binary solvent mixtures glycerol/water and 1.2-ethanediol/water show a perfect mixing behavior according to the Marcus classification, so that there is no enhancement of the water structure in any way [110]. These two mixtures are documented in Figure 7b as references. For glycerol/water and 1,2-ethanediol/water mixtures, there are perfect linear correlations of n_D^{20} as a function of $N_{avx,CH}$; see Equations (42) and (43) [89,197].

$$n_D^{20} = 2.062\ N_{av,x,CH} + 1.335,$$
$$n = 12\ (\text{glycerol/water});\ r = 0.9999. \tag{42}$$

$$n_D^{20} = 1.381\ N_{av,x,CH} + 1.333,$$
$$n = 14\ (1,2\text{-ethanediol/water});\ r = 0.997. \tag{43}$$

The larger slopes of $\Delta n_D^{20}/\Delta N_{avx,CH}$ for glycerol/water and 1,2-ethanediol/water mixtures compared to BE/water mixtures are attributed to the influence of the polarizability of the hydrogen bond network and the higher refraction of the C-O bond compared to C-H [60,62,108].

The change in the overall bulk solvent structure of 2-methyl-2-propanol/water mixtures as a function of composition is also clearly visible in the plot of the refractive index as function of $N_{av,xCH}$ (yellow dotted lines in Figures 2b and 7b), which shows a kink at about 0.035 to 0.04 mol/cm^3. Remarkably, the kink in the plot of $E_T(30)$ as a function of $N_{av,x}$ is also observed at this composition; see Figure 6a. In this concentration range, the thermodynamic changes from exothermic to endothermic with increasing $N_{av,xCH}$ for 2-methy-2-propanol/water mixtures. This thermodynamic scenario does not apply to BE/water mixtures [187,188].

Remarkably, the correlation of n_D^{20} as function of $N_{av,xCH}$ for BE/water mixtures is approximately linear over the whole composition range; see Equation (44) and Figure 7b (grey dotted line).

$$n_D^{20} = 0.876\ N_{av,x,CH} + 1.353,$$
$$n = 11\ (\text{BE/water});\ r = 0.998. \tag{44}$$

Obviously, BE as a co-solvent does not enhance the water structure, but water enhances the BE structure. That is the special thing. According to the Marcus classification, this is the reverse scenario of solvent structure enhancement. These considerations convincingly show the qualitative differences between BE/water and 2-methyl-2-propanol/water mixtures. The different solvation behavior of **B30** in BE/water mixtures compared to 2-methyl-2-propanol/water mixtures can also be supported by considering the *DI* from Equation (9). The BE/water mixtures show the greatest inhomogeneity with respect to *DI* at $x(\text{water}) = 0.85$ ($N_{av,x} = 0.029$ mol/cm^3), while the kink in the curve $E_T(30)$ as a function of $N_{av,x}$ occurs far away from this at ≈ 0.045 mol/cm^3. This is the crucial difference between BE/water mixtures and all other (monohydric) alcohol/water mixtures studied in this work. Therefore, this result could be used as a criterion to define the preferential solvation scenario of case i. However, the results do not exclude that the **B30** dye itself has an influence on the solvent cage of the BE/water mixture at low BE content, as mentioned in [192,193] for other solutes.

4. Discussion

The plot of $E_T(30)$ as a function of $N_{av,x}$ for co-solvent/water mixtures shows a different pattern depending on the co-solvent of the mixture. The scenario of each specific co-solvent/water mixture can be clearly assigned according to the Marcus classification. Four different scenarios can be identified:

A. The $E_T(30)$ increases significantly and linearly with $N_{av,x}$ (1,2-ethanediol/water, FA/water, urea/water, NMF/ water and DMSO/water mixtures) (see Figures 3a and 4a,b). These co-solvents belong to the group of solvents that do not enhance the water structure at all and form strong hydrogen bonds with water. In these cases, the $E_T(30)$ of the pure co-solvent is fitted to the linear plot.

B. The $E_T(30)$ increases asymptotically with increasing $N_{av,x}$ where the $E_T(30)$ value is always higher than with a linear dependence (1,4-dioxane/water, DMF/water and NFM/water mixtures) (see Figures 4a and 5a). In these cases, the co-solvent-rich fraction shows the non-linearity $E_T(30)$ as function of $N_{av,x}$. These co-solvents do not enhance the water structure but form weaker hydrogen bonds with water than those belonging to scenario (A).

C. The $E_T(30)$ increases as $N_{av,x}$ increases, with the $E_T(30)$ value always being lower than expected for a linear dependence (see Figure 3a). This scenario applies to methanol/water and ethanol/water mixtures. These co-solvents enhance the water structure.

D. $E_T(30)$ shows an S-shaped curve as a function of $N_{av,x}$ (see Figure 6a). With increasing $N_{av,x}$ the $E_T(30)$ value is always higher than expected with a linear dependence in the co-solvent-rich part. In the water-rich part, the $E_T(30)$ is lower than with a linear dependence according to scenario (C). The mixtures 2-propanol/water, 2-methyl-2-propanol/water and 2-butoxyethanol/water belong to this group. This scenario applies to binary solvent mixtures that interact either on the structure of the water or on the structure of the co-solvent.

In particular, these binary co-solvent/water mixtures of scenario (A), which include glycerol/water mixtures, have been shown to be robust reference liquids for contact angle measurements, as no segregation occurs when in contact with different types of surfaces [198]. This is an important result in support of the Marcus theory for the classification of aqueous mixtures.

When rapid solvent dynamics occur in a particular solvent system, thermal motion overcomes local structuring effects. An almost perfect linear relationship of $E_T(30)$ as a function of $N_{av,x}$ is then observed. This scenario is shown to hold for FA/water, DMSO/water, 1,2-ethanediol/water, urea/water and NMF/water mixtures. This interpretation is strongly supported by the results of dielectric spectroscopy and ultrafast IR experiments.

The heat of mixing of the 1,2-ethanediol/water, DMSO/water and NMF/water mixtures is exothermic over the whole composition range, whereas the heat of interaction of the FA/water mixtures is always weakly endothermic. The best fits of $E_T(30)$ as a function of $N_{av,x}$ are for 1,2-ethanediol/water, NMF/water mixtures (exothermic over the whole composition range) and FA/water mixtures (weak endothermic over the whole composition range). For all co-solvent/water mixtures with respect to scenario (A), the qualitative heat of interaction does not change as a function of composition.

With regard to scenario (B), these co-solvents also belong to the Marcus classification, which do not enhance the water structure. There is a linear dependence of $E_T(30)$ as a function of $N_{av,x}$ up to $N_{av,x} > 0.02$ mol/cm^3. However, the water fraction obviously has a greater effect on $E_T(30)$ than the average number of dipoles over the whole composition range would suggest. A linear correlation of $E_T(30)$ as a function of $N_{av,x}$ is always found in the range of higher water contents. Thus, the average molar concentration of the dipoles (water and co-solvent) acting on the probe is the dominant factor in the higher water content range. Only from $N_{av,x} > 0.0135$ is there a bend in the curve, indicating that water as a co-component loses its influence on $E_T(30)$. These co-solvents form weaker hydrogen bonds with water compared to scenario (A), indicating that the water structure changes non-linearly with composition [199–201]. These co-solvents can be classified differently depending on the criteria used by Marcus for evaluation.

For the 1,4-dioxane/water mixtures, the $E_T(30)$ is always larger than expected from the sum of water and 1,4-dioxane dipoles. The curves for 1,4-dioxane/water and DMF/water mixtures are congruent up to $N_{av,x}$ = 0.0135 mol/cm^3. However, while the heat of mixing of DMF with water is exothermic over the whole composition range, the situation is different for 1,4-dioxane/water mixtures, as discussed above. Therefore, the thermodynamic changes in the DMF/water and dioxane/water mixtures are not always captured by the correlation of $E_T(30)$ with $N_{av,x}$.

The correlation of $E_T(30)$ with x(water) gives a better indication of the thermodynamic changes at different compositions of the 1,4-dioxane/water mixture.

This detailed result is very significant because it shows the linkage of x(water) with thermodynamics but not directly with the UV/Vis shift. In summary, scenario (B) requires more detailed studies using related binary solvent mixtures. Further studies will consider the complexity of THF/water, acetone/water and acetonitrile/water mixtures and other binary solvent mixtures such as pyridine/water or piperidine/water mixtures [2] with respect to solvatochromism. These evaluations are necessary to clarify or complement some of the conclusions regarding scenario (B) of this study. As shown by Marcus [200,201], each specific mixture actually requires special treatment in order to understand the many physical effects. Therefore, this review provides only a rough overview of the overall problem using selected examples.

In scenarios (C) and (D), the co-solvents belong to the Marcus classification, which enhance the water structure. But why is the $E_T(30)$ value for methanol/water, ethanol/water, 2-propanol/water and 2-methyl-2-propanol/water mixtures lower at high water concentrations than would be expected from a linear dependence such as that observed for 1,2-ethanediol/water, as seen in Figure 3a? The answer is quite pragmatic: the hydrophobic dye **B30** is difficult to dissolve in pure water [1]. It therefore dissolves much better in the alcohol/water domain, where the partial alcohol concentration is greater than the total alcohol concentration in the initial mixture. Accordingly, a lower $E_T(30)$ is logically measured than would be expected from the total average molar concentration ($N_{av,x}$), since the average water concentration in the partial alcohol/water fraction must be lower outside the areas of enhanced water structure. The mole fraction x(water), as a measure of solvent composition in solvatochromism analysis, falsifies preferential solvation, since x is inversely proportional to $N_{av,x}$. The previously observed curved functions of $E_T(30)$ as a function of x(water) determined so far are due to the inhomogeneity (*DI*) of the solvent mixture.

This result is a significant contribution to the identification of enhanced water structures in alcohol/water mixtures. The enhanced microdomain water structure ranges of alcohol/water mixtures are apparently not recognized by **B30** for two reasons: firstly, **B30** dissolves poorly in pure water; secondly, the molar absorption coefficient of **B30** in water is rather low [1,70].

The first argument is supported by the fact that different solvatochromic dyes such as 4-nitroaniline, 4-nitrophenol, 4-nitroanisole or **B30** show qualitatively different dependencies of the UV/Vis absorption energy as a function of alcohol/water composition, especially in the water-rich range x(water) > 0.8 [18,19]. This observation holds regardless of the methanol/water, ethanol/water, 2-propanol/water or 2-methyl-2-propanol/water mixtures is considered. Obviously, hydrophilic dyes are distributed in all these fractions and hydrophobic dyes are preferentially dissolved in the alcohol/water fraction. As **B30** itself is a hydrophobic molecule, this scenario is likely.

This explanation is also consistent with the results of the BE/water mixtures. The BE/water domain captures the hydrophobic **B30** particularly well, as there is an abrupt decrease in $E_T(30)$ with increasing BE content occurs when agglomeration of n-butoxyethanol takes place.

However, it appears that **B30** and **B30**-COONa measure the water-rich fraction in the same sense [19]. Note that the full width at half maximum of the UV/Vis absorption band of **B30** measured in alcohol/water mixtures is quite broad. These very broad UV/Vis spectra with large half-widths are often measured in water/salt mixtures [202]. Unfortunately, the UV/Vis spectra are not given in [19]. Therefore, it is not possible to say whether there are superpositions of several UV/Vis bands originating from different solvation states. A definitive statement is therefore not yet possible.

The 2-propanol/water and 2-methyl-2-propanol/water mixtures belong to scenario (D). In the co-solvent rich range the **B30** is preferentially influenced by the partial water concentration of the mixture due to the larger $E_T(30)$ is measured as expected from the linearity. The situation is quite delicate because the thermodynamics of the mixture changes from exothermic to endothermic depending on the composition. Then, at $\Delta_r H_{mixing} = 0$, the plot $E_T(30)$ as function of $N_{av,x}$ shows an inflection point, as is clearly seen for 2-propanol/water, 2-methyl-2-propanol/water and 1,4-dioxane/water mixtures. However, this is only a preliminary result that needs to be confirmed by further studies.

The conclusion from these considerations is that the thermodynamics of the interaction between the solvatochromic probe and the solvent mixture is crucial. Unfortunately, the dissolution thermodynamics of **B30** in different solvents has not yet been systematically studied. There are only two papers with calorimetric results on the solvation thermodynamics of **B30** [203,204]. The ambiguous results of the two references are not consistent with the theory of exothermic solvation of the probe leading to a lowering of the ground state energy [34,35]. The dissolution process of **B30** in acetonitrile, ethyl acetate and higher alcohols is found to be endothermic, which is difficult to explain and may be due to entropic effects rather than re-association of **B30** as discussed [204]. Thus, the thermodynamics of the **B30**/HBD solvent interaction is not trivially explainable in terms of specific hydrogen bond formation. This is consistent with the results of the present study that hydrogen bond formation has no significant effect on the $E_T(30)$ value. As consequence, the influence of the thermodynamics of solvation of **B30** in terms of the real ground state energy is difficult to assess because the calorimetric studies on several **B30**/solvent systems are difficult to interpret.

There are various approaches to correlating polarity data with calorimetric results, which are partly successful, but also give very strange results [180,205]. Therefore, this approach has often not been pursued further.

The complicated situation regarding solvation thermodynamics is similar for positive solvatochromic dyes such as 4-nitroaniline [206]. The dissolution process of 4-nitroaniline in co-solvent/water mixtures is endothermic in terms of the heat of mixing in the high-water content range [180,206]. Therefore, a reinterpretation of the solvatochromic results of 4-nitroaniline and related probes in co-solvent/water mixtures from the literature [18,19,181,182] is imperative. The effect of endothermic solvation and its impact on the UV/Vis absorption energy requires more detailed analysis in future work. The idea that solvation of an electronic state leads to an energetic decrease should be abandoned.

There is another aspect to consider. Dissolved **B30** probes are statistically influenced by the dynamically moving solvent molecules. For this reason, the UV/visible spectrum only measures a snapshot of different solvation states. A superposition of many solvation states is recorded. Thus, the discrimination of domain formation in solvent mixtures by solvatochromic probes is only possible if the dynamics of the solvent is much lower than that of the optical excitation process of **B30**. This argument applies precisely to those mixtures where the co-solvent enhances the water structure. This can be explained by considering the relaxation time τ_1 measured by dielectric spectroscopy. The larger the τ_1 values, the more structural subtleties of the mixture are detected using the solvatochromic probe at ambient temperature, as shown for 2-propanol/water and 2-methyl-2-propanol/water mixtures [86,88,89]. Therefore, the results of dielectric spectroscopy in terms of relaxation time are a useful adjunct to explain the results of UV/Vis measurements.

The two groups of co-solvents, which either enhance or do not enhance the water structure, can probably be distinguished on the basis of the refractive index. When n_D^{20} is correlated as function of $N_{av,x,CH}$ as shown in Figure 2, different curves are obtained. If there is no linear correlation of n_D^{20} with $N_{av,x}$ over the entire composition range, then microdomains of water have formed in the mixture in the form of enhanced water structures. This hypothesis should be tested in further studies. However, this proposed rule does not apply to the correlation of the dielectric constant as a function of $N_{av,x}$ for DMSO/water mixtures, where a curved line is found instead of a straight line [207–209]. For this, ε_r correlates linearly with x(water), but not linearly with n_D^{20} [208]. This particular behavior of DMSO/water mixtures is due to the recently recognized unusual microheterogeneity of these particular mixtures [209]. This important detail supports the thesis that the local mass inhomogeneity in terms of DI has an influence on the result of the physical measurement. Therefore, nothing is riskier than relying on routine evaluations of measurement results as a function of the composition of co-solvent/water mixtures.

Due to the complex structure of aqueous solvent mixtures, a growing body of knowledge is emerging based on conventional and modern measurement methods that take into account the density, refractive index, heat of interaction and other properties (dielectric relaxation) of the individual solvent mixture systems. The combination of UV/Vis spectroscopic data such as $E_T(30)$ with physical properties of the solvent mixture (molar concentration of dipoles, dielectric dynamics, thermodynamics, domain size formation) has proved to be necessary but very complex.

New work on the structure of ethanol/water or 2-methyl-2-propanol/water mixtures continues to emerge, providing a refined picture of these unusual solvent systems [111,210,211]. However, the results of recent studies clearly confirm the presence of coexisting microdomains of water and alcohol/water. The considerations of this study also apply to organic solvent mixtures. As early as 1981, Langhals showed that the same relationships that apply to aqueous mixtures can also be used for binary organic solvent mixtures [212]. This motivates us to re-evaluate and classify binary organic solvent mixtures as well, in terms of the average molar concentration. This evaluation requires an extensive literature search for density data. However, the formation of hydrogen bonds on the probe can have a stronger effect in special systems [213], but the use of $N_{av,x}$ instead of x is necessary for a correct evaluation of UV/Vis spectroscopic data [67,68]

Despite the correlations found between UV/Vis data and physical solvent properties of mixtures, it must be made clear that solvatochromism is of limited use for analyzing the chemical properties of solvent mixtures, unless one simply wants to measure the composition quickly using calibration curves [6]. To understand the complex dependence of the absorption energy of a solvatochromic probe on the solvent composition, the structure of the solvent mixture must be analyzed. However, solvatochromism cannot analyze structures, as Marcus correctly concluded for acetonitrile/water mixtures [214].

5. Conclusions

The Marcus classification of aqueous solvent mixtures has proved to be very useful and explains the qualitatively different correlations of $E_T(30)$ with $N_{av,x}$ for different co-solvent/water mixtures. This should easily solve many puzzles in the literature, as all previous evaluations of solvatochromic data using the molar fraction x(water or co-solvent) as the composition variable are physically incorrect. Various linear and curvilinear relationships of $E_T(30)$ as a function of solvent composition in terms of $N_{av,x}$ have been analyzed. With increasing both the OH dipole and co-solvent dipole concentration, $E_T(30)$ increases linearly for those mixtures where the co-solvent does not enhance the water structure. This characteristic holds for FA/water, 1,2-ethandiol/water, NMF/water, urea/water and DMSO/water.

Co-solvents which enhance the water structure of aqueous mixtures (Marcus) show an S-shaped or curved curve of $E_T(30)$ as function of $N_{av,x}$. Whether a curved or an S-shaped function is obtained depends on the thermodynamics of the mixing process and the solvent dynamics in terms of the relaxation time. Even if preferential solvation of a solute occurs, this may not always be observed with solvatochromic probes if the dynamics in the binary solvent mixture are too high. An average number of solvation states is then recorded. The complexity of the structure of alcohol/water mixtures is reflected in the correlation of the refractive index or $E_T(30)$ as a function of various compositional quantities such as x(water), $N_{av,x}$, $N_{av,x,CH}$ or $N_{av,v}$.

The $N_{av,x}$ of the binary mixture is physically justified as a suitable measure of composition for the analysis of UV/Vis results of solvatochromic probes, because of its linear relationship with the UV/Vis absorption energy according to the Lorentz–Lorenz relation. The refined interpretations of $E_T(30)$ as a function of solvent composition in this work were only possible on the basis of many new and modern insights into the physical structure of solvent mixtures and the true significance of optical measurements.

In general, the significance of the various measures of the average molar mass of solvent mixtures requires further research into their relationship with physical methods of investigation and their informative value. The use of the inhomogeneity of the solvent mixtures in terms of the $M_{av,w}/M_{av,x}$ quantity should be considered for other solvent mixtures in order to support the conclusions of this study in future work.

Finally, it is incomprehensible why hardly anyone has used the average molar concentration $N_{av,z}$ as a measure of the composition of solvent mixtures. The molar fraction x is needed for thermodynamics, but not for spectroscopy. However, both x and $N_{av,z}$ are crucial in understanding the physics of mixing. Irrespective of the theoretically justified relationships of the Debye, Clausius–Mosotti or Lorentz–Lorenz equations, the use of the molar concentration is actually necessary for the evaluation of UV/vis spectroscopic data and the refractive index.

Supplementary Materials: The following supporting information can be downloaded at: https://www.mdpi.com/article/10.3390/liquids4010010/s1, References [215–222] belong to this chapter. Figure S1a. (left panel) Correlation of E_T(nile red) (kcal/mol) as a function of D_{HBD} (mol/cm^3) for protic solvents including water, methanol, ethanol, 1-propanol, 2-propanol, 1-butanol, 2-methyl-2-propanol, 2-ethanolamine, 1,2-ethandiol, 2,2,2-trifluoroethanol and 1,1,1,3,3,3-hexafluoro-2-propanol. The UV/vis-spectroscopic data are taken from ref. [72] and the D_{HBD} parameter from [62]. E_T(nile red) = $-118.8 \, D_{HBD} + 54.3$, n = 12, r = 0.924. The correlation supports the D_{HBD} parameter proposed for 2,2,2-trifluoroethanol and 1,1,1,3,3,3-hexafluoro-2-propanol from ref. [62]. Figure S1b. (right panel) Plots of $N_{av,x}$ (sum of total OH dipoles) (in mol/cm^3) as a function of x(water) for methanol/water (orange dots), ethanol/water (grey dots) 2-propanol/water (yellow dots), 2-methyl-2-propanol/water (light blue dots); 2-n-butoxyethanol/water (dark blue) mixtures, physical data from references [191,198–202]. Figure S2. Plots of $E_T(30)$ (kcal/mol) as function of n_D^{20} for 1,2-ethanediol/water (grey), DMSO/water (blue) and 1,4-dioxane/water (orange) mixtures. Figure S3. Correlation of E_T30 (in kcal/mol) as a function of N (in mol/cm^3) in the temperature range from -75 to $+75$ °C, Reichardt [1] (blue dots) and Linert [116] (orange dots). Densities see ref. [218]. Figure S4a (left panel). Comparison of $E_T(30)$ (kcal/mol) as function of $N_{av,x}$ (mol/cm^3) for ethanol/water mixtures. Data from Dimroth–Reichardt [2] (red dots) and data from Rosés [18] (blue dots), (25 °C). Figure S4b (right panel). Comparison of $E_T(30)$ (kcal/mol) as a function of x(DMSO) for DMSO/water mixtures data from [12,14,121,122] (orange dots) (25 °C) and data from Connors [15] (blue dots). Figure S5. Comparison of $E_T(30)$ (kcal/mol) as function of x(1,2-ethanediol) (mol %) for 1,2-ethanediol/water mixture, data from Kosower/Marcus [12] (orange dots) (25 °C); data from Connors [15] (blue dots). Figure S6a. $E_T(30)$ (kcal/mol) as a function of $N_{av,x}$ (mol/cm^3) for urea/water (blue) and N,N'-dimethylpropyleneurea/water (orange) mixtures; data from [143,144]. $E_T(30)$ (urea/water) = $80.3 + 58.7$; n = 8, r = 0.996; $E_T(30)$ N,N'-dimethylpropyleneurea/water) = $432 + 39.7$; n = 7, r = 0.994. Figure S6b (right panel). Overall correlation of $E_T(30)$ (kcal/mol) as a function of N for the homologous series of primary alcohols (orange dots) as well as $E_T(30)$ as function of $N_{av,v}$ (mol/cm^3) for ethanol/water mixtures (blue dots) and methanol/water mixtures (grey dots). All $E_T(30)$ data are taken from [2,18,34]. Figure S7a. Correlation of the static dielectric constant of DMSO /water mixtures [151] as a function of $N_{av,x}$ (mol/cm^3). Figure S7b. Correlation of the static dielectric constant ε_r as function of $N_{av,x}$ (mol/cm^3) for several co-solvent/water mixtures, including the pure co-solvents, 1,4-dioxane/water mixtures [162]: $\varepsilon_r = 1827 \, N_{av,x} - 25$; n = 11; r = 0.996 (grey dots). The 1,2-ethanediol/water mixtures, $\varepsilon_r = 975 \, N_{av,x} + 25$; n = 11; r = 0.994 (blue dots) and the glycerol/water mixtures, $\varepsilon_r = 908 \, N_{av,x} - 29$; n = 11; r = 0.999 (orange dots) are used as independent reference. Figure S8a (left panel) Correlation of $E_T(30)$ (kcal/mol) as a function of the static dielectric constant ε_r for DMSO/water (orange) [151] and 1,4-dioxane/water mixtures [162] (blue). Figure S8b. (right panel) Correlation of $E_T(30)$ (kcal/mol) as a function of $N_{av,x}$ (blue) and $N_{av,x}$(water) (orange) for 1,4.dioxane/water mixtures. Figure S9. Plots of Kamlet–

Taft (KAT) α HBD parameter (orange and grey dots), β (HBA) parameter) (yellow dots) and π^* dipolarity/polarizability parameter (blue dots) as a function of $N_{av,x}$ (mol/cm^3) for DMSO/water mixtures [157]. Note the maximum KAT value for π^* corresponds exactly to the kink in the curve for KAT α as a function of $N_{av,x}$ and the inflection point of β versus $N_{av,x}$ at the same composition. Figure S10. Correlations of $E_T(30)$ as a function of x(water) for methanol/water (grey), ethanol/water (orange), 2-propanol/water (blue), and 2-methyl-2-propanol/water (yellow) mixtures. Data from [2,18,19]. Figure S11. Plot of ν_{max}(Fe) [10^3 cm^{-1}] as a function of $N_{av,x}$ (mol/cm^3) for the 2-propanol/water mixtures [126]. Linear fit: ν_{max}(Fe) [10^3 cm^{-1}] = 41.95 $N_{av,x}$ + 17.05; r = 0.99, n = 11. Table S1. $E_T(30)$ values for ethanol measured at various temperatures; data from Reichardt and Linert [1,116] and densities at various temperatures of the ethanol/water mixture [218]. Table S2. Physical properties of the 1,2-ethanediol/water mixture in terms of mole fractions as well as refractive index and $E_T(30)$ values [12,77,130]. $D_{av,x,HBD}$ values are the total concentration of OH dipoles when the partial OH concentration of the 1,2-ethanediol component is taken into account according to Equation (9). Table S3. Physical data of methanol/water mixtures with respect to mole, volume and mass fraction, and $E_T(30)$ values [17]. Physical data from [215–217]. Table S4. $E_T(30)$ and $E_T(1)$ values, density, average molar masses and average molar concentrations in ethanol/water mixtures. Data from Reichardt [1,2]. Physical solvent mixture are data from [215–217]. Table S5. Physical properties and $E_T(30)$ values of the formamide/water mixtures [21,120,137,138]. Table S6. The N-methylformamide (NMF) /water mixtures. X(water), M_{avx}, $N_{av,x}$. Data from [139,219]. $E_T(30)$data from [21]. Table S7. The N,N-dimethylformamide (DMF)/water mixtures, physical data and $E_T(30)$ values [21,141,142,220]. Table S8. The N-formylmorpholine/water mixtures, $N_{av,x}$, refractive index and $E_T(30)$ values [32,221,222]. Table S9. Physical properties and $E_T(30)$ values of DMSO/water mixtures from different literature sources [12,15], Density from [146,148]. Table S10. Physical properties of the 1,4-dioxane/water mixtures in terms of mole and volume fractions as well as refractive index and $E_T(30)$ values, data from [2,159–162,165]. Table S11. Physical properties of 2-propanol/water mixtures [215–217] and $E_T(30)$ values [2,18,19]. Table S12a. $E_T(30)$, density, various average molar masses and corresponding molar concentrations of 2-methyl-2-propanol/water mixtures, data from [173,174]. $E_T(30)$ values, data from [19]. Table S12b. Refractive index, density, average molar masses and average molar concentrations of 2-methyl-2-propanol/water mixtures, data from [173,174]. $E_T(30)$ values, data from [19]. Table S13. Physical data and $E_T(30)$ values, data from [8], for the 2-Butoxyethanol (BE)/water mixtures at 25 °C, physical data from [191].

Funding: Chemnitz University of Technology provided organizational support for the work.

Acknowledgments: The author would like to thank R. Buchner, University of Regensburg, and T. G. Mayerhöfer, Friedrich Schiller University and Leibniz Institute for Photonic Technologies Jena, for helpful discussions and suggestions on dielectric spectroscopy and refractive index.

Conflicts of Interest: The author declares no conflict of interest.

References

1. Dimroth, K.; Reichardt, C.; Siepmann, T.; Bohlmann, F. On pyridinium-N-phenol-betaines and their use in characterising the polarity of solvents (original in german). *Liebigs Ann. Chem.* **1963**, *661*, 1–37. [CrossRef]
2. Dimroth, K.; Reichardt, C. Die colorimetrische Analyse binärer organischer Lösungsmittelgemische mit Hilfe der Solvatochromie von Pyridinium-N-phenolbetainen, the colourimetric analysis of binary organic solvent mixtures using the solvatochromism of pyridinium-N-phenolbetaines. *Fresenius Z. Anal. Chem.* **1966**, *215*, 344–350. [CrossRef]
3. Maksimovic, Z.B.; Reichardt, C.; Spiric, A. Determination of empirical parameters of solvent polarity E_T in binary mixtures by solvatochromic pyridinium-N-phenol betaine dyes. *Z. Anal. Chem.* **1974**, *270*, 100–104. [CrossRef]
4. Krygowski, T.M.; Wrona, P.K.; Zielkowska, U.; Reichardt, C. Empirical parameters of Lewis acidity and basicity for aqueous binary solvent mixtures. *Tetrahedron* **1985**, *41*, 4519–4527. [CrossRef]
5. Langhals, H. Die quantitative Beschreibung der Lösungsmittelpolarität binärer Gemische unter Berücksichtigung verschiedener Polaritätsskalen. *Chem. Ber.* **1981**, *114*, 2907–2913. [CrossRef]
6. Langhals, H. Polarity of binary liquid mixtures. *Angew. Chem. Int. Ed. Engl.* **1982**, *21*, 724–733. [CrossRef]
7. Langhals, H. The description of the polarity of alcohols as a function of their molar content of OH groups. *Nouv. J. Chim.* **1982**, *6*, 265–267. Available online: https://epub.ub.uni-muenchen.de/93533/ (accessed on 2 November 2022).
8. Haak, J.R.; Engberts, J.B.F.N. Solvent polarity and solvation effects in highly aqueous mixed solvents. Application of the Dimroth-Reichardt $E_T(30)$ parameter. *Recl. Trav. Chim. Pays-Bas* **1986**, *105*, 307–311. [CrossRef]
9. Michels, J.J.; Dorsey, J.G. Retention in reversed-phase liquid chromatography: Solvatochromic investigation of homologous alcohol-water binary mobile phases. *J. Chromatogr.* **1988**, *21*, 85–98. [CrossRef]

10. Chatterjee, P.; Bagchi, S. Preferential solvation of a dipolar solute in mixed binary solvent. A study by UV–visible spectroscopy. *J. Phys. Chem.* **1991**, *95*, 3311–3314. [CrossRef]
11. Bosch, E.E.; Rosés, M. Relationships between E_T Polarity and Composition in Binary Solvent Mixtures. *J. Chem. Soc. Faraday Trans.* **1992**, *88*, 3541–3546. [CrossRef]
12. Marcus, Y. The use of chemical probes for the characterization of solvent mixtures. Part 2. Aqueous mixtures. *J. Chem. Soc. Perkin Trans.* **1994**, *2*, 1751–1758. [CrossRef]
13. Kipkemboi, P.K.; Easteal, A.J. Solvent Polarity Studies of the Water+t-Butyl Alcohol and Water+t-Butylamine Binary Systems with the Solvatochromic Dyes Nile Red and Pyridinium-N-phenoxide Betaine, Refractometry and Permittivity Measurements. *Austr. J. Chem.* **1994**, *47*, 1771–1781. [CrossRef]
14. Banerjee, D.; Laha, A.K.; Bagchi, S. Preferential solvation in mixed binary solvent. *J. Chem. Soc. Faraday Trans.* **1995**, *91*, 631–636. [CrossRef]
15. Skwierczynski, R.D.; Connors, K.A. Solvent effects on chemical processes. Part 7. Quantitative description of the composition dependence of the solvent polarity measure $E_T(30)$ in binary aqueous–organic solvent mixtures. *J. Chem. Soc. Perkin Trans.* **1994**, *2*, 467–472. [CrossRef]
16. Marcus, Y. Use of chemical probes for the characterization of solvent mixtures. Part 1. Completely non-aqueous mixtures. *J. Chem. Soc. Perkin Trans.* 2 **1994**, *5*, 1015–1021. [CrossRef]
17. Rosés, M.; Ràfols, C.; José Ortega, J.; Bosch, E. Solute–solvent and solvent–solvent interactions in binary solvent mixtures. Part 1. A comparison of several preferential solvation models for describing $E_T(30)$ polarity of bipolar hydrogen bond acceptor-cosolvent mixtures. *J. Chem. Soc. Perkin Trans.* **1995**, *2*, 1607–1615. [CrossRef]
18. Bosch, E.E.; Rosés, M.; Herodes, K.; Koppel, I.; Leito, I. Solute-solvent and solvent-solvent interactions in binary solvent mixtures. Part 3. The $E_T(30)$ polarity of binary mixtures of hydroxylic solvents. *J. Chem. Soc. Perkin Trans.* **1996**, *2*, 1497–1503. [CrossRef]
19. Rosés, M.; Buhvestov, U.; Ràfols, C.; Rived, F.; Bosch, E. Solute–solvent and solvent–solvent interactions in binary solvent mixtures. Part 6. A quantitative measurement of the enhancement of the water structure in 2-methylpropan-2-ol–water and propan-2-ol–water mixtures by solvatochromic indicators. *J. Chem. Soc. Perkin Trans.* **1997**, *2*, 1341–1348. [CrossRef]
20. Buhvestov, U.; Rived, F.; Ràfols, C.; Bosch, E.; Rosés, M. Solute–solvent and solvent–solvent interactions in binary solvent mixtures. Part 7. Comparison of the enhancement of the water structure in alcohol–water mixtures measured by solvatochromic indicators. *J. Phys. Org. Chem.* **1998**, *11*, 185–196. [CrossRef]
21. Herodes, K.; Leito, I.; Koppel, I.; Rosés, M. $E_T(30)$ Polarity of Binary Mixtures of Formamides with Hydroxylic Solvent. *J. Phys. Org. Chem.* **1999**, *12*, 109–115. [CrossRef]
22. Tada, E.B.; Novaki, L.P.; El Seoud, O.A. Solvatochromism in pure and binary solvent mixtures: Effects of the molecular structure of the zwitterionic probe. *J. Phys. Org. Chem.* **2000**, *13*, 679–687. [CrossRef]
23. El Seoud, O.A. Solvation in pure and mixed solvents: Some recent developments. *Pure Appl. Chem.* **2007**, *79*, 1135–1151. [CrossRef]
24. Ghoneim, N. Study of the preferential solvation of some betaine dyes in binary solvent mixtures. *Spectrochim. Acta A Mol. Biomol. Spectrosc.* **2001**, *57*, 1877–1884. [CrossRef] [PubMed]
25. Wu, Y.G.; Tabata, M.; Takamuku, T. Preferential Solvation in Aqueous–Organic Mixed Solvents Using Solvatochromic Indicators. *J. Sol. Chem.* **2002**, *31*, 381–395. [CrossRef]
26. Bevilaqua, T.; da Silva, D.C.; Machado, V.G. Preferential solvation of Brooker's merocyanine in binary solvent mixtures composed of formamides and hydroxylic solvents. *Spectrochim. Acta Part A* **2004**, *60*, 951–958. [CrossRef] [PubMed]
27. Bentley, T.W.; Koo, I.S. Role of hydroxyl concentrations in solvatochromic measures of solvent polarity of alcohols and alcohol–water mixtures—Evidence that preferential solvation effects may be overestimated. *Org. Biomol. Chem.* **2004**, *21*, 2376–2380. [CrossRef]
28. El Seoud, O.A. Understanding solvation. *Pure Appl. Chem.* **2009**, *8*, 697–707. [CrossRef]
29. Morisue, M.; Ueno, I. Preferential Solvation Unveiled by Anomalous Conformational Equilibration of Porphyrin Dimers: Nucleation Growth of Solvent–Solvent Segregation. *J. Phys. Chem. B* **2018**, *122*, 5251–5259. [CrossRef]
30. Duereh, A.; Anantpinijwatna, A.; Latcharote, P. Prediction of Solvatochromic Polarity Parameters for Aqueous Mixed-Solvent Systems. *Appl. Sci.* **2020**, *10*, 8480. [CrossRef]
31. Pasham, F.; Jabbari, M.; Farajtabar, A. Solvatochromic Measurement of KAT Parameters and Modeling Preferential Solvation in Green Potential Binary Mixtures of N-Formylmorpholine with Water, Alcohols, and Ethyl Acetate. *J. Chem. Eng. Data* **2020**, *65*, 5458–5466. [CrossRef]
32. Sandri, J.C.; de Melo, C.E.A.; Giusti, L.A.; Rezende, M.C.; Machado, V.G. Preferential solvation index as a tool in the analysis of the behavior of solvatochromic probes in binary solvent mixtures. *J. Mol. Liq.* **2021**, *328*, 1155450. [CrossRef]
33. Pavel, C.M.; Ambrosi, E.; Dimitriu, D.G.; Dorohoi, D.O. Complex formation and microheterogeneity in water–alcohol binary mixtures investigated by solvatochromic study. *Eur. Phys. J. Spec. Top.* **2023**, *232*, 415–425. [CrossRef]
34. Reichardt, C.; Welton, T. *Solvents and Solvent Effects in Organic Chemistry*, 4th ed.; Wiley-VCH: Weinheim, Germany, 2011; ISBN 978-3-527-32473-6.
35. Reichardt, C. Solvatochromic Dyes as Solvent Polarity Indicators. *Chem. Rev.* **1994**, *94*, 2319–2358. [CrossRef]
36. Kamlet, M.-J.; Abboud, J.L.M.; Abraham, M.H.; Taft, R.W. Linear Solvation Energy Relationships. 23. A Comprehensive Collection of the Solvatochromic Parameters, π*, α, and ß, and Some Methods for Simplifying the Generalized Solvatochromic Equation. *J. Org. Chem.* **1983**, *48*, 2877–2887. [CrossRef]

37. Catalán, J.; Diaz, C. A generalized solvent acidity scale: The solvatochromism of o-tert-Butylstilbazolium Betaine dye and ist homomorph o,o'-di-tert-butylstilbazolium betaine dye. *Liebigs Ann.* **1997**, *1997*, 1941–1949. [CrossRef]
38. Catalán, J.; Díaz, C. Extending the Solvent Acidiy Scale to Highly Acidic Organic Solvents: The Unique Photophysical Behaviour of 3,6-Diethyltetrazine. *Eur. J. Org. Chem.* **1999**, *4*, 885–891. [CrossRef]
39. Catalán, J. Toward a Generalized Treatment of the Solvent Effect Based on Four Empirical Scales: Dipolarity (SdP, a New Scale), Polarizability (SP), Acidity (SA), and Basicity (SB) of the Medium. *J. Phys. Chem. B* **2009**, *113*, 5951–5960. [CrossRef] [PubMed]
40. Cerón-Carrasco, J.P.; Jacquemin, D.; Laurence, C.; Planchat, A.; Reichardt, C.; Sraïdi, K. Determination of a Solvent Hydrogen-Bond Acidity Scale by Means of the Solvatochromism of Pyridinium-N-phenolate Betaine Dye 30 and PCM-TD-DFT Calculations. *J. Phys. Chem. B* **2014**, *118*, 4605–4614. [CrossRef]
41. Laurence, C.; Mansour, S.; Vuluga, D.; Sraidi, K.; Legros, J. Theoretical, Semiempirical, and Experimental Solvatochromic Comparison Methods for the Construction of the α1 Scale of Hydrogen-Bond Donation of Solvents. *J. Org. Chem.* **2022**, *87*, 6273–6287. [CrossRef]
42. Laurence, C.; Mansour, S.; Vuluga, D.; Planchat, D.; Legros, J. Hydrogen-Bond Acceptance of Solvents: A ^{19}F Solvatomagnetic β1 Database to Replace Solvatochromic and Solvatovibrational Scales. *J. Org. Chem.* **2021**, *86*, 4143–4158. [CrossRef]
43. Spange, S.; Weiß, N. Empirical Hydrogen Bonding Donor (HBD) Parameters of Organic Solvents Using Solvatochromic Probes–A Critical Evaluation. *ChemPhysChem* **2023**, *24*, e202200780. [CrossRef]
44. Pike, S.J.; Lavagnini, E.; Varley, L.M.; Cook, J.L.; Hunter, C.A. H-Bond donor parameters for cations. *Chem. Sci.* **2019**, *10*, 5943–5951. [CrossRef] [PubMed]
45. Spange, S.; Lienert, C.; Friebe, N.; Schreiter, K. Complementary interpretation of $E_T(30)$ polarity parameters of ionic liquids. *Phys. Chem. Chem. Phys.* **2020**, *160*, 9954–9966. [CrossRef]
46. Mero, A.; Guglielmero, L.; Guazzelli, L.; D'Andrea, F.; Mezzetta, A.; Pomelli, C.S. A specific interaction between ionic liquids' cations and Reichardt's Dye. *Molecules* **2022**, *27*, 7205. [CrossRef]
47. Böttcher, C.J.F. *Theory of Electric Polarization*, 2nd ed.; Elsevier: Amsterdam, The Netherlands, 1973; ISBN 978-0-444-41019-1. [CrossRef]
48. Marcus, Y. Preferential solvation of ions in mixed solvents. Part 2—The solvent composition near the ion. *J. Chem. Soc. Faraday Trans. I* **1988**, *84*, 1465–1473. [CrossRef]
49. Marcus, Y. Preferential solvation in mixed solvents. Part 5—Binary mixtures of water and organic solvents. *J. Chem. Soc. Faraday Trans.* **1990**, *86*, 2215–2224. [CrossRef]
50. Marcus, Y. *Solvent Mixtures: Properties and Selective Solvation*; CRC Press: Boca Raton, FL, USA, 2014; pp. 9–15. ISBN 9780429175664.
51. Marcus, Y. Preferential Solvation in Mixed Solvents X. Completely Miscible Aqueous Co-Solvent Binary Mixtures at 298.15 K. *Chem. Mon.* **2001**, *132*, 1387–1411. [CrossRef]
52. Spange, S.; Lauterbach, M.; Gyra, A.K.; Reichardt, C. Über Pyridinium-N-phenolat-Betaine und ihre Verwendung zur Charakterisierung der Polarität von Lösungsmitteln, XVI. Bestimmung der empirischen Lösungsmittelpolaritäts-Parameter $E_T(30)$ und AN für 55 substituierte Phenole. *Liebigs Ann. Chem.* **1991**, 323–329. [CrossRef]
53. Coleman, C.A.; Murray, C.J. Hydrogen bonding between N-pyridinium phenolate and O-H donors in acetonitrile. *J.Org. Chem.* **1992**, *57*, 3578–3582. [CrossRef]
54. Laurence, C.; Berthelot, M.; Graton, J. Hydrogen-Bonded Complexes of Phenols. In *The Chemistry of Phenols*; Rappoport, Z., Ed.; Wiley: New York, NY, USA, 2004; ISBN 978-0-470-86945-1.
55. Cook, J.L.; Hunter, C.A.; Low, C.M.R.; Perez-Velasco, A.; Vinter, J.G. Solvent Effects on Hydrogen Bonding. *Angew. Chem. Int. Ed.* **2007**, *46*, 3706–3709. [CrossRef] [PubMed]
56. Cook, J.L.; Hunter, C.A.; Low, C.M.R.; Perez-Velasco, A.; Vinter, J.G. Preferential Solvation and Hydrogen Bonding in Mixed Solvents. *Angew. Chem. Int. Ed.* **2008**, *47*, 6275–6277. [CrossRef]
57. Reichardt, C.; Eschner, M. On pyridinium N-phenolate betaines and their use in characterising the polarity of solvents, XVIII. Synthesis and UV/Vis spectroscopic properties of a negatively solvatochromic pyridinium N-thiophenolate betaine dye (in german). *Liebigs Ann. Chem.* **1991**, *1991*, 1003–1012. [CrossRef]
58. Bolz, I.; Schaarschmidt, D.; Rüffer, T.; Lang, H.; Spange, S. A pyridinium barbiturate betaine dye with pronounced negative solvatochromism: A new approach to molecular recognition. *Angew. Chem. Int. Ed.* **2009**, *48*, 7440–7443. [CrossRef] [PubMed]
59. Plenert, C.; Mendez-Vega, E.; Sander, W. Micro- vs. Macrosolvation in Reichardt's Dyes. *J. Am. Chem. Soc.* **2021**, *143*, 13156–13166. [CrossRef]
60. Henkel, S.; Misuraca, M.C.; Troselj, P.; Davidson, J.; Hunter, C.A. Polarisation effects on the solvation properties of alcohols. *Chem. Sci.* **2018**, *9*, 88–89. [CrossRef]
61. Spange, S.; Weiß, N.; Schmidt, C.; Schreiter, K. Reappraisal of Empirical Solvent Polarity Scales for Organic Solvents. *Chem. Methods* **2021**, *1*, 42–60. [CrossRef]
62. Spange, S.; Weiß, N.; Mayerhöfer, T.G. The Global Polarity of Alcoholic Solvents and Water–Importance of the Collectively Acting Factors Density, Refractive Index and Hydrogen Bonding Forces. *Chem. Open* **2022**, *11*, e202200140. [CrossRef]
63. Lerf, C.; Suppan, P. Hydrogen Bonding and Dielectric Effects in Solvatochromic Shifts. *J. Chem. Soc. Faraday Trans.* **1992**, *88*, 963–969. [CrossRef]
64. Rezende, M.C.; Machado, V.G.; Morales, S.; Pastenes, C.; Vidal, M. Use of Nonideality Parameters for the Analysis of the Thermodynamic Properties of Binary Mixtures. *ACS Omega* **2021**, *6*, 16553–16564. [CrossRef]

65. Morales, S.; Pastenes, C.; Machado, V.G.; Rezende, M.C. Applications of a preferential–solvation index (PSI) for the comparison of binary mixtures with ionic liquids. *J. Mol. Liq.* **2021**, *343*, 117644. [CrossRef]
66. Lüdecke, C.; Lüdecke, D. Thermodynamic properties of homogeneous mixtures. In *Thermodynamik: Physikalisch-Chemische Grundlagen der Thermischen Verfahrenstechnik, Thermodynamics: Physical-Chemical Fundamentals of Thermal Process Engineering*; Springer: Berlin/Heidelberg, Germany, 2000; 424p. [CrossRef]
67. Mayerhöfer, T.G.; Dabrowska, A.; Schwaighofer, A.; Lendl, B.; Popp, J. Beyond Beer's Law: Why the Index of Refraction Depends (Almost) Linearly on Concentration. *ChemPhysChem* **2020**, *21*, 707–711. [CrossRef] [PubMed]
68. Mayerhöfer, T.G.; Popp, J. Beyond Beer's Law: Revisiting the Lorentz-Lorenz Equation. *ChemPhysChem* **2020**, *21*, 1218–1223. [CrossRef] [PubMed]
69. Mayerhöfer, T.G.; Spange, S. Understanding refractive index changes in homologous series of hydrocarbons based on Beer's law. *ChemPhysChem* **2023**, *24*, e202300430. [CrossRef] [PubMed]
70. Spange, S.; Mayerhöfer, T.G. The Negative Solvatochromism of Reichardt's Dye B30–A Complementary Study. *Chem. Phys. Chem.* **2022**, *23*, e202200100. [CrossRef] [PubMed]
71. Spange, S.; Kaßner, L.; Mayerhöfer, T.G. New insights into the negative solvatochromism of various merocyanines. *PhysChemPhys* **2023**. submitted.
72. Deye, J.F.; Berger, T.A.; Anderson, A.G. Nile Red as a Solvatochromic Dye for Measuring Solvent Strength in Normal Liquids and Mixtures of Normal Liquids with Supercritical and Near Critical Fluids. *Anal. Chem.* **1990**, *62*, 615–622. [CrossRef]
73. Kasap, S.; Capper, P. *Springer Handbook of Electronic and Photonic Materials*; Springer: Berlin/Heidelberg, Germany, 2006; p. 49. ISBN 978-0-387-26059-4.
74. Suppan, P. Polarizability of excited molecules from spectroscopic studies. *Spectrochim. Acta* **1967**, *24A*, 1161–1165. [CrossRef]
75. Suppan, P. Solvent effects on the energy of electronic transitions: Experimental observations and applications to structural problems of excited molecules. *J. Chem. Soc. A* **1968**, 3125–3133. [CrossRef]
76. Dei, L.; Grassi, S. Peculiar Properties of Water as Solute. *J. Phys. Chem. B* **2006**, *110*, 12191–12197. [CrossRef]
77. Lee, H.; Hong, W.-H.; Kim, H. Excess Volumes of Binary and Ternary Mixtures of Water, Methanol and Ethylene Glycol. *J. Chem. Eng. Data* **1990**, *35*, 371–374. [CrossRef]
78. Franks, F.; Ives, D.J.G. The structural properties of alcohol–water mixtures. *Q. Rev. Chem. Soc.* **1966**, *20*, 1–44. [CrossRef]
79. Armitage, D.A.; Blandamer, M.J.; Morcom, K.W.; Treloar, N.C. Partial Molar Volumes and Maximum Density Effects in Alcohol–Water Mixtures. *Nature* **1968**, *219*, 718–720. [CrossRef]
80. Lama, F.R.; Lu, C.Y. Excess Thermodynamic Properties of Aqueous Alcohol Solutions. *J. Chem. Eng. Data* **1965**, *10*, 216–219. [CrossRef]
81. Douheret, G.; Khadir, A.; Pal, A. Thermodynamic Characterization of the Water + Methanol System, at 298.15 K. *Thermochim. Acta* **1989**, *142*, 219–243. [CrossRef]
82. Peeters, D.; Huyskens, P. Endothermicity or exothermicity of water/alcohol mixtures. *J. Mol. Struct.* **1993**, *300*, 539–550. [CrossRef]
83. Xiao, C.; Bianchi, H.; Tremaine, P.R. Excess molar volumes and densities of (methanol+water) at temperatures between 323 K and 573 K and pressures of 7.0 MPa and 13.5 MPa. *J. Chem. Thermodyn.* **1997**, *29*, 261–286. [CrossRef]
84. Petong, P.; Pottel, R.; Kaatze, U. Water-Ethanol Mixtures at Different Compositions and Temperatures. A Dieletric Relaxation Stud. *J. Phys. Chem. A* **2000**, *104*, 7420–7428. [CrossRef]
85. Sato, T.; Chiba, A.; Nozaki, R. Hydrophobic hydration and molecular association in methanol–water mixtures studied by microwave dielectric analysis. *J. Chem. Phys.* **2000**, *112*, 2924–2932. [CrossRef]
86. Sato, T.; Buchner, R. Dielectric Relaxation Processes in Ethanol/Water Mixtures. *J. Phys. Chem. A* **2004**, *108*, 5007–5015. [CrossRef]
87. Moldoveanu, S.; David, V. Chapter 6—Solvent Extraction. In *Modern Sample Preparation for Chromatography*; Elsevier: Amsterdam, The Netherlands, 2015; pp. 131–139. [CrossRef]
88. Kaatze, U. The Dielectric Properties of Water in Its Different States of Interaction. *J. Solution Chem.* **1997**, *26*, 1049–1112. [CrossRef]
89. Behrends, R.; Fuchs, K.; Kaatze, U. Dielectric properties of glycerol/water mixtures at temperatures between 10 and 50 °C. *J. Chem. Phys.* **2006**, *124*, 144512. [CrossRef] [PubMed]
90. Dixit, S.; Crain, J.; Poon, W.C.K.; Finney, J.L.; Soper, A.K. Molecular segregation observed in a concentrated alcohol–water solution. *Nature* **2002**, *416*, 829–832. [CrossRef] [PubMed]
91. Pascal, T.A.; Goddard, W.A. Hydrophobic Segregation, Phase Transitions and the Anomalous Thermodynamics of Water/Methanol Mixtures. *J. Phys. Chem. B* **2012**, *116*, 13905–13912. [CrossRef] [PubMed]
92. Tan, M.-L.; Miller, B.T.; Te, J.; Cendagorta, J.R.; Brooks, B.R.; Ichiye, T. Hydrophobic hydration and the anomalous partial molar volumes in ethanol-water mixtures. *J. Chem. Phys.* **2015**, *142*, 064501. [CrossRef] [PubMed]
93. Moučka, C.F.; Nezbeda, I. Partial molar volume of methanol in water: Effect of polarizability. *Coll. Czech. Chem. Commun.* **2009**, *74*, 559–563. [CrossRef]
94. Besford, Q.A.; Van den Heuvel, W.; Christofferson, A.J. Dipolar Dispersion Forces in Water–Methanol Mixtures: Enhancement of Water Interactions upon Dilution Drives Self-Association. *J. Phys. Chem. B* **2022**, *126*, 6231–6239. [CrossRef]
95. Guevara-Carrion, G.; Fingerhut, R.; Vrabec, J. Density and partial molar volumes of the liquid mixture water + methanol + ethanol + 2-propanol at 298.15 K and 0.1 MPa. *J. Chem. Eng. Data* **2021**, *66*, 2425–2435. [CrossRef]
96. Han, C.; Gao, J.; Sun, W.; Han, C.; Li, F.; Li, B. Structure study of water in alcohol-water binary system based on Raman spectroscopy. *J. Phys. Conf. Ser.* **2022**, *2282*, 012021. [CrossRef]

97. Nunes, N.; Ventura, C.; Martins, F.; Elvas-Leitão, R. Modeling Preferential Solvation in Ternary Solvent Systems. *J. Phys. Chem. B* **2009**, *113*, 3071–3079. [CrossRef]
98. Nunes, N.; Elvas-Leitão, R.; Martins, F. Using solvatochromic probes to investigate intermolecular interactions in 1,4-dioxane/methanol/acetonitrile solvent mixtures. *J. Mol. Liq.* **2018**, *266*, 259–268. [CrossRef]
99. Atkins, P.W.; de Paula, J. *Physical Chemistry*, 4th ed.; Wiley-VCH: Weinheim, Germany, 2006; p. 724. ISBN 3-527-31546-2.
100. Lechner, M.D.; Gehrke, K.; Nordmeier, E.H. *Macromolecular Chemistry: A Textbook for Chemists, Physicists, Materials Scientists and Chemical Engineers*, 5th ed.; Springer: Berlin/Heidelberg, Germany, 2014; 15p, ISBN 978-3-642-41768-9. [CrossRef]
101. Stepto, T.F.T. Dispersity in Polymer Science. *Pure Appl. Chem.* **2009**, *8*, 351–353. [CrossRef]
102. Heller, W. Remarks on Refractive Index Mixture Rules. *J. Phys. Chem.* **1965**, *69*, 1123–1129. [CrossRef]
103. Tasic, A.Z.; Djordjevic, B.D.; Grozdanic, D.K.; Radojkovic, N. Use of mixing rules in predicting refractive indexes and specific refractivities for some binary liquid mixtures. *J. Chem. Eng. Data* **1992**, *37*, 310–313. [CrossRef]
104. Pretorius, F.; Focke, W.W.; Androsch, R.; du Toi, E. Estimating binary liquid composition from density and refractive index measurements: A comprehensive review of mixing rules. *J. Mol. Liq.* **2021**, *332*, 115893. [CrossRef]
105. IUPAC. *Compendium of Chemical Terminology*, 2nd ed.; McNaught, A.D., Wilkinson, A., Eds.; Blackwell Scientific Publications: Oxford, UK, 1997. [CrossRef]
106. Markgraf, H.G.; Nikuradse, A. Über den Volumeneffekt in binären Gemischen einiger organischer Flüssigkeiten. On the volume effect in binary mixtures of some organic liquids. *Z. Naturforsch. A* **1954**, *9*, 27–34. [CrossRef]
107. Redlich, O.; Kister, A.T. Algebraic Representation of Thermodynamic Properties and the Classification of Solutions. *Ind. Eng. Chem.* **1948**, *40*, 341–348. [CrossRef]
108. Speight, J.G. Atomic and Group Refractions. In *Lange's Handbook of Chemistry*, 16th ed.; McGraw-Hill Education LLC: New York, NY, USA; ISBN 10:0070163847. Available online: https://www.labxing.com/files/lab_data/1340-1625805401-Qqoazhfj.pdf (accessed on 2 November 2022).
109. Marcus, Y. Water structure enhancement in water-rich binary solvent mixtures. *J. Mol. Liq.* **2011**, *158*, 23–26. [CrossRef]
110. Marcus, Y. Water structure enhancement in water-rich binary solvent mixtures. Part II. The excess partial molar heat capacity of the water. *J. Mol. Liq.* **2012**, *166*, 62–66. [CrossRef]
111. Hsu, W.-H.; Yen, T.-C.; Chen, C.-C.; Yang, C.-W.; Fang, C.-K.; Hwang, I.-S. Observation of mesoscopic clathrate structures in ethanol-water mixtures. *J. Mol. Liq.* **2022**, *366*, 120299. [CrossRef]
112. Langhals, H. The relationship between the refractive index and the composition of binary liquid mixtures. *Z. Phys. Chem.* **1985**, *266*, 775–780. [CrossRef]
113. Herraez, J.V.; Beld, R. Refractive Indices, Densities and Excess Molar Volumes of Monoalcohols + Water. *J. Solution Chem.* **2006**, *35*, 1315–1328. [CrossRef]
114. El-Dossoki, F.I. Refractive index and density measurements for selected binary protic-protic, aprotic-aprotic, and aprotic-protic systems at temperatures from 298.15 K to 308.15 K. *J. Chin. Chem. Soc.* **2007**, *54*, 1129–1137. [CrossRef]
115. Riobóo, R.J.; Philipp, M.; Ramos, M.A.; Krüger, J.K. Concentration and temperature dependence of the refractive index of ethanol-water mixtures: Influence of intermolecular interactions. *Eur. Phys. J. E* **2009**, *30*, 19–26. [CrossRef] [PubMed]
116. Linert, W.; Jameson, R. Acceptor properties of solvents: The use of isokinetic relationships to elucidate the relationship between the acceptor number and the solvatochromism of N-phenolate betaine dyes. *J. Chem. Soc. Perkin Trans.* **1993**, *2*, 1415–1421. [CrossRef]
117. Zhao, X.; Knorr Jeanne, F.J.; Mc Hal, L. Temperature-dependent absorption spectrum of betaine-30 in methanol. *Chem. Phys. Lett.* **2002**, *356*, 214–220. [CrossRef]
118. Tada, E.B.; Silva, P.L.; El Seoud, O.A. Thermosolvatochromism of betaine dyes in aqueous alcohols: Explicit consideration of the water–alcohol complex. *J. Phys. Org. Chem.* **2003**, *16*, 691–699. [CrossRef]
119. Bosch, E.; Rosés, M.; Herodes, K.; Koppel, I.; Leito, I.; Koppel, I.; Taal, V. Solute-solvent and solvent-solvent interactions in binary solvent mixtures. 2. Effect of temperature on the $E_T(30)$ polarity parameter of dipolar hydrogen bond acceptor-hydrogen bond donor mixtures. *J. Phys. Org. Chem.* **1996**, *9*, 403–410. [CrossRef]
120. Papadakis, R. Preferential Solvation of a Highly Medium Responsive Pentacyanoferrate(II) Complex in Binary Solvent Mixtures: Understanding the Role of Dielectric Enrichment and the Specificity of Solute–Solvent Interactions. *J. Phys. Chem. B* **2016**, *120*, 9422–9433. [CrossRef]
121. Hernandez-Perni, G.; Leuenberger, H. The characterization of aprotic polar liquids and percolation phenomena in DMSO/water mixtures. *Eur J. Pharm. Biopharm.* **2005**, *61*, 201–213. [CrossRef]
122. Yalcin, A.J.D.; Christofferson, C.J.; Drummond, T.L. Greaves, Solvation properties of protic ionic liquid–molecular solvent mixtures. *Phys. Chem. Chem. Phys.* **2020**, *22*, 10995–11011. [CrossRef]
123. Kosower, E.M. The Effect of Solvent on Spectra. I. A New Empirical Measure of Solvent Polarity: Z-Values. *J. Am. Chem. Soc.* **1958**, *80*, 3253–3260. [CrossRef]
124. Kosower, E.M. The Effect of Solvent on Spectra. II. Correlation of Spectral Absorption Data with Z-Values. *J. Am. Chem. Soc.* **1958**, *80*, 3261–3267. [CrossRef]
125. Kosower, E.M.; Dodiuk, H.; Tanizawa, K.; Ottolenghi, M.; Orbach, N. Intramolecular Donor-Acceptor Systems Radiative and Nonradiative Processes for the Excited States of 2-iV-Arylamino-6-naphthalenesulfonates. *J. Am. Chem. Soc.* **1975**, *97*, 2167–2178. [CrossRef]

126. Brownstein, S. The effect of solvents upon equilibria, spectra, and reaction rates. *Can. J. Chem.* **1960**, *38*, 1590–1596. [CrossRef]
127. Gowland, J.H.; Schmid, J.H. Two linear correlations of pKa vs. solvent composition. *Can. J. Chem.* **1969**, *47*, 2953–2958. [CrossRef]
128. Brooker, L.G.S.; Arnold, C.; Craig, C.; Heseltine, D.W.; Jenkins, P.W.; Lincoln, L.L. Color and Constitution. XIII. Merocyanines as Solvent Property Indicators. *J. Am. Chem. Soc.* **1965**, *87*, 2443–2450. [CrossRef]
129. Taha, A.; Ramadan, A.A.T.; El-Behairy, M.A.; Ismaila, A.I.; Mahmoud, M.M. Preferential solvation studies using the solvatochromic dicyanobis (1, 10-phenanthroline) iron (II) complex. *New J. Chem.* **2001**, *25*, 1306–1312. [CrossRef]
130. Huot, J.Y.; Battistel, E.; Lumry, R.; Villeneuve, G.; Lavallee, J.-F.; Anusiem, A.; Jolicoeur, C. A comprehensive thermodynamic investigation of water-ethylene glycol mixtures at 5, 25, and 45° C. *J. Solution Chem* **1988**, *17*, 601–636. [CrossRef]
131. Langhals, H. Polarity of liquid mixtures with components of limited miscibility. *Tetrahedron Lett.* **1986**, *27*, 339–342. [CrossRef]
132. Langhals, H. Polarität von Flüssigkeitsgemischen mit begrenzt mischbaren Komponenten. *Z. Phys. Chem.* **1987**, *268*, 91–96. [CrossRef]
133. Jyoti, N.; Meena, A.S.; Beniwal, V. Evaluation of the Polarity in Binary Liquid Mixtures as a Function of Volume Fraction. *Asian J. Chem.* **2023**, *35*, 721–726. [CrossRef]
134. Tanaka, Y.; Kawashima, Y.; Yoshida, N.; Nakano, H. Solvatochromism and preferential solvation of Brooker's merocyanine in water–methanol mixtures. *J. Comp. Chem.* **2017**, *38*, 2411–2419. [CrossRef] [PubMed]
135. Kiss, B.; Fábián, B.; Idrissi, A.; Szőri, M.; Jedlovszky, P. Miscibility and Thermodynamics of Mixing of Different Models of Formamide and Water in Computer Simulation. *J. Phys. Chem. B* **2017**, *121*, 7147–7155. [CrossRef] [PubMed]
136. Egan, E.P., Jr.; Luff, B.B. Heat of Solution, Heat Capacity, and Density of Aqueous Formamide Solutions at 25° C. *J. Chem. Eng. Data* **1966**, *11*, 194–196. [CrossRef]
137. Campos, V.; Marigliano, A.C.G.; Sólimo, H.N. Density, Viscosity, Refractive Index, Excess Molar Volume, Viscosity, and Refractive Index Deviations and Their Correlations for the (Formamide + Water) System. Isobaric (Vapor + Liquid) Equilibrium at 2.5 kPa. *J. Chem. Eng. Data* **2008**, *53*, 211–216. [CrossRef]
138. Egorov, G.I.; Makarov, D.M. Densities and Molar Isobaric Thermal Expansions of the Water + Formamide Mixture over the Temperature Range from 274.15 to 333.15 K at Atmospheric Pressure. *J. Chem. Eng. Data* **2017**, *62*, 1247–1256. [CrossRef]
139. Perticaroli, S.; Comez, L.; Sassi, P.; Morresi, A.; Fioretto, D.; Paolanton, M. Water-like Behavior of Formamide: Jump Reorientation Probed by Extended Depolarized Light Scattering. *J. Phys. Chem. Lett.* **2018**, *9*, 120–125. [CrossRef]
140. de Visser, C.; Pel, P.; Somsen, G. Volumes and Heat Capacities of Water and N-Methylformamide in Mixtures of These Solvents. *J. Solution Chem.* **1977**, *6*, 571–580. [CrossRef]
141. Cilense, M.; Benedetti, A.V.; Vollet, D.R. Thermodynamic properties of liquid mixtures. II. Dimethylformamide-water. *Thermochim. Acta* **1983**, *63*, 151–156. [CrossRef]
142. Kota Venkata Sivakumar, K.; Murthy Neriyanuri, K.; Krishnadevaraya, S.; Surahmanyam, S.V. Excess thermodynamic functions of the systems water+ N-methyl formamide and water+ N, N-dimethyl formamide. *Acustica* **1981**, *48*, 341–345.
143. Spange, S.; Keutel, D. Untersuchungen zur Polarität wäßriger Harnstoff-und Zucker-Lösungen mit der Methode der vergleichenden Solvatochromie. *Liebigs Ann.* **1993**, *1993*, 981–985. [CrossRef]
144. Stroka, J.; Herfort, I.; Schneider, H. Dimethylpropyleneurea—Water Mixtures: 1. Physical Properties. *J. Solution Chem.* **1990**, *19*, 743–753. [CrossRef]
145. Civera, C.; del Valle, J.C.; Elorza, M.A.; Elorza, B.; Arias, C.; Díaz-Oliva, C.; Catalán, J.; Blanco, F.G. Solvatochromism in urea/water and urea-derivative/water solutions. *Phys. Chem. Chem. Phys.* **2020**, *22*, 25165. [CrossRef]
146. LeBel, R.G.; Goring, D.A.I. Density, Viscosity, Refractive Index, and Hygroscopicity of Mixtures of Water and Dimethyl Sulfoxide. *Chem. Eng. Data* **1962**, *7*, 100–101. [CrossRef]
147. Clever, H.L.; Pigott, S.P. Enthalpies of mixing of dimethylsulfoxide with water and with several ketones at 298.15 K. *J. Chem. Thermodyn.* **1971**, *3*, 221–225. [CrossRef]
148. Egorov, G.I.; Makarov, D.M. The bulk properties of the water-dimethylsulfoxide system at 278–323.15 K and atmospheric pressure. *Rus. J. Phys. Chem. A* **2009**, *83*, 693–698. [CrossRef]
149. Wong, D.B.; Sokolowsky, K.P.; El-Barghouthi, M.I.; Fenn, E.E.; Giammanco, C.H.; Sturlaugson, A.L.; Fayer, M.D. Water Dynamics in Water/DMSO Binary Mixtures. *J. Phys. Chem. B* **2012**, *116*, 5479–5490. [CrossRef] [PubMed]
150. Kaatze, U.; Pottel, R.; Schaefer, M. Dielectric spectrum of dimethyl sulfoxide/water mixtures as a function of composition. *J. Phys. Chem.* **1989**, *93*, 5623–5627. [CrossRef]
151. Płowaś, I.; Świergiel, J.; Jadżyn, J. Relative Static Permittivity of Dimethyl Sulfoxide + Water Mixtures. *J. Chem. Eng. Data* **2013**, *58*, 1741–1746. [CrossRef]
152. Kirchner, B.; Reiher, M. The Secret of Dimethyl Sulfoxide–Water Mixtures. A Quantum Chemical Study of 1DMSO–nWater Clusters. *J. Am. Chem. Soc.* **2002**, *124*, 6206–6215. [CrossRef]
153. Özal, T.A.; van der Vegt, N.F.A. Confusing cause and effect: Energy–entropy compensation in the preferential solvation of a nonpolar solute in dimethyl sulfoxide/water mixtures. *J. Phys. Chem. B* **2006**, *110*, 12104–12112. [CrossRef]
154. Idrissi, A.; Marekha, B.; Barj, M.; Jedlovszky, P. Thermodynamics of mixing water with dimethyl sulfoxide, as seen from computer simulations. *J. Phys. Chem. B* **2014**, *118*, 8724–8733. [CrossRef]
155. Catalán, J.; Dıaz, C.; Garcıa-Blanco, F. Characterization of binary solvent mixtures of DMSO with water and other cosolvents. *J. Org. Chem.* **2001**, *66*, 5846–5852. [CrossRef]

156. Inamdar, S.R.; Gayathri, B.R.; Mannekutla, J.R. Rotational diffusion of coumarins in aqueous DMSO. *J. Fluoresc.* **2009**, *19*, 693–703. [CrossRef]
157. Jabbari, M.; Khosravi, N.; Feizabadi, M.; Ajloo, D. Solubility temperature and solvent dependence and preferential solvation of citrus flavonoid naringin in aqueous DMSO mixtures: An experimental and molecular dynamics simulation study. *RSC Adv.* **2017**, *7*, 14776–14789. [CrossRef]
158. Goates, J.R.; Sullivan, R.J. Thermodynamic Properties of the System Water-p-Dioxane. *J. Phys. Chem.* **1958**, *62*, 188–190. [CrossRef]
159. Sakurai, M. Partial Molar Volumes for 1,4-Dioxane + Water. *J. Chem. Eng. Data* **1992**, *37*, 492–496. [CrossRef]
160. Schott, H. Densities, Refractive Indices, and Molar Refractions of the System Water-Dioxane at 25 °C. *J. Chem. Eng. Data* **1961**, *6*, 19–20. [CrossRef]
161. Ahn-Ercan, G.; Krienke, H.; Schmeer, G. Structural and dielectric properties of 1,4-dioxane–water mixtures. *J. Mol. Liq.* **2006**, *129*, 75–79. [CrossRef]
162. Schrödle, S.; Fischer, B.; Helm, H.; Buchner, R. Picosecond dynamics and microheterogenity of water+ dioxane mixtures. *J. Phys. Chem. A* **2007**, *111*, 2043–2046. [CrossRef]
163. Schrödle, S.; Hefter, G.; Buchner, R. Dielectric Spectroscopy of Hydrogen Bond Dynamics and Microheterogeneity of Water + Dioxane Mixtures. *J. Phys. Chem. B* **2007**, *111*, 5946–5955. [CrossRef]
164. Kumbharkhane, A.C.; Joshi, Y.S.; Mehrotra, S.C.; Yagihara, S.; Sudo, S. Study of hydrogen bonding and thermodynamic behavior in water–1,4-dioxane mixture using time domain reflectometry. *Phys. B Conden. Matter* **2013**, *421*, 1–7. [CrossRef]
165. Casassas, E.; Fonrodona, G.; de Juan, A. Solvatochromic parameters for binary mixtures and a correlation with equilibrium constants. Part I. Dioxane-water mixtures. *J. Solution Chem.* **1992**, *21*, 147–162. [CrossRef]
166. Hüttenhain, S.H.; Balzer, W. Solvatochromic Fluorescence of 8-(Phenylamino)-l-naphthalene-ammonium-sulfonate (8,1 ANS) in 1,4-Dioxane/Water Mixtures, revisited. *Z. Naturforsch. A* **1993**, *48*, 709–712. [CrossRef]
167. Raju, B.B.; Costa, S.M.B. Photophysical properties of 7-diethylaminocoumarin dyes in dioxane–water mixtures: Hydrogen bonding, dielectric enrichment and polarity effects. *Phys. Chem. Chem. Phys.* **1999**, *1*, 3539–3547. [CrossRef]
168. Sánchez, F.; Díaz, A.N.; Algarra, M.J.; Lovillo, J.; Aguilar, A. Time resolved spectroscopy of 2-(dimethylamine)fluorene. Solvent effects and photophysical behavior. *Spectrochim. Acta A Mol. Biomol. Spectrosc.* **2011**, *83*, 88–93. [CrossRef] [PubMed]
169. Sahoo, R.; Jana, D. Seth, Photophysical study of an alkaloid harmaline in 1,4-dioxane-water mixtures. *Chem. Phys. Lett.* **2018**, *706*, 158–163. [CrossRef]
170. de Visser, C.; Perron, G.; Desnoyers, J.E. The heat capacities, volumes, and expansibilities of tert-butyl alcohol–water mixtures from 6 to 65 °C. *Can. J. Chem.* **1977**, *55*, 856–862. [CrossRef]
171. Sakurai, M. Partial molar volumes in aqueous mixtures of nonelectrolytes, I. tert. butyl alcohol. *Bull. Chem. Soc. Jpn.* **1987**, *60*, 1–7. [CrossRef]
172. Bowron, D.T.; Sober, A.K.; Finney, J.L. Temperature dependence of the structure of a 0.06 mole fraction tertiary butanol-water solution. *J. Chem. Phys.* **2001**, *114*, 6203–6219. [CrossRef]
173. Egorov, G.I.; Makarov, D.M. Densities and volume properties of (water + tert-butanol) over the temperature range of (274.15 to 348.15) K at pressure of 0.1 MPa. *J. Chem. Thermodyn.* **2011**, *43*, 430–441. [CrossRef]
174. Subramanian, D.; Klauda, J.B.; Leys, J.; Anisimov, M.A. Thermodynamic anomalies and structural fluctuations in aqueous solutions of tertiary butyl alcohol. *arXiv* **2013**, arXiv:1308.3676. [CrossRef]
175. Aman-Pommier, F.; Jallut, C. Excess specific volume of water + tert-butyl alcohol solvent mixtures: Experimental data, modeling and derived excess partial specific thermodynamic quantities. *Fluid Phase Equilibria* **2017**, *439*, 43–66. [CrossRef]
176. Gavrylyak, M.S. Investigation of dynamic fluctuations of refraction index of water tertiary butanol solutions. In Proceedings of the Eighth International Conference on Correlation Optics, Chernivsti, Ukraine, 11–14 September 2007.
177. Gazi, A.H.R.; Biswas, R. Heterogeneity in Binary Mixtures of (Water + Tertiary Butanol): Temperature Dependence Across Mixture Composition. *J. Phys. Chem. A* **2011**, *115*, 2447–2455. [CrossRef] [PubMed]
178. Subramanian, D.; Anisimov, M.A. Resolving the Mystery of Aqueous Solutions of Tertiary Butyl Alcohol. *J. Phys. Chem. B* **2011**, *115*, 9179–9183. [CrossRef] [PubMed]
179. Banik, D.; Bhattacharya, S.; Kumar Datta, P.; Sarkar, N. Anomalous Dynamics in tert-Butyl Alcohol–Water and Trimethylamine N-Oxide–Water Binary Mixtures: A Femtosecond Transient Absorption Study. *ACS Omega* **2018**, *3*, 383–392. [CrossRef]
180. Arnett, E.M.; Hufford, D.; McKelvey, D.R. The Ground-State Solvation Contribution to an Electronic Spectral Shift. *J. Am. Chem. Soc.* **1966**, *88*, 3142–3148. [CrossRef]
181. Duereh, A.; Sato, Y.; Smith, R.L.; Inomata, H.; Pichierri, F. Does Synergism in Microscopic Polarity Correlate with Extrema in Macroscopic Properties for Aqueous Mixtures of Dipolar Aprotic Solvents? *J. Phys. Chem. B* **2017**, *121*, 6033–6041. [CrossRef]
182. Akay, S.; Kayan, B.; Martínez, F. Solubility, dissolution thermodynamics and preferential solvation of 4-nitroaniline in (ethanol + water) mixtures. *Phys. Chem. Liq.* **2021**, *59*, 956–968. [CrossRef]
183. Ellis, C.-M. The 2-butoxyethanol-water system: Critical solution temperatures and salting-out effects. *J. Chem. Educ.* **1967**, *44*, 405–407. [CrossRef]
184. Quirion, F.; Magid, L.J.; Drifford, M. Aggregation and Critical Behavior of 2-Butoxyethanol in Water. *Langmuir* **1990**, *6*, 244–249. [CrossRef]
185. Kaatze, U.; Menzel, K.; Pottel, R.; Schwerdtfeger, S. Microheterogeneity of 2-Butoxyethanol/Water Mixtures at Room Temperature. An Ultrasonic Relaxation Study. *Z. Phys. Chem.* **1994**, *186*, 141–170. [CrossRef]

186. Kaatze, U.; Pottel, R.; Schumacher, A. Dielectric spectroscopy of 2-butoxyethanol/water mixtures in the complete composition range. *J. Phys. Chem.* **1992**, *96*, 6017–6020. [CrossRef]
187. Siu, W.; Koga, Y. Excess partial molar enthalpies of 2-butoxyethanol and water in 2-butoxyethanol-water mixtures. *Can. J. Chem.* **1989**, *67*, 671–676. [CrossRef]
188. Andersson, B.; Olofsson, G. Partial molar enthalpies of solution as indicators of interactions in mixtures of 2-butoxyethanol and 2-butanol with water. *J. Solution Chem.* **1988**, *17*, 1169–1182. [CrossRef]
189. Yoshida, K.; Yamaguchi, T.; Nagao, M.Y.; Kawabata, Y.; Seto, H.; Takeda, T. Slow dynamics of n-butoxyethanol–water mixture by neutron spin echo technique. *Appl. Phys. A* **2002**, *74*, 386–388. [CrossRef]
190. Joshi, Y.S.; Kumbharkhan, A.C. Study of dielectric relaxation and hydrogen bonding in water + 2-butoxyethanol mixtures using TDR technique. *Fluid Phase Equilib.* **2012**, *317*, 96–101. [CrossRef]
191. Chiou, D.-R.; Chen, S.-Y.; Chen, L.J. Density, Viscosity, and Refractive Index for Water + 2-Butoxyethanol and + 2-(2-Butoxyethoxy) ethanol at Various Temperatures. *J. Chem. Eng. Data* **2010**, *55*, 1012–1016. [CrossRef]
192. Kajimoto, S.; Moria, A.; Fukumura, H. Photo-controlled phase separation and mixing of a mixture of water and 2-butoxyethanol caused by photochromic isomerisation of spiropyran. *Photochem. Photobiol. Sci.* **2010**, *9*, 208–212. [CrossRef]
193. Indra, S.; Biswas, R. Heterogeneity in (2-butoxyethanol + water) mixtures: Hydrophobicity-induced aggregation or criticality-driven concentration fluctuations? *J. Chem. Phys.* **2015**, *142*, 204501. [CrossRef] [PubMed]
194. Catalán, J.; Díaz-Oliva, C.; García-Blanco, F. On the hydrophobic effect in water–alcohol mixtures. *Chem. Phys.* **2019**, *527*, 110467. [CrossRef]
195. Li, R.; Agostino, C.; McGregor, J.; Mantle, M.D.; Zeitler, A.; Gladden, L.F. Mesoscopic Structuring and Dynamics of Alcohol/Water Solutions Probed by Terahertz Time-Domain Spectroscopy and Pulsed Field Gradient Nuclear Magnetic Resonance. *J. Phys. Chem. B* **2014**, *118*, 10156–10166. [CrossRef] [PubMed]
196. Baksi, A.; Biswas, R. Dynamical Anomaly of Aqueous Amphiphilic Solutions: Connection to Solution H-Bond Fluctuation Dynamics? *ACS Omega* **2022**, *7*, 10970–10984. [CrossRef] [PubMed]
197. Koohyar, F.; Kiani, F.; Sharifi, S.; Sharifirad, M.; Rahmanpour, S.H. Study on the Change of Refractive Index on Mixing, Excess Molar Volume and Viscosity Deviation for Aqueous Solution of Methanol, Ethanol, Ethylene Glycol, 1-Propanol and 1, 2, 3-Propantriol at T = 292.15 K and Atmospheric Pressure. *Res. J. Appl. Sci. Eng. Technol.* **2012**, *4*, 3095–3101.
198. Zhang, Z.; Wang, W.; Korpacz, A.N.; Dufour, C.R.; Weiland, Z.J.; Lambert, C.R.; Timko, M.T. Binary Liquid Mixture Contact-Angle Measurements for Precise Estimation of Surface Free Energy. *Langmuir* **2019**, *35*, 12317–12325. [CrossRef]
199. Davis, M.I. Determination and analysis of the excess molar volumes of some amide-water systems. *Thermochim. Acta* **1987**, *120*, 299–314. [CrossRef]
200. Marcus, Y. Preferential solvation in mixed solvents. 15. Mixtures of acetonitrile with organic solvents. *J. Chem. Thermodyn.* **2019**, *135*, 55–59. [CrossRef]
201. Marcus, Y. The structure of mixtures of water and acetone derived from their cohesive energy densities and internal pressures. *J. Mol. Liq.* **2020**, *320*, 112801. [CrossRef]
202. Reichardt, C. Pyridinium-N-phenolate betaine dyes as empirical indicators of solvent polarity: Some new findings. *Pure Appl. Chem.* **2008**, *80*, 1415–1432. [CrossRef]
203. Kiselev, V.D.; Kashaeva, E.A.; Luzanova, N.A.; Konovalov, A.I. Enthalpies of solution of lithium perchlorate and Reichardt' dye in some organic solvents. *Thermochim. Acta* **1997**, *303*, 225–228. [CrossRef]
204. Machado, C.; da Graça Nascimento, M.; Rezende, M.C.; Beezer, A.E. Calorimetric evidence of aggregation of the $E_T(30)$ dye in alcoholic solutions. *Thermochim. Acta* **1999**, *328*, 150–159. [CrossRef]
205. Catalán, J.; Gomez, J.; Saiz, J.L.; Couto, A.; Ferraris, M.; Laynez, J. Calorimetric quantification of the hydrogen-bond acidity of solvents and its relationship with solvent polarity. *J. Chem. Soc. Perkin Trans. 2* **1995**, *1*, 2301–2305. [CrossRef]
206. Bose, K.; Kundu, K.K. Thermodynamics of transfer of p-nitroaniline from water to alcohol + water mixtures at 25 °C and the structure of water in these media. *Can. J. Chem.* **1977**, *55*, 3961–3966. [CrossRef]
207. Cowie, J.M.G.; Toporowski, P.M. Association in the binary liquid system dimethylsulphoxide-water. *Can. J. Chem.* **1961**, *39*, 2240–2243. [CrossRef]
208. Schrader, A.M.; Donaldson, S.H., Jr.; Song, J.; Israelachvili, J.N. Correlating steric hydration forces with water dynamics through surface force and diffusion NMR measurements in a lipid–DMSO–H_2O system. *Proc. Natl. Acad. Sci. USA* **2015**, *112*, 10708–10713. [CrossRef] [PubMed]
209. Bandyopadhyay, S.N.; Singh, K.K.; Goswam, D. Sensing non-ideal microheterogeneity in binary mixtures of dimethyl sulfoxide and water. *J. Opt.* **2022**, *24*, 054001. [CrossRef]
210. Cerar, J.; Jamnik, A.; Pethes, I.; Temleitner, L.; Pusztai, L.; Tomšič, M. Structural, rheological and dynamic aspects of hydrogen-bonding molecular liquids: Aqueous solutions of hydrotropic tert-butyl alcohol. *J. Coll. Interf. Sci.* **2020**, *560*, 730–742. [CrossRef]
211. Chakraborty, S.; Pyne, P.; Mitra, R.K.; Das Mahanta, D. A subtle interplay between hydrophilic and hydrophobic hydration governs butanol (de)mixing in water. *Chem. Phys. Lett.* **2022**, *807*, 140080. [CrossRef]
212. Langhals, H. Ein neues, unkompliziertes Verfahren zur Bestimmung der Zusammensetzung binärer Flüssigkeitsgemische, A new, uncomplicated method for the determination of the composition of binary liquid mixtures. *Fresenius Z. Anal. Chem.* **1981**, *308*, 441–444. [CrossRef]

213. Królicki, R.; Jarzęba, W.; Mostafavi, M.; Lampre, I. Preferential Solvation of Coumarin 153 The Role of Hydrogen Bonding. *J. Phys. Chem. A* **2002**, *106*, 1708–1713. [CrossRef]
214. Marcus, Y. The structure of and interactions in binary acetonitrile + water mixtures. *J. Phys. Org. Chem.* **2012**, *35*, 1072–1085. [CrossRef]
215. Shalmashi, A.; Amani, F. Densities and excess molar volumes for binary solution of water + ethanol, + methanol and + propanol from (283.15 to 313.15) K). *Lat. Am. Appl. Res.* **2014**, *44*, 163–166. [CrossRef]
216. Galicia-Andrés, E.; Dominguez, H.; Pusztai, L.; Pizio, O. On the composition dependence of thermodynamic, dynamic and dielectric properties of water-methanol model mixtures. Molecular dynamics simulation results. *Condens. Matter. Phys.* **2015**, *18*, 43602. [CrossRef]
217. Martens, M.; Hadrich, M.J.; Nestler, F.; Ouda, M.; Schaad, A. Combination of Refractometry and Densimetry–A Promising Option for Fast Raw Methanol Analysis. *Chem. Ing. Tech.* **2020**, *92*, 1474–1481. [CrossRef]
218. Ortega, J. Densities and Refractive Indices of Pure Alcohols as a Function of Temperature. *J. Chem. Eng. Data* **1982**, *27*, 312–317. [CrossRef]
219. Zaichikov, A.M.; Krestyaninov, M.A. Thermodynamic characteristics of and intermolecular interactions in aqueous solutions of N-methylformamide. *Russ. J. Phys. Chem.* **2006**, *80*, 1249–1254. [CrossRef]
220. de Visser, C.; Perron, G.; Desnoyers, J.E.; Heuvelsland, W.J.M.; Somsen, G. Volumes and Heat Capacities of Mixtures of N, N-Dimethylformamide and Water at 298.15 K. *J. Chem. Eng. Data* **1977**, *22*, 74–79. [CrossRef]
221. Al-Azzawl, S.F.; Allo, E.I. Density, Viscosity, and Refractivity Data off Solutions of Potassium Iodide in /V-Formylmorpholine-Water at 25, 35, and 45 °C. *J. Chem. Eng. Data* **1992**, *37*, 158–162. [CrossRef]
222. Chen, G.; Hou, Y.; Knapp, H. Diffusion Coefficients, Kinematic Viscosities, and Refractive Indices for Heptane + Ethylbenzene, Sulfolane + 1-Methylnaphthalene, Water + N,N-Dimethylformamide, Water + Methanol, Water + N-Formylmorpholine, and Water + N-Methylpyrrolidone. *J. Chem. Eng. Data* **1995**, *40*, 1005–1010. [CrossRef]

Disclaimer/Publisher's Note: The statements, opinions and data contained in all publications are solely those of the individual author(s) and contributor(s) and not of MDPI and/or the editor(s). MDPI and/or the editor(s) disclaim responsibility for any injury to people or property resulting from any ideas, methods, instructions or products referred to in the content.

Review

Solvatochromism in Solvent Mixtures: A Practical Solution for a Complex Problem

Omar A. El Seoud [1,*], Shirley Possidonio [2] and Naved I. Malek [3]

[1] Institute of Chemistry, University of São Paulo, São Paulo 05508-000, SP, Brazil
[2] Department of Chemistry, Institute of Environmental, Chemical, and Pharmaceutical Sciences, Federal University of São Paulo, Diadema 09913-030, SP, Brazil; possidonio@unifesp.br
[3] Department of Chemistry, Sardar Vallabhbhai Nation Institute of Technology, Surat 395007, Gujarat, India; navedmalek@gmail.com
* Correspondence: elseoud.usp@gmail.com

Abstract: Many reactions are carried out in solvent mixtures, mainly because of practical reasons. For example, E2 eliminations are favored over S_N2 substitutions in aqueous organic solvents because the bases are desolvated. This example raises the question: how do we chose binary solvents to favor reaction outcomes? This important question is deceptively simple because it requires that we understand the details of all interactions within the system. Solvatochromism (solvent-dependent color change of a substance) has contributed a great deal to answer this difficult question, because it gives information on the interactions between solvents, solute-solvent, and presumably transition state-solvent. This wealth of information is achieved by simple spectroscopic measurements of selected (solvatochromic) substances, or *probes*. An important outcome of solvatochromism is that the probe solvation layer composition is almost always different from that of bulk mixed solvent. In principle, this difference can be exploited to "tune" the composition of solvent mixture to favor the reaction outcome. This minireview addresses the use of solvatochromic probes to quantify solute-solvent interactions, leading to a better understanding of the complex effects of solvent mixtures on chemical phenomena. Because of their extensive use in chemistry, we focus on *binary mixtures* containing protic-, and protic-dipolar aprotic solvents.

Keywords: binary solvent mixtures; solvatochromism; solvatochromic probes; solvation models; ester hydrolysis; biopolymer dissolution

1. Reasons for Using Mixed Solvents in Chemistry

Solvent mixtures are extensively employed in chemistry for practical reasons. For example, the solubilities of inorganic bases, such as KOH, and other electrolytes in alcohols are enhanced in presence of water [1,2]. Cellulose that is insoluble in water is, however, readily soluble in some aqueous electrolyte solutions [3], water-DMSO mixtures [4], and mixtures of ionic liquids-molecular solvents (ILs-MSs) [5–8]. In the latter example, cellulose dissolution is attributed to the disruption of the strong hydrogen-bonding (H-bonding) between the hydroxyl groups of the anhydroglucose units, as well as to the hydrophobic interactions between cellulose chains, as shown in Figure 1a (IL-DMSO). Consequently, addition of protic non-solvents to solutions of cellulose in IL-MS causes biopolymer precipitation because the non-solvent efficiently solvates the ions of the IL (Figure 1b).

In addition to enhanced biopolymer solubility, the use of mixed solvents also causes noticeable changes in the physicochemical properties, such as a reduction in viscosity, leading to better heat and mass transfer, as shown by Figure 2a,c.

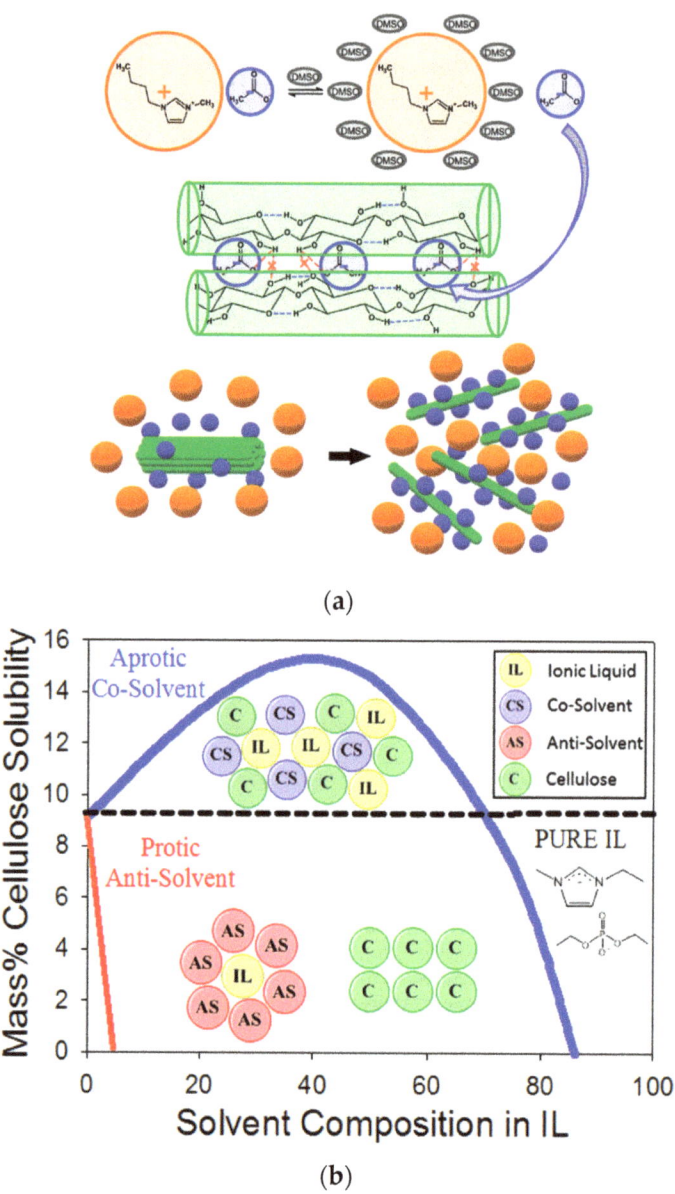

Figure 1. (a) A simplified scheme for cellulose dissolution in ionic liquid-DMSO. The biopolymer dissolution is attributed to interactions of its hydroxyl groups with the ions of the ionic liquid and the dipole of DMSO. Reproduced with permission from [9]. (b) Effects of the addition of a protic non-solvent (such as water or ethanol) on the dissolution of cellulose IL-dipolar solvent. Addition of the non-solvent leads to cellulose precipitation. Reprinted with permission from [7].

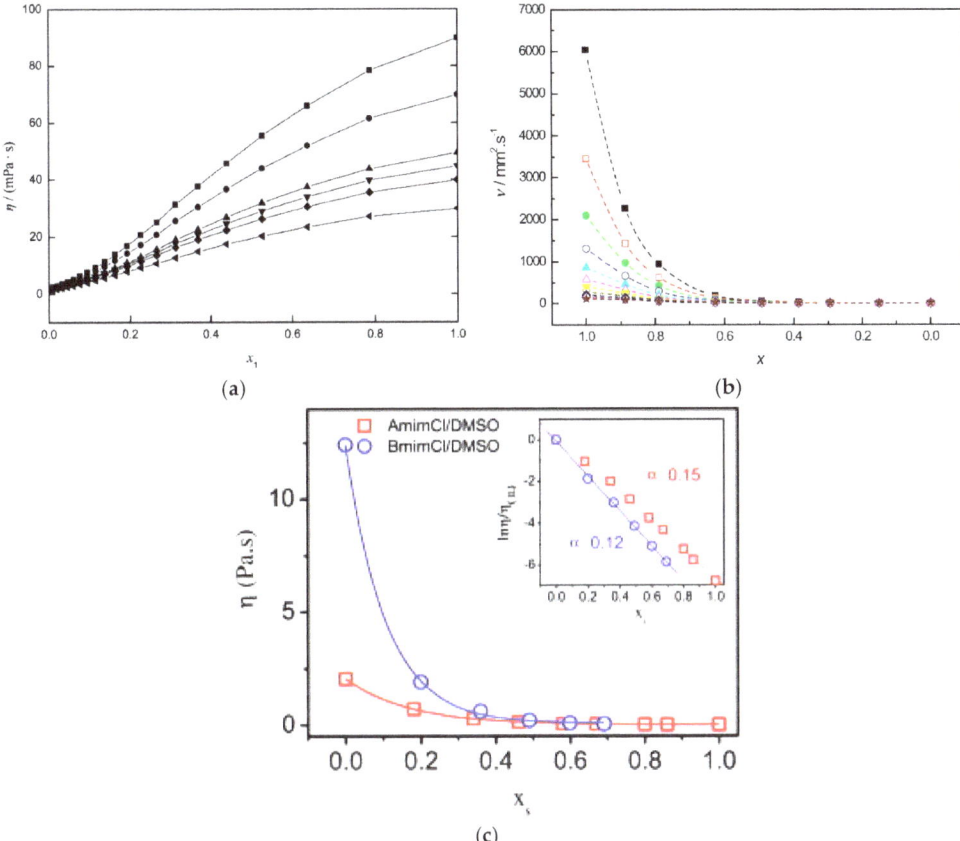

Figure 2. (a) Dependence of the viscosity (η) of PEG 400-DMSO on the mole fraction of PEG 400 at different temperatures: ■, 25 °C; ●, 30 °C; ▲, 35 °C; ▼, 40 °C; ♦, 45 °C; ◄, 50 °C. Reprinted with permission from [10]. (b) Viscosities of BuMeImCl-DMF (1-butyl-3-methylimidazolium chloride-N,N-Dimethylformamide) mixtures as a function of mole fraction of DMF: ■, 30 °C; □, 35 °C; ●, 40 °C; ○, 45 °C; ▲, 50 °C; △, 55 °C; ▼, 60 °C; ∇, 65 °C; ♦, 70 °C; ◊, 75 °C; ★, 80 °C. Reprinted with permission from [11]. (c) Effects of increasing the mole fraction of DMSO (x_S) on the viscosity of cotton cellulose in binary mixtures of DMSO with the ILs AlMeImCl (1-allyl-3-methylimidazolium chloride) and BuMeImCl. The insert is the logarithm-linear plot of the reduced viscosity ratio versus DMSO mole fraction at 25 °C Reprinted with permission from [12].

2. A Rationale for Effects of Mixed Solvents on Chemical Phenomena

How do we explain these interesting and very useful effects of binary solvent mixtures on diverse chemical phenomena? While deceptively simple, this question cannot be answered in a straightforward manner. Consider, for example, the fact that most physicochemical properties of binary mixtures are not ideal. That is, the property of the binary mixture does not vary in a simple way as a function of binary solvent composition, as shown in Figure 3a–c.

The reason behind this non-ideality is clearly the interactions between components of the binary solvent mixture. To a first approximation, one expects that the composition of the solvation layer of a dissolved substance (which we will refer to as "*probe*") should be the same as that of bulk binary mixture. Consequently, the same explanations given for bulk binary mixtures should apply to the solvation layers of the dissolved probes. This simple view, however, does not hold in most cases because probe-solvent *nonspecific and specific interactions* were not taken into consideration. These interactions change the composition of the solvation layers relative to bulk solvent mixtures, as shown below.

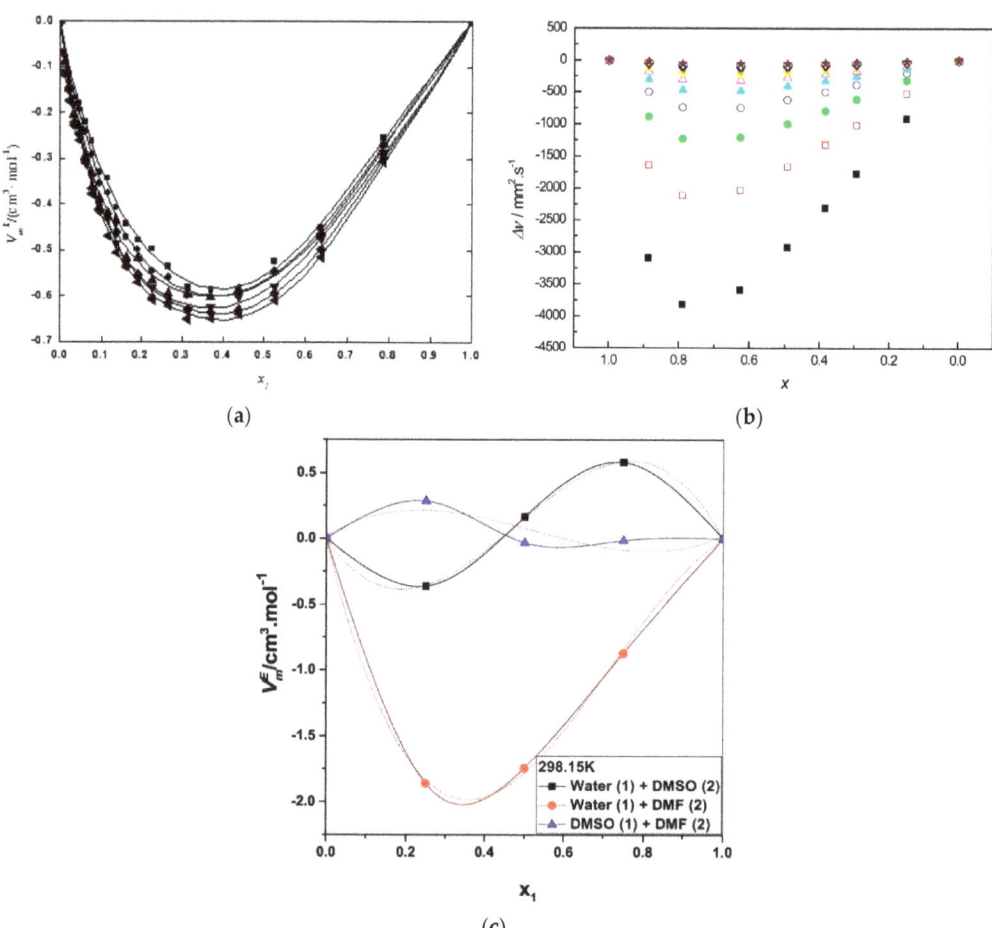

Figure 3. (a) Density-based excess molar volumes as a function of mole fraction and temperature for PEG 400-DMSO. (a) ■, 25 °C; •, 30 °C; ▲, 35 °C; ▼, 40 °C; ♦, 45 °C; ◄, 50 °C. Reprinted with permission from [10]; (b) Deviation of viscosity of mixtures of BuMeImCl-DMF from the values calculated from $\Sigma \; \chi_{component} \times V_{molar\;volume\;of\;component}$. ■, 30 °C; □, 35 °C; •, 40 °C; ○, 45 °C; ▲, 50 °C K; △, 55 °C; ▼, 60 °C; ▽, 65 °C; ♦, 70 °C; ◊, 75 °C; ★, 80 °C. Reprinted with permission from [11]; (c) Excess molar volume (V^E) of binary mixed systems for water-DMSO; water-MF, and DMSO-DMF at 25 °C. Reprinted with permission from [13].

The composition of the solvation layer of a probe may deviate from that of an (already non-ideal) bulk solvent mixture due to the so-called *"preferential solvation"* of the probe by one component of the mixture (Figure 4). In principle, this phenomenon includes contributions from probe-independent *"dielectric enrichment"*, and probe-solvent interactions. The first mechanism is operative in mixtures of nonpolar/low polar solvents, such as cyclohexane-THF (Tetrahydrofuran). It denotes enrichment of the probe solvation layer (relative to that of bulk solvent mixture) by the component of larger dielectric constant (or relative permittivity), due to non-specific probe dipole-solvent dipole interactions [14–16].

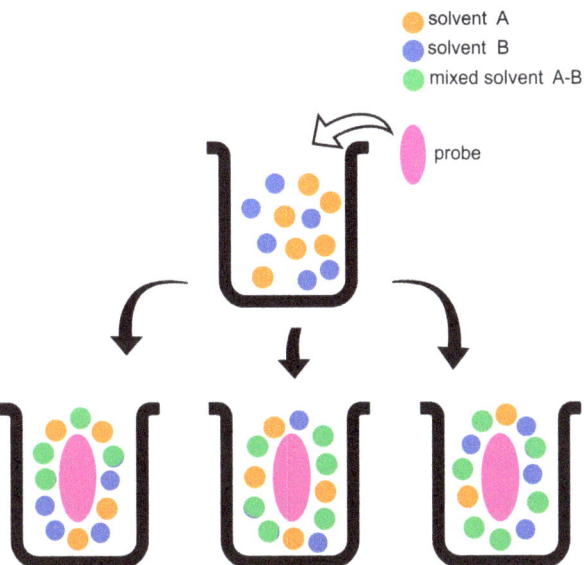

Figure 4. Schematic representation of the solvation of a solute in a binary solvent mixture composed of two solvents (A, B), and the "mixed" solvent A-B, whose formation is discussed below. The parts of the lower line represent from the left: ideal solvation, i.e., the composition of the probe solvation layer is the same as that of bulk solvent mixture; preferential solvation by the solvents (A, A-B); preferential solvation by the solvents (B, A-B).

The second solvation mechanism is dominant in protic solvents (such as aqueous alcohols) and their mixtures with strongly dipolar solvents (water-DMSO, alcohol-DMF, etc.). It is essentially due to solute-solvent H-bonding and hydrophobic interactions. One additional complication is that solvent-solvent H-bonding generates an additional or "*mixed*" solvent species that should be considered. For example, in mixtures of water (W) and alcohol (ROH), and W-DMSO, we have in solution both the parent and the mixed solvents, HO$\underline{H}$...$\underline{O}$(H)R and HO$\underline{H}$...$\underline{O}$(H)=S(CH$_3$)$_2$ [17]; this turns analysis of the solvation data more complex. In summary, most significant consequence of preferential solvation is that compositions of the solvation layers of most probes are different from those of the corresponding bulk solvent mixtures; these composition differences are probe-, and temperature-dependent [18,19].

How do we calculate the "effective" (or local) composition of the solvation layer of a probe? Several techniques were employed to solve this problem, including FTIR [20], resonance Raman spectroscopy [21], and X-ray diffraction (for solvated crystals) [22]. It is our view that the most useful approach is to use solvatochromic indicators as *models* for the compounds of interest, e.g., reactants. Solvatochromic probes are substances whose absorption or emission spectra are sensitively dependent on the solvent or the composition of solvent mixtures (Figure 5). The reason for solvatochromism is that the energy difference between the probe's ground and excited states is sensitively affected by probe-solvent interactions, leading to medium-dependent values of λ_{max}, and hence a change in solution color. For most probes, the solvatochomism is negative, meaning there is a hypsochromic shift of the longest wavelength absorption band with increasing medium polarity. The reason is that solvents stabilize the zwitterionic ground state much more than the diradical excited state (see Figure 6 for light-induced transition of the probe *t*-Bu$_5$RB). The latter corresponds to a so-called FranckCondon excited state, because the time scale of the electronic excitation (ca. 10^{-15} s) is much shorter than that required for the solvent molecule to reorient in order to stabilize the probe's excited state. The energy of this transition furnishes the solvatochromic property of interest, *vide infra*.

Figure 5. Examples of solvatochromism. Part (**a**) is for MePMBr$_2$ (2,6-dibromo-4-[(E)-2-(1-methylpyridinium-4-yl)ethenyl]) empirical polarity indicator in (from left) ethanol, water, acetone, and dichloromethane [23]. Part (**b**) is that for the empirical polarity probe t-Bu$_5$RB, Figure 6), in (mineral) diesel oil and its mixtures with 25, 50, 75% bioethanol, and in pure bioethanol, respectively. Reprinted with permission from [24]. Part (**c**) shows the dependence of solution color on the structure of the solvatochromic probe. Reprinted with permission from [25]. The structures and names of these probes are shown in Figure 7.

This approach was advanced thanks to the work of professor C. Reichardt, initially under the supervision of professor K. Dimroth at Marburg university [26,27]. The experimental part is relatively simple: register the UV-Vis spectrum of a solvatochromic probe → calculate the value of λ_{max} of a specific peak (the longest wavelength, due to intermolecular charge-transfer within the probe) → use the value of λ_{max} to calculate the desired property, or *descriptor*, of the solvent or solvent mixture. The power of solvatochromism is that it can be employed to calculate the overall (or empirical) solvent polarity scale, E_T (in kcal/mol), as well as the individual solvent descriptors that contribute to E_T, namely solvent Lewis acidity (*SA*), solvent Lewis basicity (*SB*), solvent dipolarity (*SD*), and solvent polarizability (*SP*), where S refers to solvent. Other abbreviations that were employed for designating these descriptors include SdP and SP for solvent dipolarity and polarizability, respectively. For consistency, however, we use two letters to designate each solvent descriptor.

Equation (1) shows the relationship of these solvent descriptors:

$$E_T(\text{probe}) = E_T(\text{probe})_0 + aSA + bSB + dSD + pSP \tag{1}$$

where $E_T(\text{probe})_0$ corresponds to gas phase, the descriptors (SA, SB, SD, SP) are those defined above, and (a, b, d, and p) are the corresponding regression coefficients. Figure 7 shows some solvatochromic probes used to calculate the descriptors of Equation (1). In the Taft–Kamlet–Abboud approach, similar solvatochromic parameters and different symbols were employed to describe probe–solvent interactions, α, β, and π* for solvent Lewis acidity, Lewis basicity, and (combined) dipolarity/polarizability [28]. The signs of the coefficients in Equation (1) indicate whether the property of the solvent considered increases (positive sign) or decreases (negative sign) the empirical solvent polarity [29].

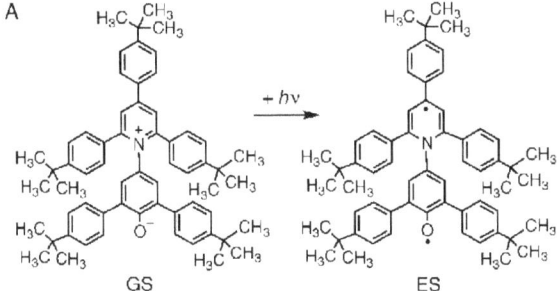

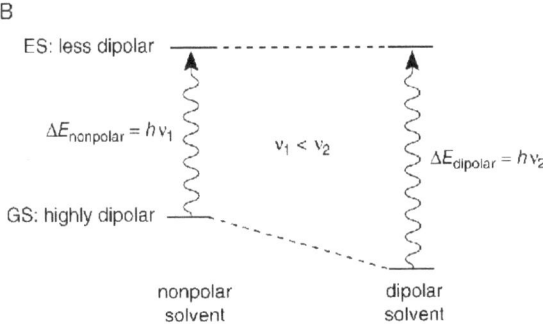

Figure 6. (**A**) The molecular structure of the solvatochromic indicator dye, t-Bu₅RB (2,6-bis [4-(t-butyl)phenyl]-4-{2,4,6-tris[4-(t-butyl)phenyl]pyridinium-1-yl}phenolate): a zwitterionic pyridinium-N-phenolate betaine dye with a highly dipolar electronic ground state (GS) and a much less dipolar first excited state (ES). (**B**) A schematic qualitative representation of the solvent influence on the differences ΔE between the energies of the GS and ES of t-Bu₅RB, dissolved in a nonpolar and a dipolar solvent, respectively. Reprinted with permission from [24].

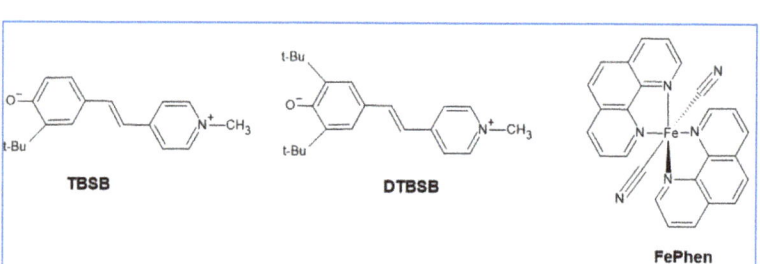
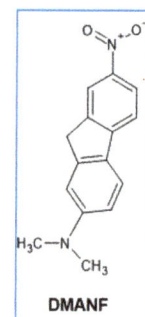

Figure 7. Structures and acronyms of some solvatochromic probes, employed to calculate the solvent descriptors shown in Equation (1).

3. A Model for the Solvation of Probes

A solvation model is required to calculate the local concentration of the species in the solvating layer of the probe. For simplicity, we consider a W-ROH mixture, containing a certain water mole fraction, χ_W. We address the question of preferential solvation by using a series of exchange equilibria between the solvent species in the bulk binary mixture and those in the probe solvation layer. Any realistic model should consider, therefore, the exchange of the three solvents (W, ROH, and W-ROH) between the bulk solvent and the probe solvation layer, where the equilibria involving the mixed solvent are given by Equations (2)–(4):

$$ROH + W \rightleftharpoons ROH - W \tag{2}$$

$$\text{Probe(ROH)}_m + m\,(ROH - W) \rightleftharpoons \text{Probe(ROH} - W)_m + m\,ROH \tag{3}$$

$$\text{Probe(W)}_m + m\,(ROH - W) \rightleftharpoons \text{Probe(ROH} - W)_m + m\,W \tag{4}$$

Note that (m) is not the solvation number of the probe; it represents the number of solvent molecules whose exchange in the probe solvation layer affects its solvatochromism; usually, the value of (m) is close to unity. For example, for the solvation of WB in mixtures of water with 4 alcohols (methanol, 1-propanol, 2-propanol, and 2-methylethanol), the calculated values of (m) range from 1.06 to 1.70 [30]. With this proviso (meaning of (m)), addressing the important point of probe-dependent volume of the solvation layer is not required for the analysis shown below. Additionally, using this model, one should be able to calculate the probe-induced preferential solvation, as expressed by the *fractionation factor* (φ), which represents the equilibrium constant for solvent exchange between the bulk binary mixture and the probe solvation layer. This model has been elaborated; Equations (5)–(7) are for W-ROH, where (Bk) refers to bulk solvent:

$$\varphi_{W/ROH} = \frac{x_W^{Probe}/x_{ROH}^{Probe}}{\left(x_W^{Bk;Effective}/x_{ROH}^{Bk;Effective}\right)^m} \tag{5}$$

$$\varphi_{ROH-W/ROH} = \frac{x_{ROH-W}^{Probe}/x_{ROH}^{Probe}}{\left(x_{ROH-W}^{Bk;Effective}/x_{ROH}^{Bk;Effective}\right)^m} \tag{6}$$

$$\varphi_{ROH-W/W} = \frac{x_{ROH-W}^{Probe}/x_W^{Probe}}{\left(x_{ROH-W}^{Bk;Effective}/x_W^{Bk;Effective}\right)^m} \tag{7}$$

In Equation (5), $\varphi_{W/ROH}$ describes the preference of (W) for the probe solvation layer relative to bulk solvent mixture. Values of $\varphi_{W/ROH} > 1$ indicate that the probe solvation layer is richer in (W) than the bulk solvent; the inverse is true for $\varphi_{W/ROH} < 1$. In absence of preferential solvation, $\varphi_{W/ROH}$ is unity, indicating that solvent composition in the probe solvation layer is the same as that of the bulk solvent. The same line of reasoning applies to $\varphi_{ROH-W/ROH}$ (mixed solvent displaces ROH) and $\varphi_{ROH-W/W}$ (mixed solvent displaces W), as depicted in Equations (6) and (7).

The use of 1:1 stoichiometry for ROH-W is an assumption that has been employed elsewhere [31–37]. Mixed solvents with a stoichiometry other than 1:1 can be regarded, to a good approximation, as mixtures of the 1:1 structure plus excess of a pure solvent. We stress that taking into account the presence of mixed solvents is more than a practical and convenient assumption; the presence of such species has been successfully employed in fitting results of spectroscopic techniques that are particularly suitable to determine the stoichiometry and association constant of solvents, e.g., NMR [38–40] and FTIR [41–43]. The observed solvatochromic property, such as the E_T^{Obs}(probe), can be then calculated by iteration from Equations (8) and (9):

$$E_T^{obs} = x_W^{Probe} E_T^W + x_{ROH}^{Probe} E_T^{ROH} + x_{ROH-W}^{Probe} E_T^{ROH-W} \tag{8}$$

$$E_T^{obs} = \frac{\left(x_{ROH}^{Bk;Effective}\right)^m E_T^{ROH} + \varphi_{W/ROH}\left(x_W^{Bk;Effective}\right)^m E_T^W + \varphi_{ROH-W/ROH}\left(x_{ROH-W}^{Bk;Effective}\right)^m E_T^{ROH-W}}{\left(x_{ROH}^{Bk;Effective}\right)^m + \varphi_{W/ROH}\left(x_W^{Bk;Effective}\right)^m + \varphi_{ROH-W/ROH}\left(x_{ROH-W}^{Bk;Effective}\right)^m} \quad (9)$$

where (m), $x_{ROH}^{Bk;Effective}$, $x_W^{Bk;Effective}$, and $x_{ROH-W}^{Bk;Effective}$ refer to the number of molecules in the probe solvation layer that affects its solvatochromic response, and effective mole fractions of the appropriate species in bulk mixed solvent, respectively. Note that these effective mole fractions differ from the analytical or starting mole fractions due to the formation of the mixed solvent, e.g., ROH-W. The input data to solve these equations include E_T^{obs}, E_T^W, E_T^{ROH}, and $x_{Species}^{Effective}$, along with initial estimates of (m), E_T^{ROH-W} and the appropriate solvent fractionation factors. The values of $x_{ROH}^{Bk;Effective}$, $x_W^{Bk;Effective}$, and $x_{ROH-W}^{Bk;Effective}$ are calculated from the dependence of a physical property (e.g., density on solution composition) by using the association model discussed by Katz et al. [44–46]; a list of the association constants of W-ROH has been published [47].

Figures 8 and 9 show the dependence of the effective concentrations of solvent species on the analytical mole fractions of the two solvents.

By using data such as those shown in Figure 8, and Equations (8) and (9), one can calculate the dependence of a solvatochromic property (e.g., solvent polarity or E_T(probe)) as a function of binary solvent composition at a fixed temperature, or at a series of temperatures (referred to as thermo-solvatochromism), as shown in Figure 9.

The preceding discussion shows that it is relatively simple to calculate medium descriptors (e.g., SA and SB) for pure and mixed solvents; the use of the appropriate solvation model permits calculation of the composition of the solvation layers of the solvatochromic probes dissolved therein. Although there are a large number of solvatochromic probes that are employed to calculate the solvent descriptors, the results are consistent when different probes were employed. This is demonstrated by application of Equation (1) to a series of ILs, where the calculated empirical solvent polarities using the RB (Reichardt betaine—2,6-diphenyl-4-(2,4,6-triphenylpyridinium-1-yl) phenolate) probe correlate linearly with the calculated values, based on the 4 descriptors (SA, SB, SD, SP), which were calculated using different solvatochromic probes (Figure 10). Values of φ were also calculated at different temperatures, showing the effect of temperature on the composition of the probe solvation layer (thermo-solvatochromism).

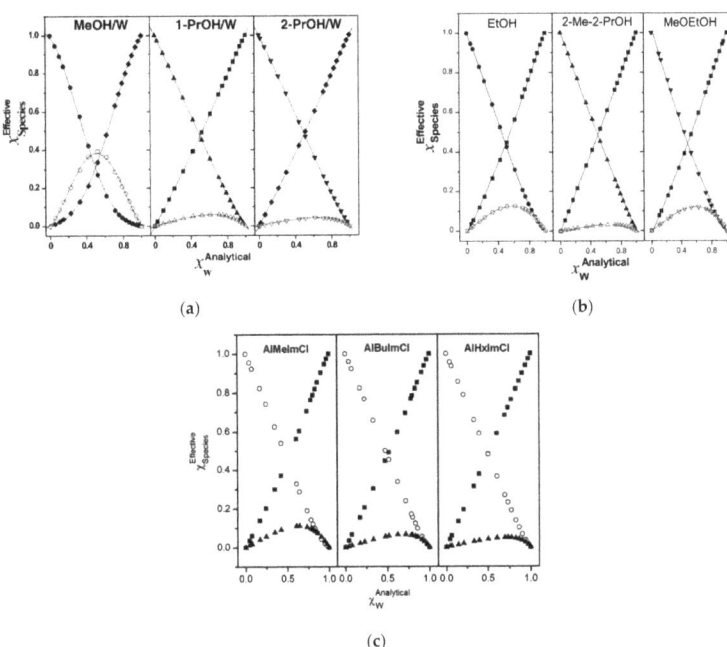

Figure 8. (**a**) Species distribution for MeOH–W, 1-PrOH–W and 2-PrOH–W mixtures at 25 °C. W (♦); ROH (•, ▲, ▼); and ROH–W (○, △, ▽). Reprinted with permission from [48]. (**b**) Species distribution for EtOH/W, 2-Me-2-PrOH (2-methyl-2-propanol)/W and MeOEtOH (2-methoxyethanol)/W mixtures, respectively, at 25 °C: W (■), ROH (•, ▲, ▼), and ROH–W (○, △, ▽). Reprinted from [30]. (**c**) Dependence of species distribution for IL-W binary mixtures on the length of R of AlRImCl (1-ally-3-methylimidazoilium chloride), at 25 °C, where R = methyl, 1-butyl, and 1-hexyl, respectively. The symbols employed are ○, ■, ▲ for IL, W, and the IL-W 1:1 complex, respectively. Reprinted with permission from [49].

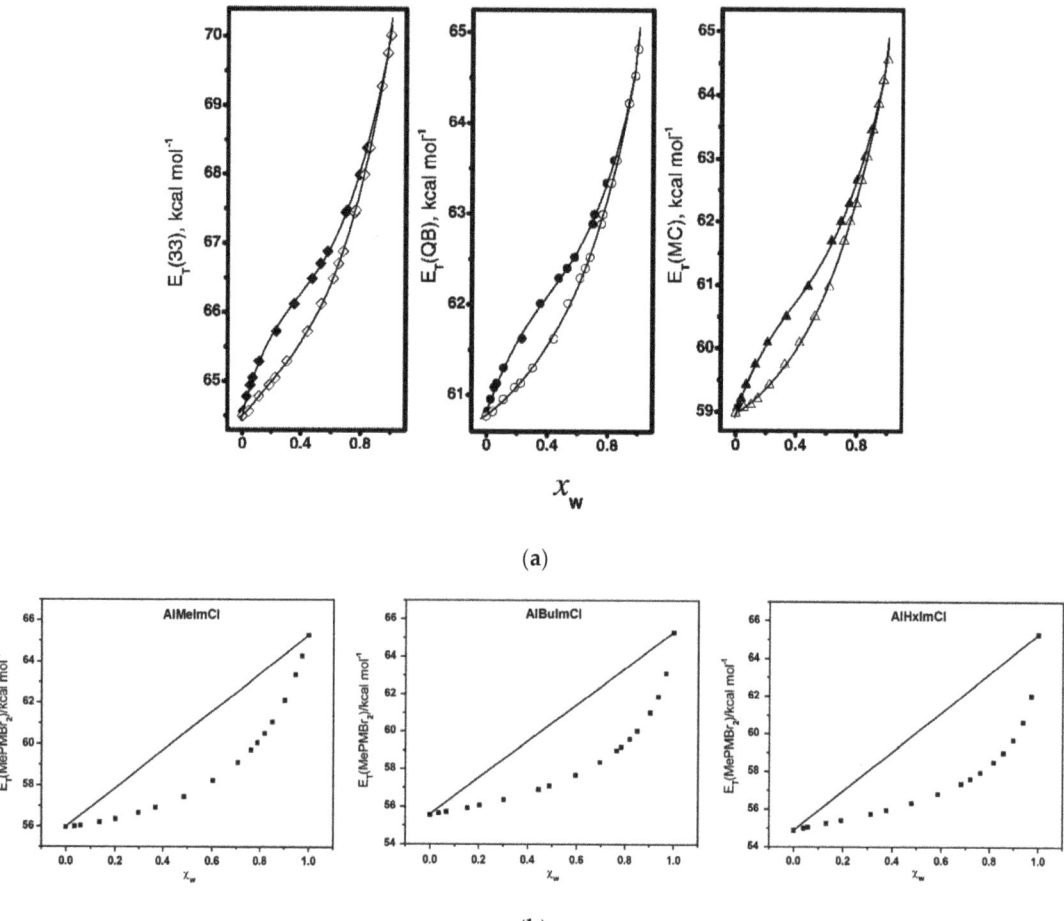

Figure 9. (a) Dependence of solvent polarity scale, E_T(probe), on analytical, $\chi_W^{Analytical}$ (open symbols), and "effective", $\chi_W^{Effective}$ (solid symbols) χ_W for MeOH-W mixtures at 25 °C. Reprinted with permission from [48]. (b) Dependence of the empirical solvent polarity parameter E_T(MePMBr$_2$) on the mole fraction of water, χ_W, at 25 °C, for mixtures of water with ILs. The straight lines connecting the polarities of the pure solvents are theoretical, plotted merely to depict the ideal solvation of the probe by the binary mixtures. Reprinted from [49].

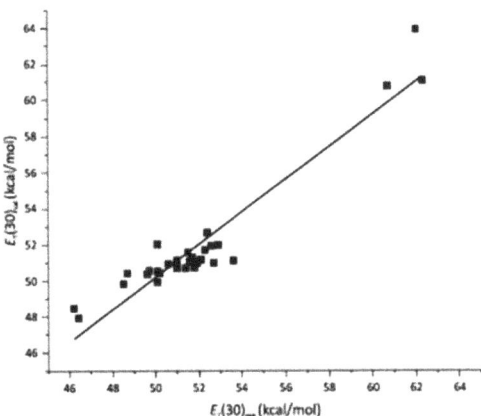

Figure 10. Correlation between experimentally determined $E_T(RB)$ and those calculated by Equation (1a) for ionic liquids. Reprinted with permission from [50].

4. Selected Examples of the Application of Solvatochromism to Understand Medium Effects on Chemical Phenomena

One of the most important applications is that the probes employed, when properly selected, can be used as models. Because most of these probes are dipolar or zwitterionic, and have varying hydrophilic/hydrophobic character, it should not be difficult to select the probes that are expected to match the reaction of interest. For a comprehensive list of solvatochromic probes, see references [19,25,27]. Consequently, the information obtained from the solvatochromic probe can be employed to understand the effects of medium composition on diverse chemical phenomena and processes.

Selected examples show the importance of the last phrase. Consider the following question: What is the medium effect on a chemical phenomenon of increasing probe hydrophobicity in a solvent mixture? An example is shown for the solvation in DMSO-W and 1-PrOH-W of 4 probes (Figure 11). Two of these are hydrophilic, 4- (pyridinium-1-yl)phenolate and 2-(pyridinium-1-yl)phenolate (p-CB and o-CB, respectively)); the other two (RB and o-RB, 2,4-dimethyl-6-(2,4,6-triphenylpyridinium-1-yl)phenolate) have more elaborate, hydrophobic structures (Figure 7); the corresponding values of φ are shown in Table 1 [25]. The straight lines in Figure 11 represent the case where there is no probe-induced preferential solvation, i.e., the composition of the probe solvation layer is equal to that of bulk solvent mixture. It is clear that this is not case; the deviation increases as a function of increasing probe hydrophobicity (RB is practically insoluble in water). Usually, more hydrophobic probes show more deviation from linearity [18].

With one exception (o-CB in DMSO-W), all values for $\varphi_{W/S}$ are <1, showing that the organic solvents displace water from the probe solvation layer. All values of $\varphi_{Mixed\ solvent/solvent}$ (mixed solvents displacing pure solvents) are > 1, i.e., the complex solvents are more efficient than the parent ones. As expected, the calculated empirical polarity of the mixed solvent is greater than that of pure DMSO or 1-PrOH, because the mixed solvent contains (more polar) water molecules. The reason for the efficiency of the mixed solvent is that probe–solvent interactions include H-bonding to the probe phenolate oxygen, as well as hydrophobic interactions. Thus, 1-PrOH/W has more sites for hydrogen-bond donation/acceptance than water or 1-PrOH, while it is also capable of solvating the probe by the hydrophobic effect due to the organic "end" of the solvent. A similar reasoning can be advanced for the efficiency of DMSO/W relative to W and DMSO. All $\varphi_{Solvent-W}$ are larger for 1-PrOH/W than DMSO/W, and for p-RB than p-CB or o-CB. The first result underlines the importance of H-bonding to solvation, whereas the second one is in agreement with the dependence of preferential solvation (hence, the values of φ) on the hydrophobicity of the probe [51].

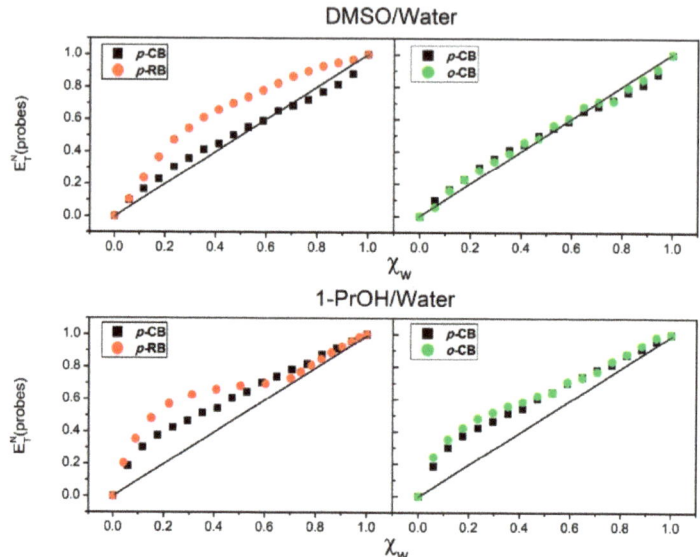

Figure 11. Dependence of E_T^N(probe) on the χ_W for W-DMSO (left) and W-1-PrOH (right) mixtures. The black and green squares are for (hydrophilic) p-CB, and o-CB, respectively. The red spheres are for (hydrophobic) p-RB (or RB). The curves clearly show that the deviation form ideality increases as a function of increasing probe hydrophobicity. Reprinted with permission from [25].

Table 1. Solvatochromism of p-RB, p-CB, and o-CB in binary mixtures of W-DMSO and W-1-PrOH at 25 °C. Reprinted with permission from [25].

	m	φ(W/DMSO)	φ(DMSO-W/DMSO)	φ(DMSO-W/W)	E_T(DMSO)	E_T(W)	E_T(W/DMSO)	$\chi^{2;\,b}$	$R^{2;\,b}$
			Solvatochromism in Water/Dimethyl Sulfoxide						
p-RB	1.32	0.49	3.29	6.71	45.3 (+0.2)[a]	63.1 (0)[a]	49.1	0.029	0.994
p-CB	0.80	0.80	1.70	2.12	58.4 (−0.4)[a]	77.9 (+0.2)[a]	67.9	0.034	0.999
o-CB	0.90	1.41	2.40	1.67	59.1 (−0.2)[a]	75.8 (+0.2)[a]	63.9	0.067	0.997
			Solvatochromism in Water/1-Propanol						
	m	φ (W/1-PrOH)	φ (1-PrOH-W/1-PrOH)	φ (1-PrOH-W/W)	E_T(1-PrOH)	E_T(W)	E_T(1-PrOH/W)	χ^2	R^2
p-RB	1.40	0.44	66.92	152.09	50.9 (+0.2)[a]	59.1 (+0.01)[a]	52.5	0.0023	0.999
p-CB	1.22	0.23	9.81	42.65	67.2 (+0.1)[a]	77.7 (0)[a]	74.0	0.006	0.999
o-CB	1.37	0.34	13.71	40.32	67.6 (+0.4)[a]	76.1 (−0.5)[a]	71.9	0.004	0.999

[a]—The numbers within parentheses refer to [calculated E_T(probe) from Equation (9)− experimentally determined E_T(probe)]. [b]—The terms χ^2 and R^2 have their usual (statistical) meaning as measure of the goodness of fit.

Likewise, for the solvation of the same probe in a series of structurally related solvents (e.g., alcohols) the fractionation factors, hence the compositions of the probe solvation layer are sensitively dependent on the hydrophobicity of ROH, as shown in Table 2, for the solvation of a hydrophobic- (WB; 2,6-dichloro-4-(2,4,6-triphenylpyridinium-1-yl)-phenolate; pKa = 4.78) and a hydrophilic probe (QB; 1-methylquinolinium-8-olate; pKa = 6.80), at 25 °C [23,52].

Table 2 shows that values of φ are probe-dependent, being larger for WB than for QB in any W-ROH mixture. Additionally, the values $\varphi_{ROH-W/ROH}$ and $\varphi_{ROH-W/W}$ increase on going from MeOH to 1-PrOH. If H-bonding to the phenolate oxygen of WB was the dominant probe-solvent interaction, then the expected order of $\varphi_{ROH-W/ROH}$ and $\varphi_{ROH-W/W}$

should have been QB (stronger base) > WB (weaker base); this is not the case. We conclude that solute-solvent hydrophobic interactions dominate the solvation of these probes. This also agrees with the dependence of φ on the hydrophobicity (or log P) of the alcohol (MeOH to 1-PrOH). Similar results were observed for the solvation of a series of merocyanine probes (see Figure 7), in binary mixture of W-ROH (MeOH, 1-PrOH) and W-MS (MeCN and DMSO) where the values of φ increase on going from MePMBr$_2$ (less hydrophobic probe) to OcPMBr$_2$ (2,6-dibromo-4-[(E)-2-(1-octylpyridinium-4-yl)ethenyl]; more hydrophobic probe) *in every binary mixture*, and from methanol to 1-PrOH *for the same probe* [23]. This type of information cannot be easily obtained by other approaches and is important, e.g., for choosing the appropriate solvent mixture for a certain application.

As in the experimental determination of the activation energy of reactions, solvation was studied as a function of temperature. An example for the dependence of E_T(probe) on T is shown in Figure 12, for the solvation of QB and MePMBr$_2$ in alkoxy-alcohols [53].

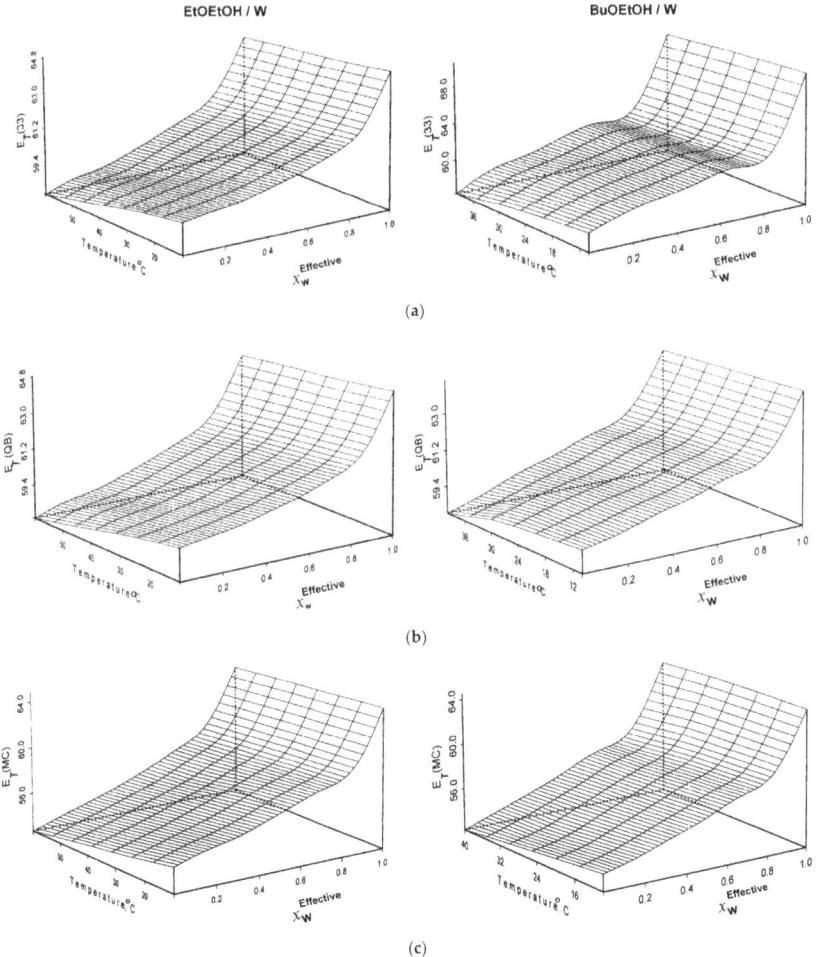

Figure 12. Dependence of the empirical solvent polarity on temperature, for the solvation of WB (a), QB (b), and MePMBr2 (c) in aqueous alcohols. Reprinted with permission from [53].

Plots (not shown) of E_T(probe)$_{Solvent}$ versus T gave excellent straight lines, the negative slopes of which are given by ΔE_T(probe)$_{Solvent}/\Delta T$ (in kcal mol^{-1} K^{-1}). These were calculated for pure solvents; the order is $|\Delta E_T(\text{probe})_{ROH}| > |\Delta E_T(\text{probe})_W|$, reflecting the greater effect of temperature on the structure of ROH. Consequently, H-bonding of water with the probe ground state is less susceptible to temperature increases than ROH. Values of $\Delta E_T(\text{probe})_{Solvent}$ for 50 °C intervals ($\Delta E_T(10\ °C)$–$\Delta E_T(60\ °C)$) for WB and MePMBr$_2$ in mixtures of water and 9 alcohols and alkoxy-alcohols were calculated, which range between 2.1 and 3.7 kcal mol^{-1}. This is a sizeable energy, relative to the activation energies of many reactions (ca. 10–12 kcal/mol). This underlines the importance of studying thermo-solvatochromism to quantify the contribution of temperature-induced changes in solvation to the energetics of reactions in solution, a quantity that cannot be calculated, e.g., from rate data.

An example that shows the usefulness of solvatochromic probes as models for chemical reactions is the pH-independent hydrolysis of esters, including the hydrophobic 2,4-dinitrophenyl carbonate [54], the relatively hydrophilic 4-nitrophenyl chloroformate (calculated log P = 1.66), and the very hydrophobic ester 4-nitrophenyl heptafluorobutyrate (calculated log P = 4.02) [55]; all reactions were studied in W-MeCN mixtures (see Scheme 1). These reactions show a complex dependence of reaction rate constants on water concentration, as shown in the left-hand parts of Figures 13 and 14.

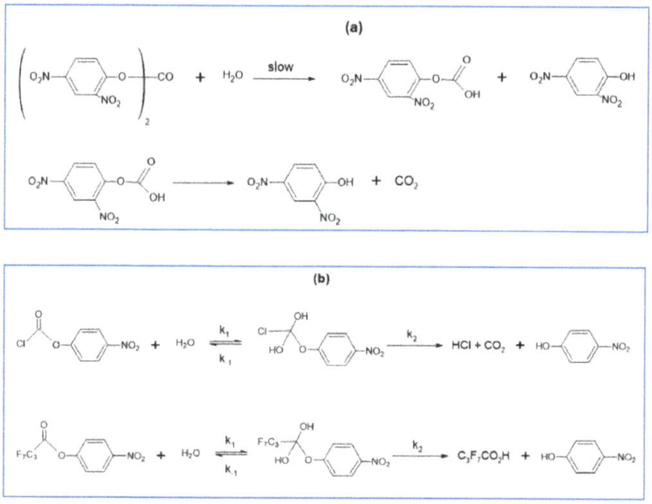

Scheme 1. Pathways for the pH-independent hydrolyses of: (a) DNPC, reprinted with permission from [54]; and (b) NPCF and NPFB, reprinted with permission from [55].

Because addition of MeCN to water induces a series of structural changes in the medium that depends on binary solvent composition [56–58], we carried out a "proton-inventory" study to probe the structures of the corresponding transition states [59]. Our results showed that the complex dependence of reaction rate constants and activation parameters on [H$_2$O] is not due to changes in the number of water molecules in the transition states. It reflects, however, the effects of acetonitrile-water interactions on solvation of reactants and transition states. We therefore expected that the dependence of kinetic data on [W] should be similar to that of model solvatochromic probes, as clearly illustrated by Figures 13 and 14. Note that these hydrolysis reactions are particularly suitable to test the potential of using solvatochromic probes as models for chemical phenomena, because complicating acid or base catalysis play no role.

Table 2. Results for the solvation of WB and QB in water-alcohol mixtures at 25 °C [30].

ROH	Log P	$K_{association}$ (L·mol^{-1})	$\varphi_{W/ROH}$	$\varphi_{ROH\text{-}W/ROH}$	$\varphi_{ROH\text{-}W/W}$
		WB			
MeOH	−0.77	173.3	0.601	2.212	3.681
EtOH	−0.31	28	0.554	11.482	20.727
1-PrOH	0.25	12.3	0.265	149.208	563.049
2-PrOH	0.05	8.1	0.551	192.625	349.592
2-Me-2-PrOH	0.35	7.0	0.484	111.267	229.890
MeO-EtOH	−0.77	32.1	0.479	5.659	11.814
		QB			
MeOH	−0.77	173.3	0.381	1.172	3.076
EtOH	−0.31	28	0.349	5.053	14.479
1-PrOH	0.25	12.3	0.305	29.599	97.046
2-PrOH	0.05	8.1	0.428	26.418	61.724
2-Me-2-PrOH	0.35	7.0	0.364	21.713	59.651
MeO-EtOH	−0.77	32.1	0.341	4.855	14.238

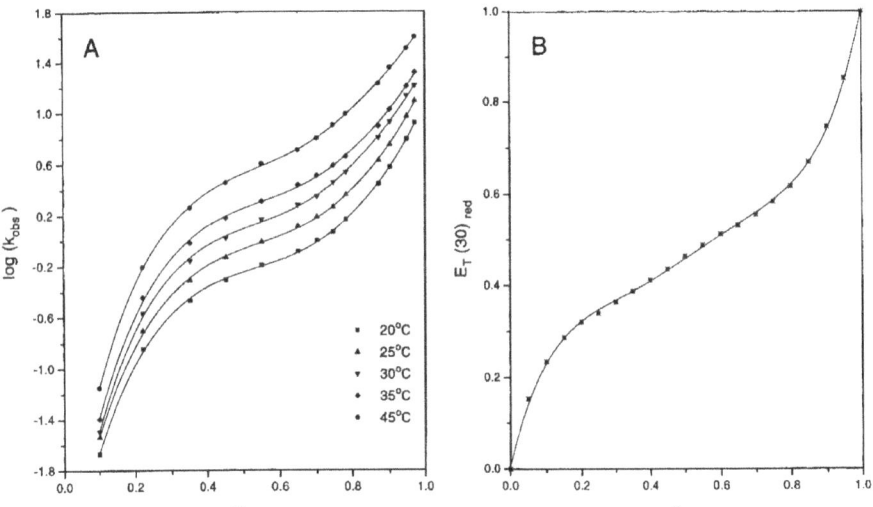

Figure 13. (**A**) Dependence of log (k_{obs}), the observed rate constant on χ_W, the mole fraction of water in aqueous acetonitrile, at different temperatures. The points are experimental, and the solid sigmoidal lines were calculated by a fourth power polynomial dependence of log (k_{obs}) o χ_W. (**B**) Dependence of the empirical solvent polarity E_T(RB) on χ_W. The solid curve was calculated from a fifth power polynomial dependence of E_T(RB) on χ_W. Reprinted with permission from [54].

Another example of the application of solvatochromic probes to elucidate chemical reactions is retro-DielsAlder reaction (RDA) of anthracene-9,10-dione in aqueous solutions (Scheme 2). The RDA reaction proceeds exceptionally fast in water compared to organic solvents (Figure 15a) because this solvent greatly accelerates pericyclic reactions through H-bonding, which stabilizes the activated complex. [60]. Figure 15b shows that this reaction has the Gibbs energy of activation clearly governed by the polarity of the solvent.

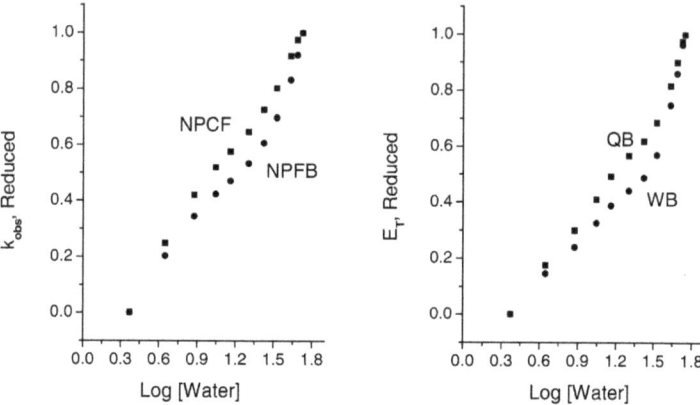

Figure 14. Plots of the dependence of log (k_{obs}) and of E_T(probe) on log [W], at 25 °C. Reduced log (k_{obs}) and reduced E_T(probe) are employed, so that the results of different species (esters, probes) may be directly compared. Values of E_T(probe) of pure water and pure MeCN are, respectively, 64.62 and 52.97 kcal mol^{-1} (QB) and 70.24 and 54.72 kcal mol^{-1} (WB). Reprinted with permission from [51].

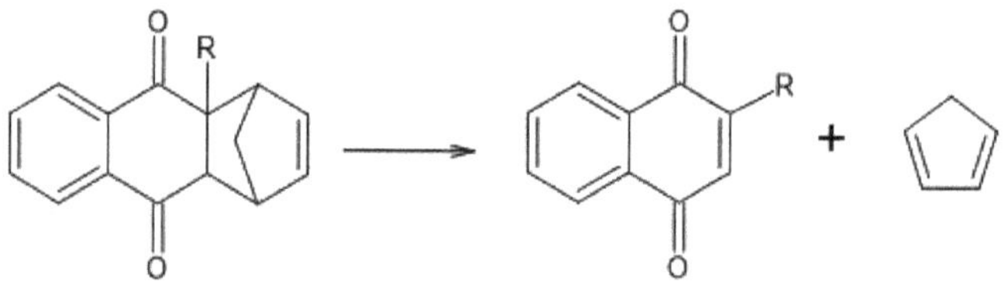

Scheme 2. RDA of anthracene-9,10-dione in aqueous solution. Reprinted with permission from [60].

We end this discussion by showing how the results of solvatochromism can be employed to determine the relative importance of solvent descriptors to cellulose dissolution. First, solvatochromism of WB was studied in several IL-DMSO mixtures, at χ_{DMSO} = 0.6 and 40 °C [61]. This study generated an equation showing the dependence of E_T(WB) on solvent descriptors SA, SB, SD and SP. In a later study, microcrystalline cellulose (MCC) was dissolved in these binary mixtures and the mass of dissolved MCC (expressed as %m) was correlated with the same solvent descriptors, calculated from the solvatochromic study [5], see Equations (10) and (11):

$$E_T(WB) = 54.61 + 2.77\ SA + 0.61\ SB - 1.06\ V_M + 0.41\ f(n) \rightarrow \text{Based on solvatochromism in 13 IL-DMSO mixtures; } R^2 = 0.939 \quad (10)$$

$$MCC\text{-}m\% = 9.49 + 3.01\ SA + 6.88\ SB - 2.99\ V_M + 2.427\ f(n) \rightarrow \text{Based on MCC dissolution in 11 IL-DMSO mixtures; } R^2 = 0.902 \quad (11)$$

where SA and SB are those defined above, V_M and $f(n)$ are the molar volume of the neat IL, and the LorentzLorenz refractive index function; these were employed instead of SD and SP, as explained elsewhere [61]. The signs of the regression coefficients in both equations are the same. That is, solvatochromism and MCC dissolution are enhanced by solvent Lewis acidity, Lewis basicity and polarizability; the inverse is true for the molar volume of the IL. The different relative importance of SA/SB in both cases can be explained because MCC is a H-bond acceptor and donor, whereas the probes are only H-bond acceptors. The relevant point, however, is that two very different chemical process can be similarly correlated with the same set of solvent descriptors. This agreement can be fruitfully employed, e.g., for screening possible candidates as solvents for cellulose and other biopolymers.

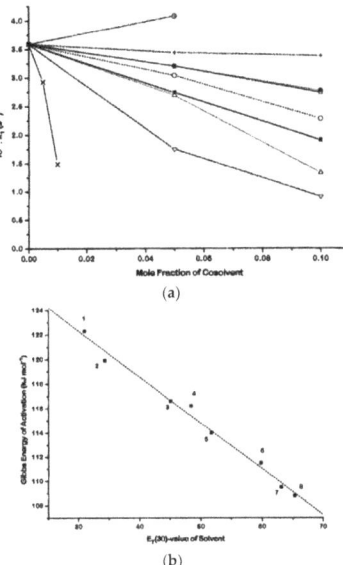

Figure 15. (a) First-order rate constant ($\times 10^6$ s^{-1}; 40 °C) of the RDA reaction of anthracene-9,10-dione in aqueous solutions versus the mole fraction of organic cosolvent: MeOH (□), EtOH (○), 1-PrOH (△), t-BuOH (▽), formamide (●), acetonitrile (■), 1-cyclohexyl-2-pyrrolidinone (×), urea (+), and glucose (⊕). (b) Correlation between the Gibbs free energy of activation of the RDA reaction of anthracene-9,10-dione and the solvatochromic parameter E_T(RB) of the solvents. Reprinted with permission from [60].

5. Conclusions

Solvent mixtures are extensively employed in chemistry for practical reasons. Their use leads to enhanced solubilities, increased rate constants, etc. These beneficial effects call for a clear understanding of solvent effects on chemical phenomena. This understanding is, however, hindered by two complications: (i) The dependencies of the physicochemical properties of the binary mixtures on its composition are usually not ideal, and (ii) The compositions of the solvation layers of the solutes are almost always different from those of the bulk solvents. The reasons for the latter differences are combinations of dielectric enrichment and, more importantly, H-bonding and solute-solvent hydrophobic interactions. This intricate situation has been greatly simplified by using solvatochromic probes as models for the reaction or process under consideration. By applying appropriate solvation models, the study of these probes furnishes a set of solvent-exchange equilibrium constants (φ) that permits calculation of the effective (or local) composition of the solvatochromic probe. If the latter is a good model for the reaction or process under consideration, one can access information (about the composition of the solvation layer) that is inaccessible by other techniques. To our view, this represents a practical solution for a complex problem. Theoretical calculations will certainly enhance our understanding of solvation, because they permit, *inter alia*, calculation of the UV-Vis spectra of probes in mixed solvents [62], and prediction of solvatochromism [63].

A list of all abbreviations employed is given below. The structures of the solvatochromic probes discussed are shown in Figure 7.

Funding: This research received no external funding.

Acknowledgments: We thank the following agencies for research project grants: O. A. El Seoud, FAPESP, grant 2014/22136-4; CNPq, grant 306108/2019-4; N. I. Malek, UGC-DAE, Collaborative Research Scheme (UDCSR/MUM/CRS-M-997/2023). O. A. El Seoud thanks Clarissa T. Martins, Carina Loffredo, Erika B. Tada, Jéssica C. de Jesus, Ludmila C. Fidale, Luzia P. Novaki, Marc Kostag, Marcella T. Dignani, Nicolas

Keppeler, Paulo R. Pires, Priscilla L. Silva, Romeu Casarano; Shirley Possidonio, and Thaís A. Bioni because their dedication and enthusiasm resulted in the solvatochromic results discussed here, and C. Reichardt for stimulating discussions on solvatochromism during his visits to Marburg.

Conflicts of Interest: The authors declare no conflict of interest.

Abbreviations and Acronyms

AlMeImCl	1-Allyl-3-methylimidazolium chloride
BuMeImCl	1-(1-butyl)-3-methylimidazolium chloride
BuPMBr$_2$	2,6-dibromo-4-[(E)-2-(1-butylpyridinium-4-yl)ethenyl]
t-Bu$_5$RB	2,6-bis[4-(t-butyl)phenyl]-4-{2,4,6-tris[4-(t-butyl)phenyl]pyridinium-1-yl}phenolate
t-BuOH	tert-Butanol
o-CB	2-(pyridinium-1-yl)phenolate
p-CB	4- (pyridinium-1-yl)phenolate
DMANF	2-(N,N-dimethylamino)-7-nitrofluorene
DMF	N,N-Dimethylformamide
DMSO	Dimethyl sulfoxide
DNPC	2,4-dinitrophenyl carbonate
DTBSB	o,o'-di-tert-butylstilbazolium betaine
E_T (probe)	Empirical solvent polarity scale using a specific probe
EtOH	Ethanol
FePhen	[Fe$_{II}$(1,10-phenanthroline)$_2$(CN)$_2$]
HxPMBr$_2$	2,6-Dibromo-4-[(E)-2-(1-hexylpyridinium-4-yl)ethenyl]
IL	Ionic liquid
Log P	Partition coefficient between two partially immiscible solvents, usually n-octanol and water
MCC	Microcrystalline cellulose
MeCN	Acetonitrile
MeOEtOH	2-Methoxyethanol
MeOH	Methanol
MePMBr$_2$	2,6-Dibromo-4-[(E)-2-(1-methylpyridinium-4-yl)ethenyl]
2-Me-2-PrOH	2-Methyl-2-propanol
MeNI	1-Methyl-5-nitroindoline
MS	Molecular solvent
NPFB	4-Nitrophenyl chloroformate
NHFB	4-Nitrophenyl heptafluorobutyrate
NI	5-Nitroindoline
OcPMBr$_2$	2,6-Dibromo-4-[(E)-2-(1-octylpyridinium-4-yl)ethenyl]
PEG	Polyethylene glycol
1-PrOH	1-Propanol
QB	1-Methylquinolinium-8-olate
RB (or p-RB)	Reichardt betaine, dye number 30 in a list of solvatochromic dyes; 2,6-diphenyl-4-(2,4,6-triphenylpyridinium-1-yl) phenolate.
o-RB	2,4-dimethyl-6-(2,4,6-triphenylpyridinium-1-yl)phenolate,
RDA	Retro-Diels—Alder reaction
SA	Solvent Lewis acidity; calculated from solvatochromic data
SB	Solvent Lewis Basicity; calculated from solvatochromic data
SD	Solvent dipolarity; calculated from solvatochromic data
Solvatochromism	Effect of the medium on the color of a solvatochromic probe
SP	Solvent polarizability; calculated from solvatochromic data
TBSB	o-tert-butylstilbazolium betaine
Thermo-solvatochromism	Effect of temperature on solvatochromism
THF	Tetrahydrofuran
WB	Wolfbeis betaine; 2,6-dichloro-4-(2,4,6-triphenylpyridinium-1-yl)-phenolate
φ	Fractionation factor: Refers to solvent exchange equilibrium constant between bulk solvent mixture and the solvation layer of the probe
χ	Mole fraction

References

1. Chiavone-Filho, O.; Rasmussen, P. Solubilities of Salts in Mixed Solvents. *J. Chem. Eng. Data* **1993**, *38*, 367–369. [CrossRef]
2. Pinho, S.P.; Macedo, E.A. Solubility of NaCl, NaBr, and KCl in Water, Methanol, Ethanol, and Their Mixed Solvents. *J. Chem. Eng. Data* **2005**, *50*, 29–32. [CrossRef]
3. Dogan, H.; Hilmioglu, N.D. Dissolution of cellulose with NMMO by microwave heating. *Carbohydr. Polym.* **2009**, *75*, 90–94. [CrossRef]
4. Kostag, M.; Jedvert, K.; Achtel, C.; Heinze, T.; El Seoud, O.A. Recent advances in solvents for the dissolution, shaping and derivatization of cellulose: Quaternary ammonium electrolytes and their solutions in water and molecular solvents. *Molecules* **2018**, *23*, 511. [CrossRef] [PubMed]
5. El Seoud, O.A.; Bioni, T.A.; Dignani, M.T. Understanding cellulose dissolution in ionic liquid-dimethyl sulfoxide binary mixtures: Quantification of the relative importance of hydrogen bonding and hydrophobic interactions. *J. Mol. Liq.* **2021**, *322*, 114848. [CrossRef]
6. Keppeler, N.; Pires, P.A.R.; Freitas, J.L.S.; El Seoud, O.A. Cellulose dissolution in mixtures of ionic liquids and molecular solvents: The fruitful synergism of experiment and theory. *J. Mol. Liq.* **2023**, *386*, 122490. [CrossRef]
7. Minnick, D.L.; Flores, R.A.; DeStefano, M.R.; Scurto, A.M. Cellulose Solubility in Ionic Liquid Mixtures: Temperature, Cosolvent, and Antisolvent Effects. *J. Phys. Chem. B* **2016**, *120*, 7906–7919. [CrossRef]
8. Zhong, C.; Cheng, F.; Zhu, Y.; Gao, Z.; Jia, H.; Wei, P. Dissolution mechanism of cellulose in quaternary ammonium hydroxide: Revisiting through molecular interactions. *Carbohydr. Polym.* **2017**, *174*, 400–408. [CrossRef]
9. El Seoud, O.A.; Keppeler, N. Education for Sustainable Development: An Undergraduate Chemistry Project on Cellulose Dissolution, Regeneration, and Chemical Recycling of Polycotton. *J. Lab. Chem. Educ.* **2020**, *8*, 11–17. [CrossRef]
10. Zhang, K.; Yang, J.; Yu, X.; Zhang, J.; Wei, X. Densities and Viscosities for Binary Mixtures of Poly(ethylene glycol) 400 + Dimethyl Sulfoxide and Poly(ethylene glycol) 600 + Water at Different Temperatures. *J. Chem. Eng. Data* **2011**, *56*, 3083–3088. [CrossRef]
11. Yang, F.; Wang, X.; Tan, H.; Liu, Z. Improvement the viscosity of imidazolium-based ionic liquid using organic solvents for biofuels. *J. Mol. Liq.* **2017**, *248*, 626–633. [CrossRef]
12. Lv, Y.; Wu, J.; Zhang, J.; Niu, Y.; Liu, C.-Y.; He, J.; Zhang, J. Rheological properties of cellulose/ionic liquid/dimethylsulfoxide (DMSO) solutions. *Polymer* **2012**, *53*, 2524–2531. [CrossRef]
13. Alam, M.S.; Ashokkumar, B.; Siddiq, A.M. The density, dynamic viscosity and kinematic viscosity of protic and aprotic polar solvent (pure and mixed) systems: An experimental and theoretical insight of thermophysical properties. *J. Mol. Liq.* **2019**, *281*, 584–597. [CrossRef]
14. Suppan, P. Local polarity of solvent mixtures in the field of electronically excited molecules and exciplexes. *J. Chem. Soc. Faraday Trans. 1 Phys. Chem. Condens. Phases* **1987**, *83*, 495. [CrossRef]
15. Suppan, P. Time-resolved luminescence spectra of dipolar excited molecules in liquid and solid mixtures. Dynamics of dielectric enrichment and microscopic motions. *Faraday Discuss. Chem. Soc.* **1988**, *85*, 173. [CrossRef]
16. LNovaki, P.; El Seoud, O.A. Solvatochromism in binary solvent mixtures: Effects of the molecular structure of the probe. *Berichte Bunsenges. Phys. Chem.* **1997**, *101*, 902–909. [CrossRef]
17. Skaf, M.S. Molecular Dynamics Study of Dielectric Properties of Water−Dimethyl Sulfoxide Mixtures. *J. Phys. Chem. A* **1999**, *103*, 10719–10729. [CrossRef]
18. El Seoud, O.A. Solvation in pure and mixed solvents: Some recent developments. *Pure Appl. Chem.* **2007**, *79*, 1135–1151. [CrossRef]
19. Machado, V.G.; Stock, R.I.; Reichardt, C. Pyridinium N-Phenolate Betaine Dyes. *Chem. Rev.* **2014**, *114*, 10429–10475. [CrossRef]
20. Bai, B.; Li, Z.; Wang, H.; Li, M.; Ozaki, Y.; Wei, J. Exploring the difference in xerogels and organogels through in situ observation. *R. Soc. Open Sci.* **2018**, *5*, 170492. [CrossRef]
21. McHale, J.L. Subpicosecond Solvent Dynamics in Charge-Transfer Transitions: Challenges and Opportunities in Resonance Raman Spectroscopy. *Acc. Chem. Res.* **2001**, *34*, 265–272. [CrossRef] [PubMed]
22. Jha, K.K.; Kumar, A.; Munshi, P. Solvatochromism and Reversible Solvent Exchange Phenomena in Solvatomorphic Organic Chromophore Crystals. *Cryst. Growth Des.* **2023**, *23*, 2922–2931. [CrossRef]
23. Martins, C.T.; Lima, M.S.; El Seoud, O.A. Thermosolvatochromism of merocyanine polarity indicators in pure and aqueous solvents: Relevance of solvent lipophilicity. *J. Org. Chem.* **2006**, *71*, 9068–9079. [CrossRef] [PubMed]
24. El Seoud, O.A.; Loffredo, C.; Galgano, P.D.; Sato, B.M.; Reichardt, C. Have biofuel, will travel: A colorful experiment and a different approach to teach the undergraduate laboratory. *J. Chem. Educ.* **2011**, *88*, 1293–1297. [CrossRef]
25. Pires, P.A.R.; El Seoud, O.A.; Machado, V.G.; de Jesus, J.C.; de Melo, C.E.A.; Buske, J.L.O.; Cardozo, A.P. Understanding Solvation: Comparison of Reichardt's Solvatochromic Probe and Related Molecular 'core' Structures. *J. Chem. Eng. Data* **2019**, *64*, 2213–2220. [CrossRef]
26. Dimroth, K.; Reichardt, C.; Siepmann, T.; Bohlmann, F. Über Pyridinium-N-phenol-betaine und ihre Verwendung zur Charakterisierung der Polarität von Lösungsmitteln. *Justus Liebigs Ann. Chem.* **1963**, *661*, 1–37. [CrossRef]
27. Reichardt, C.; Welton, T. *Solvents and Solvent Effects in Organic Chemistry*, 4th ed.; John Wiley & Sons: Hoboken, NJ, USA, 2011.
28. NWeiß; Schmidt, C.H.; Thielemann, G.; Heid, E.; Schröder, C.; Spange, S. The physical significance of the Kamlet-Taft π* parameter of ionic liquids. *Phys. Chem. Chem. Phys.* **2021**, *23*, 1616–1626. [CrossRef]
29. El Seoud, O.A.; Kostag, M.; Jedvert, K.; Malek, N.I. Cellulose in Ionic Liquids and Alkaline Solutions: Advances in the Mechanisms of Biopolymer Dissolution and Regeneration. *Polymers* **2019**, *11*, 1917. [CrossRef]

30. Tada, E.B.; Silva, P.L.; El Seoud, O.A. Thermo-solvatochromism of zwitterionic probes in aqueous alcohols: Effects of the properties of the probe and the alcohol. *Phys. Chem. Chem. Phys.* **2003**, *5*, 5378–5385. [CrossRef]
31. Antonious, M.S.; Tada, E.B.; El Seoud, O.A. Thermo-solvatochromism in aqueous alcohols: Effects of the molecular structures of the alcohol and the-solvatochromic probe. *J. Phys. Org. Chem.* **2002**, *15*, 403–412. [CrossRef]
32. Rosés, M.; Ràfols, C.; Ortega, J.; Bosch, E. Solute–solvent and solvent–solvent interactions in binary solvent mixtures. Part 1. A comparison of several preferential solvation models for describing $E_T(30)$ polarity of bipolar hydrogen bond acceptor-cosolvent mixtures. *J. Chem. Soc. Perkin Trans.* **1995**, *2*, 1607–1615. [CrossRef]
33. Bosch, E.; Herodes, K.; Koppel, I.; Leito, I.; Koppel, I.; Taal, V.; Rosés, M. Solute-solvent and solvent-solvent interactions in binary solvent mixtures. 2. Effect of temperature on the $E_T(30)$ polarity parameter of dipolar hydrogen bond acceptor-hydrogen bond donor mixtures. *J. Phys. Org. Chem.* **1996**, *9*, 403–410. [CrossRef]
34. Ortega, J.; Ràfols, C.; Bosch, E.; Rosés, M. Solute–solvent and solvent–solvent interactions in binary solvent mixtures. Part 3. The $E_T(30)$ polarity of binary mixtures of hydroxylic solvents. *J. Chem. Soc. Perkin Trans.* **1996**, *2*, 1497–1503. [CrossRef]
35. Bosch, E.; Rived, F.; Rosés, M. Solute–solvent and solvent–solvent interactions in binary solvent mixtures. Part 4. Preferential solvation of solvatochromic indicators in mixtures of 2-methylpropan-2-ol with hexane, benzene, propan-2-ol, ethanol and methanol. *J. Chem. Soc. Perkin Trans.* **1996**, *2*, 2177–2184. [CrossRef]
36. Ràfols, C.; Rosés, M.; Bosch, E. Solute–solvent and solvent–solvent interactions in binary solvent mixtures. Part 5. Preferential solvation of solvatochromic indicators in mixtures of propan-2-ol with hexane, benzene, ethanol and methanol. *J. Chem. Soc. Perkin Trans.* **1997**, *2*, 243–248. [CrossRef]
37. Buhvestov, U.; Rived, F.; Ràfols, C.; Bosch, E.; Rosés, M. Solute-solvent and solvent-solvent interactions in binary solvent mixtures. Part 7. Comparison of the enhancement of the water structure in alcohol-water mixtures measured by solvatochromic indicators. *J. Phys. Org. Chem.* **1998**, *11*, 185–192. [CrossRef]
38. Chen, J.-S.; Shiao, J.-C. Graphic method for the determination of the complex NMR shift and equilibrium constant for a hetero-association accompanying a self-association. *J. Chem. Soc. Faraday Trans.* **1994**, *90*, 429. [CrossRef]
39. Eblinger, F.; Schneider, H.-J. Self-Association of Water and Water−Solute Associations in Chloroform Studied by NMR Shift Titrations. *J. Phys. Chem.* **1996**, *100*, 5533–5537. [CrossRef]
40. Sacco, A.; De Cillis, F.M.; Holz, M. NMR Studies on hydrophobic interactions in solution Part 3 Salt effects on the self-association of ethanol in water at two different temperatures. *J. Chem. Soc. Faraday Trans.* **1998**, *94*, 2089–2092. [CrossRef]
41. Max, J.-J.; Daneault, S.; Chapados, C. 1-Propanol hydrate by IR spectroscopy. *Can. J. Chem.* **2002**, *80*, 113–123. [CrossRef]
42. Nomen, R.; Sempere, J.; Avilés, K. Detection and characterisation of water alcohol hydrates by on-line FTIR using multivariate data analysis. *Chem. Eng. Sci.* **2001**, *56*, 6577–6588. [CrossRef]
43. Ghoraishi, M.S.; Hawk, J.E.; Phani, A.; Khan, M.F.; Thundat, T. Clustering mechanism of ethanol-water mixtures investigated with photothermal microfluidic cantilever deflection spectroscopy. *Sci. Rep.* **2016**, *6*, 23966. [CrossRef] [PubMed]
44. Katz, E.D.; Ogan, K.; Scott, R.P.W. Distribution of a solute between two phases: The basic theory and its application to the prediction of chromatographic retention. *J. Chromatogr. A* **1986**, *352*, 67–90. [CrossRef]
45. Katz, E.D.; Lochmüller, C.H.; Scott, R.P.W. Methanol-Water Association and Its Effect on Solute Retention in Liquid Chromatography. *Anal. Chem.* **1989**, *61*, 349–355. [CrossRef]
46. Scott, R.P.W. The thermodynamic properties of methanol–water association and its effect on solute retention in liquid chromatography. *Analyst* **2000**, *125*, 1543–1547. [CrossRef]
47. Bastos, E.L.; Silva, P.L.; El Seoud, O.A. Thermosolvatochromism of Betaine Dyes Revisited: Theoretical Calculations of the Concentrations of Alcohol−Water Hydrogen-bonded Species and Application to Solvation in Aqueous Alcohols. *J. Phys. Chem. A* **2006**, *110*, 10287–10295. [CrossRef]
48. Tada, E.B.; Silva, P.L.; El Seoud, O.A. Thermo-solvatochromism of betaine dyes in aqueous alcohols: Explicit consideration of the water-alcohol complex. *J. Phys. Org. Chem.* **2003**, *16*, 691–699. [CrossRef]
49. Sato, B.M.; de Oliveira, C.G.; Martins, C.T.; El Seoud, O.A. Thermo-solvatochromism in binary mixtures of water and ionic liquids: On the relative importance of solvophobic interactions. *Phys. Chem. Chem. Phys.* **2010**, *12*, 1764. [CrossRef]
50. Spange, S.; Lienert, C.; Friebe, N.; Schreiter, K. Complementary interpretation of $E_T(30)$ polarity parameters of ionic liquids. *Phys. Chem. Chem. Phys.* **2020**, *22*, 9954–9966. [CrossRef]
51. Sato, B.M.; Martins, C.T.; El Seoud, O.A. Solvation in aqueous binary mixtures: Consequences of the hydrophobic character of the ionic liquids and the solvatochromic probes. *New J. Chem.* **2012**, *36*, 2353. [CrossRef]
52. Martins, C.T.; Lima, M.S.; El Seoud, O.A. A novel, convenient, quinoline-based merocyanine dye: Probing solvation in pure and mixed solvents and in the interfacial region of an anionic micelle. *J. Phys. Org. Chem.* **2005**, *18*, 1072–1085. [CrossRef]
53. Tada, E.B.; Silva, P.L.; Tavares, C.; El Seoud, O.A. Thermo-solvatochromism of zwitterionic probes in aqueous aliphatic alcohols and in aqueous 2-alkoxyethanols: Relevance to the enthalpies of activation of chemical reactions. *J. Phys. Org. Chem.* **2005**, *18*, 398–407. [CrossRef]
54. El Seoud, O.A.; El Seoud, M.I.; Farah, J.P.S. Kinetics of the pH-Independent Hydrolysis of Bis(2,4-dinitrophenyl) Carbonate in Acetonitrile−Water Mixtures: Effects of the Structure of the Solvent. *J. Org. Chem.* **1997**, *62*, 5928–5933. [CrossRef]
55. El Seoud, O.A.; Siviero, F. Kinetics of the pH-independent hydrolyses of 4-nitrophenyl chloroformate and 4-nitrophenyl heptafluorobutyrate in water-acetonitrile mixtures: Consequences of solvent composition and ester hydrophobicity. *J. Phys. Org. Chem.* **2006**, *19*, 793–802. [CrossRef]

56. Easteal, A. A Nuclear Magnetic Resonance Study of Water + Acetonitrile Mixtures. *Aust. J. Chem.* **1979**, *32*, 1379. [CrossRef]
57. Balakrishnan, S.; Easteal, A. Intermolecular interactions in water + acetonitrile mixtures: Evidence from the composition variation of solvent polarity parameters. *Aust. J. Chem.* **1981**, *34*, 943. [CrossRef]
58. Easteal, A.J.; Woolf, L.A. Measurement of (p, V, x) for (water + acetonitrile) at 298.15 K. *J. Chem. Thermodyn.* **1982**, *14*, 755–762. [CrossRef]
59. Schowen, K.B.J. Solvent Hydrogen Isotope Effects. In *Transition States of Biochemical Processes*; Springer: Boston, MA, USA, 1978; pp. 225–283.
60. Wijnen, J.W.; Engberts, J.B.F.N. Retro-Diels-Alder Reaction in Aqueous Solution: Toward a Better Understanding of Organic Reactivity in Water. *J. Org. Chem.* **1997**, *62*, 2039–2044. [CrossRef]
61. Bioni, T.A.; de Oliveira, M.L.; Dignani, M.T.; El Seoud, O.A. Understanding the efficiency of ionic liquids-DMSO as solvents for carbohydrates: Use of solvatochromic- and related physicochemical properties. *New J. Chem.* **2020**, *44*, 14906–14914. [CrossRef]
62. Lacerda, E.G.; Canuto, S.; Coutinho, K. New insights on nonlinear solvatochromism in binary mixture of solvents. In *Advances in Quantum Chemistry*; Oddershede, J., Brändas, E.J., Eds.; Jack Sabin, Scientist and Friend; Academic Press: Cambridge, MA, USA, 2022; pp. 57–79.
63. Saini, V.; Kumar, R. A machine learning approach for predicting the empirical polarity of organic solvents. *New J. Chem.* **2022**, *46*, 16981–16989. [CrossRef]

Disclaimer/Publisher's Note: The statements, opinions and data contained in all publications are solely those of the individual author(s) and contributor(s) and not of MDPI and/or the editor(s). MDPI and/or the editor(s) disclaim responsibility for any injury to people or property resulting from any ideas, methods, instructions or products referred to in the content.

Review

Solvent Replacement Strategies for Processing Pharmaceuticals and Bio-Related Compounds—A Review

Jia Lin Lee [1], Gun Hean Chong [1,*], Masaki Ota [2,3], Haixin Guo [4] and Richard Lee Smith, Jr. [2,*]

[1] Faculty of Food Science and Technology, Universiti Putra Malaysia, Serdang 43400, Selangor, Malaysia; jialee5995@gmail.com
[2] Graduate School of Environmental Studies, Tohoku University, Aramaki Aza Aoba, 468-1, Aoba-ku, Sendai 980-8572, Japan; masaki.ota.a5@tohoku.ac.jp
[3] Graduate School of Engineering, Tohoku University, Aramaki Aza Aoba, 6-6-11-403, Aoba-ku, Sendai 980-8579, Japan
[4] Agro-Environmental Protection Institute, Ministry of Agriculture and Rural Affairs, No. 31 Fukang Road, Nankai District, Tianjin 300191, China; haixin_g@126.com
* Correspondence: gunhean@upm.edu.my (G.H.C.); smith@scf.che.tohoku.ac.jp (R.L.S.J.); Tel.: +60-39679-8414 (G.H.C.); +81-22-752-2278 (R.L.S.J.)

Citation: Lee, J.L.; Chong, G.H.; Ota, M.; Guo, H.; Smith, R.L., Jr. Solvent Replacement Strategies for Processing Pharmaceuticals and Bio-Related Compounds—A Review. *Liquids* **2024**, *4*, 352–381. https://doi.org/10.3390/liquids4020018

Academic Editors: William E. Acree, Jr., Franco Cataldo and Enrico Bodo

Received: 31 December 2023
Revised: 22 February 2024
Accepted: 21 March 2024
Published: 9 April 2024

Copyright: © 2024 by the authors. Licensee MDPI, Basel, Switzerland. This article is an open access article distributed under the terms and conditions of the Creative Commons Attribution (CC BY) license (https://creativecommons.org/licenses/by/4.0/).

Abstract: An overview of solvent replacement strategies shows that there is great progress in green chemistry for replacing hazardous di-polar aprotic solvents, such as N,N-dimethylformamide (DMF), 1-methyl-2-pyrrolidinone (NMP), and 1,4-dioxane (DI), used in processing active industrial ingredients (APIs). In synthetic chemistry, alcohols, carbonates, ethers, eucalyptol, glycols, furans, ketones, cycloalkanones, lactones, pyrrolidinone or solvent mixtures, 2-methyl tetrahydrofuran in methanol, HCl in cyclopentyl methyl ether, or trifluoroacetic acid in propylene carbonate or surfactant water (no organic solvents) are suggested replacement solvents. For the replacement of dichloromethane (DCM) used in chromatography, ethyl acetate ethanol or 2-propanol in heptanes, with or without acetic acid or ammonium hydroxide additives, are suggested, along with methanol acetic acid in ethyl acetate or methyl tert-butyl ether, ethyl acetate in ethanol in cyclohexane, CO_2-ethyl acetate, CO_2-methanol, CO_2-acetone, and CO_2-isopropanol. Supercritical CO_2 (scCO_2) can be used to replace many organic solvents used in processing materials from natural sources. Vegetable, drupe, legume, and seed oils used as co-extractants (mixed with substrate before extraction) can be used to replace the typical organic co-solvents (ethanol, acetone) used in scCO_2 extraction. Mixed solvents consisting of a hydrogen bond donor (HBD) solvent and a hydrogen bond acceptor (HBA) are not addressed in GSK or CHEM21 solvent replacement guides. Published data for 100 water-soluble and water-insoluble APIs in mono-solvents show polarity ranges appropriate for the processing of APIs with mixed solvents. When water is used, possible HBA candidate solvents are acetone, acetic acid, acetonitrile, ethanol, methanol, 2-methyl tetrahydrofuran, 2,2,5,5-tetramethyloxolane, dimethylisosorbide, Cyrene, Cygnet 0.0, or diformylxylose. When alcohol is used, possible HBA candidates are cyclopentanone, esters, lactone, eucalyptol, MeSesamol, or diformylxylose. HBA—HBA mixed solvents, such as Cyrene—Cygnet 0.0, could provide interesting new combinations. Solubility parameters, Reichardt polarity, Kamlet—Taft parameters, and linear solvation energy relationships provide practical ways for identifying mixed solvents applicable to API systems.

Keywords: Reichardt polarity; Kamlet—Taft parameters; green chemistry; solvent substitution; pharmaceuticals

1. Introduction

Solvents are commonly viewed as being polar or nonpolar, depending on whether their molecular structure contains highly electronegative (N, O, S, Cl, Br, I) elements or only (C, H) elements. However, for a molecule to be polar, it must contain a polar bond and have asymmetry in its structure that causes an imbalance in charge separation between two

(+ and −) poles referred to as dipoles. The presence of an asymmetrically arranged polar bond, such as C-Cl in chloromethane (CH_3Cl), causes the molecule to be polar, whereas the presence of four symmetrically arranged C-Cl bonds in carbon tetrachloride (CCl_4) cause the molecule to be nonpolar. For two solvents to be miscible, similarity in molecular polarity is required, as given by the well-known adage, "like dissolves like", which in other words means that, for the solvation of polar molecules to occur, dipole—dipole interactions must exist, and conversely, for the solvation of nonpolar molecules to occur, dipole—dipole interactions must be absent. There are many exceptions to this adage, and certainly, system conditions (temperature, pressure) and van der Waals-London forces (dispersion) play important roles in solvation processes. Moreover, for solvent mixtures as discussed in this review, composition and interactions between hydrogen bond donor (HBD) and hydrogen bond acceptor (HBA) molecules are important.

Physical properties such as dipole moment (μ_D), dielectric constant (ε), octanol-water partition coefficient ($logK_{ow}$ or $logP$), normal boiling point (T_b), melting temperature (T_m), entropy of fusion ($\Delta_{fus}S$), Hildebrand solubility parameter, and Hansen solubility parameter help to characterize the macroscopic polarity of a molecule. On the other hand, empirical polarity scales based on solvatochromic probes (dyes), such as Reichardt $E_T(30)$ [1] and normalized E_T^N values [2], Kamlet—Taft (KT) acidity (α), basicity (β) and dipolar/polarizability (π^*) values [3,4], and Catalán parameters [5], help to characterize the microscopic polarity of a solvent [6]. In solvent selection guides developed by the industry [7,8] and chemical societies [9–12], pure component solvent properties are analyzed in detail for developing solvent replacement strategies; however, as a focus of this review, considerable opportunities exist if mixtures of two kinds of polar solvents are used to create environments of microscopic polarity. For example, mixing an HBD solvent with an HBA solvent causes complex molecules (e.g., HBD—HBA pairs) to form, such that heterogeneity (local composition) is observed for simple alcohol—water mixtures [13,14] or ethylene glycol-water mixtures [15]. In this review, the emphasis is placed on taking advantage of the local composition and microscopic polarity of a solvent mixture as opposed to the bulk properties of a pure solvent, even though temperature and pressure can also be used to vary the properties of a pure solvent.

Solutes, in the context of this review, are active pharmaceutical ingredients (APIs) and bio-related molecules that can have multiple functional groups and can contain both polar (hydrophilic) and nonpolar (hydrophobic) regions in their structure. Functional groups in the solute can interact within the molecule (intramolecular) or between neighboring molecules (intermolecular) to form associated, cyclic, complex, network, or tertiary structures, and thus, the dissolution of an API into a solvent can be the result of many different molecular interactions. The composition of a solvent mixture can be used to fine-tune dipole—dipole interactions that sometimes lead to the solubility enhancement of the API in solution that is higher than that in either of the pure mono-solvents, which is known as synergistic behavior.

2. Substances of Very High Concern (SVHC)

In the synthesis and processing of APIs, polar protic (water, alcohols, carboxylic acids), dipolar aprotic (ketones, lactones, esters, ethers), or nonpolar aprotic (hydrocarbons) solvents are used. Notably, hazardous and unsafe dipolar aprotic chemicals (e.g., N,N-dimethylformamide (DMF), 1-methyl-2-pyrrolidinone (NMP), 1,4-dioxane (DI)) account for over 40% of total solvents used in synthetic, medicine-related, and process chemistry [16], and these solvents and more than 480 others are on the candidate list of substances of very high concern (SVHC), as designated under the European Chemicals Agency (ECHA), as the European Union Registration, Evaluation Authorization and Restriction of Chemicals (REACH) guidelines limit or prohibit the use of chemicals, especially those having reproductive toxicity, carcinogenicity, or explosive decomposition properties (Table 1). Thus, the key motivation of employing mixed solvents instead of mono-solvents, new solvents, or newly developed solvents is based on environmental health and safety (EHS) guidelines for compounds with known chemical properties and conformity with the "International Council for Harmonisation of Technical Requirements for Pharmaceuticals for Human Use" (ICH). Namely, EHS and ICH should be primary factors in solvent replacement, rather than apparent greenness or economic or sustainability factors, because few newly developed solvents or solvent systems have had sufficient time for scrutiny in all areas highlighted by governmental agencies and in the solvent guides discussed below. In this review, solvent replacement strategies are analyzed with the aim of highlighting a method for identifying safe solvent mixtures for the research development and chemical processing of organic compounds.

Table 1. Selected chemicals from candidate list of substances of very high concern (SVHC) for authorization by the European Chemicals Agency (ECHA) as of 2023. Chemicals shown in various categories are for educational purposes only. Specific hazards, detailed information, case decisions, or discussion should be accessed from ECHA website [17]. LD_{50} values from PubChem or online sources based on rat/mouse oral or dermal (*d*) studies.

Chemical (CAS No.)	LD_{50} (mg/kg)	Chemical (CAS No.)	LD_{50} (mg/kg)
Carcinogenic		Respiratory Sensitizing	
1,2,3-trichloropropane (96-18-4)	120	cis-cyclohexane-1,2-dicarboxylic anhydride (13149-00-3)	-
1,2-dichloroethane (107-06-2)	670	Cyclohexane-1,2-dicarboxylic anhydride (85-42-7)	958
1,4-dioxane (123-91-1) (DI)	1550	Glutaral (111-30-8)	134
2,4-dinitrotoluene (121-14-2)	268	Toxic to Reproduction	
4,4'-Diaminodiphenylmethane (101-77-9)	120	1-Methyl-2-pyrrolidone (NMP) (872-50-4)	3914
4-aminoazobenzene (60-09-3)	200	1-vinylimidazole (1072-63-5)	180
Acrylamide (79-06-1)	170	2-ethoxyethanol (110-80-5)	2125
Anthracene oil (90640-80-5)	2000 *d*	2-ethoxyethyl acetate (111-15-9)	2700
Biphenyl-4-ylamine (92-67-1)	205	2-methoxyethanol (109-86-4)	2370
Chrysene (218-01-9)	320	2-methoxyethyl acetate (110-49-6)	2900
Furan (110-00-9)	5.2	2-methylimidazole (693-98-1)	1400
Propylene oxide (75-56-9)	1245 *d*	4,4'-sulphonyldiphenol (80-09-1)	4556
N-(hydroxymethyl)acrylamide (924-42-5)	474	Dibutyl phthalate (84-74-2) (DBP)	7499
o-aminoazotoluene (97-56-3)	300 (dog)	Dicyclohexyl phthalate (84-61-7)	30
o-toluidine (95-53-4)	670	Dihexyl phthalate (84-75-3)	29,600
Phenolphthalein (77-09-8)	>1	Diisobutyl phthalate (84-69-5)	15
Potassium dichromate (7778-50-9)	25	Diisohexyl phthalate (71850-09-4)	-
Trichloroethylene (79-01-6)	1282	Diisopentyl phthalate (605-50-5)	2000
Endocrine disruptor		Dioctyltin dilaurate (3648-18-8)	6450
2-(isononylphenoxy)ethanol (85005-55-6)	-	Formamide (75-12-7)	5577
4-(1-ethyl-1-methylhexyl)phenol (52427-13-1)	-	Methoxyacetic acid (625-45-6)	1000
4,4'-(1-methylpropylidene)bisphenol (77-40-7)	500.1	N,N-dimethylformamide (68-12-2) (DMF)	2800
4-tert-butylphenol (98-54-4)	2951	Nitrobenzene (98-95-3)	349
Isobutyl 4-hydroxybenzoate (4247-02-3)	2600	N-methylacetamide (79-16-3)	5
Nonylphenol (25154-52-3)	1200	n-pentyl-isopentyl phthalate (776297-69-9)	-
Nonylphenol, ethoxylated (9016-45-9)	1300	Perfluoroheptanoic acid (375-85-9)	500
Human health effects		Phenol, 4-dodecyl, branched (210555-94-5)	2000
Melamine (108-78-1)	3161	Phenol, tetrapropylene- (57427-55-1)	2000
Persistent, Bioaccumulative and Toxic (PBT)		Very Persistent, Very Bioaccumulative (vPvB)	
Alkanes, C14-16, chloro (1372804-76-6)	23	Phenanthrene (85-01-8)	700
Anthracene (120-12-7)	>17	Terphenyl, hydrogenated (61788-32-7)	17,500
Dodecamethylcyclohexasiloxane (540-97-6)	>50		
Octamethylcyclotetrasiloxane (556-67-2)	1540		
Pyrene (129-00-0)	2700		

3. Solvent Guides

To address the issue of the overuse of hazardous dipolar aprotic chemicals in API synthesis and processing and to improve the awareness of chemical professionals who perform solvent selection on a day-to-day basis, pharmaceutical industries have developed solvent guides with ranking systems. Chemical agencies have developed lists for solvents evaluated as hazardous that require formal authorization for use in chemical processes.

The GlaxoSmithKline (GSK) solvent guide [7,18,19] contains detailed analyses of a total of 154 small molecules (e.g., alcohols, aromatics, carbonates, chlorinated, dipolar aprotics, esters, ethers, hydrocarbons, ketones, organic acids, water) commonly used in pharmaceutical industries. The GSK solvent guide has the following categories: (i) waste (incineration, recycling, biotreatment, VOC emissions), (ii) environment (aquatic impact, air impact), (iii) human health (health hazard, exposure potential), and (iv) safety (flammability and explosion, reactivity). The GSK solvent guide allows for the quick evaluation and qualitative comparison of replacement solvents based on four primary categories that include life-cycle assessment (LCA), and it ranks solvents in their categories on a scale from 1 (major issues) to 10 (few known issues).

The European consortium and Innovative Medicines Initiative (IMI) produced CHEM21 [8], which contains guidelines and metrics for solvent usage. Byrne et al. reported environmental, health, and safety (EHS) tools and guidelines for solvents and highlighted key points in available guidelines [11]. The CHEM21 solvent guide ranks solvents in EHS categories on a scale from 1 (recommended) to 10 (hazardous), which is contrary (and opposite in order) to the scale of the GSK solvent guide. Both solvent guides provide extremely useful evaluations of solvent risks and issues and provide solvent replacement recommendations.

The American Chemical Society (ACS) Green Chemistry Institute (CGI) and pharmaceutical roundtable produced a solvent selection website (Figure 1) [10,12] dedicated to solvent usage in pharmaceutical and chemical industries and a solvent guideline [9]. Figure 1 shows a sample screen of a solvent selection tool developed for the ACS GCI Pharmaceutical Roundtable (GCIPR) that uses principle component analysis (PCA) to identify potential solvent replacements. PCA combines many physical properties, characteristics (presence of functional groups), and environmental data to generate correlations and scores according to user constraints. The solvent selection tool (Figure 1) was described by Diorazio et al. [20] and was originally designed by AstraZeneca in Spotfire, and a version was donated to GCIPR. The GCIPR solvent selection tool is useful for identifying replacement solvents based on both quantitative and qualitative characteristics (Figure 1).

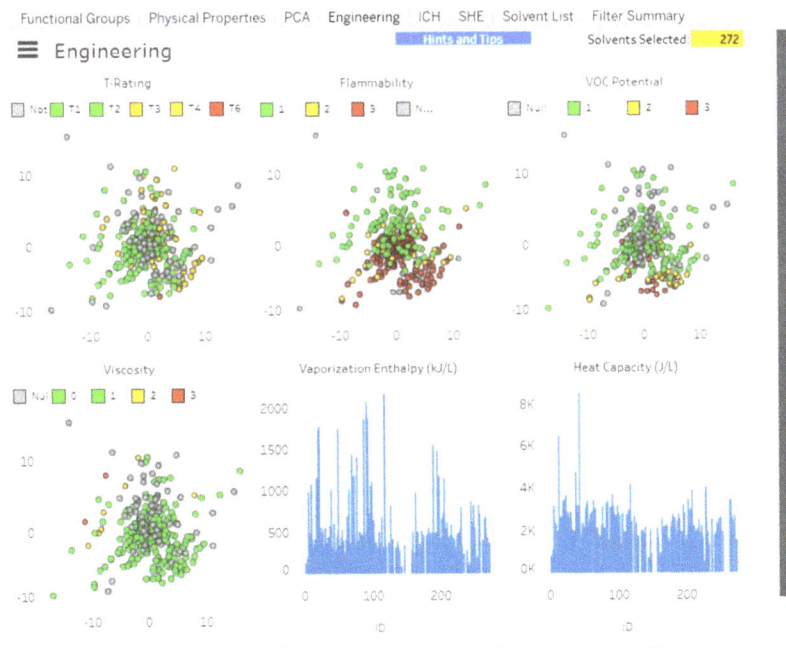

Figure 1. ACS Green Chemistry Institute and Pharmaceutical Roundtable (GCIPR) solvent selection tool https://www.acsgcipr.org/tools-for-innovation-in-chemistry/solvent-tool/ (accessed on 1 April 2024) described by Diorazio et al. [20]. Copyright ACS, 2023.

4. Replacement Solvents in Synthetic Chemistry

Syntheses of APIs are commonly performed in multistep batch processes that use hazardous or unsafe dipolar aprotic solvents in some of the key steps. Replacement strategies for non-green dipolar aprotic solvents used in reactions were suggested by Gao et al. [21].

Table 2 summarizes replacement solvents for 15 classes of synthetic reactions identified by Jordan et al. [16]. Possible replacement solvents for dipolar aprotics (Table 2) include novel water-surfactant (PS-750-M) systems that eliminate organic solvents [22], dipolar aprotic solvents with improved safety and sustainability, namely N-butyl-2-pyrrolidinone (NBP), propylene carbonate (PC), dimethylisosorbide (DMI) [23], dihydrolevoglucosenone (Cyrene) [24], eucalyptol [25], or dimethylcarbonate (DMC), or the use of mixed solvents, such as 2-methyltetrahydrofuran (2-MeTHF) with methanol (Table 2). Besides THF or DMF in Sonogashira cross-coupling reactions (Table 2), eucalyptol can possibly replace solvents such as anisole, bromobenzene, chlorobenzene, chloroform, diethyl ether (DE), N,N-dimethylacetamide (DMA), dimethyl ether (DME), DI, ethyl acetate, ethyl benzoate, and toluene [25].

Table 2. Possible replacement solvents for dipolar aprotic solvents used in synthetic chemistry transformations. Content was summarized and adapted from *Unified solvent selection guide for replacement of common dipolar aprotic solvents in synthetically useful transformations* contained in ref. [16]. Copyright ACS, 2022.

Reaction	Unsafe Dipolar Aprotics	Replacement Solvents
Amide formation	DCM; DMF	Cyrene; surfactant-water
Boc deprotection	DI	HCl in CPME; TFA in PC
Borylation chemistry	DI	2-MeTHF:MeOH (1:1); CPME; MTBE; CH
Buchwald—Hartwig amination	DI	2-MeTHF; tBuOH
Carbonylation	THF; DE	DMC
Carboxylation	THF; DE	2-MeTHF; DMI; DMC
C-H activation	THF; DMF; DI	2-MeTHF; CH
Mizoroki—Heck cross-coupling	DI; THF; DMF	NBP; DMI; PC
Nucleophilic aromatic substitution	THF; DMF; DI	2-MeTHF; PEG-400
Organometallic reaction	R-MgX; R-Li; hydrides	2-MeTHF; CPME
Solid-phase peptide synthesis	DMF; DMAc; NMP	NBP; GVL
Sonogashira cross-coupling	THF; DMF	Cyrene; NBP; DMI; Eucalyptol
Steglich Esterification	DMF	DMC
Suzuki-Miyaura cross-coupling	DI; THF; DMF	Cyrene; NBP; DMI; 2-MeTHF
Urea synthesis	DMF; THF	Cyrene

In the synthesis of APIs with solvents, the type of process employed is an important point that deserves attention. A less obvious way to lower risks associated with solvent usage in API synthesis is through continuous manufacturing (CM) [26], as opposed to batch processing. In a CM process, systems can be automated, quality can be improved, waste can be reduced, and, most importantly, solvent volumes can be greatly lowered over those quantities used in batch systems by lowering the total system volumes and by eliminating the storage of API reaction intermediates, such that overall safety of the synthesis can be improved. The number of papers published on the continuous manufacturing of APIs has roughly tripled in the past 5 years, making it a highly active research area. In CM processes, solvent selection and solvent additives play key roles in flow chemistry, product quality, system operability, economics, and sustainability. Furthermore, there are some recent new approaches for CM processes; amidation by reactive extrusion has been developed as a solventless synthesis method and has been used for the preparation of teriflunomide and moclobemide APIs [27].

5. Solubility Parameters

Solubility parameters (SP) are used to characterize substances in solvent replacement strategies. The Hildebrand SP (δ) has the basis of regular solution theory [28], and its development in solubility theory relates the cohesive energy density defined by Equation (1) to the activity coefficient [29].

$$\delta \equiv (\Delta \underline{U}^{vap}/\underline{V})^{1/2} \quad (1)$$

In Equation (1), $\underline{U}$ and $\underline{V}$ are the molar internal energy of vaporization and molar volume of the substance in its liquid state, respectively. The definition of the Hildebrand SP is typically simplified by replacing $\underline{U}$ with $(\underline{H} - P\underline{V})$ and assuming ideal gas behavior:

$$\delta = ((\Delta \underline{H}^{vap} - RT)/\underline{V})^{1/2} \quad (2)$$

Hansen [30] divided the total cohesive energy given in Equation (1) into three parts: (i) dispersion (van der Waals (London) forces) interactions (δ_d), hydrogen bonding interactions (δ_h), and polar (or dipole-dipole) interactions (δ_p). Hansen solubility parameters (HSPs) are used to determine a solubility parameter distance (Ra) between two substances "1" and "2" as follows:

$$(Ra)^2 = 4 \cdot (\delta_{d1} - \delta_{d2})^2 + (\delta_{h1} - \delta_{h2})^2 + (\delta_{p1} - \delta_{p2})^2 \quad (3)$$

where the sphere provides a region of favorable solvation for a solute "1" and solvent "2", i.e., as values of Ra become closer to zero according to a chosen solvent with given HSP values, affinity becomes higher, and the solubility of the solute in the solvent should increase. The factor of four in Equation (3) is empirical and adds statistical weighting to dispersion interactions as being most important in solvation. By taking a substance such as a polymer or biomolecule and seeing whether it dissolves into solvents with known HSP values, the radius of interaction (Ro) can be determined for that compound. Then, a relative energy difference can be defined as follows:

$$RED = Ra/Ro \quad (4)$$

and solvents or solvent mixtures that have RED < 1 are candidates that dissolve the compound. It is possible, for example, for two solvents outside of the solvation sphere to be mixed, such that they form a good solvent as mixture for a polymer. HSP theory has been used to estimate the solubilities of anti-inflammatory drugs in pure and mixed solvents [31]. Fractional HSP values, which can be plotted on ternary diagrams to facilitate the assessment of interactions, have been used to identify green extraction solvents for alkaloids [32] and to screen solvent mixtures for pharmaceutical cocrystal formation [33]. HSP is a powerful tool used for solvent screening and is especially useful for large molecules, such as polymers or biomolecules, as highlighted by Abbott [34].

In comparing the Hildebrand solubility parameter theory with that of the Hansen solubility theory, the Hildebrand solubility parameter theory has some notable failures in predicting miscibility between materials [30]. However, in a critical comparison of solvent selection for 75 polymers, both theories gave similar results in predicting polymer—solvent miscibility [35]. Namely, Hildebrand SP had a prediction accuracy of 60% for solvents and 76% for non-solvents, whereas HSP had a prediction accuracy of 67% for solvents and 76% for non-solvents [35]. On the other hand, for polar polymers, the Hildebrand SP theory gave a prediction accuracy of only 57% [35]. Both Hildebrand solubility parameters and Hansen solubility parameters are useful screening tools for solvent replacement. Hildebrand SP theory is simple and provides qualitative estimation of solvent interactions for nonpolar molecules or slightly polar molecules; Hansen SP theory accounts for detailed molecular interactions and is applicable to both nonpolar and polar molecules. HSP can be applied to complex molecules, such as lignin [36] or phytochemicals [37]; however, HSP is qualitative when hydrogen bond donor (HBD) and hydrogen bond acceptor (HBA) molecular systems are considered [38].

6. Empirical Polarity Scales

Reichardt $E_T(30)$ parameters are based on the solvatochromic properties of Betaine 30 dye and provide the sensitive characterization of solvent polarity. Reichardt E_T^N values are normalized based on the $E_T(30)$ values of water and tetramethylsilane. Reichardt parameters are firmly established in the chemical literature and form the basis of a widely used polarity scale for organic chemicals [6].

Kamlet—Taft (KT) parameters are based on the solvatochromism of dyes specific to Lewis acidity (α), Lewis basicity (β), and dipolarity/polarizability (π^*) and have independent scales that depend on reference solvents [39]. The Kamlet—Taft polarity scales are meant to have values of α, β, and π^* that are between zero and one; however, when a solvent has a Lewis acidity, Lewis basicity, or dipolarity/polarizability that is outside of the range of reference compounds, ($\pi^* = 0$ (cyclohexane) and $\pi^* = 1$ (dimethylsulfoxide)) values of KT parameters can be greater than unity or less than zero.

Catalán parameters improved the KT parameter approach by using specific dyes for solvent polarizability (SP), solvent dipolarity (SdP), solvent acidity (SA), and solvent basicity (SB) parameters rather than by average values, as in the KT approach. Catalán parameters separate the polarizability (SP) and dipolarity (SdP) contributions of the KT parameter approach. All three scales have wide use in the chemical literature, although there are issues in data reduction methods and parameter values, as pointed out by Spange et al. [40], who reanalyzed polarity scales considering molar concentrations of the solvent (N), and Spange and Weiß [41], who proposed a method to unify the acid—based (pKa) and density effects of hydrogen bond donor solvents.

According to Reichardt and Welton [6], common molecular solvents (Figure 2) can be roughly divided into three groupings: (i) dipolar protic (HBD), $E_T^N > 0.5$; (ii) dipolar aprotic (HBA), $0.3 < E_T^N < 0.5$; and (iii) apolar (non-HBD or nonpolar), $E_T^N < 0.3$. Examination of the KT dipolarity/polarizability parameters (Figure 2) shows that longer chain hydrocarbons have π^* values less than zero, and water has a π^* greater than unity, which is due to the choice of reference solvents in the KT method. Most solvent replacement strategies consider Reichardt, Kamlet—Taft or Catalán parameters in their analysis. For example, dipolar aprotic solvents generally have high KT basicity and low KT acidity (Figure 3). Direct replacement solvents for dipolar aprotics could be N-butyl-2-pyrrolidinone (NBP), Cyrene^{TM} (Cyr), γ-valerolactone (GVL), γ-butyrolactone (GBL), eucalyptol (Eupt), tetramethyloxolane (TMO), dimethyl isosorbide (DMI), or cyclopentyl methyl ether (CPME). However, many solvents have E_T^N polarity values that are much lower than that of dipolar aprotics (Figure 2) and KT acidities that are either too high or KT basicities that are too low (Figure 3) to allow direct replacement of dipolar aprotics. Nevertheless, the range of Kamlet—Taft parameters of dipolar aprotics provide valuable information for considering mixed solvents and mixed solvent composition.

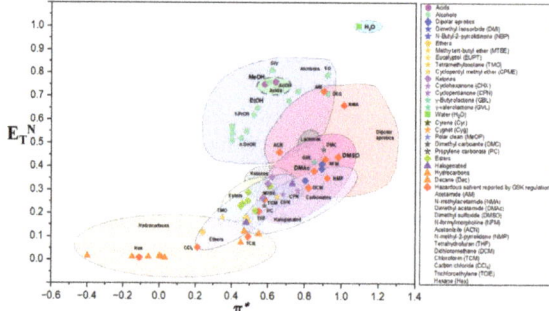

Figure 2. Reichardt E_T^N parameters plotted against Kamlet—Taft dipolarity/polarizability parameters for selected molecular solvents.

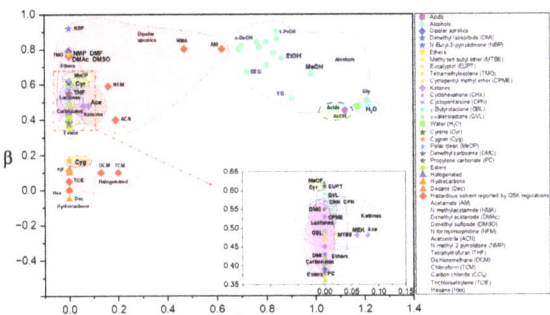

Figure 3. Kamlet—Taft basicity parameter plotted against acidity parameter for selected molecular solvents.

7. Opportunities with Mixed Solvents

Mixtures of solvents (mixed solvents) allow one to vary the chemical properties of the solution in a unique number of ways. For example, when an HBD solvent is mixed with an HBA solvent, KT parameters vary continuously with composition (Figure 4). KT parameters of mixed solvents can show synergistic behavior, which means that their β or π* values can be higher than the KT parameters of the pure solvents (Figure 4), especially when water is the HBD solvent. Duereh et al. [42] showed that there is a clear relationship between microscopic (local) polarity, complex molecule (HBD—HBA solvent pairs) interactions, and synergistic behavior in thermodynamic properties (Figure 5).

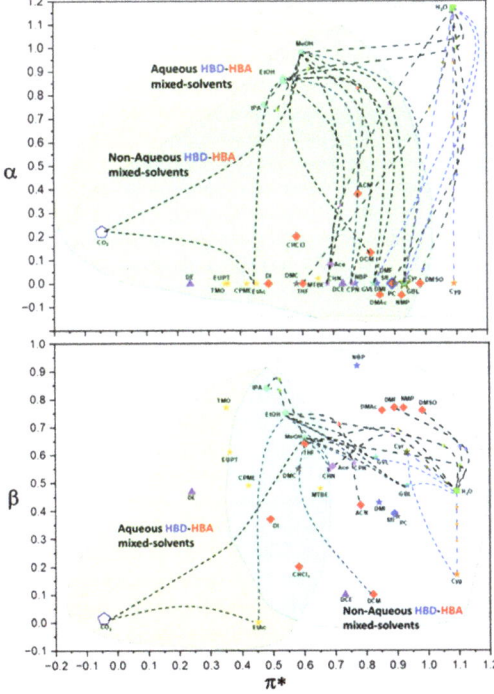

Figure 4. Kamlet—Taft acidity (α) and basicity (β) versus dipolar/polarizability (π*) for aqueous and non-aqueous mixed solvents and pure solvents. Dashed lines show approximate behavior of mixed solvent KT parameters with composition.

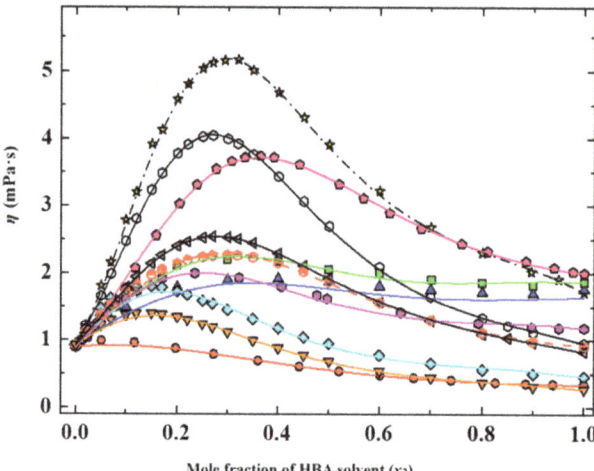

Figure 5. Dynamic viscosity (η) of water (HBD)–hydrogen bond acceptor (HBA) mixed solvent systems as a function of mole fraction of HBA solvent (x_2) at 25 °C. HBA solvents are ordered in terms of Hunter basicity (β_2^H) values (low to high): acetonitrile (● ACN), γ-valerolactone (■ GVL), γ-butyrolactone (▲ GBL), tetrahydrofuran (◆ THF), 1,4-dioxane (● DI), acetone (▼ Ace), pyridine (● PYR), N-methyl-2-pyrrolidone (✱ NMP), N,N-dimethylformamide (◀ DMF), N,N-dimethylacetamide (● DMA), and dimethyl sulfoxide (● DMSO). Reprinted with permission from [42]. Copyright American Chemical Society, 2017.

Thus, solvent composition of mixed solvents allows one to vary microscopic polarity (local composition) and the concentration of HBD—HBA complex molecules that can be used advantageously in solvent replacement schemes.

In this section, strategies for using mixed solvents to replace hazardous chemicals are highlighted for chromatography solvents, CO_2 expanded liquids, supercritical fluids, low-transition temperature mixtures, switchable solvents, and HBD—HBA mixtures of molecular solvents.

7.1. Chromatography Solvents

In chromatographic methods, great progress has been made with the introduction of mixed solvents, such as ethyl acetate (EtAc)-ethanol (EtOH) in heptanes being demonstrated as a superior replacement for dichloromethane (DCM) [43]. Mixed solvent stock solutions are marketed by leading chemical suppliers for HPLC, TLC, and flash chromatography (FC) methods [44], confirming the success of the EtAc—ethanol mixtures.

The reason why EtAc—EtOH in a heptane mixed solvent system can replace DCM can be understood by examining the variation in KT parameters of the mixture compared with the KT parameters of the DCM—MeOH system. In this case, EtOH is the HBD solvent, EtAc is the HBA solvent, and the heptanes have low overall KT acidity for the mobile phase. Composition variation of EtAc–EtOH mixtures allows for the fine control of the basicity and dipolarity/polarizability that transverse methanol KT parameters (Figure 4).

To replace hexane, CO_2–EtAc has been suggested to be applicable to thin-layer chromatography (Table 3), and CO_2–MeOH has been demonstrated to be applicable to flash chromatography [45]. The entire corporate chemistry division of Syngenta (Table 3) reduced the overall volume of seven hazardous dipolar aprotic solvents (DCM, CHCl$_3$, DCE, DI, DME, DMF, DE) by 75% over a period of two years by using solvent replacement (e.g., EtAc–EtOH mixtures for DCM) and by emphasizing reverse phase chromatography for the separation of polar compounds [46] (Table 3); however, DMF usage increased during that period. Solvent pairs, such as cyclohexanone–MeOH, cyclohexanone–EtOH, cyclopentanone–MeOH, cyclopentanone–EtOH, GBL–MeOH, GBL–EtOH, GBL–water, GVL–MeOH, GVL–EtOH, and GVL–water, have been demonstrated as replacements for NMP or DMF in polyamide synthesis and, thus, have possibilities as solvent replacements in analytical method development [47].

Improvements in high-pressure liquid chromatography (HPLC) have been made with the introduction of ultra-high-pressure liquid chromatography (UHPLC), supercritical fluid chromatography (SFC), and ultra-high-pressure supercritical fluid chromatography (UHPSFC), which reduce the amount of solvents necessary in analyses while improving resolution [48]. When UHPSFC—tandem mass spectroscopy is employed, the determination of plant hormones (cytokinins) can be analyzed in 9 min at detection limits close to 0.03 fmol [49]. ACS has introduced the analytical method greenness score (AMGS) calculator developed by Hicks et al. [48] that ranks chromatography methods according to instrument energy, solvent energy, and solvent EHS scores [10].

Table 3. Replacement solvents for dichloromethane (DCM) in high-performance liquid (HPLC), thin-layer chromatography (TLC) and flash chromatography (FC) methods. Analytes consist of neutral, basic, acidic, and polar API.

Mixed Solvent [a]	Analyte [b]	System	Ref.
EtAc:EtOH (3:1) in heptanes	Neutral	LC	[43]
EtAc:EtOH in heptanes	Neutral	LC	[43]
iPrOH in heptanes	Neutral	LC	[43]
EtAc:EtOH (3:1) in MTBE	Neutral	LC	[43]
MeOH in MTBE	Neutral	LC	[43]
EtAc:EtOH (3:1) (2% NH$_4$OH) in heptanes	Basic	LC	[43]
MeOH: NH$_4$OH (10:1) in EtAc	Basic	LC	[43]
MeOH: NH$_4$OH (10:1) in MTBE	Basic	LC	[43]
EtAc:EtOH (3:1) (2% AcOH) in heptanes	Acidic	LC	[43]
MeOH:AcOH (10:1) in EtAc	Acidic	LC	[43]
MeOH:AcOH (10:1) in MTBE	Acidic	LC	[43]
EtAc:EtOH (3:1) in cyclohexane	n.s.	LC	[46]
acetonitrile:water	Polar	LC	[46]
tert-butyl acetate	All	LC	[50]
sec-butyl acetate	All	LC	[50]
ethyl isobutyrate	All	LC	[50]
methyl pivalate	All	LC	[50]
CO$_2$:EtAc	n.s.	TLC	[51]
EtAc in heptanes	n.s.	TLC	[51]
iPrOH in heptanes	n.s.	TLC	[51]
Ace in heptanes	n.s.	TLC	[51]
CO$_2$:MeOH	Neutral	FC	[45]
CO$_2$:EtAc	n.s.	FC	[51]
CO$_2$:Ace	n.s.	FC	[51]
CO$_2$:iPrOH	n.s.	FC	[51]

[a] AcOH: acetic acid; EtAc: ethyl acetate; MTBE: methyl tert-butyl ether; [b] n.s. not specified.

7.2. Expanded Liquids and Supercritical Fluids

Chemists and chemical engineers have introduced many new types of solvents through major research initiatives. CO$_2$-expanded bio-based liquids (CXL) have been demonstrated to be favorable for enantioselective biocatalysis [52], and supercritical fluids have been shown to be able to replace the hazardous solvents used in processing APIs [53]. Supercritical carbon dioxide (scCO$_2$) has been shown to have a wide application in processing bioactive lipids [54] and bioactive-related food ingredients [55]. A comprehensive review is available on the supercritical extraction of bioactive molecules from plant matrices [56]. A less-studied methodology in the supercritical extraction of bioactive molecules from natural sources is to eliminate organic co-solvents, such as ethanol or acetone, by replacing them with co-extractants that are typically oils from plant materials (Table 4).

Table 4. Co-extractant methodology for obtaining bio-products from supercritical CO_2 extraction of natural sources. Co-extractants: vegetable, drupe, legume, or seed oils or triacylglycerols (TAGs, triglycerides). Bio-product yields shown are maximum values normalized to 100%.

Natural Source	Co-Extractant	Bio-Product	T (°C)	P (MPa)	%Yield	Ref.
Algae	Soybean oil	astaxanthin	70	40	36	[57]
Brown seaweed	Sunflower oil	carotenoids	50	30	99	[58]
Carrots	Canola oil	carotenoids	70	55	92	[59]
Mangosteen	Virgin coconut oil	xanthonoids	70	43	31	[60]
Mangosteen	Virgin coconut oil	α-mangostin	60	35	76	[61]
Marigold	Medium-chain TAGs	lutein esters	65	43	98	[62]
Marigold	Soybean oil	lutein esters	53	30	93	[63]
Propolis	Virgin coconut oil	flavonoids	50	15	25	[64]
Pumpkin	Olive oil	carotenoid	50	25	41	[65]
Red sage	Peanut oil	diterpenoids	50	38	90	[66]
Tomato	Canola oil	lycopene	40	40	86	[67]
Tomato	Hazelnut oil	lycopene	66	45	40	[68]
Tomato skin	Olive oil	lycopene	75	35	58	[69]

In co-extractant methodology (Table 4), natural source substrates (petals, pericarp, etc.) are mixed with a natural oil (co-extractant) from a vegetable, drupe, legume, or seed (or fruit) before extraction with pure $scCO_2$. The co-extractant serves to increase the mass transfer of active components from the natural source to the supercritical phase by solubilization and polarity matching, and the co-extractant properties are enhanced due to $scCO_2$ dissolution into the co-extractant phase that causes the reduction of both surface tension and viscosity while enhancing heat transfer and related properties. Thus, with co-extractant methodology (Table 4), organic co-solvents are completely eliminated in $scCO_2$ extraction such that the contamination of extracts with organic compounds is not an issue. Furthermore, with co-extractant methodology, a final product is realized directly, the cultural processing of many types of food is possible, and food safety is strictly enhanced [70].

Related to developments in supercritical fluid theory, entropy based solubility parameters have been proposed that allow the extension of traditional solubility parameter theory to chemical systems containing supercritical fluids and ethanol [71] or systems at high temperatures or high pressures [72]. Experimental systems for measuring the KT parameters of methanol, ethanol, 2-propanol, and 1,1,1,2-tetrafluoroethane (HFC134a) co-solvents in CO_2 have been developed for assessing the HBD alcohol interactions with the HBA Lewis acidity of CO_2 in the supercritical state for quantifying polarity enhancements [73].

7.3. Low Transition Temperature Mixtures

Low transition temperature mixtures (LTTMs) are special combinations of mixed solvents made up of a hydrogen bond donor (HBD) molecule and a hydrogen bond acceptor (HBA) molecule for the purpose of liquefying the mixture [74]. Ionic liquids (ILs) are combinations of discrete organic moiety containing cations and anions that are in the liquid state at room temperature. Deep eutectic solvents (DESs) are mixtures of Lewis or Brønsted acids and bases that are in the liquid state at room temperature.

The possibility of using either ILs or DESs as solvent replacements or for processing APIs allows them to have many potential innovative applications due to their solvation and tailorable properties [75]. Issues with ILs are their cost, recyclability, and relatively higher viscosity compared to molecular solvents. While DESs are inexpensive, they share some of the same issues as ILs, and in addition, their separation from chemical products may be problematic due to the formation of strong HBD—HBA complexes with the API. One innovative approach that addresses some of these issues is to incorporate the IL chemical structure into the API to improve the bioavailability in drug delivery systems [76,77]. Reviews in the area of combining HBD- or HBA-containing APIs into the structure of ILs for drug delivery systems and other purposes show that there is much activity in this research area [78,79].

7.4. Switchable Solvents

Switchable polarity solvents (SPS) [80], switchable hydrophilicity solvents (SHS) [81], switchable water (SW) [82], solvent-assisted switchable water (SASW) [83], and high-pressure switchable water (HPSW) [84] are new types of mixed solvents that can change their polarity, hydrophilicity, or characteristics through the introduction or removal of CO_2. Switchable solvent systems would seem to have many applications in processing API, and furthermore, it could be highly advantageous if APIs with an existing or added amidine group could have modified hydrophilicity with CO_2 [85] for the purposes of separation, purification, or analysis.

7.5. HBD—HBA Mixtures of Molecular Solvents

The attractiveness of using molecular solvents to form HBD—HBA mixtures is that their EHS data are available, making it possible to assess their safety. With the EHS safety of the solvents assessed, it becomes possible to focus on the technical issue of solubilizing the API in the mixed solvent for processing operations.

Duereh et al. [86] developed a methodology for replacing dipolar aprotic solvents with safe HBD—HBA solvent pairs based on solubility and Kamlet—Taft windows (Figure 6). In the methodology [86], solvent pairs are evaluated from a database with user-defined solubility parameter and KT parameter windows of an API to determine working compositions and a prioritized list of mixed solvents according to a composite GSK score. The open-access software given in ref. [86] can be extended with activity coefficient models or quantum chemistry methods to broaden the scope of the methodology.

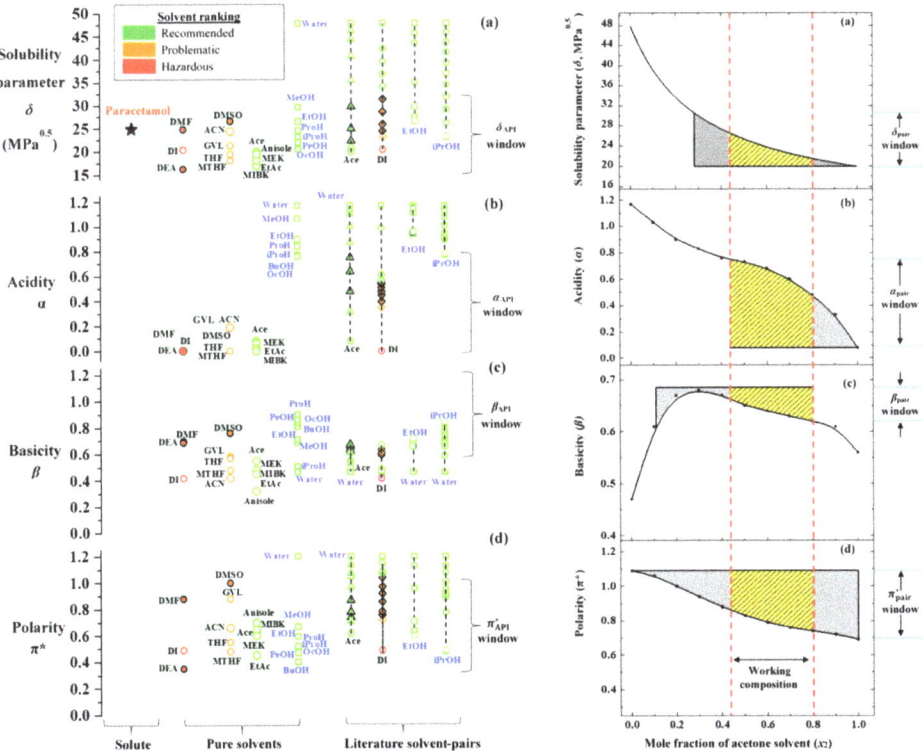

Figure 6. Concept of solubility parameter and Kamlet—Taft windows for identifying replacement solvents of an API (paracetamol): (**a**) window for solubility parameter, (**b**) window for API acidity,

(c) window for API basicity, (d) window for API dipolarity/polarizability. (**left**): Range of solubility and Kamlet—Taft parameters for dissolution of API in known solvents, including hazardous ones. (**right**): Range of solubility and Kamlet—Taft parameters superimposed onto theoretical calculations and available literature data to determine working composition ranges for a given mixed solvent pair (acetone—water). Reprinted with permission from ref. [86]. Copyright American Chemical Society, 2016.

In developing solvent replacement methodologies, physical properties can be important attributes for solvent selection. Jouyban and Acree [87] developed a single functional form for the correlation of viscosity, density, dielectric constant, surface tension, speed of sound, Reichardt E_T^N, molar volume, and isentropic compressibility of binary mixed solvents. Nazemieh et al. [88] reported data for a new set of mixed solvents, namely p-cymene with α-pinene, limonene, and citral correlated with the Jouyban—Acree model for physico-chemical properties (PCPs). Lee et al. [89] developed a local composition regular solution theory model for the correlation and prediction of API solubility in mixed solvents that had a single functional form for all compounds studied. The advantage of similar functional forms in correlative and predictive schemes for PCPs and activity coefficients is that machine learning techniques can be applied as the size of the database increases.

8. Kamlet—Taft Parameter Windows for APIs

APIs are commonly designated as being water soluble or non-water soluble. When the Reichardt E_T^N and KT parameters are plotted for mono-solvents that solvate 45 water-soluble APIs (Figure 7) and 47 water-insoluble APIs (Figure 8), the range of E_T^N and KT parameter values becomes visible, which characterizes the apparent polarity of the API. Although many water-soluble APIs are solvated by dipolar protic solvents ($E_T^N > 0.5$) and dipolar aprotic solvents ($0.3 < E_T^N < 0.5$) over a wide range of KT acidities (α), there are minimum values of π^* and β required for solvation (Figure 7). On the other hand, water-insoluble APIs are solvated by a relatively narrow range of solvent polarities ($0.2 < E_T^N < 0.75$) and KT acidities (α), in which there are maximum values of π^* and minimum values of β required for solvation (Figure 8).

When an API is dipolar protic, it interacts with basic dipolar aprotic solvents by forming hydrogen bonds with the solvent that must be stronger than those in the solid phase for solvation to occur. If the dipolar aprotic solvent is not basic or if it has insufficient basicity, then the dipolar protic API will have low solubility in the solvent, because dipole-dipole interactions generally do not have sufficient strength to break hydrogen bonds in the solid phase. On the other hand, when an API is dipolar aprotic, it interacts with dipolar aprotic solvents through dipole—dipole interactions that must be stronger than those in the solid phase for solvation to occur.

Conversely, if a solvent is dipolar protic, then the API must have sufficient basicity to accept hydrogen bonds that must be stronger than those in the solvent phase. Thus, scales for molecular basicity are extremely important in identifying potential new solvents and solvent systems. However, note that all KT α, β, and π^* parameters influence API solubility in a mixed solvent and that they depend on the mixed solvent local composition, frequently in a non-linear or synergistic way [90,91].

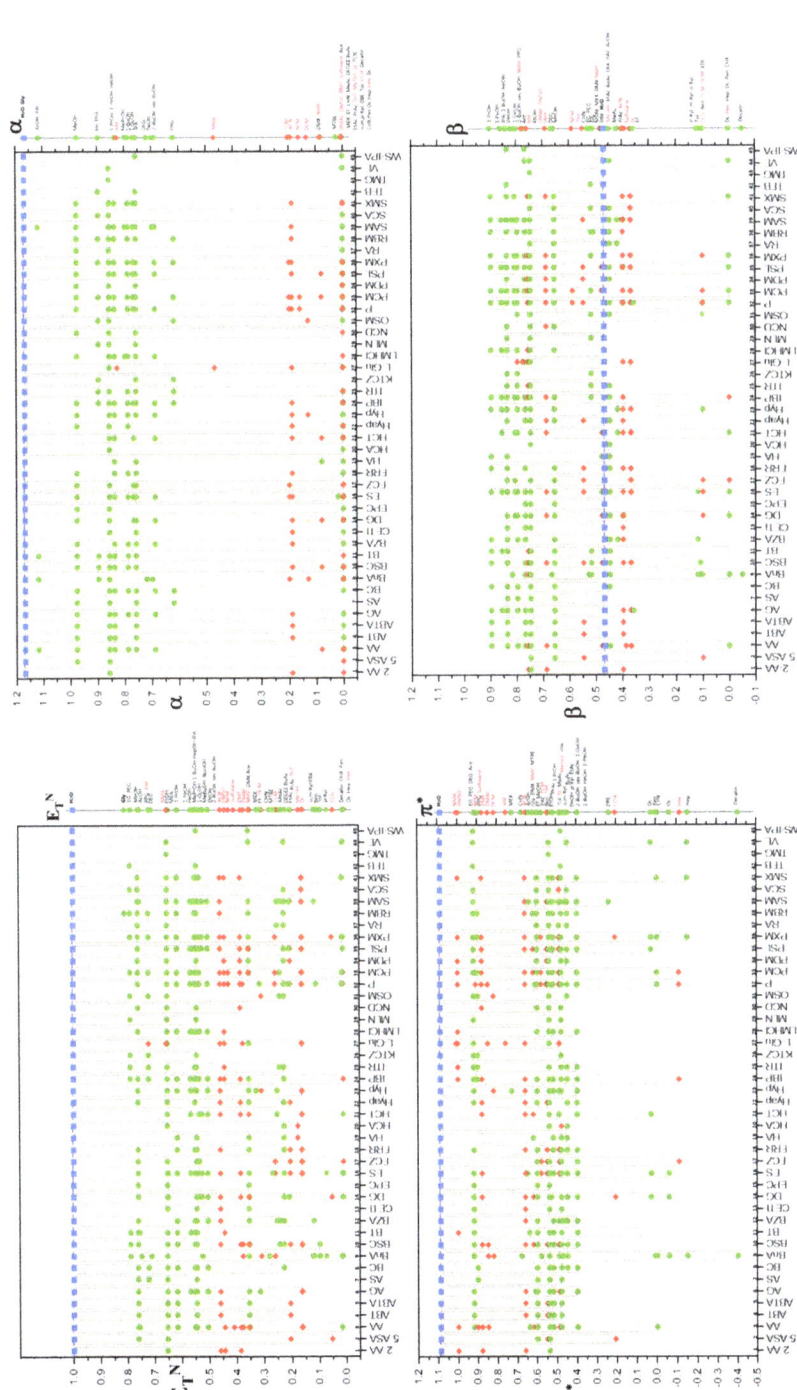

Figure 7. Reichardt E_T^N and Kamlet–Taft parameters of mono-solvents that solvate water-soluble APIs at ca. 25 °C. Data from refs. [1,92–302]. Water (Blue). Less-hazardous solvents (Green). Hazardous solvents (Red). Detailed information in Supplementary Materials.

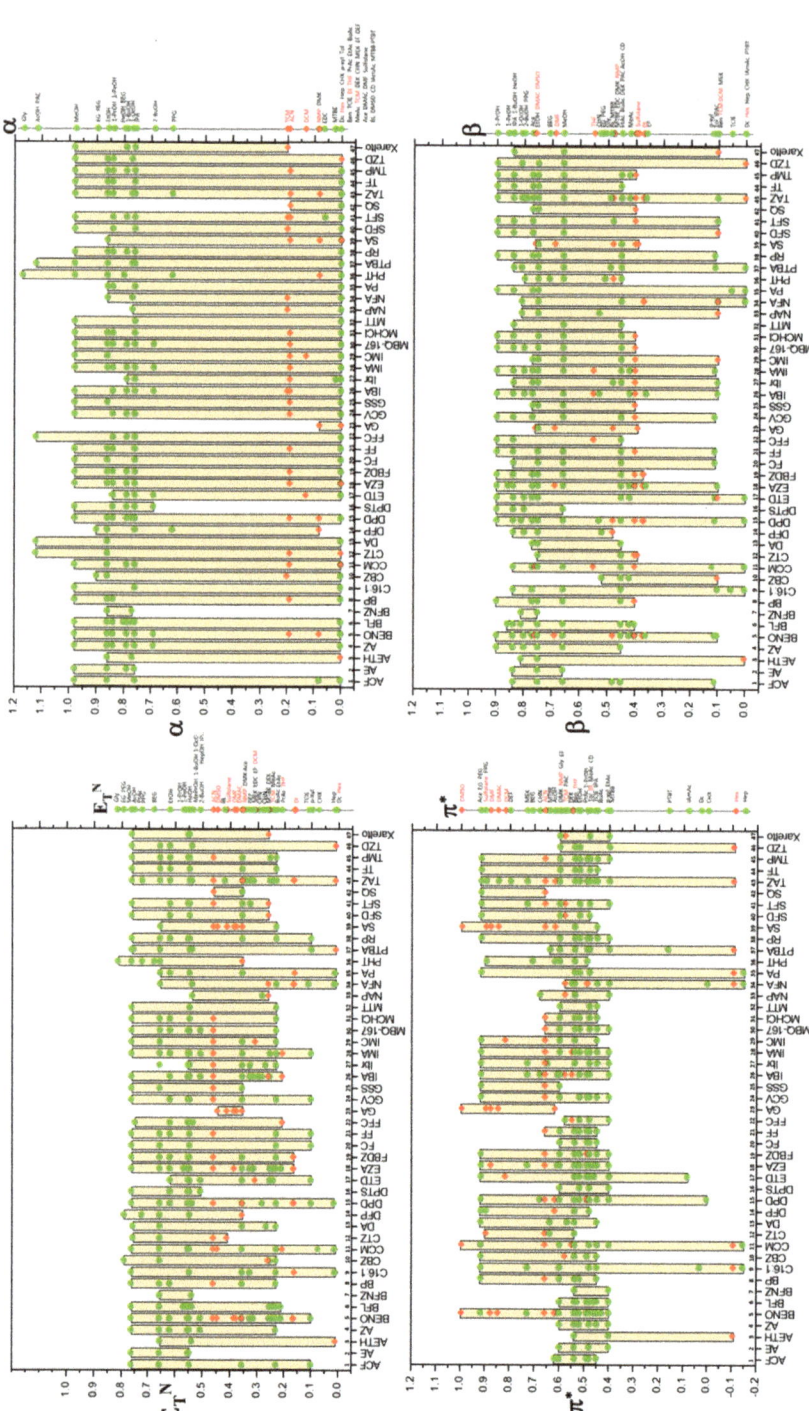

Figure 8. Reichardt E_T^N and Kamlet—Taft parameters of mono-solvents that solvate water-insoluble APIs at ca. 25 °C. Data from refs. [1,92–302]. Less-hazardous solvents (Green). Hazardous solvents (Red). Detailed information in Supplementary Materials.

Consider the HBD—HBA mixed-solvent systems shown in Figure 9. When water is used as the HBD solvent (Figure 9a–c), the HBA solvent addition (increasing x_2) lowers the KT α and generally lowers π^* values depending on the HBA polarity and causes KT β values to initially sharply increase, during which the microscopic polarity changes greatly due to the formation of complex molecules [91]. For example, water—lactone mixed solvents have been shown to exhibit synergy in KT basicity [91]. For an alcohol as the HBD solvent, the addition of an HBA solvent lowers the KT α, and it causes the KT π^* and KT β values to linearly increase or decrease with bulk composition depending on whether the pure alcohol KT π^* or KT β values are less than, equal to, or greater than those of the HBA solvent alcohol KT π^* or KT β values (Figure 9d–i). Duereh et al. showed examples of the case of ethanol (HBD) –cyclopentanone (HBA), in which mixed solvent composition can be used to favorably solvate an API (paracetamol), and they also showed a case of methanol (HBD) –cyclopentanone (HBA), in which mixed solvent composition failed to provide any solvation benefit, along with examples of 12 APIs [90].

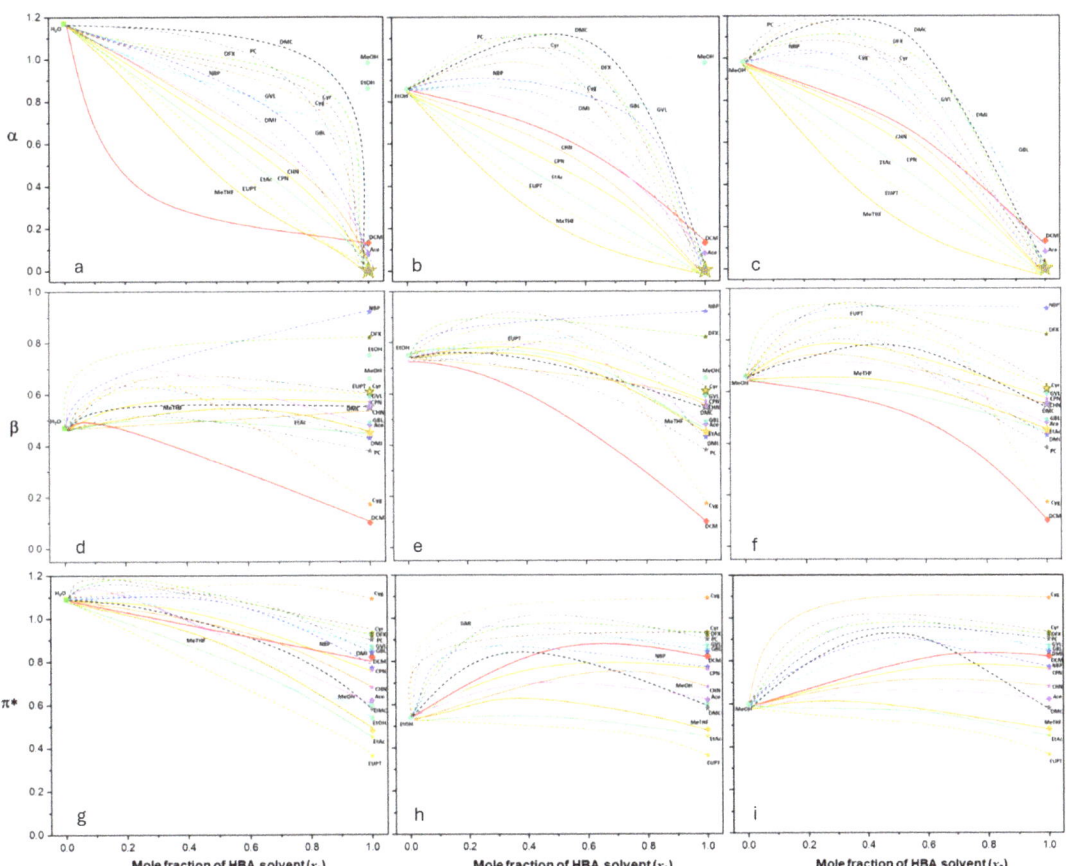

Figure 9. Kamlet—Taft acidity, basicity, and polarity for selected mixed solvents versus HBA solvent mole fraction: (**a,d,g**) water—HBA; (**b,e,h**) methanol—HBA; (**c,f,i**) ethanol—HBA. Trends shown are based on estimations (dashed lines) and actual data (solid lines) [51,86,90,91].

There are a number of HBD—HBA solvent combinations that could be replacements for hazardous solvents (Figure 9). For possible HBD solvents, water, methanol, and ethanol are good candidates. When water is the HBD solvent, possible candidate HBA solvents are acetone, acetic acid, acetonitrile, EtOH, MeOH, 2-MeTHF, water-2,2,5,5-tetramethyloxalane (TMO), DMI, Cyrene, Cygnet 0.0, or possibly diformylxylose. Safety and conditions must be considered carefully. For example, 2-MeTHF forms peroxides more rapidly than IPE, THF, or CPME when inhibitors are not present; ethereal solvents form peroxides [7]. Cygnet 0.0 is solid at room temperature [303], and 2-MeTHF in water has inverse temperature behavior up until temperatures of 340 K [304], meaning that its solubility in water decreases with increasing temperature.

When alcohols are used as the HBD solvent, cyclohexanone (CHN), cyclopentanone (CPN), many kinds of esters, GBL, GVL, eucalytol (water insoluble), or possibly MeSesamol (water insoluble) [305] or diformylxylose [306] are candidates. Furthermore, interesting HBA—HBA combinations, such as Cyrene—Cygnet 0.0, are being suggested for polymer syntheses [303] to replace hazardous dipolar aprotic solvents, and these types of HBA—HBA mixed solvents could have advantages in processing APIs.

9. Linear Solvation Energy Relationships (LSER)

Polarity parameters originally reported by Kamlet, Abboud, Abraham, and Taft were intended for use in linear solvation energy relationships (LSER) [307], expressed as follows:

$$XYZ = XYZ_0 + s(\pi^* + d\delta) + a\alpha + b\beta + h\delta_H + e\xi \tag{5}$$

where XYZ is a chemical phenomenon, XYZ_0 is a reference phenomenon, and $s, a, b, h,$ and e are descriptors that are used to correlate polarity parameters to XYZ. Many adaptations have been made of Equation (5), and a well-known one is due to Abraham [308], which expressed water—octanol partition coefficients (log P) and gas—solvent partition coefficient (log K) as follows [309]:

$$\log(P) = c + eE + sS + aA + bB + vV \tag{6}$$

$$\log(K) = c + eE + sS + aA + bB + vL \tag{7}$$

Where the bold symbols are properties of the solute related to excess molar refraction (**E**), dipolarity/polarizability (**S**), hydrogen bond acidity (**A**), Lewis basicity (**B**), McGowan's molecular volume (**V**), and gas-to-hexadecane partition coefficient (**L**). LSER models are directly applicable to predict the solubility of APIs in solvents [309]. LSER models are widely used in the field of chromatography for characterizing columns and estimating retention times [310,311] or in the analysis of petroleum distillate conditions with group contribution activity coefficient models such as UNIFAC [312], but they do not appear to have been used more broadly (in reverse) in mixed solvent replacement schemes, although environmentally related partition coefficients are incorporated into life-cycle assessment tools, such as EPA's CompTox chemical dashboard system [313] or machine learning studies for solvent characterization factors [314].

10. Conclusions

In this work, several strategies were highlighted for the replacement of hazardous dipolar aprotic solvents related to pharmaceutical and bio-related compounds. Solvent guides form the basis of solvent replacement and consider categories of safety, human health, environment, waste, and sustainability. Linking online solvent selection sites with GSK, CHEM21, ECHA, and other guidelines would allow for the efficient dissemination of solvent replacements.

An example of drop-in replacement solvents and several mixed solvent combinations for synthesizing APIs is one strategy that shows it is possible for academia and the industry to replace hazardous dipolar aprotic solvents by adopting new chemical systems that are both efficient and safe. The universal guide for the replacement of hazardous dipolar aprotic solvents in synthetic chemistry is one of the key strategies.

Mixed solvents can be used in many ways to replace hazardous solvents, often with a performance benefit. Dichloromethane can be replaced by ethanol (HBD) and ethyl acetate (HBA) mixed solvents, as is evident from marketed stock solutions by chemical companies. The use of CO_2 with esters or alcohols instead of hexane or chlorinated hydrocarbons is seen to be effective for thin-layer, flash, and supercritical chromatography, and with the introduction of marketed industrial analytical equipment, it is clear that the new technology will become established.

Expanded liquids, supercritical fluids, low-transition temperature (HBD—HBA) mixtures, and switchable solvents all offer safer chemical systems that have low energy, performance, and sustainability benefits. Chemical systems based on HBD—HBA mixtures of molecular solvents for processing APIs offer a simple way to replace hazardous solvents by considering the range of solubility parameters, Reichardt polarity, and Kamlet—Taft parameters of the pure components. Reichardt polarity and Kamlet—Taft parameters of pure components are necessary physical properties for the development of solvent replacement strategies. By using the available solubility data of APIs in mono-solvents, new mixed solvent combinations can be seen.

11. Future Outlook

Presently, there are many measurements of Reichardt polarity and Kamlet—Taft parameters of pure compounds, but far fewer measurements have been made for mixed solvent systems that can potentially replace hazardous dipolar aprotic solvents. Many new measurements are needed of Reichardt polarity and Kamlet—Taft parameters of HBD—HBA and HBA—HBA mixed solvents, especially those systems such as ethanol—ethyl acetate, to understand fundamental interactions of complex molecules with APIs.

Theoretical methods applied to HBD—HBA systems could greatly accelerate the identification of new chemical systems for processing APIs. COSMO-RS is able to quantitatively predict Kamlet—Taft parameters for both molecular solvents and deep eutectic solvents [315]. COSMO-RS gives qualitative predictions of Hansen solubility parameters [316], which is encouraging because the values of APIs could lead to a great reduction in experimental effort.

Supplementary Materials: The following supporting information can be downloaded at: https://www.mdpi.com/article/10.3390/liquids4020018/s1, Table S1. Water-soluble APIs solvated by monosolvents and their solvent polarity (E_T^N), Kamlet-Taft acidity (α), basicity (β) window, dipolarity/polarizability (π^*) and corresponding literature. Solvents listed as hazardous in GSK solvent guide are highlighted in red. Table S2. Water-insoluble APIs solvated by monosolvents and their solvent polarity (E_T^N), Kamlet-Taft acidity (α), basicity (β) window, dipolarity/polarizability (π^*) and corresponding literature. Solvents listed as hazardous in GSK solvent guide are highlighted in red.

Author Contributions: J.L.L.: data curation, investigation, formal analysis, visualization, software. G.H.C.: supervision, resources, project administration, funding acquisition, writing—reviewing and editing. M.O.: writing-reviewing and editing. H.G.: writing—reviewing and editing. R.L.S.J.: conceptualization, supervision, methodology, formal analysis, project administration, funding acquisition, writing—original draft, reviewing and editing. All authors have read and agreed to the published version of the manuscript.

Funding: The authors would like to acknowledge Trumer Medicare Sdn Bhd (Vot number: 6300827-11801) for the funding of the subscription of software.

Acknowledgments: Partial support of this project in the form of facilities and resources of the Research Center of Supercritical Fluid Technology, Tohoku University, is gratefully acknowledged.

Conflicts of Interest: The authors declare no conflicts of interest.

References

1. Reichardt, C. Empirical Parameters of Solvent Polarity as Linear Free-Energy Relationships. *Angew. Chem. Int. Ed. Engl.* **1979**, *18*, 98–110. [CrossRef]
2. Krygowski, T.M.; Wrona, P.K.; Zielkowska, U.; Reichardt, C. Empirical parameters of lewis acidity and basicity for aqueous binary solvent mixtures. *Tetrahedron* **1985**, *41*, 4519–4527. [CrossRef]
3. Reichardt, C. Solvation Effects in Organic Chemistry: A Short Historical Overview. *J. Org. Chem.* **2022**, *87*, 1616–1629. [CrossRef] [PubMed]
4. Kamlet, M.J.; Taft, R.W. The Solvatochromic Comparison Method. I. The β-Scale Of Solvent Hydrogen-Bond Acceptor (HBA) Basicities. *J. Am. Chem. Soc.* **1976**, *98*, 377–383. [CrossRef]
5. Catalán, J. On the empirical scales of organic solvents established using probe/homomorph pairs. *J. Phys. Org. Chem.* **2021**, *34*, e4206. [CrossRef]
6. Reichardt, C.; Welton, T. *Solvents and Solvent Effects in Organic Chemistry*; Wiley-VCH Verlag GmbH & Co.: Weinheim, Germany, 2011. [CrossRef]
7. Alder, C.M.; Hayler, J.D.; Henderson, R.K.; Redman, A.M.; Shukla, L.; Shuster, L.E.; Sneddon, H.F. Updating and further expanding GSK's solvent sustainability guide. *Green Chem.* **2016**, *18*, 3879–3890. [CrossRef]
8. Prat, D.; Wells, A.; Hayler, J.; Sneddon, H.; McElroy, C.R.; Abou-Shehada, S.; Dunn, P.J. CHEM21 selection guide of classical- and less classical-solvents. *Green Chem.* **2016**, *18*, 288–296. [CrossRef]
9. ACS. ACS GCI Pharmaceutical Roundtable. Collaboration to Deliver a Solvent Selection Guide for the Pharmaceutical Industry. 2018. Available online: https://www.acs.org/content/dam/acsorg/greenchemistry/industriainnovation/roundtable/solvent-selection-guide.pdf (accessed on 16 December 2023).
10. ACS. Tools for Innovation in Chemistry. 2023. Available online: https://www.acsgcipr.org/tools-for-innovation-in-chemistry/ (accessed on 16 December 2023).
11. Byrne, F.P.; Jin, S.; Paggiola, G.; Petchey, T.H.M.; Clark, J.H.; Farmer, T.J.; Hunt, A.J.; Robert McElroy, C.; Sherwood, J. Tools and techniques for solvent selection: Green solvent selection guides. *Sustain. Chem. Process.* **2016**, *4*, 7. [CrossRef]
12. Diorazio, L.J.; Richardson, P.; Sneddon, H.F.; Moores, A.; Briddell, C.; Martinez, I. Making Sustainability Assessment Accessible: Tools Developed by the ACS Green Chemistry Institute Pharmaceutical Roundtable. *ACS Sustain. Chem. Eng.* **2021**, *9*, 16862–16864. [CrossRef]
13. Dixit, S.; Crain, J.; Poon, W.C.K.; Finney, J.L.; Soper, A.K. Molecular segregation observed in a concentrated alcohol–water solution. *Nature* **2002**, *416*, 829–832. [CrossRef]
14. Ono, T.; Horikawa, K.; Ota, M.; Sato, Y.; Inomata, H. Insight into the local composition of the Wilson equation at high temperatures and pressures through molecular simulations of methanol-water mixtures. *J. Chem. Eng. Data* **2014**, *59*, 1024–1030. [CrossRef]
15. Ono, T.; Ito, Y.; Ota, M.; Takebayashi, Y.; Furuya, T.; Inomata, H. Difference in aqueous solution structure at 293.2 and 473.2 K between ethanol and ethylene glycol via molecular dynamics. *J. Mol. Liq.* **2022**, *368*, 120764. [CrossRef]
16. Jordan, A.; Hall, C.G.J.; Thorp, L.R.; Sneddon, H.F. Replacement of Less-Preferred Dipolar Aprotic and Ethereal Solvents in Synthetic Organic Chemistry with More Sustainable Alternatives. *Chem. Rev.* **2022**, *122*, 6749–6794. [CrossRef] [PubMed]
17. ECHA. Candidate List of Substances of Very High Concern for Authorisation. 2023. Available online: https://echa.europa.eu/candidate-list-table (accessed on 24 December 2023).
18. Henderson, R.K.; Jiménez-González, C.; Constable, D.J.C.; Alston, S.R.; Inglis, G.G.A.; Fisher, G.; Sherwood, J.; Binks, S.P.; Curzons, A.D. Expanding GSK's solvent selection guide—Embedding sustainability into solvent selection starting at medicinal chemistry. *Green Chem.* **2011**, *13*, 854–862. [CrossRef]
19. Jiménez-González, C.; Curzons, A.D.; Constable, D.J.C.; Cunningham, V.L. Expanding GSK's Solvent Selection Guide—Application of life cycle assessment to enhance solvent selections. *Clean Technol. Environ. Policy* **2004**, *7*, 42–50. [CrossRef]
20. Diorazio, L.J.; Hose, D.R.J.; Adlington, N.K. Toward a More Holistic Framework for Solvent Selection. *Org. Process Res. Dev.* **2016**, *20*, 760–773. [CrossRef]
21. Gao, F.; Bai, R.; Ferlin, F.; Vaccaro, L.; Li, M.; Gu, Y. Replacement strategies for non-green dipolar aprotic solvents. *Green Chem.* **2020**, *22*, 6240–6257. [CrossRef]
22. Gabriel, C.M.; Keener, M.; Gallou, F.; Lipshutz, B.H. Amide and Peptide Bond Formation in Water at Room Temperature. *Org. Lett.* **2015**, *17*, 3968–3971. [CrossRef]
23. Dalla Torre, D.; Annatelli, M.; Aricò, F. Acid catalyzed synthesis of dimethyl isosorbide via dimethyl carbonate chemistry. *Catal. Today* **2023**, *423*, 113892. [CrossRef]
24. Sherwood, J.; Constantinou, A.; Moity, L.; McElroy, C.R.; Farmer, T.J.; Duncan, T.; Raverty, W.; Hunt, A.J.; Clark, J.H. Dihydrolevoglucosenone (Cyrene) as a bio-based alternative for dipolar aprotic solvents. *Chem. Commun.* **2014**, *50*, 9650–9652. [CrossRef]
25. Campos, J.F.; Scherrmann, M.-C.; Berteina-Raboin, S. Eucalyptol: A new solvent for the synthesis of heterocycles containing oxygen, sulfur and nitrogen. *Green Chem.* **2019**, *21*, 1531–1539. [CrossRef]
26. Domokos, A.; Nagy, B.; Szilágyi, B.; Marosi, G.; Nagy, Z.K. Integrated Continuous Pharmaceutical Technologies—A Review. *Org. Process Res. Dev.* **2021**, *25*, 721–739. [CrossRef]
27. Lavayssiere, M.; Lamaty, F. Amidation by reactive extrusion for the synthesis of active pharmaceutical ingredients teriflunomide and moclobemide. *Chem. Commun.* **2023**, *59*, 3439–3442. [CrossRef] [PubMed]

28. Hildebrand, J.H. solubility. xii. Regular solutions1. *J. Am. Chem. Soc.* **1929**, *51*, 66–80. [CrossRef]
29. Prausnitz, J.M.; Lichtenthaler, R.N.; Azevedo, E.G.D. *Molecular Thermodynamics of Fluid-Phase Equilibria*, 3rd ed.; Prentice Hall PTR: London, UK, 1999.
30. Hansen, C.M. Hansen Solubility Parameters. In *A User's Handbook*, 2nd ed.; CRC Press: Boca Raton, FL, USA, 2007. [CrossRef]
31. Takebayashi, Y.; Sue, K.; Furuya, T.; Yoda, S. Solubilities of Organic Semiconductors and Nonsteroidal Anti-inflammatory Drugs in Pure and Mixed Organic Solvents: Measurement and Modeling with Hansen Solubility Parameter. *J. Chem. Eng. Data* **2018**, *63*, 3889–3901. [CrossRef]
32. Wu, Y.; Li, W.; Vovers, J.; Thuan Lu, H.; Stevens, G.W.; Mumford, K.A. Investigation of green solvents for the extraction of phenol and natural alkaloids: Solvent and extractant selection. *Chem. Eng. J.* **2022**, *442*, 136054. [CrossRef]
33. Kumar, A.; Nanda, A. In-silico methods of cocrystal screening: A review on tools for rational design of pharmaceutical cocrystals. *J. Drug Deliv. Sci. Technol.* **2021**, *63*, 102527. [CrossRef]
34. Abbott, S. Solubility, similarity, and compatibility: A general-purpose theory for the formulator. *Curr. Opin. Colloid Interface Sci.* **2020**, *48*, 65–76. [CrossRef]
35. Venkatram, S.; Kim, C.; Chandrasekaran, A.; Ramprasad, R. Critical Assessment of the Hildebrand and Hansen Solubility Parameters for Polymers. *J. Chem. Inf. Model.* **2019**, *59*, 4188–4194. [CrossRef]
36. Ma, Q.; Yu, C.; Zhou, Y.; Hu, D.; Chen, J.; Zhang, X. A review on the calculation and application of lignin Hansen solubility parameters. *Int. J. Biol. Macromol.* **2024**, *256*, 128506. [CrossRef]
37. Novaes, F.J.M.; de Faria, D.C.; Ferraz, F.Z.; de Aquino Neto, F.R. Hansen Solubility Parameters Applied to the Extraction of Phytochemicals. *Plants* **2023**, *12*, 3008. [CrossRef] [PubMed]
38. Otárola-Sepúlveda, J.; Cea-Klapp, E.; Aravena, P.; Ormazábal-Latorre, S.; Canales, R.I.; Garrido, J.M.; Valerio, O. Assessment of Hansen solubility parameters in deep eutectic solvents for solubility predictions. *J. Mol. Liq.* **2023**, *388*, 122669. [CrossRef]
39. Kamlet, M.J.; Abboud, J.L.; Taft, R.W. The solvatochromic comparison method. 6. The .pi.* scale of solvent polarities. *J. Am. Chem. Soc.* **1977**, *99*, 6027–6038. [CrossRef]
40. Spange, S.; Weiß, N.; Schmidt, C.H.; Schreiter, K. Reappraisal of Empirical Solvent Polarity Scales for Organic Solvents. *Chem. Methods* **2021**, *1*, 42–60. [CrossRef]
41. Spange, S.; Weiß, N. Empirical Hydrogen Bonding Donor (HBD) Parameters of Organic Solvents Using Solvatochromic Probes—A Critical Evaluation. *ChemPhysChem* **2023**, *24*, e202200780. [CrossRef] [PubMed]
42. Duereh, A.; Sato, Y.; Smith, R.L.; Inomata, H.; Pichierri, F. Does Synergism in Microscopic Polarity Correlate with Extrema in Macroscopic Properties for Aqueous Mixtures of Dipolar Aprotic Solvents? *J. Phys. Chem. B* **2017**, *121*, 6033–6041. [CrossRef]
43. Taygerly, J.P.; Miller, L.M.; Yee, A.; Peterson, E.A. A convenient guide to help select replacement solvents for dichloromethane in chromatography. *Green Chem.* **2012**, *14*, 3020–3025. [CrossRef]
44. Sigma-Aldrich. Greener Chromatography Solvents. 2015. Available online: https://www.sigmaaldrich.com/content/dam/sigma-aldrich/docs/Sigma-Aldrich/General_Information/1/greener-chromatography-solvents-82207.pdf (accessed on 27 December 2023).
45. McClain, R.; Rada, V.; Nomland, A.; Przybyciel, M.; Kohler, D.; Schlake, R.; Nantermet, P.; Welch, C.J. Greening Flash Chromatography. *ACS Sustain. Chem. Eng.* **2016**, *4*, 4905–4912. [CrossRef]
46. Dutta, P.; McGranaghan, A.; Keller, I.; Patil, Y.; Mulholland, N.; Murudi, V.; Prescher, H.; Smith, A.; Carson, N.; Martin, C.; et al. A case study in green chemistry: The reduction of hazardous solvents in an industrial R&D environment. *Green Chem.* **2022**, *24*, 3943–3956. [CrossRef]
47. Duereh, A.; Sato, Y.; Smith, R.L.; Inomata, H. Replacement of Hazardous Chemicals Used in Engineering Plastics with Safe and Renewable Hydrogen-Bond Donor and Acceptor Solvent-Pair Mixtures. *ACS Sustain. Chem. Eng.* **2015**, *3*, 1881–1889. [CrossRef]
48. Hicks, M.B.; Farrell, W.; Aurigemma, C.; Lehmann, L.; Weisel, L.; Nadeau, K.; Lee, H.; Moraff, C.; Wong, M.; Huang, Y.; et al. Making the move towards modernized greener separations: Introduction of the analytical method greenness score (AMGS) calculator. *Green Chem.* **2019**, *21*, 1816–1826. [CrossRef]
49. Petřík, I.; Pěnčík, A.; Stýskala, J.; Tranová, L.; Amakorová, P.; Strnad, M.; Novák, O. Rapid profiling of cytokinins using supercritical fluid chromatography coupled with tandem mass spectrometry. *Anal. Chim. Acta* **2024**, *1285*, 342010. [CrossRef]
50. Lynch, J.; Sherwood, J.; McElroy, C.R.; Murray, J.; Shimizu, S. Dichloromethane replacement: Towards greener chromatography via Kirkwood–Buff integrals. *Anal. Methods* **2023**, *15*, 596–605. [CrossRef] [PubMed]
51. Duereh, A.; Smith, R.L. Strategies for using hydrogen-bond donor/acceptor solvent pairs in developing green chemical processes with supercritical fluids. *J. Supercrit. Fluids* **2018**, *141*, 182–197. [CrossRef]
52. Hoang, H.N.; Nagashima, Y.; Mori, S.; Kagechika, H.; Matsuda, T. CO_2-expanded bio-based liquids as novel solvents for enantioselective biocatalysis. *Tetrahedron* **2017**, *73*, 2984–2989. [CrossRef]
53. Kankala, R.K.; Zhang, Y.S.; Wang, S.-B.; Lee, C.-H.; Chen, A.-Z. Supercritical Fluid Technology: An Emphasis on Drug Delivery and Related Biomedical Applications. *Adv. Healthc. Mater.* **2017**, *6*, 1700433. [CrossRef] [PubMed]
54. Temelli, F. Perspectives on supercritical fluid processing of fats and oils. *J. Supercrit. Fluids* **2009**, *47*, 583–590. [CrossRef]
55. Temelli, F. Perspectives on the use of supercritical particle formation technologies for food ingredients. *J. Supercrit. Fluids* **2018**, *134*, 244–251. [CrossRef]
56. De Melo, M.M.R.; Silvestre, A.J.D.; Silva, C.M. Supercritical fluid extraction of vegetable matrices: Applications, trends and future perspectives of a convincing green technology. *J. Supercrit. Fluids* **2014**, *92*, 115–176. [CrossRef]

57. Krichnavaruk, S.; Shotipruk, A.; Goto, M.; Pavasant, P. Supercritical carbon dioxide extraction of astaxanthin from Haematococcus pluvialis with vegetable oils as co-solvent. *Bioresour. Technol.* **2008**, *99*, 5556–5560. [CrossRef]
58. Saravana, P.S.; Getachew, A.T.; Cho, Y.-J.; Choi, J.H.; Park, Y.B.; Woo, H.C.; Chun, B.S. Influence of co-solvents on fucoxanthin and phlorotannin recovery from brown seaweed using supercritical CO_2. *J. Supercrit. Fluids* **2017**, *120*, 295–303. [CrossRef]
59. Sun, M.; Temelli, F. Supercritical carbon dioxide extraction of carotenoids from carrot using canola oil as a continuous co-solvent. *J. Supercrit. Fluids* **2006**, *37*, 397–408. [CrossRef]
60. Kok, S.L.; Lee, W.J.; Smith, R.L.; Suleiman, N.; Jom, K.N.; Vangnai, K.; Bin Sharaai, A.H.; Chong, G.H. Role of virgin coconut oil (VCO) as co-extractant for obtaining xanthones from mangosteen (*Garcinia mangostana*) pericarp with supercritical carbon dioxide extraction. *J. Supercrit. Fluids* **2021**, *176*, 105305. [CrossRef]
61. Lee, W.J.; Ng, C.C.; Ng, J.S.; Smith, R.L.; Kok, S.L.; Hee, Y.Y.; Lee, S.Y.; Tan, W.K.; Zainal Abidin, N.H.; Halim Lim, S.A.; et al. Supercritical carbon dioxide extraction of α-mangostin from mangosteen pericarp with virgin coconut oil as co-extractant and in-vitro bio-accessibility measurement. *Process Biochem.* **2019**, *87*, 213–220. [CrossRef]
62. Gao, Y.; Liu, X.; Xu, H.; Zhao, J.; Wang, Q.; Liu, G.; Hao, Q. Optimization of supercritical carbon dioxide extraction of lutein esters from marigold (*Tagetes erecta* L.) with vegetable oils as continuous co-solvents. *Sep. Purif. Technol.* **2010**, *71*, 214–219. [CrossRef]
63. Ma, Q.; Xu, X.; Gao, Y.; Wang, Q.; Zhao, J. Optimisation of supercritical carbon dioxide extraction of lutein esters from marigold (*Tagetes erect* L.) with soybean oil as a co-solvent. *Int. J. Food Sci. Technol.* **2008**, *43*, 1763–1769. [CrossRef]
64. Pattiram, P.D.; Abas, F.; Suleiman, N.; Mohamad Azman, E.; Chong, G.H. Edible oils as a co-extractant for the supercritical carbon dioxide extraction of flavonoids from propolis. *PLoS ONE* **2022**, *17*, e0266673. [CrossRef]
65. Shi, X.; Wu, H.; Shi, J.; Xue, S.J.; Wang, D.; Wang, W.; Cheng, A.; Gong, Z.; Chen, X.; Wang, C. Effect of modifier on the composition and antioxidant activity of carotenoid extracts from pumpkin (Cucurbita maxima) by supercritical CO_2. *LWT-Food Sci. Technol.* **2013**, *51*, 433–440. [CrossRef]
66. Fikri, I.; Yulianah, Y.; Lin, T.-C.; Lin, R.-W.; Chen, U.-C.; Lay, H.-L. Optimization of supercritical fluid extraction of dihydrotanshinone, cryptotanshinone, tanshinone I, and tanshinone IIA from Salvia miltiorrhiza with a peanut oil modifier. *Chem. Eng. Res. Des.* **2022**, *180*, 220–231. [CrossRef]
67. Saldaña, M.D.A.; Temelli, F.; Guigard, S.E.; Tomberli, B.; Gray, C.G. Apparent solubility of lycopene and β-carotene in supercritical CO_2, CO_2+ethanol and CO_2+canola oil using dynamic extraction of tomatoes. *J. Food Eng.* **2010**, *99*, 1–8. [CrossRef]
68. Vasapollo, G.; Longo, L.; Rescio, L.; Ciurlia, L. Innovative supercritical CO_2 extraction of lycopene from tomato in the presence of vegetable oil as co-solvent. *J. Supercrit. Fluids* **2004**, *29*, 87–96. [CrossRef]
69. Shi, J.; Yi, C.; Xue, S.J.; Jiang, Y.; Ma, Y.; Li, D. Effects of modifiers on the profile of lycopene extracted from tomato skins by supercritical CO_2. *J. Food Eng.* **2009**, *93*, 431–436. [CrossRef]
70. Yara-Varón, E.; Li, Y.; Balcells, M.; Canela-Garayoa, R.; Fabiano-Tixier, A.S.; Chemat, F. Vegetable oils as alternative solvents for green oleo-extraction, purification and formulation of food and natural products. *Molecules* **2017**, *22*, 1474. [CrossRef] [PubMed]
71. Ota, M.; Hashimoto, Y.; Sato, M.; Sato, Y.; Smith, R.L.; Inomata, H. Solubility of flavone, 6-methoxyflavone and anthracene in supercritical CO_2 with/without a co-solvent of ethanol correlated by using a newly proposed entropy-based solubility parameter. *Fluid Phase Equilibria* **2016**, *425*, 65–71. [CrossRef]
72. Ota, M.; Sugahara, S.; Sato, Y.; Smith, R.L.; Inomata, H. Vapor-liquid distribution coefficients of hops extract in high pressure CO_2 and ethanol mixtures and data correlation with entropy-based solubility parameters. *Fluid Phase Equilibria* **2017**, *434*, 44–48. [CrossRef]
73. Duereh, A.; Sugimoto, Y.; Ota, M.; Sato, Y.; Inomata, H. Kamlet-Taft Dipolarity/Polarizability of Binary Mixtures of Supercritical Carbon Dioxide with Cosolvents: Measurement, Prediction, and Applications in Separation Processes. *Ind. Eng. Chem. Res.* **2020**, *59*, 12319–12330. [CrossRef]
74. Francisco, M.; van den Bruinhorst, A.; Kroon, M.C. Low-Transition-Temperature Mixtures (LTTMs): A New Generation of Designer Solvents. *Angew. Chem. Int. Ed.* **2013**, *52*, 3074–3085. [CrossRef] [PubMed]
75. Smith, E.L.; Abbott, A.P.; Ryder, K.S. Deep Eutectic Solvents (DESs) and Their Applications. *Chem. Rev.* **2014**, *114*, 11060–11082. [CrossRef]
76. Moshikur, R.M.; Goto, M. Ionic Liquids as Active Pharmaceutical Ingredients (APIs). In *Application of Ionic Liquids in Drug Delivery*; Goto, M., Moniruzzaman, M., Eds.; Springer Singapore: Singapore, 2021; pp. 13–33. [CrossRef]
77. Md Moshikur, R.; Goto, M. Pharmaceutical Applications of Ionic Liquids: A Personal Account. *Chem. Rec.* **2023**, *23*, e202300026. [CrossRef]
78. Wu, X.; Zhu, Q.; Chen, Z.; Wu, W.; Lu, Y.; Qi, J. Ionic liquids as a useful tool for tailoring active pharmaceutical ingredients. *J. Control. Release* **2021**, *338*, 268–283. [CrossRef]
79. Moshikur, R.M.; Carrier, R.L.; Moniruzzaman, M.; Goto, M. Recent Advances in Biocompatible Ionic Liquids in Drug Formulation and Delivery. *Pharmaceutics* **2023**, *15*, 1179. [CrossRef] [PubMed]
80. Jessop, P.G.; Mercer, S.M.; Heldebrant, D.J. CO_2-triggered switchable solvents, surfactants, and other materials. *Energy Environ. Sci.* **2012**, *5*, 7240–7253. [CrossRef]
81. Cunha, I.T.; McKeeman, M.; Ramezani, M.; Hayashi-Mehedy, K.; Lloyd-Smith, A.; Bravi, M.; Jessop, P.G. Amine-free CO_2-switchable hydrophilicity solvents and their application in extractions and polymer recycling. *Green Chem.* **2022**, *24*, 3704–3716. [CrossRef]

82. Mercer, S.M.; Jessop, P.G. "Switchable water": Aqueous solutions of switchable ionic strength. *ChemSusChem* **2010**, *3*, 467–470. [CrossRef] [PubMed]
83. Liberato, V.S.; Ferreira, T.F.; MacDonald, A.R.; Dias Ribeiro, B.; Zarur Coelho, M.A.; Jessop, P.G. A CO_2-responsive method for separating hydrophilic organic molecules from aqueous solutions: Solvent-assisted switchable water. *Green Chem.* **2023**, *25*, 4705–4712. [CrossRef]
84. Cunha, I.T.; Yang, H.; Jessop, P.G. High pressure switchable water: An alternative method for separating organic products from water. *Green Chem.* **2021**, *23*, 3996–4007. [CrossRef]
85. Phan, L.; Jessop, P.G. Switching the hydrophilicity of a solute. *Green Chem.* **2009**, *11*, 307–330. [CrossRef]
86. Duereh, A.; Sato, Y.; Smith, R.L.; Inomata, H. Methodology for replacing dipolar aprotic solvents used in API processing with safe hydrogen-bond donor and acceptor solvent-pair mixtures. *Org. Process Res. Dev.* **2017**, *21*, 114–124. [CrossRef]
87. Jouyban, A.; Acree, W.E. A single model to represent physico-chemical properties of liquid mixtures at various temperatures. *J. Mol. Liq.* **2021**, *323*, 115054. [CrossRef]
88. Nazemieh, A.; Acree, W.E.; Jouyban, A. Further computations on physico-chemical properties of binary mixtures of p-cymene with α-pinene, limonene and citral. *J. Mol. Liq.* **2022**, *350*, 118211. [CrossRef]
89. Lee, J.L.; Chong, G.H.; Kanno, A.; Ota, M.; Guo, H.; Smith, R.L. Local composition-regular solution theory for analysis of pharmaceutical solubility in mixed-solvents. *J. Mol. Liq.* **2024**, *397*, 124012. [CrossRef]
90. Duereh, A.; Guo, H.; Honma, T.; Hiraga, Y.; Sato, Y.; Lee Smith, R.; Inomata, H. Solvent Polarity of Cyclic Ketone (Cyclopentanone, Cyclohexanone): Alcohol (Methanol, Ethanol) Renewable Mixed-Solvent Systems for Applications in Pharmaceutical and Chemical Processing. *Ind. Eng. Chem. Res.* **2018**, *57*, 7331–7344. [CrossRef]
91. Duereh, A.; Sato, Y.; Smith, R.L.; Inomata, H. Analysis of the Cybotactic Region of Two Renewable Lactone-Water Mixed-Solvent Systems that Exhibit Synergistic Kamlet-Taft Basicity. *J. Phys. Chem. B* **2016**, *120*, 4467–4481. [CrossRef] [PubMed]
92. Abbasi, M.; Vaez-Gharamaleki, J.; Fazeli-Bakhtiyari, R.; Martinez, F.; Jouyban, A. Prediction of deferiprone solubility in some non-aqueous binary solvent mixtures at various temperatures. *J. Mol. Liq.* **2015**, *203*, 16–19. [CrossRef]
93. Acree, W.E. IUPAC-NIST Solubility Data Series. 102. Solubility of Nonsteroidal Anti-inflammatory Drugs (NSAIDs) in Neat Organic Solvents and Organic Solvent Mixtures. *J. Phys. Chem. Ref. Data* **2014**, *43*, 023102. [CrossRef]
94. Akay, S.; Kayan, B.; Martínez, F. Solubility of fluconazole in (ethanol + water) mixtures: Determination, correlation, dissolution thermodynamics and preferential solvation. *J. Mol. Liq.* **2021**, *333*, 115987. [CrossRef]
95. Akay, S.; Kayan, B.; Peña, M.Á.; Jouyban, A.; Martínez, F. Solubility of Salicylic Acid in Some (Ethanol + Water) Mixtures at Different Temperatures: Determination, Correlation, Thermodynamics and Preferential Solvation. *Int. J. Thermophys.* **2023**, *44*, 121. [CrossRef]
96. Ali, H.S.M.; York, P.; Blagden, N.; Soltanpour, S.; Acree, W.E.; Jouyban, A. Solubility of Budesonide, Hydrocortisone, and Prednisolone in Ethanol + Water Mixtures at 298.2 K. *J. Chem. Eng. Data* **2009**, *55*, 578–582. [CrossRef]
97. Almandoz, M.C.; Sancho, M.I.; Blanco, S.E. Spectroscopic and DFT study of solvent effects on the electronic absorption spectra of sulfamethoxazole in neat and binary solvent mixtures. *Spectrochim. Acta Part A Mol. Biomol. Spectrosc.* **2014**, *118*, 112–119. [CrossRef]
98. Alsubaie, M.; Aljohani, M.; Erxleben, A.; McArdle, P. Cocrystal Forms of the BCS Class IV Drug Sulfamethoxazole. *Cryst. Growth Des.* **2018**, *18*, 3902–3912. [CrossRef]
99. Alvani-Alamdari, S.; Rezaei, H.; Rahimpour, E.; Hemmati, S.; Martinez, F.; Barzegar-Jalali, M.; Jouyban, A. Mesalazine solubility in the binary mixtures of ethanol and water at various temperatures. *Phys. Chem. Liq.* **2019**, *59*, 12–25. [CrossRef]
100. Aniya, V.; De, D.; Mohammed, A.M.; Thella, P.K.; Satyavathi, B. Measurement and modeling of solubility of para-tert-butylbenzoic acid in pure and mixed organic solvents at different temperatures. *J. Chem. Eng. Data* **2017**, *62*, 1411–1421. [CrossRef]
101. Anwer, M.K.; Mohammad, M.; Fatima, F.; Alshahrani, S.M.; Aldawsari, M.F.; Alalaiwe, A.; Al-Shdefat, R.; Shakeel, F. Solubility, solution thermodynamics and molecular interactions of osimertinib in some pharmaceutically useful solvents. *J. Mol. Liq.* **2019**, *284*, 53–58. [CrossRef]
102. Assis, G.P.; Garcia, R.H.L.; Derenzo, S.; Bernardo, A. Solid-liquid equilibrium of paracetamol in water-ethanol and water-propylene glycol mixtures. *J. Mol. Liq.* **2021**, *323*, 114617. [CrossRef]
103. Aydi, A.; Claumann, C.A.; Wüst Zibetti, A.; Abderrabba, M. Differential Scanning Calorimetry Data and Solubility of Rosmarinic Acid in Different Pure Solvents and in Binary Mixtures (Methyl Acetate + Water) and (Ethyl Acetate + Water) from 293.2 to 313.2 K. *J. Chem. Eng. Data* **2016**, *61*, 3718–3723. [CrossRef]
104. Banerjee, D.; Laha, A.K.; Bagchi, S. Preferential solvation in mixed binary solvent. *J. Chem. Soc. Faraday Trans.* **1995**, *91*, 631. [CrossRef]
105. Barzegar-Jalali, M.; Mirheydari, S.N.; Rahimpour, E.; Shekaari, H.; Martinez, F.; Jouyban, A. Experimental determination and correlation of bosentan solubility in (PEG 200+ water) mixtures at T = (293.15–313.15) K. *Phys. Chem. Liq.* **2018**, *57*, 504–515. [CrossRef]
106. Bernal-García, J.M.; Guzmán-López, A.; Cabrales-Torres, A.; Estrada-Baltazar, A.; Iglesias-Silva, G.A. Densities and viscosities of (N,N-dimethylformamide+ water) at atmospheric pressure from (283.15 to 353.15) K. *J. Chem. Eng. Data* **2008**, *53*, 1024–1027. [CrossRef]
107. Bhesaniya, K.; Nandha, K.; Baluja, S. Thermodynamics of Fluconazole Solubility in Various Solvents at Different Temperatures. *J. Chem. Eng. Data* **2014**, *59*, 649–652. [CrossRef]

108. Blanco-Márquez, J.H.; Ortiz, C.P.; Cerquera, N.E.; Martínez, F.; Jouyban, A.; Delgado, D.R. Thermodynamic analysis of the solubility and preferential solvation of sulfamerazine in (acetonitrile + water) cosolvent mixtures at different temperatures. *J. Mol. Liq.* **2019**, *293*, 111507. [CrossRef]
109. Blokhina, S.V.; Sharapova, A.V.; Ol'khovich, M.V.; Levshin, I.B.; Perlovich, G.L. Solid–liquid phase equilibrium and thermodynamic analysis of novel thiazolidine-2,4-dione derivative in different solvents. *J. Mol. Liq.* **2021**, *326*, 115273. [CrossRef]
110. Bosch, E.; Rived, F.; Rosés, M. Solute–solvent and solvent–solvent interactions in binary solvent mixtures. Part 4. Preferential solvation of solvatochromic indicators in mixtures of 2-methylpropan-2-ol with hexane, benzene, propan-2-ol, ethanol and methanol. *J. Chem. Soc. Perkin Trans.* **1996**, *2*, 2177–2184. [CrossRef]
111. Bosch, E.; Rosés, M.; Herodes, K.; Koppel, I.; Leito, I.; Koppel, I.; Taal, V. Solute-solvent and solvent-solvent interactions in binary solvent mixtures. 2. Effect of temperature on the ET(30) polarity parameter of dipolar hydrogen bond acceptor-hydrogen bond donor mixtures. *J. Phys. Org. Chem.* **1996**, *9*, 403–410. [CrossRef]
112. Calvo, B.; Cepeda, E.A. Solubilities of Stearic Acid in Organic Solvents and in Azeotropic Solvent Mixtures. *J. Chem. Eng. Data* **2008**, *53*, 628–633. [CrossRef]
113. Calvo, B.; Collado, I.; Cepeda, E.A. Solubilities of Palmitic Acid in Pure Solvents and Its Mixtures. *J. Chem. Eng. Data* **2008**, *54*, 64–68. [CrossRef]
114. Cañadas, R.; González-Miquel, M.; González, E.J.; Díaz, I.; Rodríguez, M. Evaluation of bio-based solvents for phenolic acids extraction from aqueous matrices. *J. Mol. Liq.* **2021**, *338*, 116930. [CrossRef]
115. Carmen Grande, M.d.; Juliá, J.A.; García, M.; Marschoff, C.M. On the density and viscosity of (water+dimethylsulphoxide) binary mixtures. *J. Chem. Thermodyn.* **2007**, *39*, 1049–1056. [CrossRef]
116. Castro, G.T.; Filippa, M.A.; Peralta, C.M.; Davin, M.V.; Almandoz, M.C.; Gasull, E.I. Solubility and Preferential Solvation of Piroxicam in Neat Solvents and Binary Systems. *Z. Für Phys. Chem.* **2017**, *232*, 257–280. [CrossRef]
117. Chen, F.; Zhao, M.; Feng, L.; Ren, B. Measurement and Correlation for Solubility of Diosgenin in Some Mixed Solvents. *Chin. J. Chem. Eng.* **2014**, *22*, 170–176. [CrossRef]
118. Chen, F.-X.; Zhao, M.-R.; Ren, B.-Z.; Zhou, C.-R.; Peng, F.-F. Solubility of diosgenin in different solvents. *J. Chem. Thermodyn.* **2012**, *47*, 341–346. [CrossRef]
119. Chen, S.; Liu, Q.; Dou, H.; Zhang, L.; Pei, L.; Huang, R.; Shu, G.; Yuan, Z.; Lin, J.; Zhang, W.; et al. Solubility and dissolution thermodynamic properties of Mequindox in binary solvent mixtures. *J. Mol. Liq.* **2020**, *303*, 112619. [CrossRef]
120. Chen, Y.; Xu, X.; Xie, L. Thermodynamic parameters on corresponding solid-liquid equilibrium of hydroxyapatite in pure and mixture organic solvents. *J. Mol. Liq.* **2017**, *229*, 189–197. [CrossRef]
121. Chen, Z.; Zhai, J.; Liu, X.; Mao, S.; Zhang, L.; Rohani, S.; Lu, J. Solubility measurement and correlation of the form A of ibrutinib in organic solvents from 278.15 to 323.15 K. *J. Chem. Thermodyn.* **2016**, *103*, 342–348. [CrossRef]
122. Cui, Z.; Yao, L.; Ye, J.; Wang, Z.; Hu, Y. Solubility measurement and thermodynamic modelling of curcumin in twelve pure solvents and three binary solvents at different temperature (T = 278.15–323.15 K). *J. Mol. Liq.* **2021**, *338*, 116795. [CrossRef]
123. Cysewski, P.; Jeliński, T.; Przybyłek, M.; Nowak, W.; Olczak, M. Solubility Characteristics of Acetaminophen and Phenacetin in Binary Mixtures of Aqueous Organic Solvents: Experimental and Deep Machine Learning Screening of Green Dissolution Media. *Pharmaceutics* **2022**, *14*, 2828. [CrossRef] [PubMed]
124. De la Rosa, M.V.G.; Santiago, R.; Romero, J.M.; Duconge, J.; Monbaliu, J.-C.; López-Mejías, V.; Stelzer, T. Solubility Determination and Correlation of Warfarin Sodium 2-Propanol Solvate in Pure, Binary, and Ternary Solvent Mixtures. *J. Chem. Eng. Data* **2019**, *64*, 1399–1413. [CrossRef] [PubMed]
125. del Mar Muñoz, M.; Delgado, D.R.; Peña, M.Á.; Jouyban, A.; Martínez, F. Solubility and preferential solvation of sulfadiazine, sulfamerazine and sulfamethazine in propylene glycol + water mixtures at 298.15K. *J. Mol. Liq.* **2015**, *204*, 132–136. [CrossRef]
126. Delgado, D.R.; Martínez, F. Solubility and solution thermodynamics of sulfamerazine and sulfamethazine in some ethanol+water mixtures. *Fluid Phase Equilibria* **2013**, *360*, 88–96. [CrossRef]
127. Delgado, D.R.; Martínez, F. Preferential solvation of sulfadiazine, sulfamerazine and sulfamethazine in ethanol+water solvent mixtures according to the IKBI method. *J. Mol. Liq.* **2014**, *193*, 152–159. [CrossRef]
128. Dizechi, M.; Marschall, E. Viscosity of some binary and ternary liquid mixtures. *J. Chem. Eng. Data* **1982**, *27*, 358–363. [CrossRef]
129. Domańska, U.; Pobudkowska, A.; Pelczarska, A.; Winiarska-Tusznio, M.; Gierycz, P. Solubility and pKa of select pharmaceuticals in water, ethanol, and 1-octanol. *J. Chem. Thermodyn.* **2010**, *42*, 1465–1472. [CrossRef]
130. Dong, Q.; Yu, S.; Wang, X.; Ding, S.; Li, E.; Cai, Y.; Xue, F. Solubility Measurement and Correlation of Itraconazole Hydroxy Isobutyltriazolone in Four Kinds of Binary Solvent Mixtures with Temperature from 283.15 to 323.15 K. *ACS Omega* **2023**, *8*, 39390–39400. [CrossRef] [PubMed]
131. Elizalde-Solis, O.; Arenas-Quevedo, M.G.; Verónico-Sánchez, F.J.; García-Morales, R.; Zúñiga-Moreno, A. Solubilities of Binary Systems α-Tocopherol + Capsaicin and α-Tocopherol + Palmitic Acid in Supercritical Carbon Dioxide. *J. Chem. Eng. Data* **2019**, *64*, 1948–1955. [CrossRef]
132. Fakhree, M.A.A.; Ahmadian, S.; Panahi-Azar, V.; Acree, W.E.; Jouyban, A. Solubility of 2-Hydroxybenzoic Acid in Water, 1-Propanol, 2-Propanol, and 2-Propanone at (298.2 to 338.2) K and Their Aqueous Binary Mixtures at 298.2 K. *J. Chem. Eng. Data* **2012**, *57*, 3303–3307. [CrossRef]
133. Filippa, M.A.; Gasull, E.I. Ibuprofen solubility in pure organic solvents and aqueous mixtures of cosolvents: Interactions and thermodynamic parameters relating to the solvation process. *Fluid Phase Equilibria* **2013**, *354*, 185–190. [CrossRef]

134. Gheitasi, N.; Nazari, A.H.; Haghtalab, A. Thermodynamic Modeling and Solubility Measurement of Cetirizine Hydrochloride and Deferiprone in Pure Solvents of Acetonitrile, Ethanol, Acetic Acid, Sulfolane, and Ethyl Acetate and Their Mixtures. *J. Chem. Eng. Data* **2019**, *64*, 5486–5496. [CrossRef]
135. Gonçalves Bonassoli, A.B.; Oliveira, G.; Bordón Sosa, F.H.; Rolemberg, M.P.; Mota, M.A.; Basso, R.C.; Igarashi-Mafra, L.; Mafra, M.R. Solubility measurement of lauric, palmitic, and stearic acids in ethanol, n-propanol, and 2-propanol using differential scanning calorimetry. *J. Chem. Eng. Data* **2019**, *64*, 2084–2092. [CrossRef]
136. Guo, S.; Yang, W.; Hu, Y.; Wang, K.; Luan, Y. Measurement and Correlation of the Solubility of N-Acetylglycine in Different Solvents at Temperatures from 278.15 to 319.15 K. *J. Solut. Chem.* **2013**, *42*, 1879–1887. [CrossRef]
137. Guo, Y.; He, H.; Huang, H.; Qiu, J.; Han, J.; Hu, S.; Liu, H.; Zhao, Y.; Wang, P. Solubility determination and thermodynamic modeling of n-acetylglycine in different solvent systems. *J. Chem. Eng. Data* **2021**, *66*, 1344–1355. [CrossRef]
138. Gusain, K.; Garg, S.; Kumar, R. Solubility Prediction of Pharmaceutical Compounds in Pure Solvent by Different Correlations and Thermodynamic Models. *SSRN Electron. J.* **2020**. [CrossRef]
139. Ha, E.-S.; Lee, S.-K.; Jeong, J.-S.; Sim, W.-Y.; Yang, J.-I.; Kim, J.-S.; Kim, M.-S. Solvent effect and solubility modeling of rebamipide in twelve solvents at different temperatures. *J. Mol. Liq.* **2019**, *288*, 111041. [CrossRef]
140. Hatefi, A.; Jouyban, A.; Mohammadian, E.; Acree, W.E.; Rahimpour, E. Prediction of paracetamol solubility in cosolvency systems at different temperatures. *J. Mol. Liq.* **2019**, *273*, 282–291. [CrossRef]
141. He, Q.; Zheng, M.; Zhao, H. Baicalin solubility in aqueous co-solvent mixtures of methanol, ethanol, isopropanol and n-propanol revisited: Solvent–solvent and solvent–solute interactions and IKBI preferential solvation analysis. *Phys. Chem. Liq.* **2019**, *58*, 820–832. [CrossRef]
142. Hellstén, S.; Qu, H.; Louhi-Kultanen, M. Screening of Binary Solvent Mixtures and Solvate Formation of Indomethacin. *Chem. Eng. Technol.* **2011**, *34*, 1667–1674. [CrossRef]
143. Heryanto, R.; Hasan, M.; Abdullah, E.C.; Kumoro, A.C. Solubility of Stearic Acid in Various Organic Solvents and Its Prediction using Non-ideal Solution Models. *ScienceAsia* **2007**, *33*, 469–472. Available online: https://www.scienceasia.org/2007.33.n4/v33_469_472.pdf (accessed on 30 December 2023).
144. Hu, W.; Shang, Z.; Wei, N.; Hou, B.; Gong, J.; Wang, Y. Solubility of benorilate in twelve monosolvents: Determination, correlation and COSMO-RS analysis. *J. Chem. Thermodyn.* **2021**, *152*, 106272. [CrossRef]
145. Hu, X.; Gong, Y.; Cao, Z.; Huang, Z.; Sha, J.; Li, Y.; Li, T.; Ren, B. Solubility, Hansen solubility parameter and thermodynamic properties of etodolac in twelve organic pure solvents at different temperatures. *J. Mol. Liq.* **2020**, *316*, 113779. [CrossRef]
146. Hu, X.; Tian, Y.; Cao, Z.; Sha, J.; Huang, Z.; Li, Y.; Li, T.; Ren, B. Solubility measurement, Hansen solubility parameter and thermodynamic modeling of etodolac in four binary solvents from 278.15 K to 323.15 K. *J. Mol. Liq.* **2020**, *318*, 114155. [CrossRef]
147. Imran, S.; Hossain, A.; Mahali, K.; Guin, P.S.; Datta, A.; Roy, S. Solubility and peculiar thermodynamical behaviour of 2-aminobenzoic acid in aqueous binary solvent mixtures at 288.15 to 308.15 K. *J. Mol. Liq.* **2020**, *302*, 112566. [CrossRef]
148. Ivanov, E.V.; Batov, D.V. Enthalpy-related parameters of interaction of simplest α-amino acids with the pharmaceutical mebicar (N-tetramethylglycoluril) in water at 298.15 K. *J. Chem. Thermodyn.* **2019**, *128*, 159–163. [CrossRef]
149. Jia, L.; Yang, J.; Cui, P.; Wu, D.; Wang, S.; Hou, B.; Zhou, L.; Yin, Q. Uncovering solubility behavior of Prednisolone form II in eleven pure solvents by thermodynamic analysis and molecular simulation. *J. Mol. Liq.* **2021**, *342*, 117376. [CrossRef]
150. Jiménez Cruz, J.M.; Vlaar, C.P.; López-Mejías, V.; Stelzer, T. Solubility Measurements and Correlation of MBQ-167 in Neat and Binary Solvent Mixtures. *J. Chem. Eng. Data* **2021**, *66*, 832–839. [CrossRef]
151. Jiménez, D.M.; Cárdenas, Z.J.; Martínez, F. Solubility and solution thermodynamics of sulfadiazine in polyethylene glycol 400 + water mixtures. *J. Mol. Liq.* **2016**, *216*, 239–245. [CrossRef]
152. Jouyban, A.; Acree, W.E.; Martínez, F. Dissolution thermodynamics and preferential solvation of ketoconazole in some {ethanol (1) + water (2)} mixtures. *J. Mol. Liq.* **2020**, *313*, 113579. [CrossRef]
153. Jouyban, A.; Mazaher Haji Agha, E.; Rahimpour, E.; Acree, W.E., Jr. Further computation and some comments on "Stearic acid solubility in mixed solvents of (water + ethanol) and (ethanol + ethyl acetate): Experimental data and comparison among different thermodynamic models". *J. Mol. Liq.* **2020**, *310*, 113228. [CrossRef]
154. Jouyban, K.; Mazaher Haji Agha, E.; Hemmati, S.; Martinez, F.; Kuentz, M.; Jouyban, A. Solubility of 5-aminosalicylic acid in N-methyl-2-pyrrolidone + water mixtures at various temperatures. *J. Mol. Liq.* **2020**, *310*, 113143. [CrossRef]
155. Jouyban-Gharamaleki, V.; Jouyban, A.; Kuentz, M.; Hemmati, S.; Martinez, F.; Rahimpour, E. A laser monitoring technique for determination of mesalazine solubility in propylene glycol and ethanol mixtures at various temperatures. *J. Mol. Liq.* **2020**, *304*, 112714. [CrossRef]
156. Jouyban-Gharamaleki, V.; Jouyban, A.; Martinez, F.; Zhao, H.; Rahimpour, E. A laser monitoring technique for solubility study of ketoconazole in propylene glycol and 2-propanol mixtures at various temperatures. *J. Mol. Liq.* **2020**, *320*, 114444. [CrossRef]
157. Kalam, M.A.; Alshehri, S.; Alshamsan, A.; Haque, A.; Shakeel, F. Solid liquid equilibrium of an antifungal drug itraconazole in different neat solvents: Determination and correlation. *J. Mol. Liq.* **2017**, *234*, 81–87. [CrossRef]
158. Kandi, S.; Charles, A.L. Measurement, correlation, and thermodynamic properties for solubilities of bioactive compound (−)-epicatechin in different pure solvents at 298.15 K to 338.15 K. *J. Mol. Liq.* **2018**, *264*, 269–274. [CrossRef]
159. Karpiuk, I.; Wilczura-Wachnik, H.; Myśliński, A. α-Tocopherol/AOT/alkane/water system. *J. Therm. Anal. Calorim.* **2017**, *131*, 2885–2892. [CrossRef]

160. Khajir, S.; Shayanfar, A.; Acree, W.E.; Jouyban, A. Effects of N-methylpyrrolidone and temperature on phenytoin solubility. *J. Mol. Liq.* **2019**, *285*, 58–61. [CrossRef]
161. Kuhs, M.; Svärd, M.; Rasmuson, Å.C. Thermodynamics of fenoxycarb in solution. *J. Chem. Thermodyn.* **2013**, *66*, 50–58. [CrossRef]
162. Kumari, A.; Kadakanchi, S.; Tangirala, R.; Thella, P.K.; Satyavathi, B. Measurement and modeling of solid–liquid equilibrium of para-tert-butylbenzoic acid in acetic acid/methanol+ water and acetic acid+ para-tert-butyltoluene binary systems at various temperatures. *J. Chem. Eng. Data* **2016**, *62*, 87–95. [CrossRef]
163. Lange, L.; Heisel, S.; Sadowski, G. Predicting the Solubility of Pharmaceutical Cocrystals in Solvent/Anti-Solvent Mixtures. *Molecules* **2016**, *21*, 593. [CrossRef] [PubMed]
164. Lee, S.-K.; Sim, W.-Y.; Ha, E.-S.; Park, H.; Kim, J.-S.; Jeong, J.-S.; Kim, M.-S. Solubility of bisacodyl in fourteen mono solvents and N-methyl-2-pyrrolidone + water mixed solvents at different temperatures, and its application for nanosuspension formation using liquid antisolvent precipitation. *J. Mol. Liq.* **2020**, *310*, 113264. [CrossRef]
165. Li, A.; Si, Z.; Yan, Y.; Zhang, X. Solubility and thermodynamic properties of hydrate lenalidomide in phosphoric acid solution. *J. Mol. Liq.* **2021**, *330*, 115446. [CrossRef]
166. Li, M.; Liu, S.; Li, S.; Yang, Y.; Cui, Y.; Gong, J. Determination and Correlation of Dipyridamole p-Toluene Sulfonate Solubility in Seven Alcohol Solvents and Three Binary Solvents. *J. Chem. Eng. Data* **2017**, *63*, 208–216. [CrossRef]
167. Li, R.; Fu, L.; Zhang, J.; Wang, W.; Chen, X.; Zhao, J.; Han, D. Solid-liquid equilibrium and thermodynamic properties of dipyridamole form II in pure solvents and mixture of (N-methyl pyrrolidone + isopropanol). *J. Chem. Thermodyn.* **2020**, *142*, 105981. [CrossRef]
168. Li, R.; Jin, Y.; Yu, B.; Xu, Q.; Chen, X.; Han, D. Solubility determination and thermodynamic properties calculation of macitentan in mixtures of ethyl acetate and alcohols. *J. Chem. Thermodyn.* **2021**, *156*, 106344. [CrossRef]
169. Li, R.; Liu, L.; Khan, A.; Li, C.; He, Z.; Zhao, J.; Han, D. Effect of Cosolvents on the Solubility of Lenalidomide and Thermodynamic Model Correlation of Data. *J. Chem. Eng. Data* **2019**, *64*, 4272–4279. [CrossRef]
170. Li, R.; Yan, H.; Wang, Z.; Gong, J. Correlation of Solubility and Prediction of the Mixing Properties of Ginsenoside Compound K in Various Solvents. *Ind. Eng. Chem. Res.* **2012**, *51*, 8141–8148. [CrossRef]
171. Li, R.; Yin, X.; Jin, Y.; Chen, X.; Zhao, B.; Wang, W.; Zhong, S.; Han, D. The solubility profile and dissolution thermodynamic properties of minocycline hydrochloride in some pure and mixed solvents at several temperatures. *J. Chem. Thermodyn.* **2021**, *157*, 106399. [CrossRef]
172. Li, R.; Zhao, B.; Chen, X.; Zhang, J.; Liu, Z.; Zhu, X.; Han, D. Solubility and apparent thermodynamic analysis of pomalidomide in (acetone + ethanol/isopropanol) and (ethyl acetate + ethanol/isopropanol) and its correlation with thermodynamic model. *J. Chem. Thermodyn.* **2021**, *154*, 106345. [CrossRef]
173. Li, W.; Yuan, J.; Wang, X.; Shi, W.; Zhao, H.; Xing, R.; Jouyban, A.; Acree, W.E. Bifonazole dissolved in numerous aqueous alcohol mixtures: Solvent effect, enthalpy–entropy compensation, extended Hildebrand solubility parameter approach and preferential solvation. *J. Mol. Liq.* **2021**, *338*, 116671. [CrossRef]
174. Li, X.; Du, C.; Cong, Y.; Zhao, H. Solubility determination and thermodynamic modelling of 3-amino-1,2,4-triazole in ten organic solvents from T = 283.15 K to T = 318.15 K and mixing properties of solutions. *J. Chem. Thermodyn.* **2017**, *104*, 189–200. [CrossRef]
175. Li, X.; Ma, M.; Du, C.; Zhao, H. Solubility of cetilistat in neat solvents and preferential solvation in (acetone, isopropanol or acetonitrile) + water co-solvent mixtures. *J. Mol. Liq.* **2017**, *242*, 618–624. [CrossRef]
176. Li, X.; Wang, M.; Du, C.; Cong, Y.; Zhao, H. Preferential solvation of rosmarinic acid in binary solvent mixtures of ethanol + water and methanol + water according to the inverse Kirkwood–Buff integrals method. *J. Mol. Liq.* **2017**, *240*, 56–64. [CrossRef]
177. Li, Y.; Wang, Y.; Ning, Z.; Cui, J.; Wu, Q.; Wang, X. Solubilities of Adipic Acid and Succinic Acid in a Glutaric Acid + Acetone or n-Butanol Mixture. *J. Chem. Eng. Data* **2014**, *59*, 4062–4069. [CrossRef]
178. Lin, L.; Zhao, K.; Yu, B.; Wang, H.; Chen, M.; Gong, J. Measurement and Correlation of Solubility of Cefathiamidine in Water + (Acetone, Ethanol, or 2-Propanol) from (278.15 to 308.15) K. *J. Chem. Eng. Data* **2015**, *61*, 412–419. [CrossRef]
179. Liu, J.-Q.; Wang, Y.; Tang, H.; Wu, S.; Li, Y.-Y.; Zhang, L.-Y.; Bai, Q.-Y.; Liu, X. Experimental Measurements and Modeling of the Solubility of Aceclofenac in Six Pure Solvents from (293.35 to 338.25) K. *J. Chem. Eng. Data* **2014**, *59*, 1588–1592. [CrossRef]
180. Liu, W.; Bao, Z.; Shen, Y.; Yao, T.; Bai, H.; Jin, X. Solubility measurement and thermodynamic modeling of carbamazepine (form III) in five pure solvents at various temperatures. *Chin. J. Chem. Eng.* **2021**, *33*, 231–235. [CrossRef]
181. Liu, Y.; Wang, Y.; Liu, Y.; Xu, S.; Chen, M.; Du, S.; Gong, J. Solubility of L-histidine in different aqueous binary solvent mixtures from 283.15 K to 318.15 K with experimental measurement and thermodynamic modelling. *J. Chem. Thermodyn.* **2017**, *105*, 1–14. [CrossRef]
182. Lou, Y.; Wang, Y.; Li, Y.; He, M.; Su, N.; Xu, R.; Meng, X.; Hou, B.; Xie, C. Thermodynamic equilibrium and cosolvency of florfenicol in binary solvent system. *J. Mol. Liq.* **2018**, *251*, 83–91. [CrossRef]
183. Mabhoot, A. Jouyban, A. Solubility of Sodium Phenytoin in Propylene Glycol + Water Mixtures in the Presence of B-Cyclodextrin. *Pharm. Sci.* **2015**, *21*, 152–156. [CrossRef]
184. Mahali, K.; Guin, P.S.; Roy, S.; Dolui, B.K. Solubility and solute–solvent interaction phenomenon of succinic acid in aqueous ethanol mixtures. *J. Mol. Liq.* **2017**, *229*, 172–177. [CrossRef]
185. Marcus, Y. The use of chemical probes for the characterization of solvent mixtures. Part 2. Aqueous mixtures. *J. Chem. Soc. Perkin Trans.* **1994**, *2*, 1751. [CrossRef]

186. Marcus, Y. Use of chemical probes for the characterization of solvent mixtures. Part 1. Completely non-aqueous mixtures. *J. Chem. Soc. Perkin Trans.* **1994**, *2*, 1015. [CrossRef]
187. Matsuda, H.; Kaburagi, K.; Matsumoto, S.; Kurihara, K.; Tochigi, K.; Tomono, K. Solubilities of Salicylic Acid in Pure Solvents and Binary Mixtures Containing Cosolvent. *J. Chem. Eng. Data* **2008**, *54*, 480–484. [CrossRef]
188. McHedlov-Petrosyan, N.O. Book review: Christian Reichardt and Thomas Welton, Solvents and Solvent Effects in Organic Chemistry (Fourth Edition, Updated and Enlarged; Wiley-VCH Verlag & Co. KGaA, Weinheim, 2011; 718 p. Hardcover). *Russ. J. Phys. Chem. A* **2011**, *85*, 1482. [CrossRef]
189. Mealey, D.; Svärd, M.; Rasmuson, Å.C. Thermodynamics of risperidone and solubility in pure organic solvents. *Fluid Phase Equilibria* **2014**, *375*, 73–79. [CrossRef]
190. Mirmehrabi, M.; Rohani, S. Measurement and Prediction of the Solubility of Stearic Acid Polymorphs by the UNIQUAC Equation. *Can. J. Chem. Eng.* **2004**, *82*, 335–342. [CrossRef]
191. Mo, F.; Ma, J.; Zhang, P.; Zhang, D.; Fan, H.; Yang, X.; Zhi, L.; Zhang, J. Solubility and thermodynamic properties of baicalein in water and ethanol mixtures from 283.15 to 328.15 K. *Chem. Eng. Commun.* **2019**, *208*, 183–196. [CrossRef]
192. Mohamadian, E.; Hamidi, S.; Martínez, F.; Jouyban, A. Solubility prediction of deferiprone in N-methyl-2-pyrrolidone+ ethanol mixtures at various temperatures using a minimum number of experimental data. *Phys. Chem. Liq.* **2017**, *55*, 805–816. [CrossRef]
193. Mohammadian, E.; Jouyban, A.; Barzegar-Jalali, M.; Acree, W.E.; Rahimpour, E. Solubilization of naproxen: Experimental data and computational tools. *J. Mol. Liq.* **2019**, *288*, 110985. [CrossRef]
194. Mohammadzade, M.; Barzegar-Jalali, M.; Jouyban, A. Solubility of naproxen in 2-propanol+water mixtures at various temperatures. *J. Mol. Liq.* **2015**, *206*, 110–113. [CrossRef]
195. Moodley, K.; Rarey, J.; Ramjugernath, D. Experimental solubility of diosgenin and estriol in various solvents between T = (293.2–328.2)K. *J. Chem. Thermodyn.* **2017**, *106*, 199–207. [CrossRef]
196. Moodley, K.; Rarey, J.; Ramjugernath, D. Experimental solubility data for prednisolone and hydrocortisone in various solvents between (293.2 and 328.2) K by employing combined DTA/TGA. *J. Mol. Liq.* **2017**, *240*, 303–312. [CrossRef]
197. Mora, C.P.; Martínez, F. Solubility of naproxen in several organic solvents at different temperatures. *Fluid Phase Equilibria* **2007**, *255*, 70–77. [CrossRef]
198. Moradi, M.; Mazaher Haji Agha, E.; Hemmati, S.; Martinez, F.; Kuentz, M.; Jouyban, A. Solubility of 5-aminosalicylic acid in {N-methyl-2-pyrrolidone + ethanol} mixtures at T = (293.2 to 313.2) K. *J. Mol. Liq.* **2020**, *306*, 112774. [CrossRef]
199. Ning, L.; Gong, X.; Li, P.; Chen, X.; Wang, H.; Xu, J. Measurement and correlation of the solubility of estradiol and estradiol-urea co-crystal in fourteen pure solvents at temperatures from 273.15 K to 318.15 K. *J. Mol. Liq.* **2020**, *304*, 112599. [CrossRef]
200. Noda, K.; Aono, Y.; Ishida, K. Viscosity and Density of Ethanol-Acetic Acid-Water Mixtures. *Kagaku Kogaku Ronbunshu* **1983**, *9*, 237–240. [CrossRef]
201. Noda, K.; Ohashi, M.; Ishida, K. Viscosities and densities at 298.15 K for mixtures of methanol, acetone, and water. *J. Chem. Eng. Data* **1982**, *27*, 326–328. [CrossRef]
202. Noubigh, A. Stearic acid solubility in mixed solvents of (water + ethanol) and (ethanol + ethyl acetate): Experimental data and comparison among different thermodynamic models. *J. Mol. Liq.* **2019**, *296*, 112101. [CrossRef]
203. Oliveira, G.; Bonassoli, A.B.G.; Rolemberg, M.P.; Mota, M.A.; Basso, R.C.; Soares, R.d.P.; Igarashi-Mafra, L.; Mafra, M.R. Water Effect on Solubilities of Lauric and Palmitic Acids in Ethanol and 2-Propanol Determined by Differential Scanning Calorimetry. *J. Chem. Eng. Data* **2021**, *66*, 2366–2373. [CrossRef]
204. Ortiz, C.P.; Cardenas-Torres, R.E.; Martínez, F.; Delgado, D.R. Solubility of Sulfamethazine in the Binary Mixture of Acetonitrile + Methanol from 278.15 to 318.15 K: Measurement, Dissolution Thermodynamics, Preferential Solvation, and Correlation. *Molecules* **2021**, *26*, 7588. [CrossRef]
205. Osorio, I.P.; Martínez, F.; Peña, M.A.; Jouyban, A.; Acree, W.E. Solubility, dissolution thermodynamics and preferential solvation of sulfadiazine in (N-methyl-2-pyrrolidone + water) mixtures. *J. Mol. Liq.* **2021**, *330*, 115693. [CrossRef]
206. Pabba, S.; Kumari, A.; Ravuri, M.G.; Thella, P.K.; Satyavathi, B.; Shah, K.; Kundu, S.; Bhargava, S.K. Experimental determination and modelling of the co-solvent and antisolvent behaviour of binary systems on the dissolution of pharma drug; L-aspartic acid and thermodynamic correlations. *J. Mol. Liq.* **2020**, *314*, 113657. [CrossRef]
207. Pacheco, D.P.; Martínez, F. Thermodynamic analysis of the solubility of naproxen in ethanol + water cosolvent mixtures. *Phys. Chem. Liq.* **2007**, *45*, 581–595. [CrossRef]
208. Padervand, M.; Naseri, S.; Boroujeni, H.C. Preferential solvation of pomalidomide, an anticancer compound, in some binary mixed solvents at 298.15 K. *Chin. J. Chem. Eng.* **2020**, *28*, 2626–2633. [CrossRef]
209. Pasham, F.; Jabbari, M.; Farajtabar, A. Solvatochromic Measurement of KAT Parameters and Modeling Preferential Solvation in Green Potential Binary Mixtures of N-Formylmorpholine with Water, Alcohols, and Ethyl Acetate. *J. Chem. Eng. Data* **2020**, *65*, 5458–5466. [CrossRef]
210. Patel, A.; Vaghasiya, A.; Gajera, R.; Baluja, S. Solubility of 5-Amino Salicylic Acid in Different Solvents at Various Temperatures. *J. Chem. Eng. Data* **2010**, *55*, 1453–1455. [CrossRef]
211. Przybyłek, M.; Miernicka, A.; Nowak, M.; Cysewski, P. New Screening Protocol for Effective Green Solvents Selection of Benzamide, Salicylamide and Ethenzamide. *Molecules* **2022**, *27*, 3323. [CrossRef]

212. Qiu, J.; Huang, H.; He, H.; Liu, H.; Hu, S.; Han, J.; Guo, Y.; Wang, P. Measurement and Correlation of *trans*-4-Hydroxyl-proline Solubility in Sixteen Individual Solvents and a Water + Acetonitrile Binary Solvent System. *J. Chem. Eng. Data* **2020**, *66*, 575–587. [CrossRef]
213. Radmand, S.; Rezaei, H.; Zhao, H.; Rahimpour, E.; Jouyban, A. Solubility and thermodynamic study of deferiprone in propylene glycol and ethanol mixture. *BMC Chem* **2023**, *17*, 37. [CrossRef]
214. Ràfols, C.; Rosés, M.; Bosch, E. Solute–solvent and solvent–solvent interactions in binary solvent mixtures. Part 5. Preferential solvation of solvatochromic indicators in mixtures of propan-2-ol with hexane, benzene, ethanol and methanol. *J. Chem. Soc. Perkin Trans.* **1997**, *2*, 243–248. [CrossRef]
215. Rani, R.S.; Rao, G.N. Stability of binary complexes of L-aspartic acid in dioxan–water mixtures. *Bull. Chem. Soc. Ethiop.* **2013**, *27*, 367–376. [CrossRef]
216. Rashid, A.; White, E.T.; Howes, T.; Litster, J.D.; Marziano, I. Effect of Solvent Composition and Temperature on the Solubility of Ibuprofen in Aqueous Ethanol. *J. Chem. Eng. Data* **2014**, *59*, 2699–2703. [CrossRef]
217. Rathi, P.B.; Kale, M.; Soleymani, J.; Jouyban, A. Solubility of Etoricoxib in Aqueous Solutions of Glycerin, Methanol, Polyethylene Glycols 200, 400, 600, and Propylene Glycol at 298.2 K. *J. Chem. Eng. Data* **2018**, *63*, 321–330. [CrossRef]
218. Ren, J.; Chen, D.; Yu, Y.; Li, H. Solubility of dicarbohydrazide bis[3-(5-nitroimino-1,2,4-triazole)] in common pure solvents and binary solvents at different temperatures. *R. Soc. Open Sci.* **2019**, *6*, 190728. [CrossRef]
219. Rezaei, H.; Rahimpour, E.; Martinez, F.; Jouyban, A. Measurement and correlation of solubility data for deferiprone in propylene glycol and 2-propanol at different temperatures. *Heliyon* **2023**, *9*, e17402. [CrossRef]
220. Rezaei, H.; Rezaei, H.; Rahimpour, E.; Martinez, F.; Jouyban, A. Solubility profile of phenytoin in the mixture of 1-propanol and water at different temperatures. *J. Mol. Liq.* **2021**, *334*, 115936. [CrossRef]
221. Rodríguez, A.; Trigo, M.; Aubourg, S.P.; Medina, I. Optimisation of Low-Toxicity Solvent Employment for Total Lipid and Tocopherol Compound Extraction from Patagonian Squid By-Products. *Foods* **2023**, *12*, 504. [CrossRef] [PubMed]
222. Rosales-García, T.; Rosete-Barreto, J.M.; Pimentel-Rodas, A.; Davila-Ortiz, G.; Galicia-Luna, L.A. Solubility of Squalene and Fatty Acids in Carbon Dioxide at Supercritical Conditions: Binary and Ternary Systems. *J. Chem. Eng. Data* **2017**, *63*, 69–76. [CrossRef]
223. Roses, M.; Ortega, J.; Bosch, E. Variation of E T(30) polarity and the Kamlet-Taft solvatochromic parameters with composition in alcohol-alcohol mixtures. *J. Solut. Chem.* **1995**, *24*, 51–63. [CrossRef]
224. Ruidiaz, M.A.; Delgado, D.R.; Martínez, F. Indomethacin solubility estimation in 1,4-dioxane + water mixtures by the extended hildebrand solubility approach. *Química Nova* **2011**, *34*, 1569–1574. [CrossRef]
225. Sajedi-Amin, S.; Barzegar-Jalali, M.; Fathi-Azarbayjani, A.; Kebriaeezadeh, A.; Martínez, F.; Jouyban, A. Solubilization of bosentan using ethanol as a pharmaceutical cosolvent. *J. Mol. Liq.* **2017**, *232*, 152–158. [CrossRef]
226. Serna-Carrizales, J.C.; Zárate-Guzmán, A.I.; Aguilar-Aguilar, A.; Forgionny, A.; Bailón-García, E.; Flórez, E.; Gómez-Durán, C.F.A.; Ocampo-Pérez, R. Optimization of Binary Adsorption of Metronidazole and Sulfamethoxazole in Aqueous Solution Supported with DFT Calculations. *Processes* **2023**, *11*, 1009. [CrossRef]
227. Sha, J.; Ma, T.; Huang, Z.; Hu, X.; Zhang, R.; Cao, Z.; Wan, Y.; Sun, R.; He, H.; Jiang, G.; et al. Corrigendum to "Solubility determination, model evaluation, Hansen solubility parameter and thermodynamic properties of benorilate in six pure solvents and two binary solvent mixtures". *J. Chem. Thermodyn.* **2021**, *158*, 106365. [CrossRef]
228. Sha, J.; Yang, X.; Hu, X.; Huang, Z.; Cao, Z.; Wan, Y.; Sun, R.; Jiang, G.; He, H.; Li, Y.; et al. Solubility determination, model evaluation, Hansen solubility parameter and thermodynamic properties of benflumetol in pure alcohol and ester solvents. *J. Chem. Thermodyn.* **2021**, *154*, 106323. [CrossRef]
229. Sha, J.; Yang, X.; Ji, L.; Cao, Z.; Niu, H.; Wan, Y.; Sun, R.; He, H.; Jiang, G.; Li, Y.; et al. Solubility determination, model evaluation, Hansen solubility parameter, molecular simulation and thermodynamic properties of benflumetol in four binary solvent mixtures from 278.15 K to 323.15 K. *J. Mol. Liq.* **2021**, *333*, 115867. [CrossRef]
230. Shakeel, F.; Iqbal, M.; Ezzeldin, E.; Haq, N. Thermodynamics of solubility of ibrutinib in ethanol+water cosolvent mixtures at different temperatures. *J. Mol. Liq.* **2015**, *209*, 461–464. [CrossRef]
231. Shao, D.; Yang, Z.; Zhou, G.; Chen, J.; Zheng, S.; Lv, X.; Li, R. Improving the solubility of acipimox by cosolvents and the study of thermodynamic properties on solvation process. *J. Mol. Liq.* **2018**, *262*, 389–395. [CrossRef]
232. Sharapova, A.; Ol'khovich, M.; Blokhina, S.; Perlovich, G. Solubility and vapor pressure data of bioactive 6-(acetylamino)-N-(5-ethyl-1,3,4-thiadiazol-2-yl) hexanamide. *J. Chem. Thermodyn.* **2019**, *135*, 35–44. [CrossRef]
233. Shen, B.; Wang, Q.; Wang, Y.; Ye, X.; Lei, F.; Gong, X. Solubilities of Adipic Acid in Acetic Acid + Water Mixtures and Acetic Acid + Cyclohexane Mixtures. *J. Chem. Eng. Data* **2013**, *58*, 938–942. [CrossRef]
234. Sheng, X.; Luo, W.; Wang, Q. Determination and Correlation for the Solubilities of Succinic Acid in Cyclohexanol + Cyclohexanone + Cyclohexane Solvent Mixtures. *J. Chem. Eng. Data* **2018**, *63*, 801–811. [CrossRef]
235. Shi, S.; Yan, M.; Tao, B.; Luo, W. Measurement and correlation for solubilities of succinic acid, glutaric acid and adipic acid in five organic solvents. *J. Mol. Liq.* **2020**, *297*, 111735. [CrossRef]
236. Singh, S. Studies on the Interactions of Paracetamol in Water and Binary Solvent Mixtures at T = (298.15–313.15) K: Viscometric and Surface Tension Approach. *Biointerface Res. Appl. Chem.* **2021**, *12*, 2776–2786. [CrossRef]
237. Smirnov, V.I. Thermochemical investigation of L-glutamine dissolution processes in aqueous co-solvent mixtures of acetonitrile, dioxane, acetone and dimethyl sulfoxide at T = 298.15 K. *J. Chem. Thermodyn.* **2020**, *150*, 106227. [CrossRef]

238. Smirnov, V.I.; Badelin, V.G. Similarity and differences of the thermochemical characteristics of l-glutamine dissolution in aqueous solutions of some acetamides and formamides at T = 298.15 K. *J. Mol. Liq.* **2019**, *285*, 84–88. [CrossRef]
239. Soltanpour, S.; Gharagozlu, A. Piroxicam Solubility in Binary and Ternary Solvents of Polyethylene Glycols 200 or 400 with Ethanol and Water at 298.2 K: Experimental Data Report and Modeling. *J. Solut. Chem.* **2015**, *44*, 1407–1423. [CrossRef]
240. Soltanpour, S.; Jouyban, A. Solubility of Acetaminophen and Ibuprofen in Binary and Ternary Mixtures of Polyethylene Glycol 600, Ethanol and Water. *Chem. Pharm. Bull.* **2010**, *58*, 219–224. [CrossRef] [PubMed]
241. Soltanpour, S.; Nazemi, V. Solubility of Ketoconazole in Binary and Ternary Solvents of Polyethylene Glycols 200, 400 or 600 with Ethanol and Water at 298.2 K. Data Report and Analysis. *J. Solut. Chem.* **2018**, *47*, 65–79. [CrossRef]
242. Soltanpour, S.; Shekarriz, A.-H. Naproxen solubility in binary and ternary solvents of polyethylene glycols 200, 400 or 600 with ethanol and water at 298.2 K—Experimental data report and modelling. *Phys. Chem. Liq.* **2015**, *53*, 748–762. [CrossRef]
243. Solymosi, T.; Tóth, F.; Orosz, J.; Basa-Dénes, O.; Angi, R.; Jordán, T.; Ötvös, Z.; Glavinas, H. Solubility Measurements at 296 and 310 K and Physicochemical Characterization of Abiraterone and Abiraterone Acetate. *J. Chem. Eng. Data* **2018**, *12*, 4453–4458. [CrossRef]
244. Sun, H.; Liu, B.; Liu, P.; Zhang, J.; Wang, Y. Solubility of Fenofibrate in Different Binary Solvents: Experimental Data and Results of Thermodynamic Modeling. *J. Chem. Eng. Data* **2016**, *61*, 3177–3183. [CrossRef]
245. Sun, J.; Liu, X.; Fang, Z.; Mao, S.; Zhang, L.; Rohani, S.; Lu, J. Solubility Measurement and Simulation of Rivaroxaban (Form I) in Solvent Mixtures from 273.15 to 323.15 K. *J. Chem. Eng. Data* **2015**, *61*, 495–503. [CrossRef]
246. Swinerd, G.G. *Orbital Mechanics: Theory and Applications*; Logsdon, T., Ed.; John Wiley and Sons Limited: Chichester, UK, 1998. [CrossRef]
247. Tang, W.; Wang, Z.; Feng, Y.; Xie, C.; Wang, J.; Yang, C.; Gong, J. Experimental Determination and Computational Prediction of Androstenedione Solubility in Alcohol + Water Mixtures. *Ind. Eng. Chem. Res.* **2014**, *53*, 11538–11549. [CrossRef]
248. Tang, W.; Xie, C.; Wang, Z.; Wu, S.; Feng, Y.; Wang, X.; Wang, J.; Gong, J. Solubility of androstenedione in lower alcohols. *Fluid Phase Equilibria* **2014**, *363*, 86–96. [CrossRef]
249. Teutenberg, T.; Wiese, S.; Wagner, P.; Gmehling, J. High-temperature liquid chromatography. Part II: Determination of the viscosities of binary solvent mixtures—Implications for liquid chromatographic separations. *J. Chromatogr. A* **2009**, *1216*, 8470–8479. [CrossRef]
250. Thati, J.; Nordström, F.L.; Rasmuson, Å.C. Solubility of Benzoic Acid in Pure Solvents and Binary Mixtures. *J. Chem. Eng. Data* **2010**, *55*, 5124–5127. [CrossRef]
251. Torres, N.; Escalera, B.; Martínez, F.; Peña, M.Á. Thermodynamic Analysis of Etoricoxib in Amphiprotic and Amphiprotic: Aprotic Solvent Mixtures at Several Temperatures. *J. Solut. Chem.* **2020**, *49*, 272–288. [CrossRef]
252. Valavi, M.; Ukrainczyk, M.; Dehghani, M.R. Prediction of solubility of active pharmaceutical ingredients by semi- predictive Flory Huggins/Hansen model. *J. Mol. Liq.* **2017**, *246*, 166–172. [CrossRef]
253. Vargas-Santana, M.S.; Cruz-González, A.M.; Ortiz, C.P.; Delgado, D.R.; Martínez, F.; Peña, M.Á.; Acree, W.E.; Jouyban, A. Solubility of sulfamerazine in (ethylene glycol + water) mixtures: Measurement, correlation, dissolution thermodynamics and preferential solvation. *J. Mol. Liq.* **2021**, *337*, 116330. [CrossRef]
254. Vieira, A.W.; Molina, G.; Mageste, A.B.; Rodrigues, G.D.; de Lemos, L.R. Partitioning of salicylic and acetylsalicylic acids by aqueous two-phase systems: Mechanism aspects and optimization study. *J. Mol. Liq.* **2019**, *296*, 111775. [CrossRef]
255. Volkova, T.V.; Levshin, I.B.; Perlovich, G.L. New antifungal compound: Solubility thermodynamics and partitioning processes in biologically relevant solvents. *J. Mol. Liq.* **2020**, *310*, 113148. [CrossRef]
256. Wang, H.; Yao, G.; Zhang, H. Measurement and Correlation of the Solubility of Baicalin in Several Mixed Solvents. *J. Chem. Eng. Data* **2019**, *64*, 1281–1287. [CrossRef]
257. Wang, S.; Chen, N.; Qu, Y. Solubility of Florfenicol in Different Solvents at Temperatures from (278 to 318) K. *J. Chem. Eng. Data* **2011**, *56*, 638–641. [CrossRef]
258. Wang, S.; Chen, Y.; Gong, T.; Dong, W.; Wang, G.; Li, H.; Wu, S. Solid-liquid equilibrium behavior and thermodynamic analysis of dipyridamole in pure and binary solvents from 293.15 K to 328.15 K. *J. Mol. Liq.* **2019**, *275*, 8–17. [CrossRef]
259. Wang, S.; Li, Q.-S.; Su, M.-G. Solubility of 1*H*-1,2,4-Triazole in Ethanol, 1-Propanol, 2-Propanol, 1,2-Propanediol, Ethyl Formate, Methyl Acetate, Ethyl Acetate, and Butyl Acetate at (283 to 363) K. *J. Chem. Eng. Data* **2007**, *52*, 856–858. [CrossRef]
260. Wang, S.; Qin, L.; Zhou, Z.; Wang, J. Solubility and Solution Thermodynamics of Betaine in Different Pure Solvents and Binary Mixtures. *J. Chem. Eng. Data* **2012**, *57*, 2128–2135. [CrossRef]
261. Wang, S.; Song, Z.; Wang, J.; Dong, Y.; Wu, M. Solubilities of Ibuprofen in Different Pure Solvents. *J. Chem. Eng. Data* **2010**, *55*, 5283–5285. [CrossRef]
262. Wang, S.; Zhang, Y.; Wang, J. Solubility Measurement and Modeling for Betaine in Different Pure Solvents. *J. Chem. Eng. Data* **2014**, *59*, 2511–2516. [CrossRef]
263. Wang, X.; Zhang, D.; Liu, S.; Chen, Y.; Jia, L.; Wu, S. Thermodynamic Study of Solubility for Imatinib Mesylate in Nine Monosolvents and Two Binary Solvent Mixtures from 278.15 to 318.15 K. *J. Chem. Eng. Data* **2018**, *63*, 4114–4127. [CrossRef]
264. Wang, Z.; Xu, Z.; Xu, X.; Yang, A.; Luo, W.; Luo, Y. Solubility of benzoic acid in twelve organic solvents: Experimental measurement and thermodynamic modeling. *J. Chem. Thermodyn.* **2020**, *150*, 106234. [CrossRef]
265. Watterson, S.; Hudson, S.; Svärd, M.; Rasmuson, Å.C. Thermodynamics of fenofibrate and solubility in pure organic solvents. *Fluid Phase Equilibria* **2014**, *367*, 143–150. [CrossRef]

266. Wei, H.; Gao, N.; Dang, L. Solubility and Thermodynamic Properties of Sulfamethazine–Saccharin Cocrystal in Pure and Binary (Acetonitrile + 2-Propanol) Solvents. *Trans. Tianjin Univ.* **2020**, *27*, 460–472. [CrossRef]
267. Wu, J.; Gu, L.; Wang, H.; Tao, L.; Wang, X. Solubility of Baicalein in Different Solvents from (287 to 323) K. *Int. J. Thermophys.* **2014**, *35*, 1465–1475. [CrossRef]
268. Wu, K.; Li, Y. Solubility and solution thermodynamics of isobutyramide in 15 pure solvents at temperatures from 273.15 to 324.75 K. *J. Mol. Liq.* **2020**, *311*, 113294. [CrossRef]
269. Wu, S.; Shi, Y.; Zhang, H. Solubility Measurement and Correlation for Amrinone in Four Binary Solvent Systems at 278.15–323.15 K. *J. Chem. Eng. Data* **2020**, *65*, 4108–4115. [CrossRef]
270. Wu, Y.; Ren, M.; Zhang, X. Solubility Determination and Model Correlation of Benorilate between T = 278.18 and 318.15 K. *J. Chem. Eng. Data* **2020**, *65*, 3690–3695. [CrossRef]
271. Wu, Y.; Wu, C.; Yan, S.; Hu, B. Solubility of Bisacodyl in Pure Solvent at Various Temperatures: Data Correlation and Thermodynamic Property Analysis. *J. Chem. Eng. Data* **2019**, *65*, 43–48. [CrossRef]
272. Wu, Y.; Wu, J.; Wang, J.; Gao, J. Effect of Solvent Properties and Composition on the Solubility of Ganciclovir Form I. *J. Chem. Eng. Data* **2019**, *64*, 1501–1507. [CrossRef]
273. Wüst Zibetti, A.; Aydi, A.; Claumann, C.A.; Eladeb, A.; Adberraba, M. Correlation of solubility and prediction of the mixing properties of rosmarinic acid in different pure solvents and in binary solvent mixtures of ethanol + water and methanol + water from (293.2 to 318.2) K. *J. Mol. Liq.* **2016**, *216*, 370–376. [CrossRef]
274. Xia, Q.; Chen, S.-N.; Chen, Y.-S.; Zhang, M.-S.; Zhang, F.-B.; Zhang, G.-L. Solubility of decanedioic acid in binary solvent mixtures. *Fluid Phase Equilibria* **2011**, *304*, 105–109. [CrossRef]
275. Xu, R.; Han, T.; Shen, L.; Zhao, J.; Lu, X.a. Solubility Determination and Modeling for Artesunate in Binary Solvent Mixtures of Methanol, Ethanol, Isopropanol, and Propylene Glycol + Water. *J. Chem. Eng. Data* **2019**, *64*, 755–762. [CrossRef]
276. Marcus, Y. Preferential Solvation of Drugs in Binary Solvent Mixtures. *Pharm. Anal. Acta* **2017**, *10*, 4172. [CrossRef]
277. Yan, M.; Li, X.; Tao, B.; Yang, L.; Luo, W. Solubility of succinic acid, glutaric acid and adipic acid in propionic acid + ε-caprolactone mixtures and propionic acid + cyclohexanone mixtures: Experimental measurement and thermodynamic modeling. *J. Mol. Liq.* **2018**, *272*, 106–119. [CrossRef]
278. Yang, H.; Rasmuson, Å.C. Solubility of Butyl Paraben in Methanol, Ethanol, Propanol, Ethyl Acetate, Acetone, and Acetonitrile. *J. Chem. Eng. Data* **2010**, *55*, 5091–5093. [CrossRef]
279. Yang, H.; Zhang, T.; Xu, S.; Han, D.; Liu, S.; Yang, Y.; Du, S.; Li, M.; Gong, J. Measurement and Correlation of the Solubility of Azoxystrobin in Seven Monosolvents and Two Different Binary Mixed Solvents. *J. Chem. Eng. Data* **2017**, *62*, 3967–3980. [CrossRef]
280. Yang, L.; Zhang, Y.; Cheng, J.; Yang, C. Solubility and thermodynamics of polymorphic indomethacin in binary solvent mixtures. *J. Mol. Liq.* **2019**, *295*, 111717. [CrossRef]
281. Yang, Z.; Shao, D.; Zhou, G. Analysis of solubility parameters of fenbendazole in pure and mixed solvents and evaluation of thermodynamic model. *J. Chem. Thermodyn.* **2020**, *140*, 105876. [CrossRef]
282. Yang, Z.; Shao, D.; Zhou, G. Solubility profile of imatinib in pure and mixed solvents and calculation of thermodynamic properties. *J. Chem. Thermodyn.* **2020**, *144*, 106031. [CrossRef]
283. Yang, Z.; Shao, D.; Zhou, G. Improvement of solubility and analysis thermodynamic properties of β tegafur in pure and mixed organic solvents. *J. Chem. Thermodyn.* **2020**, *146*, 106090. [CrossRef]
284. Yaws, C.L. Physical Properties—Inorganic Compounds. In *The Yaws Handbook of Physical Properties for Hydrocarbons and Chemicals*; Elsevier: Amsterdam, The Netherlands, 2015; pp. 684–810.
285. Yu, Q.; Ma, X.; Xu, L. Solubility, dissolution enthalpy and entropy of l-glutamine in mixed solvents of ethanol+water and acetone+water. *Thermochim. Acta* **2013**, *558*, 6–9. [CrossRef]
286. Zadaliasghar, S.; Jouyban, A.; Martinez, F.; Barzegar-Jalali, M.; Rahimpour, E. Solubility of ketoconazole in the binary mixtures of 2-propanol and water at different temperatures. *J. Mol. Liq.* **2020**, *300*, 112259. [CrossRef]
287. Zhang, C.-L.; Li, B.-Y.; Wang, Y. Solubilities of Sulfadiazine in Methanol, Ethanol, 1-Propanol, 2-Propanol, Acetone, and Chloroform from (294.15 to 318.15) K. *J. Chem. Eng. Data* **2010**, *55*, 2338–2339. [CrossRef]
288. Zhang, F. Commentary on the "Solubility of l-histidine in different aqueous binary solvent mixtures from 283.15 K to 318.15 K with experimental measurement and thermodynamic modelling". *J. Chem. Thermodyn.* **2018**, *124*, 98–100. [CrossRef]
289. Zhang, H.; Yin, Q.; Liu, Z.; Gong, J.; Bao, Y.; Zhang, M.; Hao, H.; Hou, B.; Xie, C. Measurement and correlation of solubility of dodecanedioic acid in different pure solvents from T = (288.15 to 323.15)K. *J. Chem. Thermodyn.* **2014**, *68*, 270–274. [CrossRef]
290. Zhang, J.; Huang, C.; Chen, J.; Xu, R. Equilibrium Solubility Determination and Modeling of Fenbendazole in Cosolvent Mixtures at (283.15–328.15) K. *J. Chem. Eng. Data* **2019**, *64*, 4095–4102. [CrossRef]
291. Zhang, J.; Huang, C.; Xu, R. Solubility of Bifonazole in Four Binary Solvent Mixtures: Experimental Measurement and Thermodynamic Modeling. *J. Chem. Eng. Data* **2019**, *64*, 2641–2648. [CrossRef]
292. Zhang, J.; Huang, C.; Xu, R. Solubility Determination and Mathematical Modeling of Nicorandil in Several Aqueous Cosolvent Systems at Temperature Ranges of 278.15–323.15 K. *J. Chem. Eng. Data* **2020**, *65*, 4063–4070. [CrossRef]
293. Zhang, J.; Song, X.; Xu, R. Solubility Determination and Modeling for Milrinone in Binary Solvent Mixtures of Ethanol, Isopropanol, Ethylene Glycol, and N,N-Dimethylformamide + Water. *J. Chem. Eng. Data* **2020**, *65*, 4100–4107. [CrossRef]

294. Zhang, J.; Zhang, H.; Xu, R. Solubility Determination and Modeling for Tirofiban in Several Mixed Solvents at 278.15–323.15 K. *J. Chem. Eng. Data* **2020**, *65*, 4071–4078. [CrossRef]
295. Zhang, N.; Li, S.; Yang, H.; Li, M.; Yang, Y.; Tang, W. Measurement and Correlation of the Solubility of Tetramethylpyrazine in Nine Monosolvents and Two Binary Solvent Systems. *J. Chem. Eng. Data* **2019**, *64*, 995–1006. [CrossRef]
296. Zhang, P.; Sha, J.; Wan, Y.; Zhang, C.; Li, T.; Ren, B. Apparent thermodynamic analysis and the dissolution behavior of levamisole hydrochloride in three binary solvent mixtures. *Thermochim. Acta* **2019**, *681*, 178375. [CrossRef]
297. Zhang, P.; Wan, Y.; Zhang, C.; Zhao, R.; Sha, J.; Li, Y.; Li, T.; Ren, B. Solubility and mixing thermodynamic properties of levamisole hydrochloride in twelve pure solvents at various temperatures. *J. Chem. Thermodyn.* **2019**, *139*, 105882. [CrossRef]
298. Zhang, P.; Zhang, C.; Zhao, R.; Wan, Y.; Yang, Z.; He, R.; Chen, Q.; Li, T.; Ren, B. Measurement and Correlation of the Solubility of Florfenicol Form A in Several Pure and Binary Solvents. *J. Chem. Eng. Data* **2018**, *63*, 2046–2055. [CrossRef]
299. Zhang, X.; Chen, J.; Hu, J.; Liu, M.; Cai, Z.; Xu, Y.; Sun, B. The solubilities of benzoic acid and its nitro-derivatives, 3-nitro and 3,5-dinitrobenzoic acids. *J. Chem. Res.* **2021**, *45*, 1100–1106. [CrossRef]
300. Zhang, X.; Cui, P.; Yin, Q.; Zhou, L. Measurement and Correlation of the Solubility of Florfenicol in Four Binary Solvent Mixtures from T = (278.15 to 318.15) K. *Crystals* **2022**, *12*, 1176. [CrossRef]
301. Zhu, Y.; Yang, H.; Si, Z.; Zhang, X. Solubility and thermodynamics of l-hydroxyproline in water and (methanol, ethanol, n-propanol) binary solvent mixtures. *J. Mol. Liq.* **2020**, *298*, 112043. [CrossRef]
302. Zorrilla-Veloz, R.I.; Stelzer, T.; López-Mejías, V. Measurement and Correlation of the Solubility of 5-Fluorouracil in Pure and Binary Solvents. *J. Chem. Eng. Data* **2018**, *63*, 3809–3817. [CrossRef]
303. Milescu, R.A.; Zhenova, A.; Vastano, M.; Gammons, R.; Lin, S.; Lau, C.H.; Clark, J.H.; McElroy, C.R.; Pellis, A. Polymer Chemistry Applications of Cyrene and its Derivative Cygnet 0.0 as Safer Replacements for Polar Aprotic Solvents. *ChemSusChem* **2021**, *14*, 3367–3381. [CrossRef]
304. Glass, M.; Aigner, M.; Viell, J.; Jupke, A.; Mitsos, A. Liquid-liquid equilibrium of 2-methyltetrahydrofuran/water over wide temperature range: Measurements and rigorous regression. *Fluid Phase Equilibria* **2017**, *433*, 212–225. [CrossRef]
305. Dargo, G.; Kis, D.; Gede, M.; Kumar, S.; Kupai, J.; Szekely, G. MeSesamol, a bio-based and versatile polar aprotic solvent for organic synthesis and depolymerization. *Chem. Eng. J.* **2023**, *471*, 144365. [CrossRef]
306. Komarova, A.O.; Dick, G.R.; Luterbacher, J.S. Diformylxylose as a new polar aprotic solvent produced from renewable biomass. *Green Chem.* **2021**, *23*, 4790–4799. [CrossRef]
307. Kamlet, M.J.; Abboud, J.L.M.; Abraham, M.H.; Taft, R.W. Linear solvation energy relationships. 23. A comprehensive collection of the solvatochromic parameters,. pi.*,. alpha., and. beta., and some methods for simplifying the generalized solvatochromic equation. *J. Org. Chem.* **1983**, *48*, 2877–2887. [CrossRef]
308. Abraham, M.H. Scales of solute hydrogen-bonding: Their construction and application to physicochemical and biochemical processes. *Chem. Soc. Rev.* **1993**, *22*, 73–83. [CrossRef]
309. Liu, X.; Acree, W.E.; Abraham, M.H. Descriptors for some compounds with pharmacological activity; calculation of properties. *Int. J. Pharm.* **2022**, *617*, 121597. [CrossRef]
310. Vitha, M.; Carr, P.W. The chemical interpretation and practice of linear solvation energy relationships in chromatography. *J. Chromatogr. A* **2006**, *1126*, 143–194. [CrossRef]
311. West, C.; Lesellier, E. Characterisation of stationary phases in subcritical fluid chromatography with the solvation parameter model: III. Polar stationary phases. *J. Chromatogr. A* **2006**, *1110*, 200–213. [CrossRef]
312. Efimov, I.; Povarov, V.G.; Rudko, V.A. Comparison of UNIFAC and LSER Models for Calculating Partition Coefficients in the Hexane–Acetonitrile System Using Middle Distillate Petroleum Products as an Example. *Ind. Eng. Chem. Res.* **2022**, *61*, 9575–9585. [CrossRef]
313. EPA. CompTox Chemicals Dashboard v2.3.0. 2023. Available online: https://comptox.epa.gov/dashboard/ (accessed on 31 December 2023).
314. Hou, P.; Jolliet, O.; Zhu, J.; Xu, M. Estimate ecotoxicity characterization factors for chemicals in life cycle assessment using machine learning models. *Environ. Int.* **2020**, *135*, 105393. [CrossRef]
315. Wojeicchowski, J.P.; Abranches, D.O.; Ferreira, A.M.; Mafra, M.R.; Coutinho, J.A.P. Using COSMO-RS to Predict Solvatochromic Parameters for Deep Eutectic Solvents. *ACS Sustain. Chem. Eng.* **2021**, *9*, 10240–10249. [CrossRef]
316. Wojeicchowski, J.P.; Ferreira, A.M.; Okura, T.; Pinheiro Rolemberg, M.; Mafra, M.R.; Coutinho, J.A.P. Using COSMO-RS to Predict Hansen Solubility Parameters. *Ind. Eng. Chem. Res.* **2022**, *61*, 15631–15638. [CrossRef]

Disclaimer/Publisher's Note: The statements, opinions and data contained in all publications are solely those of the individual author(s) and contributor(s) and not of MDPI and/or the editor(s). MDPI and/or the editor(s) disclaim responsibility for any injury to people or property resulting from any ideas, methods, instructions or products referred to in the content.

Article

Conventional and Green Rubber Plasticizers Classified through Nile Red [E(NR)] and Reichardt's Polarity Scale [$E_T(30)$]

Franco Cataldo

Actinium Chemical Research Institute, Via Casilina 1626A, 00133 Rome, Italy; franco.cataldo@fastwebnet.it

Abstract: After a survey on polymer plasticization theories and conventional criteria to evaluate polymer–plasticizer compatibility through the solubility parameter, an attempt to create a polymer–plasticizer polarity scale through solvatochromic dyes has been made. Since Reichardt's $E_T(30)$ dye is insoluble in rubber hydrocarbon polymers like polyisoprene, polybutadiene and styrene–butadiene copolymers and is not useful for the evaluation of the hydrocarbons and ester plasticizers, the Nile Red solvatochromic dye was instead used extensively and successfully for this class of compounds. A total of 53 different compounds were evaluated with the Nile Red dye and wherever possible also with Reichardt's $E_T(33)$ dye. A very good correlation was then found between the Nile Red scale E(NR) and Reichardt's $E_T(30)$ scale for this class of compounds focusing on diene rubbers and their typical hydrocarbons and new ester plasticizers. Furthermore, the E(NR) scale also shows a reasonable correlation with the total solubility parameter calculated according to the Van Krevelen method. Based on the above results, some conclusion was made about the compatibility between the diene rubbers and the conventional plasticizers, as well as a new and green plasticizer proposed for the rubber compounds.

Keywords: rubbers; plasticizers; solvatochromic dyes; Nile Red dye; Reichardt's dye $E_T(30)$; Reichardt's dye $E_T(33)$; compatibility; solubility parameter

Citation: Cataldo, F. Conventional and Green Rubber Plasticizers Classified through Nile Red [E(NR)] and Reichardt's Polarity Scale [$E_T(30)$]. *Liquids* 2024, 4, 305–321. https://doi.org/10.3390/liquids4020015

Academic Editors: Enrico Bodo and Anton Airinei

Received: 13 January 2024
Revised: 13 February 2024
Accepted: 20 March 2024
Published: 31 March 2024

Copyright: © 2024 by the author. Licensee MDPI, Basel, Switzerland. This article is an open access article distributed under the terms and conditions of the Creative Commons Attribution (CC BY) license (https://creativecommons.org/licenses/by/4.0/).

1. Introduction

Plasticizers are liquids at ambient temperature with a relatively high molecular weight or, less frequently, low-melting-point solids, which are added to a polymer matrix to change its viscoelastic properties [1–9]. The changes in the physical properties of the polymer matrix have many consequences starting from an improved processability, passing through an increased flexibility of the resulting polymer compound, and leading to improved low-temperature performances [1–9]. The latter result is achieved because the plasticizer is often a liquid characterized by a lower glass transition temperature (T_g) than the guest polymer matrix, leading to a shift of the compound T_g toward lower temperatures. Indeed, the plasticizer efficiency is often measured by the degree of glass transition temperature shift toward lower temperatures of the resulting plasticized compound (T_g^c) with respect to the glass transition of the raw polymer (T_g^p) so that $\Delta T_g = T_g^p - T_g^c$ [1–9]. However, the ΔT_g is most pronounced in polymers with rigid chains (e.g., PVC), whereas the plasticizer causes a shift of the order of 100–160 °C. On the other hand, the effect of plasticizers on already flexible and rubber-like polymers is much less pronounced and limited to a ΔT_g shift to just a few tens of °C toward lower temperatures [1–9]. For rubbers already characterized by low glass transition temperature values (e.g., natural rubber $T_g = -72$ °C or high cis-polybutadiene with Tg = -105 °C), the plasticizer effect can cause a $\Delta T_g \approx -10$ °C [10,11]. For each given type of polymer or polymer blend, the efficiency of a plasticizer is measured by the degree of ΔT_g it is able to cause with respect to another plasticizer. Furthermore, at least for polar polymers (and with a series of limitations), a simple relationship has been found: $\Delta T_g = kn$ with k being a proportionality constant and n being the moles of plasticizer added, suggesting that the glass transition temperature shift is directly proportional to the amount of the plasticizer added [3]. On the other hand, for apolar polymers, it holds a similar relationship: $\Delta T_g = k'\varphi$, where φ is the volume fraction of the plasticizer [3]. Another simple relationship for the determination of the compound glass transition is the following [8]:

$$1/T_g^c = \omega_1/T_1 + \omega_2/T_2 \qquad (1)$$

where ω_1 and ω_2 are the weight fractions of polymer and plasticizer, respectively, while T_1 and T_2 values are the respective glass transition temperatures. Other relationships for the estimation of the glass transition of a plasticized compound can be found in ref. [8].

At this point, it is worth very shortly surveying the three main theories of plasticization mechanisms: the lubricity theory, the gel theory and the free volume theory. It is also interesting to note that the latter theory also includes the other two [1–9].

The lubricity theory [1–9] starts from the observation that in a pure polymer, the resistance to deformation and flow derives from the intermolecular friction between adjacent polymer chains which are in direct contact. The introduction of the plasticizer molecules between the polymer chains facilitates the movement of the chain segments through a slippage mechanism provided by the lubricating action of the plasticizer.

The gel theory [1–9] represents a further step ahead with respect to the lubricity theory. It starts from the idea that in amorphous polymers, the resistance to deformation derives from a model structure of the polymer intended to be a three-dimensional or honeycomb-type structure, which can also be defined as a gel structure. Such a gel structure is conceived as derived from loose attachments or secondary forces that occur at rather regular intervals along the polymer chains and hence in the matrix. The introduction of a plasticizer in such honeycomb structures has the effect of breaking the loose attachments and masking the center of forces by preventing the reformation of the three-dimensional macromolecular interaction. It is admitted that the masking action of the force centers derives from the fact that the polymer chain segments are solvated by the plasticizer. Solvation is always intended in a dynamic way so that each chain segment is solvated and de-solvated and, sometimes, the masking action is lost for a while. This implies an increased flexibility and flowability of the polymer chain and the resulting plasticized compound. Naturally, the chemical nature of each plasticizer exerts a different effect on the masking effect of the secondary force centers, leading to the individual effect of each plasticizer.

The free volume theory [1–9] starts from the observation that the free volume in a polymer is a measure of the internal available space for the motion of the chain segments, the chain ends and the side groups. The free volume reaches the minimum value, say 2.5% of the volume of the given polymer body, when it is cooled below its glass transition temperature. Indeed, below the Tg, because of the limited free volume available, the mentioned chain, end and side group movements are frozen and the polymer is in a glassy state. An increase in the motion of these moieties of the polymer can be achieved by (1) heating, (2) the addition of a plasticizer (which, being a low-molecular-weight molecule, increases the number of chain ends dramatically), (3) introducing branching or bulky side groups to the main polymer chain, and (4) inserting more flexible chain segments into the main polymer chain. Thus, any action that leads to an increase in the free volume of a polymer is measurable by a shift of the glass transition toward lower temperatures. The simplest action, which does not cause a chemical modification of the polymer, is just the physical addition of a plasticizer. This is known as external plasticization to distinguish it from internal plasticization, which instead is due to the chemical modification of the polymer. It is evident that the free volume theory is the most convincing and includes all the notions of the lubricity and gel theories. In fact, the plasticizer fills the available free volume and creates extra free volume; the first molecular layer of the plasticizer is adsorbed on the polymer chain segments and provides a certain degree of solvation, shielding the force centers of interaction between chains and preventing polymer networking reformation on cooling. The excess plasticizer molecules, not interacting directly with the polymer chain segments, act as a volume filler of the free volume created by the first layer of molecules solvating the chain segments and may provide a lubricating effect under deformation and flow. Additional discussion about the free volume theory and recent developments can be found in ref. [8].

The key practical problem in the use of plasticizers is the evaluation of the compatibility between the polymer, the polymer blend and the plasticizers. A general and updated survey can be found in ref. [9]. However, the most popular and accessible approach to evaluating the polymer–plasticizer compatibility involves the use of the solubility parameter either in rubber compounds [12] or in plastics [13].

The solubility parameter has been defined by Hildebrand and Scott [12] as follows:

$$\delta = [(\Delta H_{vap} - RT)/V_m]^{0.5} \qquad (2)$$

The evaporation enthalpy ΔH_{vap} was taken as the parameter of the cohesion energy between molecules minus the thermal energy needed to separate them (RT) divided by the molar volume V_m. Equation (2) can be re-written as

$$\delta = [(E_{coh})/V_m]^{0.5} \qquad (3)$$

The cohesive energy E_{coh} of a substance in a condensed state is defined as the increase in internal energy ΔU per mole of substance if all the intermolecular forces are eliminated.

Hansen [13] showed that the solubility parameter proposed by Hildebrand and Scott does not take into account the contribution of polar forces and hydrogen bonding; therefore, a more complex solubility parameter has been proposed:

$$\delta^2 = \delta_d^2 + \delta_p^2 + \delta_h^2 \qquad (4)$$

derived from the contribution of three components of the cohesive energy:

$$E_{coh} = E_d + E_p + E_h \qquad (5)$$

which are due to the contribution of dispersion and polar forces plus a hydrogen bonding contribution, respectively.

It is possible to calculate the solubility parameters and the solubility parameter components of almost all molecules and polymers by a group contribution method [14]. For this purpose, as explained by Van Krevelen [14], it is useful to introduce the molar attraction constant simply defined as

$$\varphi = (E_{coh} V_m)^{0.5} \qquad (6)$$

A set of equations was proposed by Van Krevelen [14] for the calculation of the solubility parameter components using molar attraction by a group contribution methodology:

$$\delta_d = (\sum \varphi_d)/V_m \qquad (7)$$

$$\delta_p = (\sum \varphi_p^2)^{0.5}/V_m \qquad (8)$$

$$\delta_h = [(\sum E_h)/V_m]^{0.5} \qquad (9)$$

The total solubility parameter can be calculated as follows:

$$\delta_t = (\delta_d^2 + \delta_p^2 + \delta_h^2)^{0.5} \qquad (10)$$

It can be observed from Equation (9) that the hydrogen bond parameter δ_h cannot be calculated from the molar attraction, but directly from the hydrogen bonding energy E_h [14]. There are numerous ways of evaluating the solubility of a given polymer P in a given solvent S; Van Krevelen [14] suggests the criteria imposed by the following equation:

$$\Delta\delta = [(\delta_{d,P} - \delta_{d,S})^2 + (\delta_{p,P} - \delta_{p,S})^2 + (\delta_{h,P} - \delta_{h,S})^2]^{0.5} \qquad (11)$$

To predict solubility,

$$\Delta\delta \leq 5 \tag{12}$$

Alternatively, Hansen [13] proposed a more sophisticated and relatively complex approach for the evaluation of the solubility of a polymer in a solvent.

A simpler and more practical approach regards the direct adoption of the total solubility parameter δ_t, eventually determined according to Equation (10), to evaluate the solubility between a polymer and a plasticizer or a solvent, which conforms to the criteria imposed by the following equation:

$$|\Delta\delta| = (\delta_{t,P} - \delta_{t,S}) \leq 3.0 \tag{13}$$

a criterion proposed by Brydson [12].

By using the Van Krevelen methodology [14] in previous works, we have calculated the solubility parameters of fullerenes [15] and their solubility in fatty acid esters and glycerides [16]. Similarly, it was through the calculated solubility parameters according to the Van Krevelen methodology [14] that the compatibility between biodiesel and diene rubbers as well as other typical petroleum-derived plasticizers used in rubber compounding was calculated [17].

In the present work, we wish to show an alternative or complementary approach to the solubility parameter to evaluate in a practical way the compatibility between a plasticizer and a polymer matrix (in particular a rubber polymer) by also classifying the plasticizers and the rubbers through Reichardt's polarity scale $E_T(30)$ or through a complementary scale. After all, Reichardt's polarity scale was also successfully applied for the first time in the selection of a bonding agent for a rocket propellant composite [18].

2. Materials and Methods

2.1. Materials

All plasticizers, solvents and polymers, unless otherwise stated, were purchased from Merck-Aldrich (Darmstadt, Germany– St.Louis, MO, USA). The petroleum-based plasticizers or extenders used in the rubber (tire) industry were commercially available products sourced from the market of rubber chemicals and additives. Namely, these plasticizers were T-DAE (treated distillate aromatic extract) and MES (mild extract solvate). Only the product Nytex BIO 6200 (a blend of naphthenic oil and fatty acids from renewable sources) was obtained from Nynas (Stockholm, Sweden). The methyl ester of rapeseed oil was a commercial biodiesel we used as a rubber compound plasticizer in our earlier work [17]. Another commercially available plasticizer of this study was the methyl ester of coconut oil. These methyl esters are prepared on an industrial scale by transesterification of the corresponding glycerides with methanol. The sunflower oil (high oleic content) and the soybean oil of the present work were industrial products obtained from the market of rubber chemicals. There is a trend in the patent [19,20] and in the open literature, e.g., [21], to adopt vegetable oils in tire treads as a substitute for petroleum-based plasticizers to produce greener tires. The most interesting vegetable oils appear to be soybean [19] and sunflower oil [20], also including a combination of both oils [21] as plasticizers in tire treads. Regarding the rubber samples, cis-1,4-polybutadiene (Europrene Neocis 60) was obtained from Versalis (Ravenna, Italy), synthetic cis-1,4-polyisoprene sample was SKI-5PM produced by JSC Sterlitamak Petrochemical Plant (Russia) and solution styrene–butadiene rubber (S-SBR) was obtained from Arlanxeo (Dormagen, Germany). The S-SBR was FX5000 grade, characterized by 50% vinyl content and 20% styrene content. Epoxidized natural rubber (ENR-25) was supplied by Ekoprena (Kuala Lumpur, Malaysia). The nitrile rubber sample was sourced from Versalis (Ravenna, Italy), and it was the Europrene N3345 with 33% acrylonitrile content. All the other polymers including the liquid polymer samples were purchased from Merck-Aldrich (Darmstadt, Germany– St.Louis, MO, USA).

The solvatochromic dyes $E_T(30)$ and Nile Red, whose chemical structures are shown in Scheme 1, were purchased from Merck-Aldrich (Germany–USA). The solvatochromic dye $E_T(33)$ was obtained from Fluka (Buchs, Switzerland).

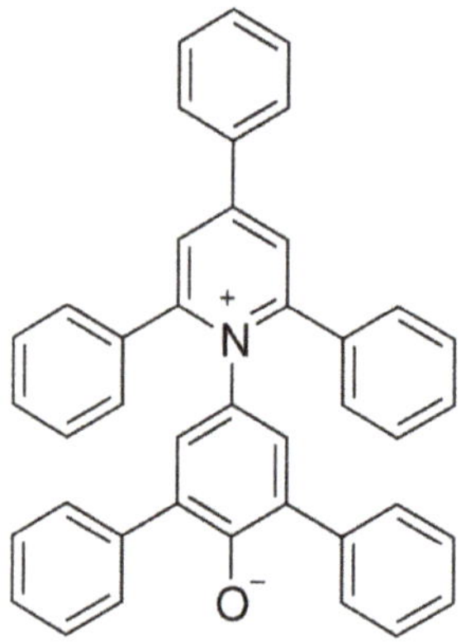

2,6-diphenyl-4-(2,4,6-triphenyl-1-pyridinio)phenolate
$E_T(30)$ dye

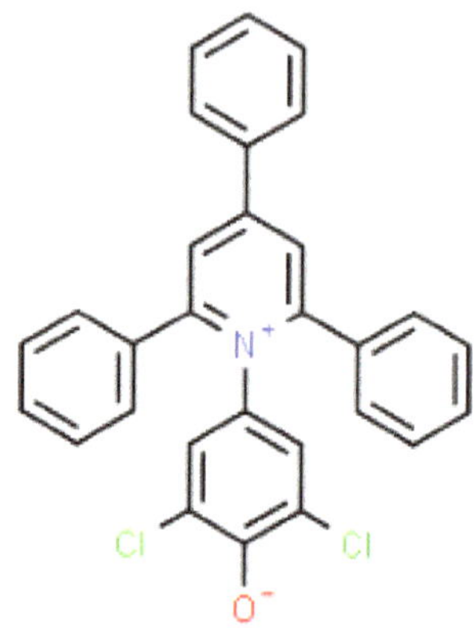

2,6-dichloro-4-(2,4,6-triphenyl-1-pyridinio)phenolate
$E_T(33)$ dye

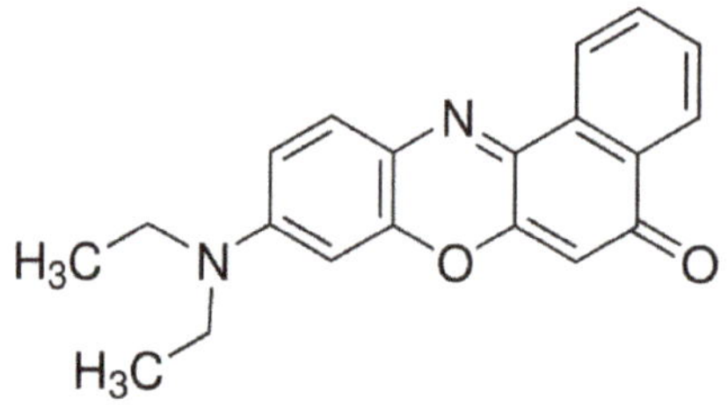

9-diethylamino-5H-benzo[a]phenoxazinone
Nile Red dye

Scheme 1. The solvatochromic dyes used in the present work.

2.2. Determination of the Maximum Absorbance with the Solvatochromic Dyes in Liquid Samples

For all plasticizers, solvents and liquid polymers, the dissolution of the minimal quantity of the selected solvatochromic dye was made in a beaker by stirring. For viscous liquids, gentle heating was applied to accelerate the dissolution of the dye. The typical volume of each sample under analysis was 10 mL and the dye added was much less than a spatula tip. Before making any spectrophotometric measurement on our Shimadzu UV2450 equipped with thermostated cells kept at 25 °C, a back correction was made with quartz cuvettes filled with the pure liquid sample under study. This operation is crucial, especially with yellow or even brown liquid samples, for instance, the plasticizers used in the tire industry, to reduce or minimize the color interference of the sample. After the back-correction step, the electronic absorption spectrum of the solvatochromic dye in the selected liquid was recorded using the reference cuvette filled with the reference liquid without any dye.

2.3. Determination of the Maximum Absorbance with the Solvatochromic Dyes in Solid Thin Films of Polymer Sample

Regarding the polymer thin films with the solvatochromic dye (typically Nile Red) embedded inside, the sample preparation involved the following steps. The selected polymer or rubber cut into the smallest possible pieces was weighed in a flask (typically 500–600 mg) and dissolved in about 13 g of dichloromethane (CH_2Cl_2). Only for epoxidized natural rubber (ENR-25) and for nitrile rubber (Europrene N3345) was it necessary to use tetrahydrofuran (THF) as a solvent instead of CH_2Cl_2. Once the polymer was fully dissolved, less than a tip of a spatula of Nile Red was added to the solution and stirred till a complete dissolution of the dye was achieved. The resulting homogeneous solution was poured into an optical glass dish with a diameter of 11 cm and the solvent was allowed to evaporate slowly under a fume hood. After the complete evaporation of the solvent, the resulting dyed polymer film in the dish was heated to 60–70 °C to permit the complete evaporation of any solvent trace. After cooling, the electronic absorption spectrum of the dyed polymer film was directly collected through the optical glass dish. Only in a couple of cases, i.e., with polystyrene and poly(lactic acid), was it possible to separate the polymer film from the glass dish as a free-standing thin solid film. In the latter two cases, the electronic absorption spectra were recorded directly on the free-standing thin solid films.

It is known from the literature [22] that the solvatochromic dye may separate into the form of microcrystals on the polymer surface leading to non-useful absorption maxima readings. This potentially undesired effect was minimized by using the minimum possible amount of dye during the preparation of the composite. Furthermore, Nile Red dye is characterized by a high solubility in hydrocarbon polymers (in contrast to the $E_T(30)$ and $E_T(33)$ dyes). Consequently, the potential blooming was therefore completely suppressed.

The potential solvent effect was checked in the case of polybutadiene. This polymer was dissolved both in CH_2Cl_2 and in THF. The resulting two different polymer films with Nile Red dye prepared on glass dishes were analyzed spectrophotometrically, obtaining the same maximum absorption value. Of course, the final "drying" step of the film under moderate heating is crucial to remove all the solvent traces, which otherwise may affect the position of the maximum absorption.

3. Results

3.1. General Aspects of Nile Red Solvatochromic Dye with Respect to $E_T(30)$ Dye

The challenge of this work is to evaluate, through a solvatochromic scale, the potential compatibility between polymers and plasticizers with special attention to rubber polymers and plasticizers used in the rubber and tire industry. This new approach could be a complementary and experimental way with respect to the evaluation and estimation of the solubility parameter of each component as conducted, for example, in a previous work [17] using the Van Krevelen methodology [14]. The key problem faced by the present work is represented by the fact that the fundamental Reichardt's dye, $E_T(30)$ of Scheme 1, is insoluble in hydrocarbons, and it is also not suitable for ester plasticizers either because of the relatively low solubility but also because the small residual acidity in the esters causes the protonation of the phenolate oxygen anion, leading to the disappearance of the long-wavelength solvatochromic charge transfer (CT) band [23]. Indeed, the $E_T(30)$ values of hydrocarbons were all determined through another complementary penta-t-butyl-substituted betaine dye called $E_T(45)$ [24], which is not easily accessible and furthermore, it is exactly as sensitive to protonation as the $E_T(30)$ dye. Thus, the Reichardt dye suitable for polar media with weak acidity is the $E_T(33)$ dye (see Scheme 1) [23], which in fact was adopted in the present work. The symmetrical chlorine substitution in the ortho-position to the phenolate group changes the pKa of the conjugated acid of the dye $E_T(33)$ by four orders of magnitude with respect to $E_T(30)$, making the former dye suitable for weakly acidic media [23]. Since the $E_T(30)$ but also the $E_T(33)$ values expressed in kcal/mol are both determined through the well-known equation [23,24]

$$E_T = 28{,}591\, \lambda^{-1} \tag{14}$$

where λ is the maximum absorption of the long-wavelength CT band of the pyridinium-N-phenolate betaine dye in a given liquid medium; the conversion from $E_T(33)$ values to $E_T(30)$ can be achieved through the following equation [23]:

$$E_T(30) = 0.9953\, E_T(33) - 8.1132 \tag{15}$$

Despite the availability of $E_T(33)$, the potential measurement of the polarity in certain hydrocarbon-based plasticizers, in certain apolar polymers and in certain esters remains unaffordable.

In a seminal paper belonging to this Special Issue of "*Liquids*", Acree and Lang [25] have shown an interesting and sophisticated approach that permits the estimation of the $E_T(30)$ values of certain "difficult" substrates where the direct polarity measurement is hindered for a series of reasons, as in the present case for apolar polymers (e.g., rubbers) and certain plasticizers.

An alternative to the above approach is to resume the solvatochromic dye known as Nile Red. The chemical structure of Nile Red is shown in Scheme 1 and the full chemical name of the dye is 9-diethylamino-5H-benzo[α]phenoxazinone [26]. It is a very stable fluorescent dye used for staining biological tissues, being highly lipophilic [27]. Certain structural analogies between Nile Red and phenol blue have been known for a long time, including the interesting solvatochromic behavior of the former dye [28]. Indeed, Nile Red is not only soluble in hydrocarbons but also in fats, where $E_T(30)$ is insoluble. In acid media, Nile Red does not lose its solvatochromic behavior, in contrast with the $E_T(30)$ dye [28]. In these instances, Nile Red is completely complementary to $E_T(30)$ and it is the ideal dye for the present work, where the substrates are apolar polymers (e.g., rubbers), hydrocarbons and ester plasticizers, but also triglycerides.

Deye, Berger and Anderson [28,29] have already performed a systematic study of Nile Red in comparison to $E_T(30)$ dye, showing that with the former dye, it is possible to measure the solvent polarity in many more liquids than those directly accessible by the $E_T(30)$, including fluorinated molecules and supercritical CO_2. As in the case of $E_T(30)$, the electronic absorption maximum of Nile Red can be expressed in kcal/mol according to Equation (14), so that an E(NR), i.e., Nile Red, scale for solvent can be constructed. A recognized disadvantage of the E(NR) scale with respect to the $E_T(30)$ scale regards the fact that the band shift of the Nile Red dye in media of different polarities is less pronounced than the case of the band shift offered by Reichardt's dye [28]. This implies an intrinsic lower sensitivity of the E(NR) to polarity change with respect to the $E_T(30)$. Furthermore, as can be seen in Figures 1 and 2, the absorption maxima of Nile Red and Reichardt's dyes vary in the opposite direction to the polarity of the medium, i.e., Nile Red is a positively solvatochromic dye, whereas Reichardt's dye is a negatively solvatochromic probe [30].

Figure 1 shows that the correlation between E(NR) and $E_T(30)$ is not linear but quite complex when considering all data available for any type of solvent (excluding the fluorinated molecules):

$$E(NR) = -0.0012\,[E_T(30)]^3 + 0.1728\,[E_T(30)]^2 - 8.1534\,[E_T(30)] + 182.17 \quad (16)$$

with a correlation coefficient $R^2 = 0.888$.

However, when considering only certain homogenous classes of solvents, for example, non-HBD solvents, simpler and more linear correlations are obtained, as will be shown later.

It is also interesting to note that the attention toward the Nile Red dye as a solvatochromic probe alternative or complementary to $E_T(30)$ is steady from a theoretical point of view but also from a practical point of view. The theoretical analysis focuses on the nature of the charge transfer transition of Nile Red, which makes it an effective probe. A twisted intramolecular charge transfer transition was advocated for this dye [31–37], and the ground and excited dipole moments of the dye were also estimated. The Nile Red dye is steadily employed in many measurements including probing hydrocarbon liquids [36], absorption and fluorescence in a series of alcohols [37], analysis of green chemistry solvents' polarity [38], study of anisotropic liquids [39], in zeolites [40], detection of lipid order heterogeneity in cells [41], and detection of hydrogen bonding strength in microenvironments [42], limiting the numerous fields of current applications.

Table 1. Measured or calculated ET(30) and E(NR) values of plasticizers, rubber and polymers.

PLASTICIZER, SOLVENT OR POLYMER	$E_T(30)$ Kcal/mol	E(NR) Kcal/mol	References or Notes on the $E_T(30)$ Values	References or Notes on the E(NR) Values	Solubility Parameter δ_t in $(MPa)^{0.5}$ Calc. According to Ref. [14]
Isooctane	30.7	58.77	ref. [24]	this work	14.6
N-hexane and cyclohexane	31.0	58.75	ref. [24]	this work	15.1
Tetradecane	31.0	58.21	ref. [24]	this work	16.2
SQUALANE	31.0	58.12	estimated	this work	15.6
Decalin	31.2	57.87	ref. [24]	this work	16.9
SQUALENE	36.5	57.64	this work fm $E_T(33)$	this work	16.0
Liquid cis-POLYISOPRENE (Liq-IR)	35.4	57.55	this work fm $E_T(33)$	this work	16.6
Liquid 1,2-POLYBUTADIENE	33.0	57.32	calculated (*)	this work	17.1
Liquid cis-POLYBUTADIENE (Liq-BR)	33.2	57.25	calculated (*)	this work	17.1
Cyclohexene	32.2	57.19	ref. [24]	this work	15.4
POLYBUTADIENE thin film (BR)	34.2	56.96	calculated (*)	this work	17.1
POLYISOPRENE thin film (IR)	34.6	56.82	calculated (*)	this work	16.6
Oleyl Oleate	36.2	56.27	this work fm $E_T(33)$	this work	16.3
S-SBR with styrene 21% & vinyl 50%	36.8	56.17	calculated (*)	this work	17.4
T-DAE oil	36.8	56.2	this work fm $E_T(33)$	calculated (**)	17.3
POLYSTYRENE thin film	37.8	55.84	calculated (*)	this work	17.5
Di-n-butyl PHTHALATE	39.5	55.3	ref. [24]	calculated (**)	17.8
Dimethyl PHTHALATE	40.7	55.0	ref. [24]	calculated (**)	18.0
Diisododecyl ADIPATE	36.4	54.98	this work fm $E_T(33)$	this work	17.5
Ethyl OLEATE	40.5	54.85	this work fm $E_T(30)$	this work	17.3
MES oil	37.3	54.83	this work fm $E_T(33)$	this work	16.7
Biodiesel fm rapeseed oil	41.4	54.77	calculated (*)	this work	16.6

Table 1. Cont.

PLASTICIZER, SOLVENT OR POLYMER	$E_T(30)$ Kcal/mol	E(NR) Kcal/mol	References or Notes on the $E_T(30)$ Values	References or Notes on the E(NR) Values	Solubility Parameter δ_t in $(MPa)^{0.5}$ Calc. According to Ref. [14]
Diethylhexyl SEBACATE	36.3	54.76	this work fm $E_T(33)$	this work	17.0
Coconut methyl ester	41.6	54.70	calculated (*)	this work	16.7
Dimethyl SEBACATE	41.8	54.7	this work fm $E_T(33)$	calculated (**)	18.1
Diethylhexyl ADIPATE (DOA)	36.4	54.67	calculated (*)	this work	17.8
Dibutyl SEBACATE	42.3	54.5	this work fm $E_T(33)$	calculated (**)	17.8
Soybean oil	42.1	54.55	calculated (*)	this work	17.2
Nytex BIO 6200	42.1	54.55	calculated (*)	this work	16.7
Epoxidized natural rubber (ENR-25)	42.9	54.32	calculated (*)	this work	17.2
Dioctylterephthalate (DOTP)	43.0	54.27	calculated (*)	this work	17.4
Sunflower oil (high oleic content)	43.5	54.13	calculated (*)	this work	16.7
Ethyl PALMITATE	43.9	54.0	this work fm $E_T(30)$	calculated (**)	16.8
Diethyl AZELATE	45.1	54.08	this work fm $E_T(33)$	this work	17.3
Methyl undecenoate	44.0	53.99	calculated (*)	this work	17.9
PEG dioleate	44.6	53.79	calculated (*)	this work	17.7
PEG monooleate	45.2	53.60	calculated (*)	this work	17.6
Polymethylmethacrylate film (PMMA)	45.8	53.44	calculated (*)	ref. [43]	
Poly(Lactic acid) film (PLLA)	45.8	53.44	calculated (*)	this work	19.5
Bis(THFA) ADIPATE	46.2	53.30	this work fm $E_T(33)$	calculated (**)	19.0
Ethyl levulinate	46.2	53.30	calculated (*)	this work	18.9
Dioctylphthalate (ethylhexyl) (DOP)	46.4	53.24	calculated (*)	this work	18.2
Nitrile Rubber with 33% ACN film	47.0	53.07	calculated (*)	this work	19.4
Bis(THFA) SEBACATE	47.1	53.0	this work fm $E_T(33)$	calculated (**)	18.4
Bis(THFA) AZELATE	47.3	53.0	this work fm $E_T(33)$	calculated (**)	18.6

Table 1. Cont.

PLASTICIZER, SOLVENT OR POLYMER	$E_T(30)$ Kcal/mol	$E(NR)$ Kcal/mol	References or Notes on the $E_T(30)$ Values	References or Notes on the $E(NR)$ Values	Solubility Parameter δ_t in $(MPa)^{0.5}$ Calc. According to Ref. [14]
THFA OLEATE	47.4	53.0	this work fm $E_T(33)$	calculated (**)	17.8
THFA LAURATE (30 °C)	48.0	52.8	this work fm $E_T(33)$	calculated (**)	17.2
THFA PELARGONATE (30 °C)	49.3	52.4	this work fm $E_T(33)$	calculated (**)	18.9
Tetrahydrofurfuryl alcohol	49.9	52.2	this work fm $E_T(30)$	calculated (**)	
L-(−) ethyl lactate	51.1	52.16	ref. [24]	this work	
Polyethylene glycol (PEG-400)	49.7	52.15	this work fm $E_T(30)$	this work	19.7
Polytetrahydrofuran (PolyTHF]low Mw	49.3	52.12	this work fm $E_T(30)$	this work	18.9
Water	63.1	48.21	ref. [24]	this work	

(*) Calculated with $E(NR) = -0.303\ E_T(30) + 67.31$; (**) Calculated with $E_T(30) = [E(NR) - 67.31]/-0.303$; THFA = Tetrahydrofurfuryl.

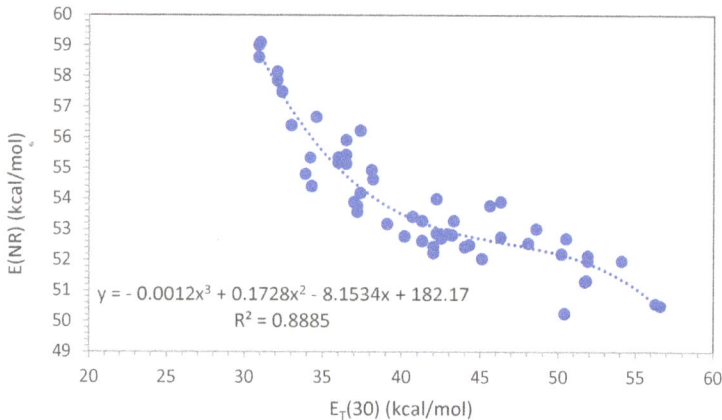

Figure 1. Correlation between E(NR) and $E_T(30)$ using the data of ref. [28] and excluding the fluorinated molecules.

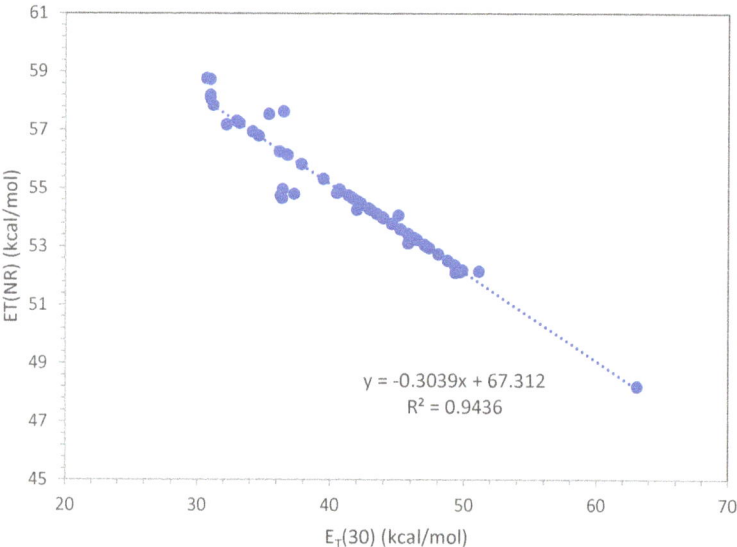

Figure 2. Correlation between E(NR) and $E_T(30)$ using all the experimental data of Table 1.

3.2. Determination of Rubber and Polymer Polarity with Nile Red Solvatochromic Dye

As detailed in Section 2.3, the micropolarity of the rubber polymers was determined with Nile Red solvatochromic dye either using model compounds or liquid polymers where the dye was soluble or using thin solid-state polymer films where the dye was embedded in the polymer matrix with the aid of a solvent, which was then removed completely. The basic idea of the latter approach derives from the seminal work of Dutta et al. [43], who used Nile Red as a micropolarity probe for plastics in polymethylmethacrylate (PMMA) and polyvinylalcohol films. We also extended the same approach to rubber polymers like cis-1,4-polyisoprene, cis-1,4-polybutadiene, the styrene–butadiene copolymer, epoxidized natural rubber and nitrile rubber. The micropolarity of polystyrene and poly(lactic acid) films was also studied, whereas the E(NR) value reported by Dutta et al. [43] was adopted for PMMA.

Table 1 reports the experimental E(NR) values for all the above-mentioned rubbers and other liquids. In Table 1, the apolar compounds are at the top of the table and the most polar at the bottom of it.

Regarding the rubber and other polymers studied, there is a logical trend from apolar to polar compounds starting with the liquid model compounds squalane and squalene (squalane is the liquid analog of the ethylene–propylene copolymer, while squalene can be considered a liquid model compound of natural rubber or cis-1,4-polyisoprene as discussed elsewhere [44]) as the most apolar, followed by the liquid polyisoprene and polybutadiene and then by the thin solid films of polybutadiene (BR) and polyisoprene (IR). The copolymer styrene–butadiene rubber (S-SBR) with 21% styrene content is found to be more polar than either BR or IR as expected. S-SBR shows E(NR) values in the same range as polystyrene thin solid film and the petroleum-based rubber plasticizer T-DAE, whose aromatic content is in the range of 20–25%.

As expected, polar polymers are appropriately positioned in the E(NR) scale, with epoxidized natural rubber (ENR-25) showing an E(NR) value of 54.32 kcal/mol with respect to 57.55 kcal/mol of liquid IR and 56.82 kcal/mol of the solid film of IR.

Even more polar than ENR-25 are polymethylmethacrylate (PMMA) and poly(lactic acid) (PLLA), both with E(NR) = 53.44 kcal/mol, followed by nitrile rubber (NBR) with 33% acrylonitrile (ACN) content and a Nile Red electronic transition at 53.07 kcal/mol. The most polar polymers (among those studied in the present work) were two oligomers: polyethylene glycol (PEG-400) and polytetrahydrofuran liquid oligomer (polyTHF), both at 52.1 kcal/mol. Being oligomers, the latter two compounds have their micropolarity affected by the OH end groups which, of course, play an important role in affecting their polarity.

3.3. Determination of Rubber Plasticizer Polarity with Nile Red Solvatochromic Dye

Treated distillate aromatic extract (T-DAE) and mild extract solvate (MES) represent the most-used petroleum-based plasticizers in rubber compounds since 2010 when the use of distillate aromatic extract (DAE) was banned. T-DAE has an aromatic content in the range of 20–25%, while the aromatic content of MES is limited to 10–15%. Unfortunately, T-DAE is too dark to measure its E(NR) value but it was possible to perform a measurement on it with the $E_T(33)$ dye. On the other hand, MES is much more clear in color than T-DAE and it was possible to make a measurement both with $E_T(33)$ and Nile Red dyes as summarized in Table 1. From these data, T-DAE is appropriately positioned in Table 1 just between the S-SBR rubber and the polystyrene polymer. Surprisingly, the MES, which is prevalently aliphatic and naphthenic in its chemical nature, is positioned not among the hydrocarbons at the top of Table 1 but rather in the upper-mid part of Table 1, sharing the same E(NR) value as biodiesel i.e., the methyl ester of fatty acids from rapeseed oil, which instead is appropriately positioned in Table 1. There are probably components and impurities in the MES (which is a mixture of hydrocarbons) that affect its real polarity value. The presence of water and heteroatoms (e.g., N, S, Cl) may significantly alter the polarity in these industrial compounds.

As already stated in Section 2.1, there is a trend to use vegetable oils as rubber plasticizers, and the most popular are soybean and sunflower oils [19–21]. In Table 1, both oils were tested with the Nile Red dye, and the soybean oil appears to be slightly less polar with an E(NR) = 54.55 kcal/mol with respect to the sunflower oil (high oleic content), which has an E(NR) = 54.13 kcal/mol. It is also interesting that a commercial product Nytex Bio 6200, which is a blend of petroleum-based naphthenic oil and fatty acids, is classified by the E(NR) scale to be just in the middle between soybean oil and sunflower oil.

On the other hand, the methyl esters of fatty acids derived from rapeseed oil (biodiesel) or from coconut oil are slightly less polar than triglycerides. In fact, for the methyl esters of fatty acids, E(NR) = 54.7 kcal/mol, while the triglycerides show E(NR) = 54.1–54.5 kcal/mol. Thus, at least in terms of compatibility with rubbers for tire application, based on the above data, there are no practical differences between the methyl esters of fatty acids and the triglycerides, as already disclosed some time ago [17].

In the past, phthalates were used as plasticizers for winter tire treads. However, at present, there are a number of concerns regarding the environmental and health impact of phthalates [45–47] so they must be replaced by safer plasticizers. The typical plasticizers used in place of phthalates are adipates and sebacates. In our E(NR) scale, it is possible to see that the phthalates are located at E(NR) = 55.0–55.3 kcal/mol, although DOP (diethylhexylphthalate), the most used in the past, was found at 53.2 kcal/mol. It is interesting to note that the adipates are found at about 55 kcal/mol, with diethylhexyladipate (DOA) being the most common at 54.67 kcal/mol. The same comments apply to sebacates, which, although more expensive than adipates, display similar E(NR) values to adipates and DOA in particular.

Another trend regarding the substitution of DOP involves the use of terephthalic acid derivatives, and in particular, the use of dioctylterephthalate (DOTP), whose E(NR) in Table 1 was found to be 54.27 kcal/mol, which was less polar, as expected, than DOP at 53.24 kcal/mol.

Tetrahydrofurfuryl alcohol (THFA) has been proposed as an ideal alcohol from renewable sources. In fact, it derives from furfurol obtained from mineral acid treatment of biomasses. Furfurol is fully hydrogenated to THFA. We have synthesized several esters of THFA as part of another project [48] and in Table 1, we report the polarity values of the adipate, sebacate, oleate, laurate and pelargonate measured with the $E_T(33)$ dye; in the Discussion section, it will be shown how the $E_T(30)$ values correlate linearly with the E(NR) values for the class of compounds considered in the present paper. The THFA esters appear too polar to have good compatibility with common rubbers such as IR, BR and S-SBR but may be more than suitable as plasticizers for nitrile rubber (NBR) with E(NR) = 53.07 kcal/mol and 52-53 kcal/mol for the THFA esters.

Polyethylene glycol is used in certain rubber compounds as a compatibilizer aid between silica filler and rubber. The E(NR) value for PEG-400 was found at 52.15 kcal/mol, while the E(NR) value of raw silica surfaces is not known, but a work on modified silica surfaces reports E(NR) ≈ 48-7–52.6 kcal/mol (depending from the type and degree of modification of the surface) [49]. Based on these data, PEG-400 is really suitable for interacting with silica surfaces. In Table 1, it is shown that polytetrahydrofuran (polyTHF) oligomer is also a suitable liquid for the compatibilization of silica surfaces with rubber, since polyTHF shows the same E(NR) value as PEG-400. To reduce the polarity of PEG-400, it is possible to esterificate the OH end groups of the glycol with oleic acid to obtain either PEG monooleate with E(NR) = 53.6 kcal/mol or PEG dioleate with 53.8 kcal/mol.

4. Discussion

4.1. Correlation between the E(NR) Scale and the $E_T(30)$ Scale

The results have shown that Nile Red is an excellent and complementary solvatochromic dye with respect to $E_T(30)$. It is complementary because it is soluble in hydrocarbons and it is not sensitive to the residual acidity that may be present in ester plasticizers. Furthermore, Nile Red is also soluble in the same solvents as $E_T(30)$ and $E_T(33)$. Without the use of Nile Red, the present work could not have been conducted since we dealt with hydrocarbon rubber polymers and plasticizers where $E_T(30)$ was insoluble or problematic but where $E_T(33)$ also had problems. Furthermore, $E_T(33)$ is no longer easily commercially available, while the easy availability of Nile Red is due to the fact that it is also used as a staining dye for biological tissues and cells [27].

In this work, 53 compounds, either polymers or plasticizers, were studied with Nile Red and wherever possible also with the $E_T(33)$ dye. The data in Table 1 obtained with the latter dye were converted to the $E_T(30)$ scale using Equation (15). Only in a couple of cases was it possible to use $E_T(30)$ dye directly. The experimental data in Table 1 were then put in a graph as shown in Figure 2, so that the following equation was obtained with a very good correlation coefficient $R^2 = 0.944$:

$$E(NR) = -0.3039\, E_T(30) + 67.312 \tag{17}$$

and the reverse equation,

$$E_T(30) = [E(NR) - 67.312]/(-0.3039) \quad (18)$$

Equation (17) was then used to calculate the E(NR) values which were not measured with Nile Red dye. The calculated values are shown in italic characters in Table 1. Similarly, for the $E_T(30)$ values in Table 1, use was made of Equation (18), and the calculated values are shown in italic characters as well.

4.2. Correlation between the Solubility Parameter δ_t and the E(NR) Scale

In the introduction, we presented the classical approach to determine the compatibility between a polymer and a plasticizer. The classical approach involves the determination of the solubility parameter of both components, by finding these values in published tables (e.g., Hansen's book [13]) or calculating the solubility parameter of each component by group increments through the Van Krevelen method [14] or other similar approaches. Then, the solubility criteria are shown in Equations (11)–(13).

It is interesting at this point to evaluate how the total solubility parameter δ_t correlates with the E(NR) scale. Our first approach was to use the Hansen solubility parameter δ_t derived from the data tabulated in ref. [13]. However, the correlation with the E(NR) values reported in Table 1 was not satisfactory.

The following step was to calculate the δ_t value for each solvent, plasticizer and polymer reported in Table 1 according to the Van Krevelen method [14]. The calculation results are shown in the last column on the right of Table 1. These calculated δ_t values were then put in a graph against the E(NR) values of Table 1 and the results are shown in Figure 3. Although this time, the correlation coefficient $R^2 = 0.64$ is not as good as the E(NR) vs ET(30) correlation shown in Figure 1, an evident trend can be observed so that at least a rough estimation of E(NR) can be derived from δ_t according to the following:

$$\delta_t = -0.472\ E(NR) + 43.33 \quad (19)$$

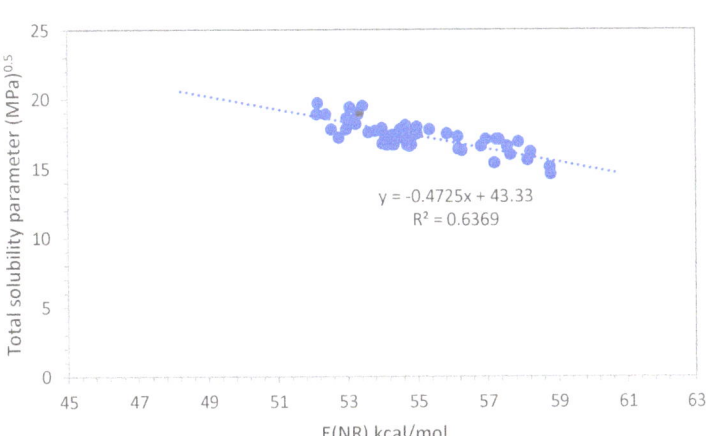

Figure 3. Correlation between the solubility parameter δ_t and the E(NR) scale.

In the correlation between the δ_t values of Table 1 and the E(NR) values, we were forced to exclude the δ_t values of tetrahydrofurfuryl alcohol, ethyl lactate and the polymer polymethylmethacrylate (PMMA).

4.3. Some Reflections on the Diene Rubber Compatibility Based on the E(NR) Scale

The main purpose of the present work was to create a polarity scale for hydrocarbon rubbers and conventional and new plasticizers. As shown in Table 1, it was possible to measure the E(NR) value of the key diene rubbers polyisoprene (IR), polybutadiene (BR) and the copolymer styrene–butadiene (S-SBR). From this measurement, it turned out that the liquid oligomers of IR and BR were adequate model compounds of the higher polymers IR and BR since very similar E(NR) values were measured. The same reasoning also applies for squalene as a model compound of IR polymer. Regarding the conventional hydrocarbon plasticizers of diene rubber, from Table 1, it is immediately evident that the T-DAE plasticizer with its relatively high aromatic content is fully compatible with S-SBR, since both compounds share the same E(NR) value. Furthermore, it is also confirmed that T-DAE is also more compatible with IR and BR than MES, which, although less aromatic, evidently contains some structural feature that increases its polarity to the same level as biodiesel (the methyl ester of rapeseed oil tested previously with success as a rubber compound plasticizer [17]).

The phthalate-based plasticizers used in the past in passenger winter tire treads result in the E(NR) scale being less polar and more compatible with the diene rubber than the currently used ester plasticizers like adipates and sebacates.

"Green" plasticizers are those produced from renewable sources. Thus, emerging green plasticizers for the tire industry are soybean and sunflower oils as well as the semi-green blend of naphthenic oil with fatty acids (Nytex Bio 6200). As expected, these green plasticizers are in the middle of Table 1, and the compatibility with the diene rubbers like IR, BR and S-SBR is probably moderate. On the other hand, in Table 1, it is immediately evident that such mentioned green plasticizers have excellent compatibility with more polar rubbers like epoxidized natural rubber (ENR) and nitrile rubber (NBR).

Tetrahydrofurfuryl alcohol (THFA) is considered a green alcohol fully derived from renewable sources. Thus, its esters with carboxylic acids also from renewable sources are 100% green. However, such esters of THFA are all found at the bottom of Table 1, suggesting high polarity and poor compatibility with IR, BR and S-SBR. Instead, the THFA esters are certainly suitable as plasticizers for ENR and NBR.

5. Conclusions

The evaluation of the polymer–plasticizer compatibility through an $E_T(30)$ scale is hindered by the insolubility of the $E_T(30)$ dye in hydrocarbon polymers like IR, BR and S-SBR. Furthermore, $E_T(30)$ dye may have limited solubility in plasticizers and is sensitive to their weak acidity. In fact, protonation of the $E_T(30)$ dye destroys the CT band of this solvatochromic dye. To circumvent such a situation, other authors [25] have proposed an interesting, original and sophisticated approach that permits the estimation of the $E_T(30)$ values of certain "difficult" substrates where the direct polarity measurement is hindered. The alternative approach proposed in the present work is to use Nile Red dye as a solvatochromic dye. Nile Red has the advantage of being soluble in hydrocarbons and in polar solvents and it is not sensitive to protonation.

In Table 1, a series of 53 different compounds including rubbers, plasticizers, hydrocarbon solvents and even some polar polymers (like PMMA, PLLA, PEG-400 and polyTHF) were tested with the Nile Red probe. For the evaluation of liquid polarity, Nile Red was dissolved in the selected liquid and the spectrum was recorded, while for polymers, Nile Red was embedded in a thin solid film of the selected polymer and the spectrum was recorded on the solid-state film. Wherever possible, the liquid or plasticizer was also evaluated with the $E_T(33)$ dye (seldom with the $E_T(30)$ dye) and the resulting value converted into the $E_T(30)$ scale through Equation (15).

Thus, Table 1, in a certain number of cases, reports both the E(NR) value and the $E_T(30)$ value. These data were used to derive a correlation between E(NR) scale and $E_T(30)$, limited to a selected number of polymers and plasticizers, as shown in Figure 2 and displayed by Equations (17) and (18) with a very good correlation coefficient.

Furthermore, the total solubility parameter δ_t as defined by Equation (10) has also been calculated for each compound according to the Van Krevelen method [14] and reported in

Table 1. A reasonable correlation was also found between the solubility parameter and the E(NR) scale, as shown in Figure 3 and described by Equation (19).

Thus, with Nile Red dye, it is possible to study the polarity of hydrocarbon polymers (like rubbers) and plasticizers and to connect the results either with Reichardt's $E_T(30)$ scale or with the total solubility parameter.

Funding: This research received no external funding.

Data Availability Statement: This work will be available on the Researchgate website on the author's page.

Conflicts of Interest: The author declares no conflicts of interest.

References

1. Sears, J.K.; Darby, J.R. *The Technology of Plasticizers*; John Wiley & Sons: New York, NY, USA, 1982.
2. Stepek, J.; Daoust, H. *Additives for Plastics*; Springer Science: New York, NY, USA, 1983; pp. 7–33.
3. Tager, A. *The Physical Chemistry of Polymers*; Mir Publishers: Moscow, Russia, 1978; pp. 547–567.
4. Cadogan, D.; Howick, C.J. Plasticizers. In *Ullmann's Encyclopedia of Industrial Chemistry*, 5th ed.; Elvers, B., Hawkins, S., Schultz, Eds.; VCH: Weinheim, Germany, 1992; Volume A20, pp. 439–458.
5. Sears, J.F.; Touchette, N.W. Plasticizers. In *Concise Encyclopedia of Polymer Science and Engineering*; Kroschwitz, J., Ed.; John Wiley & Sons: New York, NY, USA, 1999; pp. 734–744.
6. Howick, C.J. Plasticizers. In *Kirk-Othmer Encyclopedia of Chemical Technology*, 5th ed.; Ley, C., Ed.; John Wiley & Sons: New York, NY, USA, 2007.
7. Godwin, A.D. Plasticizers. In *Applied Plastics and Engineering Handbook*; Kutz, M., Ed.; Elsevier: Amsterdam, The Netherlands, 2011; pp. 487–502.
8. Marcilla, A.; Breltràn, A. Mechanism of plasticizers action. In *Handbook of Plasticizers*, 4th ed.; Wypich, G., Ed.; ChemTec Publishing: Toronto, ON, Canada, 2023; pp. 139–158.
9. Wypich, A. Compatibility of plasticizers. In *Handbook of Plasticizers*, 4th ed.; Wypich, G., Ed.; ChemTec Publishing: Toronto, ON, Canada, 2023; pp. 159–180.
10. Morris, G. Plasticizers. In *Developments in Rubber Technology—1*; Whelan, A., Lee, K.S., Eds.; Elsevier Applied Science: London, UK, 1979; pp. 207–225.
11. Crouter, B.G. Processing aids and plasticizers. In *Developments in Rubber Technology—4*; Whelan, A., Lee, K.S., Eds.; Elsevier Applied Science: London, UK, 1987; pp. 119–158.
12. Brydson, J.A. *Rubber Chemistry*; Applied Science Publishers Ltd.: London, UK, 1978.
13. Hansen, C.M. *Hansen Solubility Parameters: A User's Handbook*; CRC Press/Taylor & Francis: Boca Raton, FL, USA, 2007.
14. Van Krevelen, D.W. *Properties of Polymers. Their Correlation with Chemical Structure, Their Numerical Estimation and Prediction from Additive Group Contributions*, 3rd ed.; Elsevier: Amsterdam, The Netherlands, 1990.
15. Cataldo, F. On the solubility parameter of C_{60} and higher fullerenes. *Fuller. Nanot. Carbon Nanostruct.* **2009**, *17*, 79–84. [CrossRef]
16. Cataldo, F. Solubility of fullerenes in fatty acids esters: A new way to deliver in vivo fullerenes. Theoretical calculations and experimental results. In *Medicinal Chemistry and Pharmacological Potential of Fullerenes and Carbon Nanotubes*; Cataldo, F., Da Ros, T., Eds.; Springer Science: Dordrecht, The Netherlands, 2008; Chapter 13.
17. Cataldo, F.; Ursini, O.; Angelini, G. Biodiesel as a plasticizer of a SBR-based tire tread formulation. *ISRN Polym. Sci.* **2013**, *2013*, 340426. [CrossRef]
18. Cataldo, F. Application of Reichardt's solvent polarity scale ($E_T(30)$) in the selection of bonding agents for composite solid rocket propellants. *Liquids* **2022**, *2*, 289–302. [CrossRef]
19. Sandstrom, P.H.; Rodewald, S.; Ramanathan, A. Rubber Composition and Tire with Component Comprised of Polyisoprene Rubber and Soybean Oil. U.S. Patent N°20140135424A1, 15 May 2014.
20. Brunelet, T.; Dinh, M.; Favrot, J.M.; Labrunie, P.; Lopitaux, G.; Royet, J.G. Plasticizing System for Rubber Composition. U.S. Patent N°7834074B2, 22 August 2004.
21. Mohamed, N.R.; Othman, N.; Shuib, R.K. Synergistic effect of sunflower oil and soybean oil as alternative processing oil in the development of greener tyre tread compound. *J. Rubber Res.* **2022**, *25*, 239–249. [CrossRef]
22. Khristenko, I.V.; Kholin, Y.V.; Mchedlov-Petrossyan, N.O.; Reichardt, C.; Zaitsev, V.N. Probing of chemically modified silica surfaces by solvatochromic pyridinium N-phenolate betaine indicators. *Colloid J.* **2006**, *68*, 511–517. [CrossRef]
23. Reichardt, C. Pyridinium-N-phenolate betaine dyes as empirical indicators of solvent polarity: Some new findings. *Pure Appl. Chem.* **2008**, *80*, 1415–1432. [CrossRef]
24. Reichardt, C.; Welton, T. *Solvents and Solvent Effects in Organic Chemistry*; John Wiley & Sons: New York, NY, USA, 2011.
25. Acree Jr, W.E.; Lang, A.S. Reichardt's Dye-Based Solvent Polarity and Abraham Solvent Parameters: Examining Correlations and Predictive Modeling. *Liquids* **2023**, *3*, 303–313. [CrossRef]
26. Davis, M.M.; Helzer, H.B. Titrimetric and Equilibrium Studies Using Indicators Related to Nile Blue A. *Anal. Chem.* **1966**, *38*, 451–461. [CrossRef]

27. Green, F.J. *The Sigma-Aldrich Handbook of Stains, Dyes and Indicators*; Aldrich Chemical Company, Inc.: Milwaukee, WI, USA, 1990; pp. 519–520.
28. Deye, J.F.; Berger, T.A.; Anderson, A.G. Nile Red as a solvatochromic dye for measuring solvent strength in normal liquids and mixtures of normal liquids with supercritical and near critical fluids. *Anal. Chem.* **1990**, *62*, 615–622. [CrossRef]
29. Deye, J.F.; Berger, T.A.; Anderson, A.G. Errata Corrige. *Anal. Chem.* **1990**, *62*, 1552. [CrossRef]
30. Nigam, S.; Rutan, S. Principles and applications of solvatochromism. *Appl. Spectrosc.* **2001**, *55*, 362A–370A. [CrossRef]
31. Ghoneim, N. Photophysics of Nile red in solution: Steady state spectroscopy. *Spectrochim. Acta Part A Mol. Biomol. Spect.* **2000**, *56*, 1003–1010. [CrossRef] [PubMed]
32. Dias, L.C., Jr.; Custodio, R.; Pessine, F.B. Investigation of the Nile Red spectra by semi-empirical calculations and spectrophotometric measurements. *Int. J. Quantum Chem.* **2006**, *106*, 2624–2632. [CrossRef]
33. Kawski, A.; Bojarski, P.; Kukliński, B. Estimation of ground-and excited-state dipole moments of Nile Red dye from solvatochromic effect on absorption and fluorescence spectra. *Chem. Phys. Lett.* **2008**, *463*, 410–412. [CrossRef]
34. Guido, C.A.; Mennucci, B.; Jacquemin, D.; Adamo, C. Planar vs. twisted intramolecular charge transfer mechanism in Nile Red: New hints from theory. *Phys. Chem. Chem. Phys.* **2010**, *12*, 8016–8023. [CrossRef] [PubMed]
35. Ya. Freidzon, A.; Safonov, A.A.; Bagaturyants, A.A.; Alfimov, M.V. Solvatofluorochromism and twisted intramolecular charge-transfer state of the nile red dye. *Int. J. Quantum Chem.* **2012**, *112*, 3059–3067. [CrossRef]
36. Yablon, D.G.; Schilowitz, A.M. Solvatochromism of Nile Red in nonpolar solvents. *Appl. Spectrosc.* **2004**, *58*, 843–847. [CrossRef] [PubMed]
37. Zakerhamidi, M.S.; Sorkhabi, S.G. Solvent effects on the molecular resonance structures and photo-physical properties of a group of oxazine dyes. *J. Lumin.* **2015**, *157*, 220–228. [CrossRef]
38. Jessop, P.G.; Jessop, D.A.; Fu, D.; Phan, L. Solvatochromic parameters for solvents of interest in green chemistry. *Green Chem.* **2012**, *14*, 1245–1259. [CrossRef]
39. Gilani, A.G.; Moghadam, M.; Zakerhamidi, M.S. Solvatochromism of Nile red in anisotropic media. *Dyes Pigment.* **2012**, *92*, 1052–1057. [CrossRef]
40. Sarkar, N.; Das, K.; Nath, D.N.; Bhattacharyya, K. Twisted charge transfer processes of Nile red in homogeneous solutions and in faujasite zeolite. *Langmuir* **1994**, *10*, 326–329. [CrossRef]
41. Kreder, R.; Pyrshev, K.A.; Darwich, Z.; Kucherak, O.A.; Mély, Y.; Klymchenko, A.S. Solvatochromic Nile Red probes with FRET quencher reveal lipid order heterogeneity in living and apoptotic cells. *ACS Chem. Biol.* **2015**, *10*, 1435–1442. [CrossRef] [PubMed]
42. Cser, A.; Nagy, K.; Biczók, L. Fluorescence lifetime of Nile Red as a probe for the hydrogen bonding strength with its microenvironment. *Chem. Phys. Lett.* **2002**, *360*, 473–478. [CrossRef]
43. Dutta, A.K.; Kamada, K.; Ohta, K. Spectroscopic studies of nile red in organic solvents and polymers. *J. Photochem. Photobiol. A Chem.* **1996**, *93*, 57–64. [CrossRef]
44. Cataldo, F. Aminoxyl (nitroxyl or nitroxide) radical formation by the action of ozone on squalene containing secondary aromatic amine antioxidants. *J. Vinyl. Addit. Technol.* **2022**, *28*, 379–389. [CrossRef]
45. Mariana, M.; Castelo-Branco, M.; Soares, A.M.; Cairrao, E. Phthalates' exposure leads to an increasing concern on cardiovascular health. *J. Hazard. Mater.* **2023**, *457*, 131680. [CrossRef] [PubMed]
46. Arrigo, F.; Impellitteri, F.; Piccione, G.; Faggio, C. Phthalates and their effects on human health: Focus on erythrocytes and the reproductive system. *Compar. Biochem. Physiol. Part C Toxicol. Pharmacol.* **2023**, *270*, 109645. [CrossRef] [PubMed]
47. Luís, C.; Algarra, M.; Câmara, J.S.; Perestrelo, R. Comprehensive insight from phthalates occurrence: From health outcomes to emerging analytical approaches. *Toxics* **2021**, *9*, 157. [CrossRef]
48. Cataldo, F. unpublished results.
49. Moreno, E.M.; Levy, D. Role of the comonomer GLYMO in ORMOSILs as reflected by nile red spectroscopy. *Chem. Mater.* **2000**, *12*, 2334–2340. [CrossRef]

Disclaimer/Publisher's Note: The statements, opinions and data contained in all publications are solely those of the individual author(s) and contributor(s) and not of MDPI and/or the editor(s). MDPI and/or the editor(s) disclaim responsibility for any injury to people or property resulting from any ideas, methods, instructions or products referred to in the content.

 *liquids*

Article

Exploring Solvation Properties of Protic Ionic Liquids by Employing Solvatochromic Dyes and Molecular Dynamics Simulation Analysis

Stuart J. Brown, Andrew J. Christofferson, Calum J. Drummond, Qi Han and Tamar L. Greaves *

School of Science, STEM College, RMIT University, Melbourne, VIC 3000, Australia; s3399895@student.rmit.edu.au (S.J.B.); andrew.christofferson@rmit.edu.au (A.J.C.); calum.drummond@rmit.edu.au (C.J.D.); qi.han@rmit.edu.au (Q.H.)
* Correspondence: tamar.greaves@rmit.edu.au

Abstract: Solvation properties are key for understanding the interactions between solvents and solutes, making them critical for optimizing chemical synthesis and biochemical applications. Designable solvents for targeted optimization of these end-uses could, therefore, play a big role in the future of the relevant industries. The tailorable nature of protic ionic liquids (PILs) as designable solvents makes them ideal candidates. By alteration of their constituent structural groups, their solvation properties can be tuned as required. The solvation properties are determined by the polar and non-polar interactions of the PIL, but they remain relatively unknown for PILs as compared to aprotic ILs and their characterization is non-trivial. Here, we use solvatochromic dyes as probe molecules to investigate the solvation properties of nine previously uncharacterized alkyl- and dialkylammonium PILs. These properties include the Kamlet–Aboud–Taft (KAT) parameters: π* (dipolarity/polarizability), α (H-bond acidity) and β (H-bond basicity), along with the $E_T(30)$ scale (electrophilicity/polarizability). We then used molecular dynamics simulations to calculate the radial distribution functions (RDF) of 21 PILs, which were correlated to their solvation properties and liquid nanostructure. It was identified that the hydroxyl groups on the PIL cation increase α, π* and $E_T(30)$, and correspondingly increase the cation–anion distance in their RDF plots. The hydroxyl group, therefore, reduces the strength of the ionic interaction but increases the polarizability of the ions. An increase in the alkyl chain length on the cation led to a decrease in the distances between cations, while also increasing the β value. The effect of the anion on the PIL solvation properties was found to be variable, with the nitrate anion greatly increasing π*, α and anion–anion distances. The research presented herein advances the understanding of PIL structure–property relationships while also showcasing the complimentary use of molecular dynamics simulations and solvatochromic analysis together.

Keywords: ionic liquids; protic ionic liquids; KAT; molecular dynamics; solvation

Citation: Brown, S.J.; Christofferson, A.J.; Drummond, C.J.; Han, Q.; Greaves, T.L. Exploring Solvation Properties of Protic Ionic Liquids by Employing Solvatochromic Dyes and Molecular Dynamics Simulation Analysis. *Liquids* **2024**, *4*, 288–304. https://doi.org/10.3390/liquids4010014

Academic Editors: William E. Acree, Jr., Franco Cataldo and Enrico Bodo

Received: 15 January 2024
Revised: 1 February 2024
Accepted: 5 March 2024
Published: 20 March 2024

Copyright: © 2024 by the authors. Licensee MDPI, Basel, Switzerland. This article is an open access article distributed under the terms and conditions of the Creative Commons Attribution (CC BY) license (https://creativecommons.org/licenses/by/4.0/).

1. Introduction

Ionic liquids (ILs) are a unique class of solvents possessing tailorable physicochemical properties through varying their chemical structure [1–3]. They are often defined as liquid salts with melting points ≤100 °C, though MacFarlane et al. (2018) proposed expanding this definition to include IL-solvent mixtures [4]. In both industry and research applications, the beneficial properties that can be provided by ILs include low viscosity, negligible vapor pressure, control over thermal phase transitions, support for amphiphilic self-assembly and even biocompatibility and biodegradation [4–9]. A key property of many ILs is the ability to solvate a combination of polar and non-polar compounds due to containing both ionic components and organic alkyl groups [5,10–12]. Despite this duality, the majority of ILs are considered polar solvents based on a generalization of IL solvent properties. This highlights the complexity of solvent properties, and their challenging nature to define and characterize. Specifically, the International Union of Pure and Applied Chemistry (IUPAC) defines polarity as 'the action of all possible intermolecular interactions between solute ions or molecules and solvent molecules, excluding interactions leading to definite chemical alterations of the ions or molecules of the solute' [13]. Not surprisingly, the literature, therefore, lacks a comprehensive understanding of select structure–property relationships describing the solvent properties of ILs.

Characterization of solvent–solute interactions is a non-trivial task, and many different techniques have been used for this purpose. These include but are not limited to, ^{1}H and ^{13}C NMR [14], UV/vis absorption and fluorescence [15–19], octanol-water partition [20], chromatography [20–22] and analyzing the solvent effect based on standard chemical reactions [23–28]. Physical properties can also provide insight into the macroscopic polar properties of a solvent through for example, refractive index and molar refractivity [29,30], but do not provide detail on intermolecular electrostatic and polarization forces between the solute and solvent. The technique that is arguably most effective at comparing solvent properties across ILs is the use of UV/visible spectroscopy with optimized solvatochromic/solvatofluorochromic dyes [23,30–32]. The absorbance/fluorescence shift of these dyes occurs due to their interactions with the solvent and can be measured via spectroscopy and calculated on an empirical basis [33–37]. They have previously been used to develop the Kamlet–Aboud–Taft (KAT) parameters as a multi-parameter approach that has been shown to be suitable for ILs [30–32,38]. Using this method, the solvation properties of a solvent are described using three unique parameters [33–37]. These include, π^* (dipolarity/polarizability), α (H-bond donating (HBD) acidity) and β (H-bond accepting (HBA) basicity). Another well-utilised empirical scale of polarity is the $E_T(30)$ electronic transition scale [29,39]. It is calculated from the maximum absorbance wavelength shift of Reichardt's betaine dye or Reichardt's dye 30 and is a result of solvent dipolarity/polarizability and H-bond acidity. However, we have previously noted significant solubility and protonation issues when applying Reichardt's dye 30 during experimentation with ILs. Therefore, Reichardt's dye 33 (RD33) [40] has been used as a substitute, producing an $E_T(33)$ value that can be correlated to $E_T(30)$ via a linear relationship. Combined, these four parameters account for the diverse interactions possible between an IL solvent and solute.

Protic ILs (PILs) are an easy-to-synthesize, non-aqueous solvent class with applications in a variety of areas, such as biomolecule stability and activity and chemical catalysis [11,41–45]. PILs differ from aprotic ILs due to the proton transfer that occurs during synthesis from the neutralization of a Brønsted acid by a Brønsted base, producing a solvent capable of hydrogen bonding. Their overall polarity is, therefore, based on the combination of hydrogen bonds, dipole–dipole interactions and electrostatic interactions. These multiple contributing factors and the non-specific nature of polarity mean that the use of a multi-parameter analysis method is a necessity. Previously, 11 solvatochromic dyes were trialled for the characterization of PIL solvation properties, identifying N,N-diethyl-4-nitroaniline (DE4A), 4-nitroaniline (4NA) and RD33 as the optimal choice [38]. To date, solvation properties of only 16 alkylammonium nitrate, formate, acetate and thiocyanate PILs have previously been fully reported [21,38]. Comparatively, alkylammonium cations were shown to have weaker H-bond basicity and elevated π^* as compared to tetraalkylammonium sulfonate aprotic ILs [21,46]. These literature studies concluded that H-bond acidity was shown to be a result of both cation and anion structural groups, while H-bond basicity was dominated by the anion. The nitrate anion was identified as a stronger H-bond anion and more polarizable than the organic carboxylate anions. Additionally, a decrease in π^* with increasing alkyl chain length has been observed for PILs. However, the impact of hydroxyl groups on the anion remains an unexplored area of PIL solvation properties, and it is unknown how robust these trends are across ion series.

Molecular dynamics (MD) simulations provide additional and complementary information to the experimental investigation of molecular interactions of a solvent, such as information on the specific interactions of PILs based on their structure. The literature on MD of ILs has primarily focused on interactions and the structure of aprotic ILs. For example, Miao et al. (2022) modeled the liquid structure of choline-amino acid-based ILs through their internal H-bonds and atom–atom pair correlation functions then correlated this with their ability to solvate lignin [47]. Previously, Eyckens et al. (2016) combined KAT parameters with the modeled structure of tri- and tetraglyme with lithium bis(trifluoromethyl)sulfonimide [48]. The low β value of the system was attributed to the chelation of each glyme to the bis(trifluoromethyl)sulfonimide anion and confirmed with MD simulations. More recently, MD simulations were compared to diffusion coefficients of guanidinium-based PILs. This study found that H-bonding was the main interaction between cation and anion and specifically, anions with high proton affinities showed a clear localization of the acidic proton of the cation [49]. Therefore, the KAT multi-parameter method and MD analysis of IL solvent properties is a potent combination but has not been explored for PILs.

Herein, we combine experimentally determined KAT parameters with radial distribution functions (RDFs) from MD simulations for the expansion of our understanding of PIL properties. First, we present the KAT and $E_T(30)$ solvation parameters of twelve PILs, nine of which were previously uncharacterized, using the dyes DE4A, 4NA and RD33 (Figure 1a). The focus on PILs as a subclass of ILs in this study is due to a lack of representation in the literature. The molecular structures of these PILs are presented in Figure 1b. These PILs have been selected to investigate the effect of hydroxyl groups, alkyl chain length, cation substitution and choice of anion on PIL solvation properties. Each of these structural properties has shown considerable impact on PIL physicochemical properties observed in previous studies [50]. MD simulations were then conducted for the nine PILs in combination with a further twelve PILs (Figure 1c) where their KAT and $E_T(30)$ solvation parameters had been previously reported in the literature [38]. In total, 21 sets of RDF plots have been calculated for these PILs and compared to their experimentally determined solvation properties. This combination of experimental and computational analysis allows for a deeper understanding of the structure–property trends of alkyl- and dialkylammonium PILs liquid structure and solvation properties.

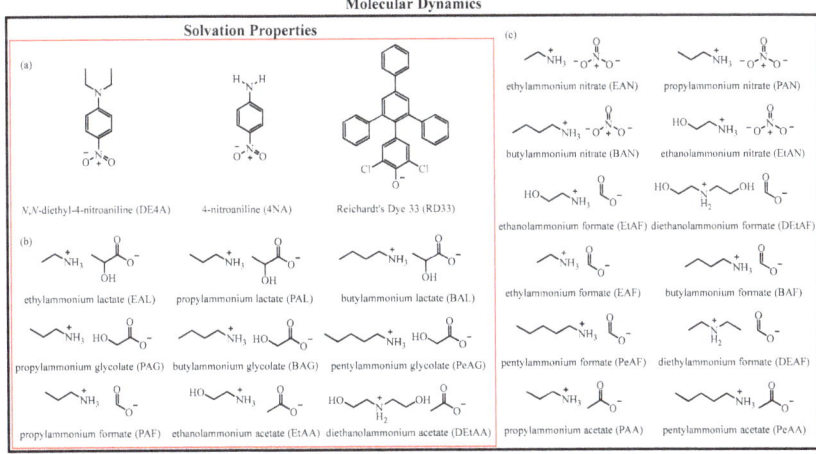

Figure 1. The chemical structures, names and abbreviations of (**a**) the solvatochromic dyes used for calculation of KAT parameters and the $E_T(30)$ scale, (**b**) previously uncharacterized PILs that have been characterized for their solvation properties via solvatochromic dye absorbance analysis in this study (in the red frame), and (**c**) PILs from the literature previously characterized for their solvation properties and have been analyzed in this study via MD simulations for their respective RDF plots.

2. Method

2.1. PIL Synthesis

All reagents for the synthesis of ILs were used as received. The precursors included diethanolamine (98.0%), ethanolamine (99.5%), ethylamine (66.0% in water), propylamine (98.0%), butylamine (99.5%), pentylamine (99.0%), acetic acid (99.0%), glycolic acid (99.0%) and lactic acid (85.0%), purchased from Sigma-Aldrich (St. Louis, MO, USA). Formic acid (98.0%) was purchased from Merck (Darmstadt, Germany). The molecular solvents were used as received and include dimethylsulfoxide (DMSO) (>99.9%) and methanol (MeOH) (99.8%) from Sigma-Aldrich, acetonitrile (ACN) (>99.9%) from Merck and MilliQ water from a Merck Synergy system with UV, Type 1 water (18.2 MΩ·cm at 25 °C ultrapure water).

Each PIL was synthesized via the dropwise addition of a Brønsted acid to a Brønsted base for stoichiometric neutralization in an ethanol bath (<5 °C) according to the literature method reported [1,50]. Each synthesis was kept below 10 °C to avoid the amide side reaction in batches of 20 g per PIL (102–220 mmol of reagent). After synthesis, each PIL was dried to remove excess water using a rotary evaporator for 24 h followed by approximately 72 h below 0.3 mbar on a freeze dryer. The water content was then measured using a combination of coulometric Karl-Fischer titration for <1 wt% water content and volumetric Karl-Fischer titration for PILs with >1 wt% water content. The final water content of each of the PILs is provided in Table S1 of the ESI.

2.2. UV/Visible Spectroscopy Analysis

The dye molecules used were N,N-diethyl-4-nitroaniline (99%, Santa Cruz Biotechnology, Dallas, TX, USA), 4-nitroaniline (99%, Fluka, Thermo Fisher Scientific, Waltham, MA, USA), and Reichardt's Dye 33 (99%, Aurora Fine Chemicals, San Diego, CA, USA). The chemical structures of these dyes are shown in Figure 1a.

Solutions of solvatochromic dyes in PILs were prepared according to our previously recorded method [38]. This was accomplished by serial dilution of each dye in methanol, and then the methanol was evaporated via a vacuum oven. Once the methanol was completely evaporated, each solvent of interest was added for a final dye concentration of 0.014 mM, 0.02 mM and 0.63 mM of DE4A, 4NA and RD33, respectively. It should be noted, that for PILs with particularly high viscosity, a combination of manual stirring, a benchtop vortex and time (approximately 12 h) was used to ensure full dispersion of the solubilized dye. The PILs with high viscosity were EAL, EAH, EtAA, EtAL, PAG, PAL, BAL, BAG, PeAG, PeAL and DEtAA.

The spectroscopic measurements were performed using a PerkinElmer EnSight Multi-mode plate reader (PerkinElmer, Waltham, MA, USA) and the spectral range for absorbance measurements was 300–600 nm with a bandwidth of 1 nm for all absorbance measurements.

2.3. KAT Formulation

The $E_T(33)$ parameter was calculated using Equation (1), where λ_{max} is the wavelength corresponding to the maximum absorbance of the solvatochromic dye RD33. $E_T(33)$ was then calculated from $E_T(30)$ using the linear relationship shown in Equation (2) [38]. For a more detailed description of the inception and application of the equations used here, consult the original papers as referenced here [29,33,34,36].

$$E_T(33)\left(\text{kcal·mol}^{-1}\right) = \frac{28591.5}{\lambda_{max}} \quad (1)$$

$$E_T(30) = 0.9442 E_T(33) - 5.7329 \quad (2)$$

The KAT parameter π^* was calculated using Equation (3). The calculation of π^* uses the maximum absorbance frequency of the dye DE4A (v_{max}), where $s = -3.182$ and $v_O = 27.52$ kK where each are constants that have been reported previously in the literature [36,38].

$$v_{max} = v_O + s\pi^* \quad (3)$$

The parameter α is calculated from a combination of the $E_T(30)$ and π^* values in Equation (4).

$$\alpha = 0.0649(E_T(30)) - 0.72\pi^* - 2.03 \quad (4)$$

Finally, parameter β was obtained using Equation (5) where λ_{DE4A} is the maximum absorbance wavelength of the dye DE4A and λ_{4NA} is the maximum absorbance wavelength of the dye 4NA.

$$\beta = \left(1.035\left(\frac{10^4}{\lambda_{DE4A}}\right) - \left(\frac{10^4}{\lambda_{4NA}}\right) + 2.64\right) \bigg/ 2.80 \quad (5)$$

The PILs of EAN, PAN and EAF, as well as the molecular solvents DMSO, ACN, water and MeOH, were used as controls to compare to previous studies. All calculated results, including wavelength of maximum absorbance, are reported in Table 1.

2.4. Molecular Dynamics Simulations

All systems comprised 500 cations and 500 anions randomly packed into a 60 × 60 × 60 Å³ unit cell using PACKMOL [51]. Initial atomic partial charges were calculated using Gaussian 16 [52], and the general amber force field (GAFF) [53] standard protocol for partial charge calculation (HF/6-31G*) was applied to all atoms using the Antechamber program of the AMBER 20 package [54]. All MD simulations were performed using the GROMACS 2019.3 software [55]. ACEPYPE [56] was used to convert the topology from AMBER format to GROMACS format. Prior to the MD simulation, a molecular mechanics minimization was performed on each structure employing the steepest descent method, with a maximum force convergence criterion of 20 kJ mol^{-1} nm^{-1}. Each simulation was equilibrated by 500 ps of annealing where the temperature was increased linearly from 298.15 K to 600 K in the first 250 ps, then reduced to 298.15 K in the final 250 ps. Production MD simulations, with atomic coordinates saved every 10 ps, were run for 100 ns in the NPT ensemble at 298.15 K and 1 bar with the Nose–Hoover thermostat and Parrinello–Raman barostat. The LINCS algorithm was applied to all bonds to allow a 2 fs timestep, and a 10 Å cutoff was applied to electrostatic and van der Waals interactions, with the particle-mesh Ewald scheme applied to long-range electrostatics. Analysis was carried out using VMD 1.9.3 [57].

3. Results

3.1. KAT Parameter Characterisation of PILs

The maximum absorbance wavelengths of DE4A, 4NA and RD(33) were obtained for 12 PILs and four molecular solvents and are provided in Table 1. From each of these the KAT parameters π* (polarizability), α (H-bond acidity) and β (H-bond basicity) in conjunction with the $E_T(30)$ scale were calculated and these values are also provided in Table 1. By using a consistent method and dyes, it is, therefore, possible to compare to our previous work.

Table 1. The maximum absorbance wavelength of the solvatochromic dyes in each of the PILs and their corresponding calculated solvation parameters.

		Max Absorbance Wavelength (nm)			π* (±0.01)	α (±0.03)	β (±0.03)	$E_T(30)$ (±0.3)
		DE4A (±1)	4NA (±1)	RD(33) (±1)				
Novel	PAF	410	385	426	0.95	1.03	0.68	57.6
	PAG	401	383	440	0.78	1.02	0.84	55.6
	BAG	397	382	444	0.70	1.04	0.90	55.1
	PeAG	395	382	447	0.67	1.04	0.95	54.7
	EAL	407	385	433	0.89	1.00	0.75	56.6
	PAL	398	383	440	0.72	1.06	0.91	55.6
	BAL	395	383	443	0.67	1.07	0.98	55.2
	EtAA	414	388	430	1.02	0.94	0.67	57.0
	DEtAA	412	385	426	0.98	1.00	0.64	57.6
Control	DMSO	413	390	529	1.00	0.19	0.74	45.30
	acetonitrile	402	368	-	0.80	-	0.43	-
	methanol	396	370	442	0.69	1.07	0.62	55.34
	water	429	377	406	1.27	1.00	0.09	60.76
	EAN	417	380	-	1.07	-	0.41	-
	PAN	414	383	411	1.02	1.13	0.55	59.95
	EAF	411	385	423	0.96	1.05	0.66	58.09

The KAT parameters (π*, α and β) are a unitless scale relative to each dye while $E_T(30)$ is measured as kcal/mol.

3.1.1. $E_T(30)$ Scale

The $E_T(30)$ scale or electrophilicity is a measure of the energy required to transfer charge through the PILs. The dominating charged contributions to this measurement are the H-bond donors (acidity), dipole–dipole and dipole/induced dipole interactions of the PILs. The $E_T(30)$ values are shown in Figure 2 where the newly characterized PILs and data from this study are presented as circles, and the previously reported values as triangles. The two molecular solvents presented here are water and methanol with relatively high and low $E_T(30)$ values, respectively. The $E_T(30)$ of the PILs studied here is approximately within the two molecular solvents and ranges from 54.7 to 57.6 kcal/mol.

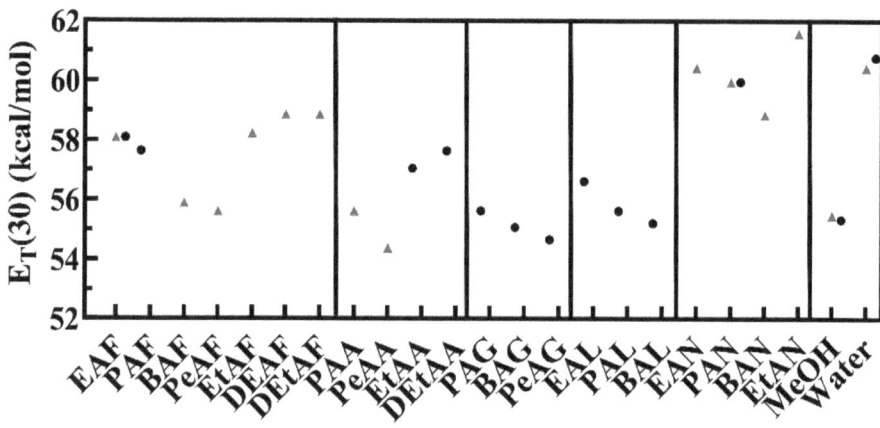

Figure 2. The calculated $E_T(30)$ values of PILs separated by anion. The experimental values determined in this study are denoted by a circle (•) and the reported values from Yalcin et al. [38] are denoted by a triangle (▲). The error bars are smaller than the symbols, and the values are provided in Table 1.

Structural changes of the PIL cation of alkyl chain length, hydroxyl groups and primary or secondary substitution all had an effect on the $E_T(30)$ values. Decreasing the alkyl chain length induced an increase in $E_T(30)$ values, as seen with the glycolate and lactate alkylammonium series, with the order of PAG > BAG > PeAG and EAL > PAL > BAL. There was a noticeable increase in $E_T(30)$ when substitution increased from primary to secondary ammonium cations, shown by DEAF > EAF, DEtAF > EtAF, and DEtAA > EtAA, likely due to the increased number of functional groups. This is consistent with the previous study where the $E_T(30)$ of DEtAF was slightly higher than EtAF [38]. However, from literature results of similar ILs, tertiary and quaternary substituted alkylammonium ILs generally have lower $E_T(30)$ values relative to primary and secondary [46]. This may be due to the increased steric hindrance caused by additional structures, reducing the number and/or strength of the interactions of the cation.

The effect of the anion can be seen in Figure 2. For the ethylammonium cation, there is a noticeable difference between anions in the order nitrate > formate > lactate ≈ glycolate ≈ acetate. Notably, nitrate-containing PILs consistently have the highest $E_T(30)$. Increasing the alkyl chain length from formate to acetate led to a decrease in $E_T(30)$. Interestingly, PILs containing the glycolate or lactate anions had similar $E_T(30)$ values, when paired with the same cation. This suggests that the effect of the hydroxyl group on a small-chained carboxylate anion is independent of its position. In contrast to the cation, there was only a minor change in the $E_T(30)$ values of the lactate and glycolate PILs as compared to the anion counterparts without a hydroxyl group, e.g., PAG ≈ PAA and PeAG only 0.3 kcal/mol greater than PeAA. We conclude that the hydroxyl groups present on the anion are less capable of acting as H-bond donors as compared to those on the cation [38].

3.1.2. π^* (Polarizability)

The polarizability of a solvent depends on multiple solvent interactions, including π–π stacking when available, along with non-specific dipole–dipole interactions and dispersive forces. Since the PILs used in this study consisted of alkylammonium and dialkylammonium cations they are incapable of π–π stacking, and hence the polarizability is characterizing the PILs dipole formation and ion–ion interactions. Values of π^* are normalized to DMSO, where DMSO = 1 to ensure consistency with the literature.

The π^* values of the PILs in this study are presented in Figure 3, along with those in the literature for PILs, DMSO, ACN and MeOH. A good consistency of π^* values with the literature was obtained for most of the ILs, though we note some variability in the measurement of π^* in the case of EAF. By order of the trend 'decreasing π^* with increasing alkyl chain length', the experimental value of PAF (0.95) would be expected to have a π^* between the literature values of EAF (0.90) and BAF (0.70). However, the determined value of EAF (0.96) here was higher than the literature value; therefore, PAF aligned with said trend in this study. Contributions from possible water content variation or stoichiometric variance in the PILs may cause shifts in π^* as seen in their ability to alter the physical properties of PILs [50,58,59]. In general, increasing the alkyl chain length of the cation led to a decrease in π^*, while the presence of a hydroxyl group on the cation led to an increase.

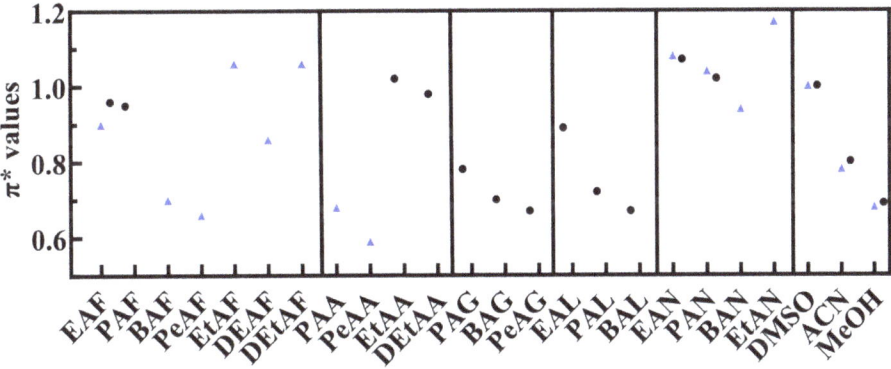

Figure 3. The calculated π^* values of PILs separated by anion. The experimental values determined in this study are denoted by a circle (•) and the reported values from Yalcin et al. [38] are denoted by a triangle (▲). The error bars are smaller than the symbols, and the values are provided in Table 1.

Similarly, the addition of a hydroxyl group on the anion of the PIL increases π^*. A comparison between the glycolate/lactate PILs to acetate or formate shows a consistent increase in π^*, e.g., PAA has a π^* value of 0.68 as compared to PAG at 0.78 and PAL at 0.72. Similarly, the hydroxyl group on the cation increases π^*, though to a larger degree as previously reported [38]. Increasing the alkyl chain length of the anion from formate to acetate anion shows a decrease in π^*, consistent with results from the literature [38].

3.1.3. α H-Bond Acidity

The α value of a solvent is defined as its ability to donate a H-bond or its H-bond acidity. It was found that α was the KAT parameter with the highest variance between those in this study and those in the literature, with an average variation of only 0.03. This variation may be due to its calculation from not one but two dyes, compounding any variation observed.

All experimental and literature α values are presented in Figure 4. With a range of 0.94 to 1.07 for α, most PILs are higher than water and all are much higher than the molecular solvents DMSO and ACN [38]. Only EtAA was found to have an α less than water, showing that PILs have similar H-bond donating properties to water. This is not surprising as some PILs have been shown to form H-bonding networks similar to that of water [60,61]. The effect of the alkyl chain on the cation shows an increase with increasing alkyl chain length for the glycolate and lactate series. Comparatively, the formate and nitrate series each only vary marginally with no definitive trend. There was an apparent decrease in α with the presence of a hydroxyl group on the cation; however, an additional hydroxyl group supplied by increased substitution of the ammonium group increases α, e.g., EtAF (0.96) < DEtAF (1.00) < EAF (1.05) and EtAA (0.94) < Delta (1.00).

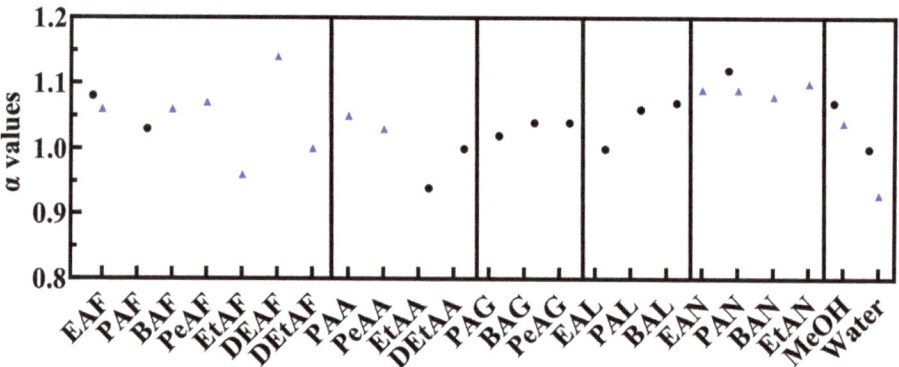

Figure 4. The calculated α values of PILs separated by anion. The experimental values determined in this study are denoted by a circle (•) and the reported values from Yalcin et al. [38] are denoted by a triangle (▲). The error bars are smaller than the symbols, and the values are provided in Table 1.

In regard to the anions, the nitrate anion led to PILs with the overall highest α of the anions collated here. As the nitrate group [NO_3^-] is unable to donate H-bonds due to its structure, this is likely due to the high ionizability of the anion, enhancing the H-bond donating ability of the partnered cation [50]. The hydroxyl group on the anion showed a variable effect on α, where the lactate series was generally higher than the glycolate. However, the formate and nitrate anions had consistently higher α than acetate, glycolate and lactate. While the α results here show higher variance as compared to the rest of the solvation properties, the trends presented are consistent with the literature and are a result of both anion and cation contributions [38].

3.1.4. β H-Bond Basicity

As α is the ability of a solvent to donate H-bonds, β is a measure of its ability to accept H-bonds or its H-bond basicity. Figure 5 presents the calculated β values of this study in conjunction with previously reported β from Yalcin et al. [38] As opposite properties, the order of PILs is naturally inverse when comparing β to α, as are their trends. Specifically, the effect of the cation appears varied, as the alkyl chain length of the cation generally increased β, while hydroxyl groups showed a decrease in β. This is not surprising as the molecular structure of alkylammonium cations lends itself to H-bond donating with the ammonium (–NH_3^+) group. Additional research into cation structures may further shed light on the relationship between cation structure and PILs β.

The presence of a hydroxyl group on the anion appears to generally increase the β of PILs as compared to its effect when on the cation. This can be seen in Figure 5 via a comparison of the glycolate and lactate anions with the ethanolammonium and diethanolammonium cations. Where β values of the hydroxylated anions range from 0.75–0.98 and the hydroxylated cations range from 0.45–0.67 [38]. This work shows the strong influence of anion structure on the β of PILs and its inverse relationship to α, most visible in the nitrate PILs, each with relatively high α, but comparatively low β. Consistent results were obtained for the PILs EAF and PAN with some variation in the EAN β value, potentially due to changes in water content. The molecular solvents across studies had high accuracy with almost identical values for DMSO, ACN and MeOH. Noticeably, the variation between carboxylate-based PIL α and β values are significantly less than that of the molecular solvents as well as the nitrate series of PILs. Therefore, it indicates the contribution of both cation and anion to these properties, specifically $-NH_3^+$ to α and $-COO^-$ to β.

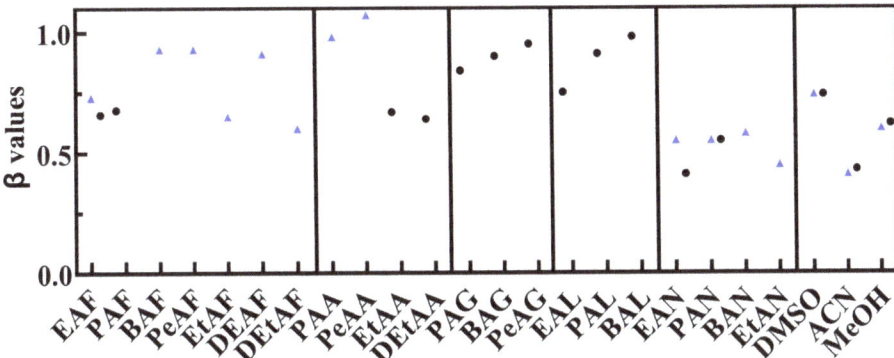

Figure 5. The calculated β values of PILs separated by anion. The experimental values determined in this study are denoted by a circle (•) and the reported values from Yalcin et al. [38] are denoted by a triangle (▲). The error bars are smaller than the symbols, and the values are provided in Table 1.

3.2. Radial Distribution Functions of PILs

MD simulations were performed on the 21 alkylammonium PILs shown in Figure 1, and from these, their RDFs were calculated. These 21 PILs include the 9 PILs from this study and the previous 12 where the $E_T(30)$ and KAT parameters have been reported. This series of PILs provides systematic structural variation to enable insight into the effect of hydroxyl groups on either the cation or the anion, increased alkyl chain length of the cation, increased branching of the ammonium cation, and variations in the anion between nitrate and carboxylates.

The RDF plots for the PILs containing a formate anion are visualized in Figures 6 and 7 to demonstrate the effect of changes in cation structure on solvation properties. Similarly, the RDF plots for PILs containing the propylammonium cation are included in Figure 8 for determining the effect of the anion structure. The remaining 11 RDF plots are presented in Figures S1–S11. This analysis has used the N^+ of the respective cation or the negatively charged O^- of the anion as the representative position of each ion in the solution and consequently their solvation environment.

3.2.1. The Effect of Cation Structure

The RDFs for the ILs with a formate anion (EAF, PAF, BAF, PeAF, EtAF and DEtAF) are presented in Figure 6, where each structure is overlayed with the atoms of interest and their labels within the RDF analysis. A solvation map of the distances between atoms according to their peaks in the RDF plots is presented in Figure 7. Though not to scale, these plots allow for a visual comparison of the distances between ions and the effects of their different structural groups.

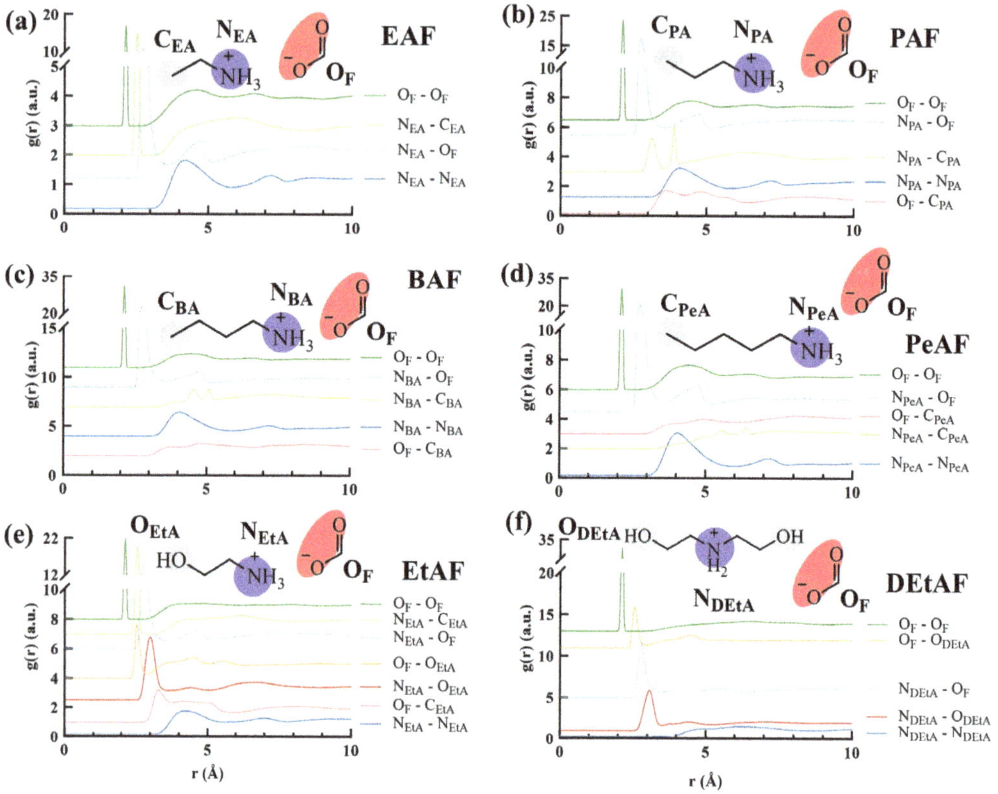

Figure 6. The radial distribution functions of the formate series of PILs (**a**) EAF, (**b**) PAF, (**c**) BAF, (**d**) PeAF, (**e**) EtAF and (**f**) DEtAF. The RDF plots are inset with the molecular structure of each PIL, where the ions are highlighted, N^+ in blue with the cation subscript e.g., N_{EA} is the N^+ atom on ethylammonium, O^- in red with the anion subscript e.g., O_F is the O^- atoms of formate. Additional functional groups are highlighted in grey. The terminal carbon in the cation alkyl chain e.g., C_{PeA} is the terminal carbon of the pentylammonium cation, and hydroxyl groups e.g., O_{EtA} is the oxygen atom of the hydroxyl group in the ethanolammonium cation.

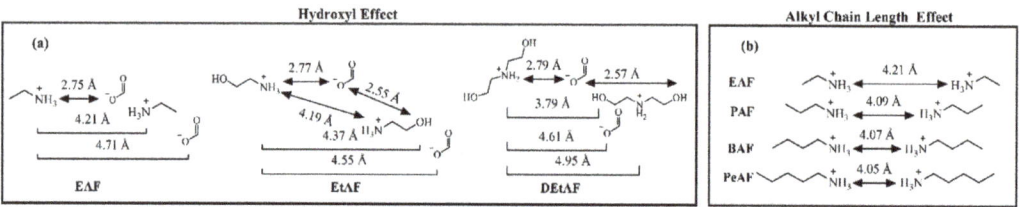

Figure 7. Solvation maps developed based on the RDF plots presented in Figure 6 (not to scale). The effect of hydroxyl groups on interatomic distances is presented in (**a**) for EAF, EtAF and DEtAF. (**b**) The effect of the alkyl chain on N–N distances is inversely proportional to its length, where increasing alkyl chain length decreases N–N distance.

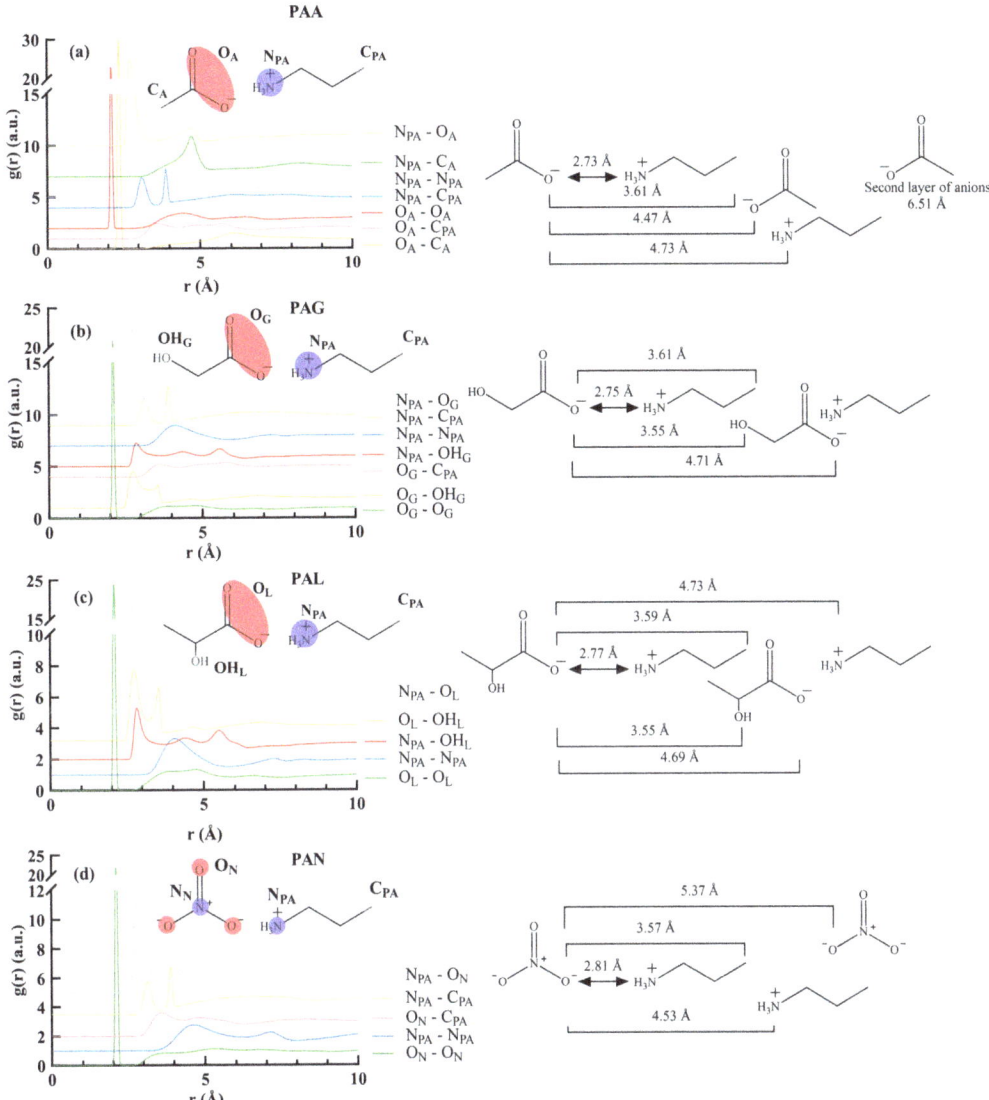

Figure 8. The RDF plots of (**a**) PAA, (**b**) PAG, (**c**) PAL and (**d**) PAN inlayed with its molecular structure with N⁺ ions highlighted in blue and O⁻ ions in red. Each RDF plot (**left**) has its corresponding solvation map adjacent (**right**) with a visualization of the peaks seen in each plot. The solvation maps are not to scale.

The distance between cation and anion as represented by the distances between N_{cation} and O_F was consistent and was 2.75 Å for the ethyl- to pentylammonium PILs (Figure 6). With the introduction of a hydroxyl group on the cation (EtAF) the cation–anion distance increases to 2.77 Å and then increases further with DEtAF to 2.79 Å (Figure 7a). This indicates the slight rearrangement of charge upon the cation with the presence of hydroxyl groups, possibly due to increased H-bond formation within the PIL which can be seen as an increase in α from EtAF to DEtAF. The distances of H-bonds between the carboxylate anion and the cations hydroxyl groups slightly increased (from 2.55 Å to 2.57 Å) as well as cation–cation (from 4.19 Å to 4.95 Å) and secondary cation–anion distances (from 4.55 Å to 4.61 Å) for EtAF as compared to DEtAF (Figure 7a). It may be possible the branching of DEtAF increased steric hindrance of the structural packing within the solvation environment of the PIL, causing weaker interactions despite the increased availability of hydroxyl groups.

The liquid nanostructure of PILs is a reported phenomenon describing the weak structuring of ions based on the segregation of hydrophilic and hydrophobic domains [62]. Experimentally, this can be observed using small angle scattering, where the peak position measured in inverse angstroms can be used to estimate the correlation distance, e.g., PeAF has a correlation distance of 15.4 Å [50]. This represents the distance between the repeating groups in solution, i.e., the distance between cations segregated by non-polar groups. Liquid nanostructure, therefore, does not describe an ion's solvation environment but can be correlated to the trends observed. Here it is observed that by increasing the alkyl chain length, a decrease in the N–N distance occurs, indicating the ammonium cations are closer (Figure 7b). This is due to the increased structuring of the system and closer packing of ions [2]. Analysis of EAF N_{EA}–C_{EA} distances shows a sharp peak at 2.55 Å for the alkyl chain of the same molecule; however, beyond that is only a broad increase in g(r) before reaching bulk EAF. This is representative of the weak liquid nanostructure of EAF which is caused by the short chain length (weak hydrophobicity) on the cation and has been observed by previous small angle X-ray scattering experiments (SAXS) [62]. Upon increasing the alkyl chain length, two peaks can be observed for N–C of each cation. The first peak is indicative of the alkyl chain of the same molecule and the second is its neighbor. However, the second peak is closer than the length of a second alkyl chain and, therefore, indicates the grouping of alkyl chains from nearby cations. Representative of the liquid nanostructure observed by SAXS. These alkyl chains form the non-polar domain of their liquid nanostructure, where increasing alkyl chain length increases the distance at which these peaks are observed. Notably, these RDF peaks for N–C are not observed for the ethanolammonium cation and is as expected from the previously reported SAXS analysis of ethanolammonium based PILs [62].

3.2.2. The Effect of Anion Structure

The RDFs for the PILs with a propylammonium cation (PAA, PAG, PAL, PAN) are presented in Figure 8, where each structure is inlaid with the atoms of interest and their labels within the RDF analysis. Through analysis of the RDF plots in Figure 8, individual solvation maps of the PILs were developed and are presented adjacent to their corresponding RDF.

The most noticeable change caused by the anion is the change in distance between anions (e.g., O_A–O_A for PAA). This difference can be grouped by the type of anion, where PAF and PAA have an O–O distance of 4.45 Å (Figure 6b) and 4.47 Å (Figure 8a), respectively. PAG and PAL have distances of 4.71 Å and 4.69 Å, respectively (Figure 8b,c) and the inorganic anion of PAN has a distance of 5.37 Å (Figure 8d). This increase in O–O distance indicates an increase in the repulsive forces of the anion. Despite increased O-O distance for the glycolate and lactate anions, the distance between the hydroxyl group and the anion carboxylate (O_G–OH_G and O_L–OH_L, respectively) does not change as seen in their solvation maps. The RDF of O–OH has two distinct peaks, only 0.8 Å apart. The first peak is representative of O–OH on the same molecule (2.75 Å), while the second peak is the O–OH of an adjacent anion molecule (3.55 Å). Therefore, it is likely that the hydroxyl group is oriented towards the anions to enable H-bonding between anions and help form a polar domain within the liquid structure of the PIL. Comparatively, the distance between the hydroxyl group in EtAF and the formate anion is 2.55 Å (Figure 7), revealing the stronger H-bond donating ability when present on the cation, as compared to the anion. This is supported by the α results obtained during the experimental KAT analysis.

There is an increase in the cation to anion distance in the order PAA (2.73 Å) < PAF (2.75 Å) = PAG (2.75 Å) < PAL (2.77 Å) < PAN (2.81 Å). The overall change in distance from PILs containing the acetate or nitrate anion is 0.8 Å, which indicates slightly weaker ionic interactions from PAA to PAN. This is likely related to the relatively high ΔpK_a of nitrate-based PILs and indicative of more strongly dissociated ions in solution [50]. Although it is noted that acid-base dissociation in neat PILs will be energetically different from that in water, so ΔpK_a based on acid-base equilibria in water should be employed with caution.

4. Discussion

Structural feature trends can provide design rules for tailored solvents and so the continued characterization of PIL properties remains an important area of study. While there has been much research into aprotic IL solvation properties, many PILs have not been characterized, despite their cheap precursors and ease of synthesis [1]. Due to the nature of PILs and ILs in general, the repeatability of characterization studies is variable as many factors can influence the PILs' physical and chemical properties [50]. Notably water content, IL ionizability and cation–anion stoichiometry influence these properties [4,59,63], and it is essential for water content to be reported. Here, the water content of all PILs used has been presented in Table S1.

The characterization of PIL solvation properties through the multi-parameter KAT method provides detailed information on the possible interactions of PILs as well as the interaction strengths. However, the solvation parameters of π^*, α, β as well as $E_T(30)$ are dependent on the chosen solvatochromic dyes, and while they should be representative of interactions with other solutes, variations are expected between studies using different dyes, and for solutes of different sizes. Importantly, while these provide important insights into solvent–solute interactions, they are constructed from specific dyes and will not be transferrable to all solutes. The inclusion of MD simulation data for PILs is advantageous for understanding their solvent structure by investigating the short-range structures of PIL via RDF plots to provide a visualization of the structure surrounding the charged moieties of either cation or anion.

Increasing the length of the cation alkyl chain was influential on all solvation properties, and typically led to an increase in β, and reduction in α, π^* and $E_T(30)$. In addition, it led to a decrease in the N–N distances in the RDF plots. This is consistent with segregation of polar and non-polar moieties of the ions into separate domains. The increase in β with increasing alkyl chain length is consistent with an increase in the non-polar solvent environment for solutes, due to the increasing proportion of non-polar domains, thus, reducing the number of polarizable interactions with solutes.

The presence of hydroxyl groups in PILs has been shown to be one of the most important structural moieties in relation to the effect of a chemical group on their physicochemical properties [50]. This is also true for their solvation properties [38] and their ability to act as successful solvents for biological molecules [64]. The hydroxyl group on the cation significantly increases α, π^* and $E_T(30)$ values while decreasing the β value. This solvation change is seen as an increase in the cation–anion distance from the RDF plots, where the structure change allows for weaker ionic interactions and greater polarizability. For DEtAF where there were two hydroxyl groups present, there was an even larger increase in cation–anion distances as well as polarizability. This increase in PIL polarizability is noticeable as an increase in cohesive forces within the PIL, while simultaneously altering the physicochemical properties of the PIL by increasing density, viscosity, surface tension, glass transition temperature and refractive index [50].

The influence of the anions on the solvation properties of the PILs had a different impact compared to the cations, for the ions used in this study. The anion impacted the $E_T(30)$ values, with decreasing $E_T(30)$ as nitrate > formate > acetate, which was consistent with our previous study [38]. We see this order again in the specific distance from the MD simulations between the cation and anion (N_{PA}–O) of PAN > PAF > PAA. This follows the order of PIL ionicity where the ionicity can be estimated by the ΔpK_a of the acid/base pair, with nitric acid and propylamine having the greatest ΔpK_a and acetic acid and propylamine having the lowest. Overall PILs containing the nitrate anion led to notably higher π^*

and α with lower β, than those with formate or acetate anions. This is highly important for the application of PILs towards biomolecules, where the solvent β has been correlated with IL hydrophobicity [65] and IL effect on protein stability [66], and previous studies have shown that the low β of nitrate anions and short cation alkyl chains preserve native protein structures at low PIL concentrations [43,64,67]. In addition, the presence of the hydroxyl group on carboxylate anions showed variability, with comparable π^*, α and β relative to the formate and acetate PILs, although it increased β as compared to cation-based hydroxyl groups, revealing β to be predominantly governed by anion structure.

5. Conclusions

The solvation properties of nine PILs were characterized using the KAT multi-parameter method to obtain π^*, α, and β, along with $E_T(30)$ values. These were combined with literature data of 12 related PILs to develop structure–property relationships. RDF plots from MD simulations were obtained for all 21 PILs to gain insight into the atomic distances and solvation environments of individual ions. Increasing the cation alkyl chain length increased β, while decreasing α, π^* and $E_T(30)$ and was observed to decrease cation–cation (N–N) distances in the RDF plots. Hydroxyl groups on the cation generally increased α, π^* and $E_T(30)$ and led to an increase in cation–anion (N–O) distance in their respective RDF plots. Further analysis correlated anions with their polarizability in the order of nitrate > formate > acetate, and the increase in cation–anion distances calculated by RDF plots followed the same order of anions. Overall, nitrate was the most influential and polarizable anion for the PIL solvation properties, followed by glycolate/lactate and then formate/acetate anions. This order correlated to anion–anion distances from the RDF plots, where nitrate had the greatest distance between anions. The research herein is an extension of the current literature on PIL solvent properties, and to the best of our knowledge, is the first combination of RDF analysis and solvatochromic dyes applied to the study of PIL structure–property trends. Further work in this area can be used to investigate the poorly understood effect of non-stoichiometry in PILs. Additionally, reported solvation properties of PIL mixtures in the literature are lacking but would be highly beneficial to applications requiring select biomolecular solvation.

Supplementary Materials: The following supporting information can be downloaded at: https://www.mdpi.com/article/10.3390/liquids4010014/s1, Table S1: Water content of each PIL; Figures S1–S11: RDF plots of the PILs.

Author Contributions: S.J.B.: Conceptualization, Methodology, Formal Analysis, Investigation, Writing—Original Draft, Visualization. A.J.C.: Investigation, Methodology, Writing—Review and Editing. C.J.D.: Resources, Supervision, Writing—Review and Editing. Q.H.: Conceptualization, Formal Analysis, Investigation, Supervision, Writing—Review and Editing, Visualization. T.L.G.: Conceptualization, Resources, Supervision, Writing—Review and Editing, Project Administration. All authors have read and agreed to the published version of the manuscript.

Funding: The authors acknowledge funding support from the Australian Research Council under its Discovery grant Scheme for this work (DP230101712).

Data Availability Statement: Data are contained within the article or Supplementary Materials.

Acknowledgments: S.J.B. acknowledges the financial support of an RMIT PhD Stipend, and a Postgraduate Research Award (PGRA) granted by the Australian Institute of Nuclear Science and Engineering (AINSE). This work was supported by computational resources provided by the Australian Government through the National Computational Infrastructure (NCI) and Pawsey Supercomputing Research Centre under the National Computational Merit Allocation Scheme (project kl59 and resource grant ne25). The authors would also like to recognize the fundamental contributions made to this field by Professor Christian Reichardt. Wherein this work would not be possible without his pioneering research using solvatochromic dyes to characterize solvation properties.

Conflicts of Interest: The authors declare no conflict of interest.

References

1. Greaves, T.L.; Drummond, C.J. Protic Ionic Liquids: Properties and Applications. *Chem. Rev.* **2008**, *108*, 206–237. [CrossRef] [PubMed]
2. Greaves, T.L.; Drummond, C.J. Solvent nanostructure, the solvophobic effect and amphiphile self-assembly in ionic liquids. *Chem. Soc. Rev.* **2013**, *42*, 1096–1120. [CrossRef] [PubMed]
3. Hayes, R.; Warr, G.G.; Atkin, R. Structure and Nanostructure in Ionic Liquids. *Chem. Rev.* **2015**, *115*, 6357–6426. [CrossRef] [PubMed]
4. MacFarlane, D.R.; Chong, A.L.; Forsyth, M.; Kar, M.; Vijayaraghavan, R.; Somers, A.; Pringle, J.M. New dimensions in salt–solvent mixtures: A 4th evolution of ionic liquids. *Faraday Discuss.* **2018**, *206*, 9–28. [CrossRef] [PubMed]
5. Welton, T. Room-temperature ionic liquids. Solvents for synthesis and catalysis. *Chem. Rev.* **1999**, *99*, 2071–2084. [CrossRef] [PubMed]
6. Forsyth, S.A.; Pringle, J.M.; MacFarlane, D.R. Ionic Liquids; An Overview. *Aust. J. Chem.* **2004**, *57*, 113–119. [CrossRef]
7. Ghandi, K. A Review of Ionic Liquids, Their Limits and Applications. *Green Sustain. Chem.* **2014**, *4*, 44–53. [CrossRef]
8. Greaves, T.L.; Drummond, C.J. Protic Ionic Liquids: Evolving Structure–Property Relationships and Expanding Applications. *Chem. Rev.* **2015**, *115*, 11379–11448. [CrossRef]
9. Ohno, H.; Yoshizawa-Fujita, M.; Kohno, Y. Functional Design of Ionic Liquids: Unprecedented Liquids that Contribute to Energy Technology, Bioscience, and Materials Sciences. *Bull. Chem. Soc. Jpn.* **2019**, *92*, 852–868. [CrossRef]
10. Greaves, T.L.; Schaffarczyk McHale, K.S.; Burkart-Radke, R.F.; Harper, J.B.; Le, T.C. Machine learning approaches to understand and predict rate constants for organic processes in mixtures containing ionic liquids. *Phys. Chem. Chem. Phys.* **2021**, *23*, 2742–2752. [CrossRef]
11. Hallett, J.P.; Welton, T. Room-Temperature Ionic Liquids: Solvents for Synthesis and Catalysis. 2. *Chem. Rev.* **2011**, *111*, 3508–3576. [CrossRef] [PubMed]
12. Sedov, I.A.; Magsumov, T.I.; Salikov, T.M.; Solomonov, B.N. Solvation of apolar compounds in protic ionic liquids: The non-synergistic effect of electrostatic interactions and hydrogen bonds. *Phys. Chem. Chem. Phys.* **2017**, *19*, 25352–25359. [CrossRef] [PubMed]
13. Perrin, C.L.; Agranat, I.; Bagno, A.; Braslavsky, S.E.; Fernandes, P.A.; Gal, J.-F.; Lloyd-Jones, G.C.; Mayr, H.; Murdoch, J.R.; Nudelman, N.S.; et al. Glossary of terms used in physical organic chemistry (IUPAC Recommendations 2021). *Pure Appl. Chem.* **2022**, *94*, 353–534. [CrossRef]
14. Guan, W.; Chang, N.; Yang, L.; Bu, X.; Wei, J.; Liu, Q. Determination and Prediction for the Polarity of Ionic Liquids. *J. Chem. Eng. Data* **2017**, *62*, 2610–2616. [CrossRef]
15. Deye, J.F.; Berger, T.A.; Anderson, A.G. Nile Red as a solvatochromic dye for measuring solvent strength in normal liquids and mixtures of normal liquids with supercritical and near critical fluids. *Anal. Chem.* **1990**, *62*, 615–622. [CrossRef]
16. Webb, M.A.; Morris, B.C.; Edwards, W.D.; Blumenfeld, A.; Zhao, X.; McHale, J.L. Thermosolvatochromism of Phenol Blue in Polar and Nonpolar Solvents. *J. Phys. Chem. A* **2004**, *108*, 1515–1523. [CrossRef]
17. Cave, R.J.; Castner, E.W. Time-Dependent Density Functional Theory Investigation of the Ground and Excited States of Coumarins 102, 152, 153, and 343. *J. Phys. Chem. A* **2002**, *106*, 12117–12123. [CrossRef]
18. Ando, Y.; Homma, Y.; Hiruta, Y.; Citterio, D.; Suzuki, K. Structural characteristics and optical properties of a series of solvatochromic fluorescent dyes displaying long-wavelength emission. *Dye. Pigment.* **2009**, *83*, 198–206. [CrossRef]
19. Jin, H.; Baker, G.A.; Arzhantsev, S.; Dong, J.; Maroncelli, M. Solvation and Rotational Dynamics of Coumarin 153 in Ionic Liquids: Comparisons to Conventional Solvents. *J. Phys. Chem. B* **2007**, *111*, 7291–7302. [CrossRef]
20. Shetty, P.H.; Youngberg, P.J.; Kersten, B.R.; Poole, C.F. Solvent properties of liquid organic salts used as mobile phases in microcolumn reversed-phase liquid chromatography. *J. Chromatogr. A* **1987**, *411*, 61–79. [CrossRef]
21. Poole, S.K.; Shetty, P.H.; Poole, C.F. Chromatographic and spectroscopic studies of the solvent properties of a new series of room-temperature liquid tetraalkylammonium sulfonates. *Anal. Chim. Acta* **1989**, *218*, 241–264. [CrossRef]
22. Anderson, J.L.; Ding, J.; Welton, T.; Armstrong, D.W. Characterizing Ionic Liquids on the Basis of Multiple Solvation Interactions. *J. Am. Chem. Soc.* **2002**, *124*, 14247–14254. [CrossRef] [PubMed]
23. Bini, R.; Chiappe, C.; Mestre, V.L.; Pomelli, C.S.; Welton, T. A rationalization of the solvent effect on the Diels–Alder reaction in ionic liquids using multiparameter linear solvation energy relationships. *Org. Biomol. Chem.* **2008**, *6*, 2522–2529. [CrossRef] [PubMed]
24. Jeličić, A.; García, N.; Löhmannsröben, H.-G.; Beuermann, S. Prediction of the Ionic Liquid Influence on Propagation Rate Coefficients in Methyl Methacrylate Radical Polymerizations Based on Kamlet–Taft Solvatochromic Parameters. *Macromolecules* **2009**, *42*, 8801–8808. [CrossRef]
25. Crowhurst, L.; Falcone, R.; Lancaster, N.L.; Llopis-Mestre, V.; Welton, T. Using Kamlet–Taft Solvent Descriptors To Explain the Reactivity of Anionic Nucleophiles in Ionic Liquids. *J. Org. Chem.* **2006**, *71*, 8847–8853. [CrossRef] [PubMed]
26. Ranieri, G.; Hallett, J.P.; Welton, T. Nucleophilic Reactions at Cationic Centers in Ionic Liquids and Molecular Solvents. *Ind. Eng. Chem. Res.* **2008**, *47*, 638–644. [CrossRef]
27. Hawker, R.R.; Haines, R.S.; Harper, J.B. The effect of varying the anion of an ionic liquid on the solvent effects on a nucleophilic aromatic substitution reaction. *Org. Biomol. Chem.* **2018**, *16*, 3453–3463. [CrossRef]

28. Butler, B.J.; Harper, J.B. The effect of the structure of the anion of an ionic liquid on the rate of reaction at a phosphorus centre. *J. Phys. Org. Chem.* **2019**, *32*, e3819. [CrossRef]
29. Reichardt, C. Polarity of ionic liquids determined empirically by means of solvatochromic pyridinium N-phenolate betaine dyes. *Green Chem.* **2005**, *7*, 339–351. [CrossRef]
30. Chiappe, C.; Pomelli, C.S.; Rajamani, S. Influence of Structural Variations in Cationic and Anionic Moieties on the Polarity of Ionic Liquids. *J. Phys. Chem. B* **2011**, *115*, 9653–9661. [CrossRef]
31. Ab Rani, M.A.; Brant, A.; Crowhurst, L.; Dolan, A.; Lui, M.; Hassan, N.H.; Hallett, J.P.; Hunt, P.A.; Niedermeyer, H.; Perez-Arlandis, J.M.; et al. Understanding the polarity of ionic liquids. *Phys. Chem. Chem. Phys.* **2011**, *13*, 16831–16840. [CrossRef] [PubMed]
32. Spange, S.; Lungwitz, R.; Schade, A. Correlation of molecular structure and polarity of ionic liquids. *J. Mol. Liq.* **2014**, *192*, 137–143. [CrossRef]
33. Kamlet, M.J.; Taft, R.W. The solvatochromic comparison method. I. The β-scale of solvent hydrogen-bond acceptor (HBA) basicities. *J. Am. Chem. Soc.* **1976**, *98*, 377–383. [CrossRef]
34. Taft, R.W.; Kamlet, M.J. The solvatochromic comparison method. 2. The α-scale of solvent hydrogen-bond donor (HBD) acidities. *J. Am. Chem. Soc.* **1976**, *98*, 2886–2894. [CrossRef]
35. Yokoyama, T.; Taft, R.W.; Kamlet, M.J. The solvatochromic comparison method. 3. Hydrogen bonding by some 2-nitroaniline derivatives. *J. Am. Chem. Soc.* **1976**, *98*, 3233–3237. [CrossRef]
36. Kamlet, M.J.; Abboud, J.L.; Taft, R.W. The solvatochromic comparison method. 6. The π^* scale of solvent polarities. *J. Am. Chem. Soc.* **1977**, *99*, 6027–6038. [CrossRef]
37. Minesinger, R.R.; Jones, M.E.; Taft, R.W.; Kamlet, M.J. The solvatochromic comparison method. 5. Spectral effects and relative strengths of the first and second hydrogen bonds by 4-nitroaniline to hydrogen bond acceptor solvents. *J. Org. Chem.* **1977**, *42*, 1929–1934. [CrossRef]
38. Yalcin, D.; Drummond, C.J.; Greaves, T.L. Solvation properties of protic ionic liquids and molecular solvents. *Phys. Chem. Chem. Phys.* **2020**, *22*, 114–128. [CrossRef]
39. Reichardt, C. Pyridinium N-phenolate betaine dyes as empirical indicators of solvent polarity: Some new findings. *Pure Appl. Chem.* **2004**, *76*, 1903–1919. [CrossRef]
40. Pardo, R.; Zayat, M.; Levy, D. ET(33) dye as a tool for polarity determinations: Application to porous hybrid silica thin-films. *J. Photochem. Photobiol. A Chem.* **2010**, *210*, 17–22. [CrossRef]
41. Han, Q.; El Mohamad, M.; Brown, S.; Zhai, J.; Rosado, C.; Shen, Y.; Blanch, E.W.; Drummond, C.J.; Greaves, T.L. Small angle X-ray scattering investigation of ionic liquid effect on the aggregation behavior of globular proteins. *J. Colloid Interface Sci.* **2023**, *648*, 376–388. [CrossRef] [PubMed]
42. Han, Q.; Binns, J.; Zhai, J.; Guo, X.; Ryan, T.M.; Drummond, C.J.; Greaves, T.L. Insights on lysozyme aggregation in protic ionic liquid solvents by using small angle X-ray scattering and high throughput screening. *J. Mol. Liq.* **2022**, *345*, 117816. [CrossRef]
43. Han, Q.; Brown, S.J.; Drummond, C.J.; Greaves, T.L. Protein aggregation and crystallization with ionic liquids: Insights into the influence of solvent properties. *J. Colloid Interface Sci.* **2022**, *608*, 1173–1190. [CrossRef]
44. Han, Q.; Smith, K.M.; Darmanin, C.; Ryan, T.M.; Drummond, C.J.; Greaves, T.L. Lysozyme conformational changes with ionic liquids: Spectroscopic, small angle X-ray scattering and crystallographic study. *J. Colloid Interface Sci.* **2021**, *585*, 433–443. [CrossRef] [PubMed]
45. Han, Q.; Ryan, T.M.; Rosado, C.J.; Drummond, C.J.; Greaves, T.L. Effect of ionic liquids on the fluorescence properties and aggregation of superfolder green fluorescence protein. *J. Colloid Interface Sci.* **2021**, *591*, 96–105. [CrossRef]
46. Poole, C.F. Chromatographic and spectroscopic methods for the determination of solvent properties of room temperature ionic liquids. *J. Chromatogr. A* **2004**, *1037*, 49–82. [CrossRef] [PubMed]
47. Miao, S.; Imberti, S.; Atkin, R.; Warr, G. Nanostructure in amino acid ionic molecular hybrid solvents. *J. Mol. Liq.* **2022**, *351*, 118599. [CrossRef]
48. Eyckens, D.J.; Demir, B.; Walsh, T.R.; Welton, T.; Henderson, L.C. Determination of Kamlet–Taft parameters for selected solvate ionic liquids. *Phys. Chem. Chem. Phys.* **2016**, *18*, 13153–13157. [CrossRef]
49. Rauber, D.; Philippi, F.; Becker, J.; Zapp, J.; Morgenstern, B.; Kuttich, B.; Kraus, T.; Hempelmann, R.; Hunt, P.; Welton, T.; et al. Anion and ether group influence in protic guanidinium ionic liquids. *Phys. Chem. Chem. Phys.* **2023**, *25*, 6436–6453. [CrossRef]
50. Brown, S.J.; Yalcin, D.; Pandiancherri, S.; Le, T.C.; Orhan, I.; Hearn, K.; Han, Q.; Drummond, C.J.; Greaves, T.L. Characterising a protic ionic liquid library with applied machine learning algorithms. *J. Mol. Liq.* **2022**, *367*, 120453. [CrossRef]
51. Martínez, L.; Andrade, R.; Birgin, E.G.; Martínez, J.M. PACKMOL: A package for building initial configurations for molecular dynamics simulations. *J. Comput. Chem.* **2009**, *30*, 2157–2164. [CrossRef] [PubMed]
52. Frisch, M.J.; Trucks, G.W.; Schlegel, H.B.; Scuseria, G.E.; Robb, M.A.; Cheeseman, J.R.; Scalmani, G.; Barone, V.; Petersson, G.A.; Nakatsuji, H.; et al. *Gaussian 16 Rev. C.01*; Gaussian Inc.: Wallingford, CT, USA, 2016.
53. Wang, J.; Wolf, R.M.; Caldwell, J.W.; Kollman, P.A.; Case, D.A. Development and testing of a general amber force field. *J. Comput. Chem.* **2004**, *25*, 1157–1174. [CrossRef] [PubMed]
54. Case, D.A.; Aktulga, H.M.; Belfon, K.; Ben-Shalom, I.Y.; Brozell, S.R.; Cerutti, D.S.; Cheatham, I.T.E.; Cisneros, G.A.; Cruzeiro, V.W.D.; Darden, T.A.; et al. *Amber 2021*; University of California: San Francisco, CA, USA, 2021.

55. Abraham, M.J.; Murtola, T.; Schulz, R.; Páll, S.; Smith, J.C.; Hess, B.; Lindahl, E. GROMACS: High performance molecular simulations through multi-level parallelism from laptops to supercomputers. *SoftwareX* **2015**, *1*, 19–25. [CrossRef]
56. Sousa da Silva, A.W.; Vranken, W.F. ACPYPE—AnteChamber PYthon Parser interfacE. *BMC Res. Notes* **2012**, *5*, 367. [CrossRef] [PubMed]
57. Humphrey, W.; Dalke, A.; Schulten, K. VMD: Visual molecular dynamics. *J. Mol. Graph.* **1996**, *14*, 33–38. [CrossRef]
58. Yalcin, D.; Christofferson, A.J.; Drummond, C.J.; Greaves, T.L. Solvation properties of protic ionic liquid–molecular solvent mixtures. *Phys. Chem. Chem. Phys.* **2020**, *22*, 10995–11011. [CrossRef] [PubMed]
59. Yalcin, D.; Drummond, C.J.; Greaves, T.L. High throughput approach to investigating ternary solvents of aqueous non-stoichiometric protic ionic liquids. *Phys. Chem. Chem. Phys.* **2019**, *21*, 6810–6827. [CrossRef] [PubMed]
60. Fumino, K.; Wulf, A.; Ludwig, R. Hydrogen Bonding in Protic Ionic Liquids: Reminiscent of Water. *Angew. Chem. Int. Ed.* **2009**, *48*, 3184–3186. [CrossRef] [PubMed]
61. Evans, D.F.; Chen, S.-H.; Schriver, G.W.; Arnett, E.M. Thermodynamics of solution of nonpolar gases in a fused salt. Hydrophobic bonding behavior in a nonaqueous system. *J. Am. Chem. Soc.* **1981**, *103*, 481–482. [CrossRef]
62. Greaves, T.L.; Kennedy, D.F.; Mudie, S.T.; Drummond, C.J. Diversity Observed in the Nanostructure of Protic Ionic Liquids. *J. Phys. Chem. B* **2010**, *114*, 10022–10031. [CrossRef]
63. MacFarlane, D.R.; Forsyth, M.; Izgorodina, E.I.; Abbott, A.P.; Annat, G.; Fraser, K. On the concept of ionicity in ionic liquids. *Phys. Chem. Chem. Phys.* **2009**, *11*, 4962–4967. [CrossRef] [PubMed]
64. Brown, S.J.; Ryan, T.M.; Drummond, C.J.; Greaves, T.L.; Han, Q. Lysozyme aggregation and unfolding in ionic liquid solvents: Insights from small angle X-ray scattering and high throughput screening. *J. Colloid Interface Sci.* **2024**, *655*, 133–144. [CrossRef] [PubMed]
65. Han, Q.; Wang, X.; Bynre, N. Utilizing Water Activity as a Simple Measure to Understand Hydrophobicity in Ionic Liquids. *Front. Chem.* **2019**, *7*, 112. [CrossRef] [PubMed]
66. Han, Q.; Wang, X.; Byrne, N. Understanding the Influence of Key Ionic Liquid Properties on the Hydrolytic Activity of Thermomyces lanuginosus Lipase. *ChemCatChem* **2016**, *8*, 1551–1556. [CrossRef]
67. Han, Q.; Su, Y.; Smith, K.M.; Binns, J.; Drummond, C.J.; Darmanin, C.; Greaves, T.L. Probing ion-binding at a protein interface: Modulation of protein properties by ionic liquids. *J. Colloid Interface Sci.* **2023**, *650*, 1393–1405. [CrossRef]

Disclaimer/Publisher's Note: The statements, opinions and data contained in all publications are solely those of the individual author(s) and contributor(s) and not of MDPI and/or the editor(s). MDPI and/or the editor(s) disclaim responsibility for any injury to people or property resulting from any ideas, methods, instructions or products referred to in the content.

Communication

The Photophysics of Diphenyl Polyenes Analyzed by Their Solvatochromism

Javier Catalán [1],* and Henning Hopf [2]

[1] Departamento de Química Física Aplicada, Universidad Autónoma de Madrid, 28049 Madrid, Spain
[2] Institut für Organische Chemie, Technische Universitat Braunschweig, Hagenring 30, D-38106 Braunschweig, Germany; h.hopf@tu-braunschweig.de
* Correspondence: javier.catalan@uam.es

Abstract: The solvent-dependent intensity changes in the first UV/Vis absorption band of the three polyenes DPH, DPHb, and ttbP3 dissolved in a hydrocarbon solvent with temperature allow for the conclusion that, at temperatures above 233 K, the two phenyl groups of DPH are rotated out-of-plane to yield a non-coplanar molecular structure. This leads to the conclusion that DPH becomes increasingly less coplanar as the temperature rises above 233 K. When the phenyl groups rotate out-of-plane, the polarizability decreases, and the energy of the first electronic transition increases by an extra value. Therefore, below 233 K, the correlation lines between the absorption energy of the 0–0 component of the UV/Vis absorption band and the solvent polarizability, as measured by the *SP* values, show bilinear behavior. The unexpected behavior shown by DPH dissolved in tetrachloro- and dichloromethane is discussed. We dedicate this research as a tribute to the very important contribution to the solvent effect made by Prof. Christian Reichardt and also to his generous and altruistic scientific help that he has always shown.

Keywords: solvent effect on the spectra of organic compounds; solvatochromism of diphenylpolyenes; on the polarizability of diphenylpolyenes in Cl_4C and in Cl_2CH_2

Citation: Catalán, J.; Hopf, H. The Photophysics of Diphenyl Polyenes Analyzed by Their Solvatochromism. *Liquids* **2024**, *4*, 278–287. https://doi.org/10.3390/liquids4010013

Academic Editors: William E. Acree, Jr., Franco Cataldo and Enrico Bodo

Received: 2 November 2023
Revised: 20 February 2024
Accepted: 4 March 2024
Published: 13 March 2024

Copyright: © 2024 by the authors. Licensee MDPI, Basel, Switzerland. This article is an open access article distributed under the terms and conditions of the Creative Commons Attribution (CC BY) license (https://creativecommons.org/licenses/by/4.0/).

1. Introduction

Polyene compounds are part of the chromophores involved in important biochemical processes such as vision, the coloration of relevant fruits and vegetables, the generation of vitamins, and protection mechanisms against sunlight. In these biochemical processes, the photophysics of these compounds plays a very important role, and therefore, it has been imperative to know their photophysical functioning.

Thus, Hausser et al. [1,2] synthesized in 1935 seven diphenylpolyenes, $[C_6H_5\text{-}(CH=CH)_n\text{-}C_6H_5$ with n = 1–7]. From their photophysical studies, carried out at 293 and 77 K, they established for the first time a series of guidelines on the photophysical properties of these compounds [2,3]. From these studies, the authors were able to conclude that the bathochromic shift of the first UV/Vis absorption band of these vinylogous diphenyl polyenes by the successive addition of a vinylene group is significantly large. They also found that this band shift does not depend on (a) the solvent's dipole moment or (b) the solvent's relative permittivity, but (c) it does depend on the solvent's refractive index. In summary, these authors established in 1935 that the vinylogous batahochromism of diphenylpolyenes depends only on the solvent's polarizability.

Hausser et al. also confirmed that (a) the solvatochromism of the emission band of these diphenylpolyenes is almost negligible, as compared to that shown by their UV/Vis absorption, and (b) that these compounds show a significant Stokes band shift, which increases with increasing polyene chain length. These findings supported the hypothesis, which was later widespread, that the electronic states involved in the UV/vis absorption and the emission process in these compounds must be of a different nature.

On the other hand, it is important to point out that from theoretical calculations [4–6] obtained at a temperature of 0 K, it was assumed that these diphenylpolyenes have a coplanar molecular structure in their fundamental electronic state. The evidence obtained by X-ray diffraction [7–9] allowed for the conclusion that all these compounds have flat polyene chains.

However, Drente et al. [9,10] already indicated that, although in a crystalline state these polyenes adopt a coplanar conformation with their phenyl rings, in the case of 1,8-diphenyl-1,3,5,7-octatetraene (DPO), the phenyl groups appear rotated by 7.5° and 5.4° at 293 K and 173 K, respectively [10]. On the other hand, Hall et al. [11] found that for 1,6-diphenyl-1,3,5-hexatriene (DPH) in crystal form at 293 K, its phenyl groups are rotated by only 1.9° with respect to its coplanar polyene chain.

One should keep in mind that in a crystalline state, the diphenylpolyene molecules are arranged in a way that forces the molecules to adopt a coplanar structure, particularly at very low temperatures. However, it cannot be ruled out that this coplanar arrangement can change in solutions at higher temperatures.

There is also evidence from the ^{13}C NMR spectra of diphenylpolyenes, measured in a thermotropic liquid crystal ZLI-1167 between 354 and 307 K, showing a loss of planarity due to the rotation of the two phenyl groups, as reported by Benzi et al. [12], and the spectra of DPH, dissolved in 2-methyl-tetrahydrofuran and measured at 323, 293, 268, 238, 208, and 188 K, as reported by Catalán et al. [13].

Let us now focus on the solvatochromism of diphenylpolyenes, which, according to Hausser et al. [1,2], depends solely on the polarizability of the dissolving medium. It is already accepted that the polarizability of α-ω-diphenylpolyenes increases with the elongation of their flat molecular structure. In addition, Bramley and Fèbre [14] indicate that this increase in polarizability is larger than expected by applying the additivity rule to these compounds.

It should be noted that, upon raising the temperature and the connected out-of-plane rotation of the two phenyl groups of α-ω-diphenylpolyene, a decrease in the size of the planar area of the compound would inevitably lead to a significant reduction in its sensitivity to the solvent's polarizability.

At low temperatures, the electronic transition energy of the first UV/Vis absorption band should follow a linear correlation with the corresponding values of the solvent's polarizability. But this linear behavior should change with the increasing rotation of the two phenyl groups, causing a loss of planarity and of the polarizability of the diphenylpolyene.

It is well established by the Mulliken–Rieke rule [15] that the intensity of an electronic transition is independent of temperature, and if the intensity is not kept constant with temperature, it must be concluded that the compound under consideration is changing its molecular structure. We already showed [13,16] that by raising the solution temperature, there comes a point at which the intensity of the first UV/Vis absorption band of the diphenylpolyenes decreases.

The aforementioned behavior can be confirmed using a solution of the corresponding DPP in a solvent that allows us to measure the absorption band of the compound over a wide range of temperatures. For this, a suitable solvent is a 1/1 mixture by volume of decaline and methylcyclohexane, which we will call DEMCH. We have shown that in this solvent, we can measure the absorption of these compounds even at 77 K.

It seems logical to expect that if a diphenylpolyene, above a certain temperature, rotates its phenyl groups, not only will its polarizability decrease, but if not, it will also cease to be nonpolar. Consequently, at high temperatures, its solvatochromism will decrease with the polarizability of the solvent and increase with the dipolarity of the solvent. This would allow us to understand why Ponder and Mathies [17] found that diphenylpolyenes had a non-zero dipole moment in 1,4-dioxane at a temperature of 20.5 °C.

In this work, we will also show the thermochromic behavior of diphenylpolyenes, which are considered non-acidic and non-basic compounds, when dissolving them in a solvent such as Cl_4C, which only shows polarizability (SP) (SP = 0.768, SdP = 0.00) [18], and in Cl_2CH_2, which shows both polarizability (SP) and dipolarity (SP) (SP = 761 and SdP = 0.769) [19].

Consequently, the peaks of the first absorption band of the polyene compound dissolved in Cl_4C should show linear behavior at low temperatures, and when reaching temperatures at which the diphenylpolyene compound rotates its phenyl groups, it should follow another type of linear behavior as the polarizability of the diphenylpolyene compound decreases. That is to say, in the wide range of temperatures studied, the energy of one of its peaks in the first absorption band should show bilinear behavior with temperature [20].

However, when dissolved in Cl_2CH_2, the behavior will be different since, while the polarizability contribution will decrease as the compound rotates its phenyl groups, the contribution due to the dipolarity of the compound will increase, possibly compensating for the previous effect caused by the polarizability of the solvent. Consequently, in these compounds, the energy of their peaks will tend to remain linear with temperature.

All this will be analyzed from the corresponding spectroscopic data obtained with DPH, ttbP3, and DPHb in the solvents indicated above. It is interesting to note that a thorough and critical review of the knowledge of the photophysics of diphenylpolyenes has been recently published [21].

In Scheme 1, the molecular structures of the polyenes used in the present work and available in our laboratory are shown: all-trans-1,6-diphenyl-1,3,5-hexatriene (DPH), all-trans-diindanylidene-2-butene (DPHb), and 3,8-di-tert-butyl-2,2,9,9-tetramethyl-1,3,5,7-decatriene (ttbP3).

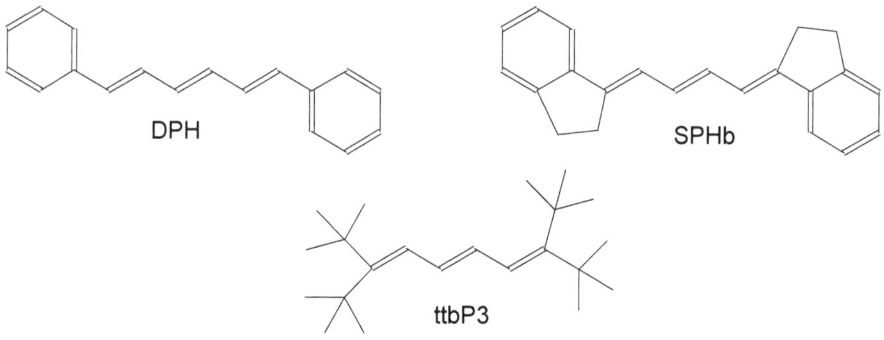

Scheme 1. Molecular structures of the polyenes utilized in this work.

2. Experimental Section

Spectrophotometric-grade methylcyclohexane and decaline from Sigma-Aldrich (St. Louis, MO, USA) are of 99 and 98% purity. Di- and tetrachloromethane are Uvasol for Spectroscopy solvents from Merck Uvasol (Rahway, NJ, USA) with purities larger than 99.9%.

Solution temperatures were controlled with an Oxford DN1704 cryostat that was purged with dried nitrogen (99.99% pure) and equipped with an ITC4 controller interfaced to the spectrometer. All UV/Vis absorption spectra were recorded on a Cary-5 spectrophotometer at variable temperatures, using suprasil quartz cells with a 1 cm path length that were fixed to the cryostat.

3. Results and Discussion

3.1. Changes in the Polyene's Molecular Structure with Temperature

In Figures 1–3, we present the first UV/Vis absorption band, respectively, of the three polyenes studied dissolved in decaline/methylcyclohexane (1:1 by volume; a solvent which we will call DEMCHEM), obtained in a temperature range between 93 and 363 K.

Figure 4 represents the areas of the first UV/Vis absorption bands of the three polyenes studied dissolved in DEMCHEM and measured at constant concentrations between 93 and 363 K; all areas are referred to as the area values measured at 93 K.

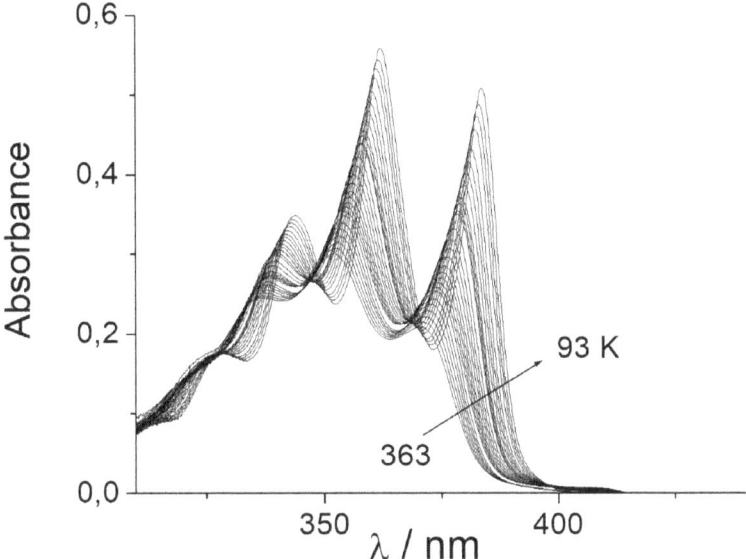

Figure 1. First UV/Vis absorption band of a solution of DPH dissolved in DEMCHEM at temperatures from 363 K to 93 K.

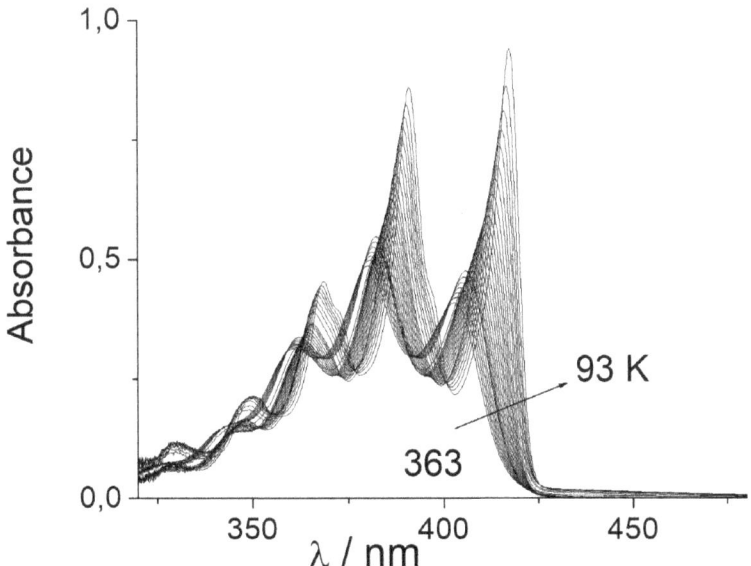

Figure 2. First UV/Vis absorption band of a solution of DPHb dissolved in DEMCHEM at temperatures from 363 K to 93 K.

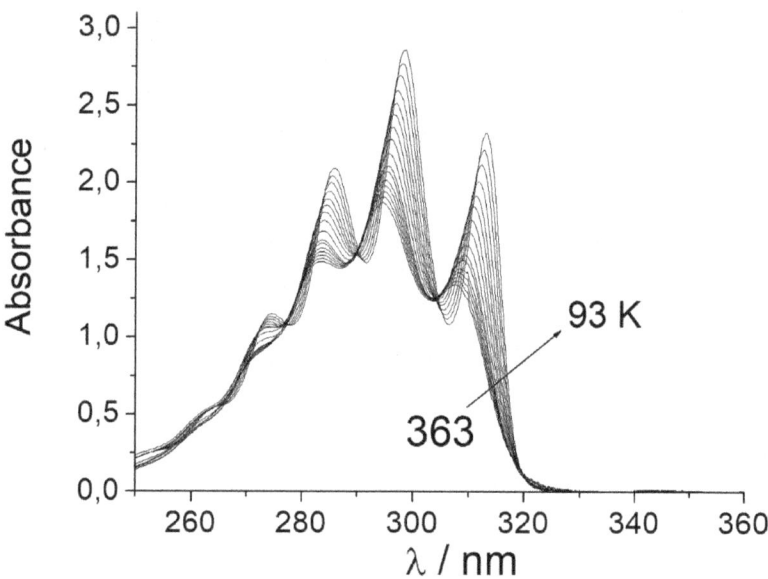

Figure 3. First UV/Vis absorption band of a solution of ttbP3 dissolved in DEMCHEM at temperatures from 363 K to 93 K.

The data for DPH shown in Figure 4a allow for the clear conclusion that (a) DPH studied in a temperature range of 93 to 223 K maintains its molecular structure, and (b) above 223 K, its molecular structure experiences increasing deformation with increasing temperature.

The data for DPHb shown in Figure 4b exhibit that hardly any molecular structure deformation occurs within the temperature range studied. The area of this first absorption band measured between 93 and 363 K maintains its values, referring to the corresponding value at 93 K within 1 ± 0.01.

This behavior of DPH and DPHb dissolved in DEMCHEM is in incontestable agreement with the supposition that DPH experiences a twisting of its two phenyl groups, whereas this twisting does not occur in DPHb because its two phenyl groups are blocked and cannot rotate (see Scheme 1).

The corresponding areas of the first absorption band of ttbP3 dissolved in DEMCHEM are shown in Figure 4c, clearly confirming the previously mentioned supposition.

The results achieved in ttbP3 indicate that in this triene, the small temperature-induced deformations, which may be due to small changes in the *tert*-butyl groups, will barely be translated into small changes in the intensity of its first absorption band. The measured values of the corresponding band areas, referring to the corresponding value at 93 K, are within 1 ± 0.02 throughout the whole temperature range studied.

3.2. Temperature-Dependent Solvatochromic Changes in Polyenes

As was previously shown (see Figure 4a), by increasing the temperature of the DPH solution in DEMCHEM above 233 K, as shown in Figure 4a, the compound begins to rotate its phenyl groups, and the rotation of these groups will increase with an increase in the temperature of the solution. As a consequence of this rotational movement, the coplanar surface of the DPH will begin to decrease, and therefore, its polarizability will decrease below an *SP* of 0.71.

Figure 5 presents the correlation between the temperature-dependent energy of the 0–0 component of the first UV/Vis absorption band of DPH dissolved in DEMCHEM and the corresponding values of the empirically determined solvent polarizability parameter *SP* of DEMCHEM. The correlation clearly shows bilinear behavior with respect to *SP*: for *SP* values larger than 0.7, a correlation line corresponding to $\bar{v}_{abs} = (-5300 \pm 135)\, SP + (31{,}806 \pm 99)$ with n = 14 and r = 0.996 can be seen, and for SP values smaller than 0.7, a correlation line of $\bar{v}_{abs} = (-5671 \pm 89)\, SP + (32{,}081 \pm 63)$ with n = 10 and r = 0.997 is found.

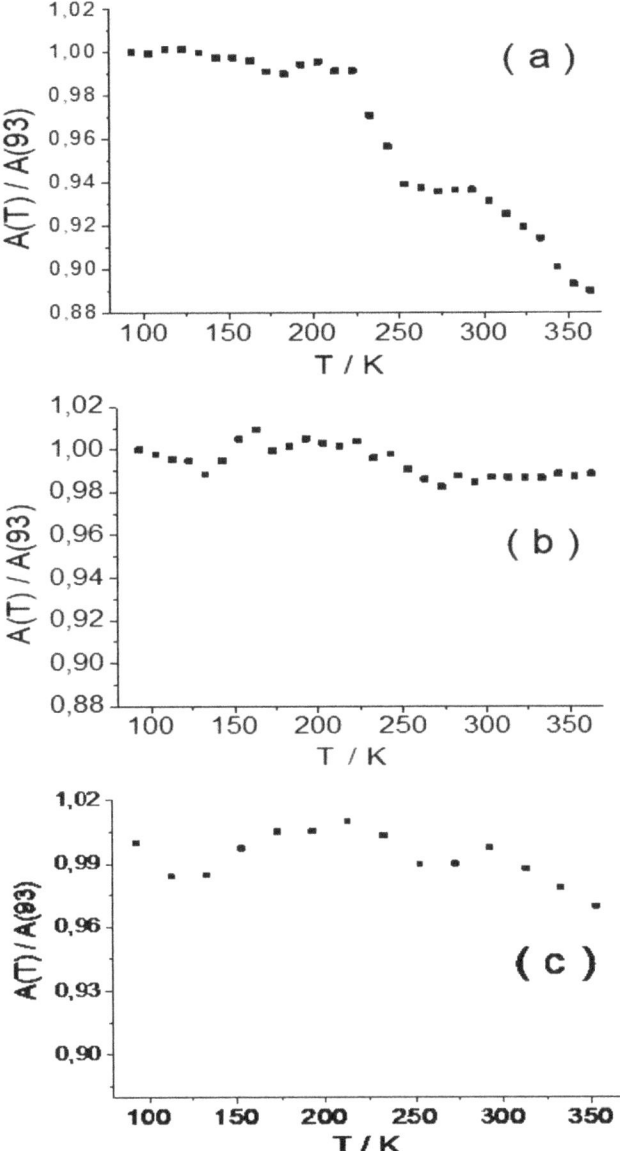

Figure 4. Temperature-dependent areas of the first UV/Vis absorption band of (**a**) DPH, (**b**) DPHb, and (**c**) ttbP3 dissolved in DEMCHEM at a constant concentration divided by the respective area measured at 93 K.

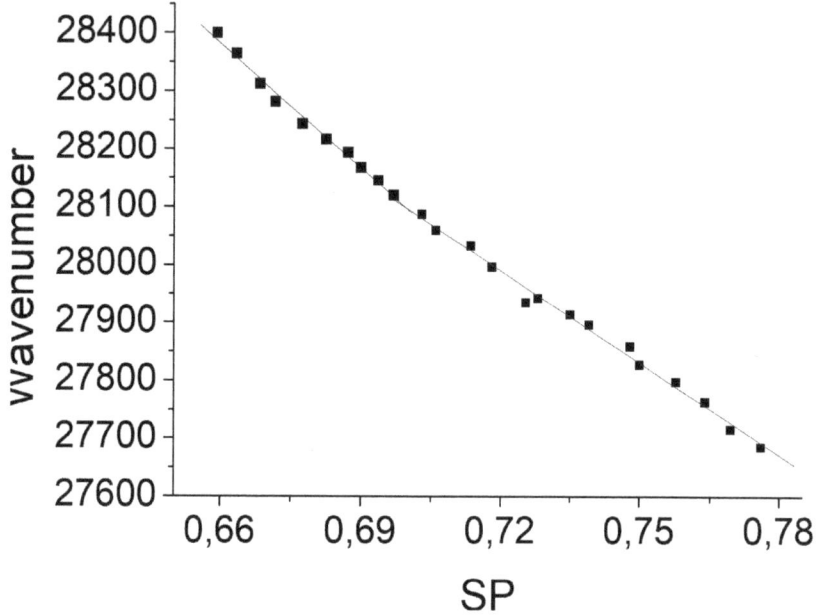

Figure 5. Wavenumber (in cm^{-1}) of the 0–0 component of DPH in DEMCHEM versus the polarizability (*SP*) of the solvent.

Figure 6 shows the correlation between the temperature-dependent energy of the 0–0 component of the first UV/Vis absorption band of DPHb dissolved in DEMCHEM and the corresponding *SP* values. The results show a clear linear correlation line for the entire *SP* range, which can be adjusted by $\bar{\nu}_{abs} = (-6229 \pm 43)\, SP + (28{,}856 \pm 30)$, with n = 26 and r = 0.999.

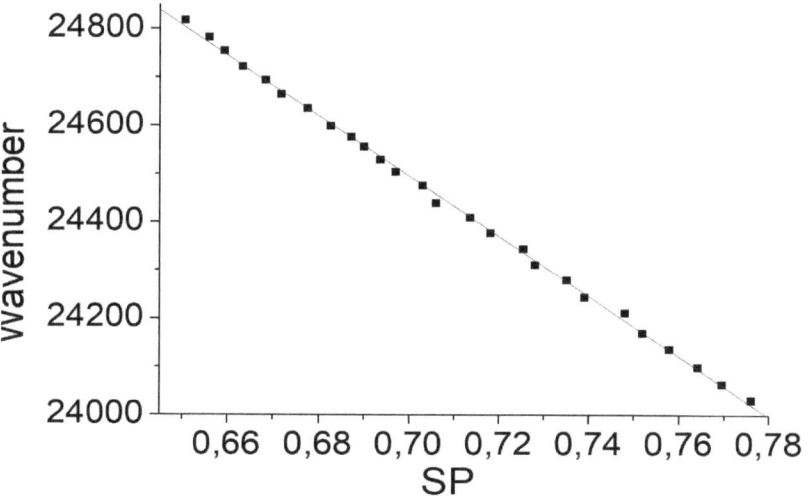

Figure 6. Wavenumber (in cm^{-1}) of the 0–0 component of DPHb in DEMCHEM versus the polarizability (*SP*) of the solvent.

Figure 7 presents the correlation between the temperature-dependent energy of the 0–0 component of the first UV/Vis absorption band of ttbP3 in DEMCHEM and the corresponding SP values. Again, a clear linear correlation line for the entire SP range is found, which can be described by the correlation equation $\bar{\nu}_{abs} = (-3946 \pm 56)\, SP + (35{,}055 \pm 40)$, with n = 17 and r = 0.998.

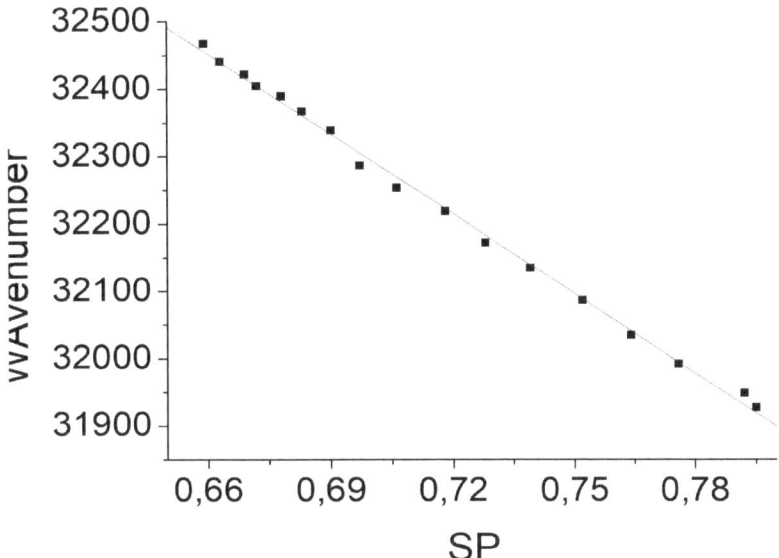

Figure 7. Wavenumber (in cm^{-1}) of the 0–0 component of ttbP3 in DEMCHEM versus the polarizability (SP) of the solvent.

Figure 8 describes the correlation between the temperature-dependent energy of the 0–0 component of the first UV/Vis absorption band of DPH dissolved in Cl$_4$C and the corresponding SP values. A clearly bilinear behavior is shown for SP values larger than 0.78, with a corresponding correlation of $\bar{\nu}_{abs} = (-3023 \pm 127)\, SP + (30{,}295 \pm 103)$ with n = 11 and r = 0.992, and for SP values smaller than 0.78, the fitting is expressed by $\bar{\nu}_{abs} = (-4848 \pm 352)\, SP + (31{,}724 \pm 265)$ with n = 7 and r = 0.987.

Finally, Figure 6 shows the correlation between the temperature-dependent energy of the 0–0 component of the first UV/Vis absorption band of DPH dissolved in Cl$_2$CH$_2$ and the corresponding SP values. A clear linear behavior is evident, in agreement with the correlation equation $\bar{\nu}_{abs} = (-3882 \pm 69)\, SP + (30{,}971 \pm 56)$, with n = 15 and r = 0.998.

The results given in Figures 8 and 9 do not only highlight the particular solvatochromic influence of these two solvents, apolar Cl$_4$C and dipolar Cl$_2$CH$_2$, but they are also in agreement with the accepted model for the solvatochromism of diphenylpolyenes. As already mentioned in the Introduction, it is commonly accepted that diphenylpolyenes are polarizable planar compounds which are non-polar, non-acidic, and non-basic. CL$_4$C and CL$_2$CH$_2$ are solvents with high polarizability and have, with SP = 0.768 and 0.761, nearly the same polarizability SP values. However, they differ in their other empirical solvatochromic parameters: for Cl$_4$C, the solvent dipolarity parameter SdP = 0.000, the solvent basicity parameter SB = 0.044, and the solvent acidity parameter SA = 0.000 [19], whereas for Cl$_2$CH$_2$, the corresponding solvent parameters are SdP = 0.769, SB = 0.0178, and SA = 0.040 [19].

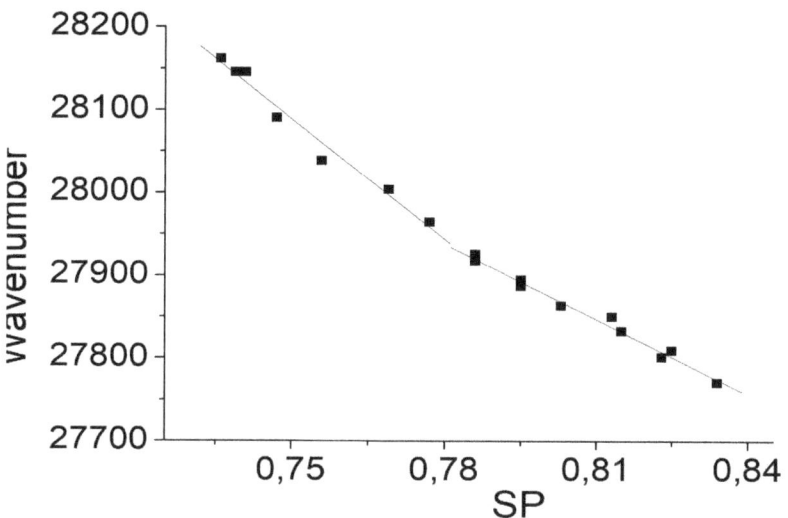

Figure 8. Wavenumber (in cm^{-1}) of the 0–0 component of DPH in CL$_4$C versus the polarizability (*SP*) of the solvent.

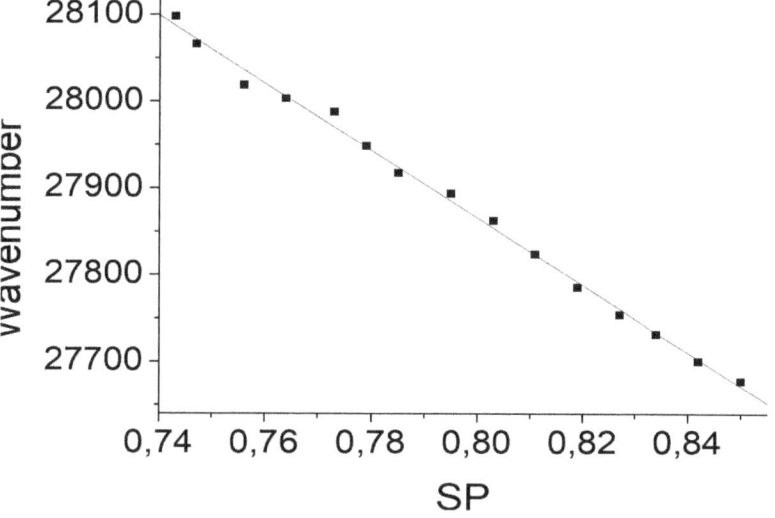

Figure 9. Wavenumber (in cm^{-1}) of the 0–0 component of DPH in CL$_2$CH$_2$ versus the polarizability (*SP*) of the solvent.

Consequently, DPH sees its bilinear behavior revealed when dissolved in CL$_4$C due to its high polarizability (see Figure 8), while in CL$_2$CH$_2$, the destabilization created by its polarizability is practically compensated for by the stabilization generated by its dipolarity, thus explaining the linear correlation behavior shown in Figure 9.

4. Conclusions

In this work, an adequate analysis of the solvatochromism of three carefully selected polyene compounds, measured in well-selected suitable solvents, was carried out, allowing for a better understanding of their photophysical behavior.

Author Contributions: Data collection J.C., used products H.H., Both authors: wrote and editing the manuscript. All authors have read and agreed to the published version of the manuscript.

Funding: This research received no external funding.

Data Availability Statement: Data are contained within the article.

Conflicts of Interest: The authors declare that the research was conducted in absence of any commercial or financial relations that could be construed as a potential conflict of interest.

References

1. Hausser, K.W.; Kuhn, R.; Smakula, A.Z. Lichtabsorption und Doppelbindung. IV. Diphenylpolyene. *Phys. Chem.* **1935**, *29B*, 384–389.
2. Hausser, K.W.; Kuhn, R.; Seitz, A.Z. Lichtabsorption und Doppelbindung. Über die Absorption von Verbindungen mit konjugierten Kohlenstoffdoppelbindungen bei tiefer Temperatur. *Phys. Chem.* **1935**, *29*, 391–416.
3. Hausser, K.W.; Kuhn, R.; Kuhn, E.Z. Über die Flaurescenz der Diphenylpolyene. *Phys. Chem.* **1935**, *29B*, 417–454.
4. Catalán, J. On the inversion of the 1 B u and 2 A g electronic states in α,ω-diphenylpolyenes. *J. Chem. Phys.* **2003**, *119*, 1373–1385. [CrossRef]
5. Ye, J.F.; Chen, H.; Note, R.; Mizuseki, H.; Kawazoe, Y. Excess polarizabilities upon excitation from the ground state to the first dipole-allowed excited state of diphenylpolyenes. *Int. J. Quantum. Chem.* **2007**, *107*, 2006–2014. [CrossRef]
6. Falková, K.; Cajzi, R.; Burda, J.V. The vertical excitation energies and a lifetime of the two lowest singlet excited states of the conjugated polyenes from C2 to C22: Ab initio, DFT, and semiclassical MNDO-MD simulations. *J. Comput. Chem.* **2022**, *29*, 777–787.
7. Hengstenberg, J.; Kuhn, R. 19. Die Kristallstruktur der Diphenylpolyene. *Zeitschrift für Kristallographie-Cryst. Mater.* **1930**, *75*, 301–310. [CrossRef]
8. Drenth, W.; Wiebenga, E.H. Structure of α,ω-diphenylpolyenes.: I. crystal data of 1,4-diphenyl-1,3-butadiene,1,6-diphenyl-1,3,5-hexatriene and 1,8-diphenyl-1,3,5,7-octatetraene. *Recueil* **1953**, *72*, 39–43. [CrossRef]
9. Drenth, W.; Wiebenga, E.H. Structure of α.ω-diphenylpolyenes: II. Crystal structure of the monoclinic and orthorhombic modification of 1,10-diphenyl-1,3,5,7,9-decapentaene. *Recueil* **1954**, *73*, 218–228. [CrossRef]
10. Drenth, W.; Wiebenga, E.H. Structure of α,ω-diphenyl-polyenes. IV. Crystal and molecular structure of 1,8-diphenyl-1,3,5,7-octatetraene. *Acta Cryst.* **1955**, *8*, 755–760. [CrossRef]
11. Hall, T.; Bachrach, S.M.; Spangler, C.W.; Sapochak, L.S.; Lin, C.T.; Guan, H.W.; Rogers, R.D. Structure of all-*trans*-1,6-diphenyl- (A) and all-trans-1,6-bis(*o*-methoxyphenyl)-1,3,5-hexatriene (B). *Acta Cryst.* **1989**, *C45*, 1541–1543. [CrossRef]
12. Benzi, C.; Barone, V.; Tarroni, R.; Zannoni, C. Order parameters of α,ω-diphenylpolyenes in a nematic liquid crystal from an integrated computational and C13 NMR spectroscopic approach. *J. Chem. Phys.* **2006**, *125*, 174904. [CrossRef]
13. Catalán, J.; Diaz-Oliva, C.; del Valle, J.C. On the first electronic transitions in molecular spectra of conjugated diphenylpolyenes: A reappraisal. *Chem. Phys.* **2019**, *525*, 110422. [CrossRef]
14. Bramley, R.; Le Fèvre, R.J.W. Molecular polarisability. The anisotropies of diphenylpolyenes. *J. Chem. Soc.* **1960**, 1820–1824. [CrossRef]
15. Mulliken, R.S.; Rieke, C.A. Molecular electronic spectra, dispersion and polarization: The theoretical interpretation and computation of oscillator strengths and intensities. *Rep. Progr. Phys.* **1941**, *8*, 231–273. [CrossRef]
16. Catalán, J. Molecular structure distortions and the Mulliken-Rieke rule: The case of t-stilbene. *Chem. Phys. Lett.* **2005**, *416*, 165–170. [CrossRef]
17. Ponder, M.; Mathies, R. Excited-state polarizabilities and dipole moments of diphenylpolyenes and retinal. *J. Phys. Chem.* **1983**, *87*, 5090–5098. [CrossRef]
18. Catalán, J.; Hopf, H. Empirical treatment of the inductive and dispersive components of solute—Solvent interactions: The solvent polarizability (SP) scale. *Eur. J. Org. Chem.* **2004**, *22*, 4694–4702. [CrossRef]
19. Catalán, J. Toward a generalized treatment of the solvent effect based on four empirical scales: Dipolarity (SdP, a new scale), polarizability (SP), acidity (SA), and basicity (SB) of the medium. *J. Phys. Chem. B* **2009**, *113*, 5951–5960. [CrossRef] [PubMed]
20. Catalán, J. On the temperature—Dependent isomerization of all-trans-1,6-diphenyl-1,3,5-hexatriene in solution: A reappraisal. *J. Phys. Org. Chem.* **2022**, *35*, e4336. [CrossRef]
21. Catalán, J. On the photophysics of polyene compounds. *Trends Phys. Chem.* **2022**, *22*, 17–23.

Disclaimer/Publisher's Note: The statements, opinions and data contained in all publications are solely those of the individual author(s) and contributor(s) and not of MDPI and/or the editor(s). MDPI and/or the editor(s) disclaim responsibility for any injury to people or property resulting from any ideas, methods, instructions or products referred to in the content.

Article

Solvatochromic and Computational Study of Some Cycloimmonium Ylids

Daniela Babusca [1,2,*], Andrei Vleoanga [3] and Dana Ortansa Dorohoi [1,*]

1. Faculty of Physics, Alexandru Ioan Cuza University, 11 Carol I Blvd., 700506 Iasi, Romania
2. Alexandru Vlahuta School, 10 Buridava Street, 700432 Iasi, Romania
3. Faculty of General Medicine, Grigore T. Popa University of Medicine and Pharmacy, 16 Universității Street, 700115 Iasi, Romania; mg-rom-32917@students.umfiasi.ro
* Correspondence: daniela.babusca@vlahuta.ro (D.B.); ddorohoi@uaic.ro (D.O.D.)

Abstract: This article contains a comparative spectral analysis corroborated with the quantum mechanical computations of four cycloimmonium ylids. The spectral shift of the visible electronic absorption band of the studied molecules in 20 solvents with different empirical parameters is expressed by linear multi-parametric dependences that emphasize the intramolecular charge transfer (ICT) process. The nature of molecular interactions and their contribution to the spectral shift of the visible ICT band of solutes are also established in this manuscript. The results of the statistical analysis are used to estimate the cycloimmonium ylids' excited dipole moment by the variational method, using the hypothesis of McRae. The importance of the structure of both the heterocycle and carbanion substituents to the stability and reactivity of the studied cycloimmonium ylids is underlined by the quantum mechanical computations of the molecular descriptors.

Keywords: cycloimmonium ylids; solvatochromic study; nature and strength of molecular interactions; excited-state dipole moment

1. Introduction

Cycloimmonium ylids [1–3] are N-ylids with separated charges on a nitrogen belonging to a heterocycle and on a negative α-exo-cycle carbon, named the carbanion. Two highly electronegative atomic substituents are bonded to the carbanion in carbanion-disubstituted methylids, while in carbanion-monosubstituted methylids, one electronegative atomic group and a hydrogen atom are bonded to the carbanion.

In the second half of the 20th century, Professors I. Zugravescu and M. Petrovanu with their teams, working in the Faculty of Chemistry of Alexandru Ioan Cuza University of Iasi, made a great contribution by obtaining and characterizing new molecules belonging to the cycloimmonium ylid class [2].

The studied cycloimmonium ylids contain heterocycles (pyridinium, iso-quinolinium, pyridazinium and benzo-[f]-quinolinium) and the carbanion's substituents (carbethoxy, acetyl and benzoyl) [2].

The stability of cycloimmonium ylids is influenced by both the electronegativity of the carbanion substituents and the electron-withdrawing ability of the heterocycles [2,3]. For example, the carbanion-disubstituted cycloimmonium ylids are more stable than those that are carbanion-monosubstituted, and iso-quinolinium ylids are more stable than pyridinium ylids.

The most common method for obtaining cycloimmonium ylids is the "salt method" from the quaternary halogenures of heterocycles in basic solutions [2].

All cycloimmonium ylids are characterized by a high-wavelength electronic absorption band (usually placed in the visible range) that is attributed to an internal charge transfer (ICT) from the carbanion toward the heterocycle and is very sensitive to the solvent nature [3–8]. This band shifts to high wavenumbers when the ylid is passed from nonpolar

Citation: Babusca, D.; Vleoanga, A.; Dorohoi, D.O. Solvatochromic and Computational Study of Some Cycloimmonium Ylids. *Liquids* **2024**, *4*, 171–190. https://doi.org/10.3390/liquids4010009

Academic Editors: William E. Acree, Jr. and Enrico Bodo

Received: 19 November 2023
Revised: 28 December 2023
Accepted: 29 January 2024
Published: 12 February 2024

Copyright: © 2024 by the authors. Licensee MDPI, Basel, Switzerland. This article is an open access article distributed under the terms and conditions of the Creative Commons Attribution (CC BY) license (https://creativecommons.org/licenses/by/4.0/).

or non-protic solvents to polar and protic ones; it disappears or its intensity diminishes in acid media [3,7].

Cycloimmonium ylids are used as acid–basic indicators due to their change in color in the presence of acids and bases [2,3]. They are also used as semiconductors [9], in obtaining thin conducting films [10], or as chemical precursors in reactions for new heterocycles [11,12]. The 2,3 dipolar cycloadditions of cycloimmonium ylids to dienophiles have been used to synthesize steroid analogs with biological activity [13], including antiviral [14], anticancer [15], antifungal [16,17] and antituberculosis [18] effects.

Due to their high reactivity, some cycloimmonium ylids with monosubstituted carbanions are used as precursors in synthesis reactions [19,20].

As dipolar and polarizable molecules, cycloimmonium ylids are used, in small concentrations, as sounders in liquids in order to obtain information about the nature and the strength of the molecular interactions in solutions [5–8,21].

In order to illustrate the importance of solvents' empirical scales in characterizing the nature and strength of molecular interactions and in estimating the excited-state dipole moment of the solute molecules, we have chosen four cycloimmonium ylids: two with common heterocycle and two pairs with common carbanions. These molecules were previously studied separately from a spectral point of view in binary and ternary solutions [5–8,22–24]. The chosen cycloimmonium molecules show electronic absorption bands in the UV range of the $\pi - \pi^*$ type and a visible band of the n–π^* type [3]. The UV bands are of high intensity and low sensitivity to the solvent nature, but the visible absorption band is of low intensity, disappears in acid media and shifts to blue when the solute molecules pass through a non-polar/aprotic solvent to a polar/protic solvent. Based on these characteristics, the electronic visible band was attributed to an intramolecular charge transfer (ICT) transition from the ylid carbanion toward its heterocycle [3,6–8].

The mechanism of the visible ICT band of cycloimmonium ylids is similar to that based on the definition of the solvent polarity scales ($E_T(30)$ and E_T^N) introduced by Cristian Reichardt and co-workers [25–27] using betaine dye derivatives as spectrally active molecules.

The empirical mono-parameter solvent scales classify liquids regarding their solvation ability but do not allow us to establish the contribution of each type of molecular interaction to the solvation energy. They measure only the difference between the solvation energies in the electronic states responsible for the electronic (absorption/emission) band appearance according to its position in the electronic spectrum relative to the solute gaseous phase.

The theoretical relationships [28–30] established for describing the influence of dielectric continuous liquids on the electronic bands of the solutes, as well as the new empirical parameters [31] introduced by Kamlet, Abboud and Taft (KAT) and by Catalan [32] in order to consider the ability of solvents to form hydrogen bonds by receiving or donating protons, were used, with better results, in establishing both the nature and contribution of each type of interactions to the total spectral shift in a given solvent and in estimating the dipole moment in the excited state of the studied cycloimmonium ylids.

In all solutions, interactions between the components can modify some chemical descriptors of the molecules. Both specific (hydrogen bonds, change of electrons or protons, dimerization and clusterization) and non-specific (orientation, induction, polarization and dispersive) interactions can take place between the solution components. Intermolecular interactions induce changes in spectral tools of solutions, such as spectral shifts, variations in intensities and/or polarization of the spectral bands.

Solvatochromism describes the changes in the position/intensity of molecular UV–Vis bands of the solute induced by the solvent polarity. Some researchers previously used the notion of the solvent polarity. For example, Brooker [33] suggested the term solvent polarity in 1951, and Kosower realized a solvent scale based on the energy at the maximum of a visible electronic absorption band in 1958 [34]. But the term solvent polarity was recognized by IUPAC in 1994 when it was introduced in the Glossary of terms used in physical organic chemistry and defined as the "solvation capability of solvent depending on all intermolecular solute/solvent interactions which do not change the chemical nature of the solution components". When a chemical reaction occurs in situ, it must be ensured that the reaction components of the chosen solvent have good solubility and do not change their chemical structure.

In the middle of the 20th century, the chemists used the term solvent polarity (suggested by Brooker in 1951 [33]) to characterize liquid properties. Kosower [34,35] was the first to determine the positions of the visible electronic band of the complex 1-ethyl-4-carbomethoxy pyridinium iodide in 21 solvents and classified them based on the energy at the maximum of intermolecular charge transfer band between the iodine and pyridine derivative, listing a Z-scale, or solvent polarity scale.

The complex structure and the charge transfer mechanism studied by Kosower are shown in Figure 1. Due to the change in the orientation of the complex dipole moment caused by the visible photon absorption (see Figure 1a), the spectral position of the visible absorption band of the complex depends only on the ground dipole moment of complex (by the solvation energy of the complex in its non-excited electronic state). The dipole moment in the excited electronic state of the complex is perpendicular on the ground state dipole moment. The interaction with the reactive field created by the solvent molecules surrounding the solute becomes null after the visible photon absorption. Therefore, the parameter Z introduced by Kosower measures the solvation energy of the considered complex in its ground electronic state. The solvent polarity scale is based on Equation (1) in which the wavelength at the maximum of the electronic absorption band is expressed in cm and the solvent polarity Z in kcal/mol.

$$Z\left(\frac{\text{kcal}}{\text{mol}}\right) = \frac{2.8591 \cdot 10^{-3}}{\lambda(\text{cm})} \qquad (1)$$

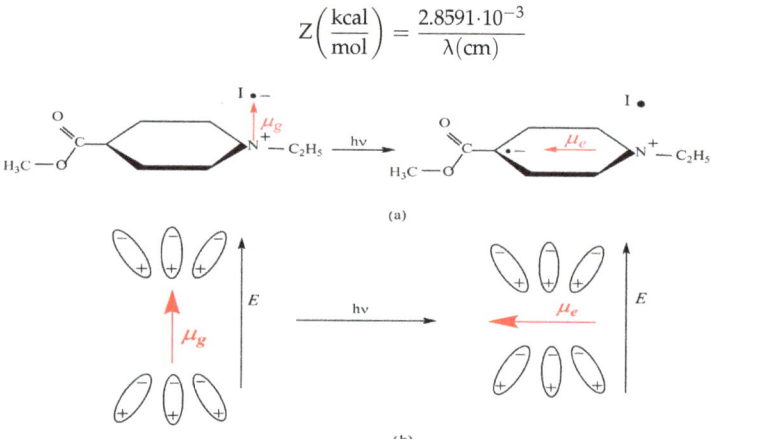

Figure 1. Chemical structure of 1-ethyl-4-carbomethoxy-pyridinium iodide complex and the mechanism of ICT of the visible electronic band appearance. (**a**) Electron transfer from iodine to heterocycle. (**b**) Relative orientation of the solvent molecules and the dipole moments of solute on its ground and excited states.

Some inconvenient factors, such as insolubility of this complex in the non-polar solvents and high concentration at which the spectral recordings must be carried out, make it impossible to measure Z values for a great number of solvents.

A large number of solvent scales have been defined. More than 30 empirical polarity solvent scales have been published to date [36], and the existence of linear correlations between them was emphasized by different authors.

The Dimroth and Reichardt [25–27] empirical scale of solvent polarity is usually applied to characterize the solvatochromic effects of solvents. It is constructed based on the spectral characteristics of betaine 1 (2,6-bis[4-(t-butyl)phenyl]-4-[2,4,6-tris[4-(t-butyl)phenyl]pyridine-1-yl]phenolate) [25] with the chemical structure illustrated in Figure 2a.

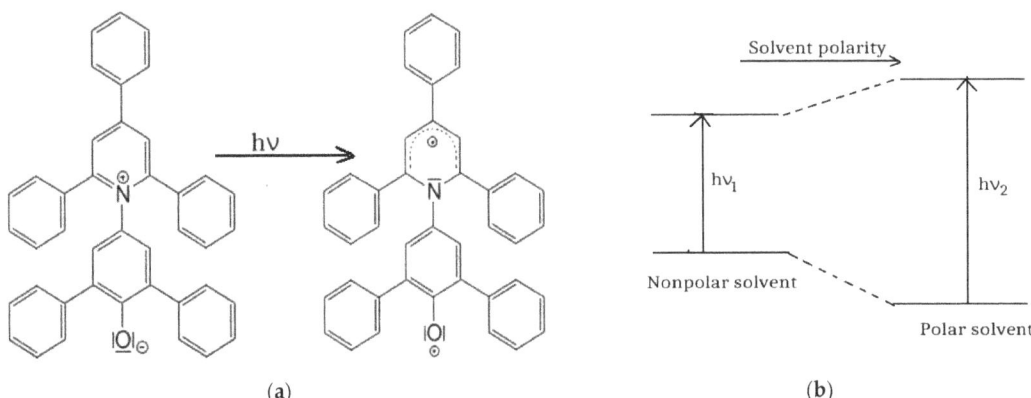

Figure 2. (a) Chemical structure of betaine 1 dye and mechanism of the ICT in this molecule; (b) hypsochromic shift of the visible band of betaine 1 in polar solvents compared with non-polar ones.

Betaine 1 dye (Figure 2a) shows a visible electronic absorption band due to an intramolecular charge transfer (ICT) through oxygen towards the heterocycle; this process causes the decrease in the dye excited-state dipole moment and the change in its sensitivity. So, the solvation energy in the ground state of the solute determines its stabilization, while in the excited state, the local field created by the solvent becomes antiparallel to the molecular dipole moment, and the energy corresponds to a destabilization process, as Figure 2b suggests.

Betaine 1 dye is used to measure the solvent polarity due to its high solubility in the majority of liquids and due to the great difference between its dipole moments in the ground ($\mu_g = 15D$) and in the first excited ($\mu_e = 6D$) electronic states, inducing high values of the spectral shifts in the visible range for the polar solvents compared with the non-polar ones.

Betaine 1 dye is at the basis of two solvent polarity scales, $E_T(30)$ and E_T^N, with the following definitions.

$$E_T(30)\frac{kcal}{mol} = hcN_A\bar{\nu} = 2.8591\cdot 10^{-3}\bar{\nu} = 2.8951\cdot 10^{-3}/\lambda_{max}(cm) \qquad (2)$$

In Equation (2), $\bar{\nu}$ is the wavenumber at the maximum of the electronic absorption band, measured in cm^{-1}, and λ_{max} is the wavelength (expressed in cm) corresponding to the same maximum.

The second empirical solvent scale is named E_T^N, and it is defined using the values of $E_T(30)$ in the respective solvent and in water and TMS, corresponding with Equation (3).

$$E_T^N = \frac{E_T(solvent) - E_T(TMS)}{E_T(water) - E_T(TMS)} = \frac{E_T(solvent) - 30.7}{32.4} \qquad (3)$$

The last scale was introduced in order to avoid the use of non-IS unity kcal/mol and then transform it in kJ/mol. The parameter E_T^N is adimensional.

Reichardt contributed to the proposal of these empirical scales and to the increase in the solvent number characterized by parameter E_T^N, considering the electronic absorption spectra of other betaine dyes [26] as standard dyes for the solvents in which betaine 1 is not soluble. A complex study of the solvent influence on organic molecules was realized by prof. C. Reichardt in [27].

These two solvent polarity scales are known for more 400 pure and mixed liquids. The increasing the solvent polarity is in the following order:

- Non-polar non hydrogen bond donor (HBD) (cyclohexane, benzene, THF and dichloromethane);

- Dipolar non-polar HBD (aprotic) solvents (acetone, DMF and DMSO);
- Dipolar protic solvents (alcohols and acids).

The empirical scales of the solvents [36] arrange the liquids using the spectral data and can be applied to the solute molecules showing electronic spectra and developing the same type of interactions with a given solvent. In the solution of a solute, molecular interactions differ according to the standard molecule, and deviations in the linear dependence of the spectral data vs. empirical parameters are observed. The presence of the aberrant points in graphs of experimental wavenumbers vs. empirical solvent parameters indicates the presence of supplemental interactions between the solute and the solvent molecules.

The statistical analysis of experimental data shows the linearity between the experimental wavenumbers at the maximum of the electronic band of the solute and the empirical polarity parameter of the solvent according to Equation (4).

$$\nu = mP_p + n \qquad (4)$$

In Equation (4), the following notation is used: ν is the wavenumber at the maximum of the electronic band, and P_p is the empirical polarity parameter of the solvent. The correlation coefficients m and n can be statistically established using experimental spectral data and the values of the solvent polarity parameters. Due to the linear relationships between various polarity scales according to [36], in the present paper, we used an equation of the type (4) both for Z and $E_T(30)$ parameters.

The empirical solvent scales based on multiple parameters have the advantage of separately considering the effect of the different forces on the solute molecules.

Some theoretical representations [28–30] were concomitantly developed in order to compute the contributions of each type of solvent to the electronic bands of the solute. But, as Reichardt comments [26,27], these theories take into consideration only universal interactions, considering the solvent a continuum dielectric acting on the valence electrons of the solute molecules. The solvation energy in the electronic states responsible for the appearance of the electronic transition in absorption or emission processes is computed using macroscopic parameters n (refractive index) and ε (electric permittivity) of the solvent in the existent theories. The specific interactions of the solute/solvent type are neglected by the existent theories.

The impossibility to develop a unitary theory describing the complex phenomena from the liquid phase results from the multitude of the interactions developed between its components and from the very close values of the interaction energies and the energy of the thermal motion, which determine only a partial order in liquids.

In this situation, the researchers combined [5,8,37] the theoretical and empirical correlations using multi-parametric linear relations between the spectral characteristics of the solute and the solvent parameters in order to assure good correspondence between the experimental and computed results. Usually, equations of the type (5) are used to describe the solvent influence on the electronic bands of the solutes.

$$\nu_{calc.} = \nu_0 + C_1 f(n) + C_2 f(\varepsilon) + C_3 \alpha + C_4 \beta \qquad (5)$$

In Equation (5), the first term signifies the wavenumber at the maximum of the electronic band of the solute measured in its gaseous phase; the two following terms describe the universal interactions as functions of the macroscopic parameters of the solvent (refractive index, n, and electric permittivity, ε), and the terms $C_3\alpha$ and $C_4\beta$ describe the specific interactions in which the solvent molecules donate and receive protons, respectively.

This combination of the theoretical representations and the empirical treatment permit us to use the theoretical expressions of the correlation coefficients C_1 and C_2 [6,7] in order to compute some molecular parameters of the solute in its excited state of electronic transition.

Equations of the types (6) and (7) express the correlation coefficients (determined by statistical means) as functions of microscopic parameters of the solute (molecular radius, a; dipole moment, μ; polarizability, α; and ionization potentials, I, of solute (u) and solvent (v) in the solute ground (g) and excited (e) states of the electronic transition). The corresponding parameters of the solute in its ground state can be computed by quantum mechanical procedures.

$$C_1 = \frac{2\mu_g(\mu_g - \mu_e \cos\varphi)}{a^3} + 3kT\frac{\alpha_g - \alpha_e}{a^3} \tag{6}$$

$$C_2 = \frac{\mu_g - \mu_e}{a^3} - \frac{2\mu_g(\mu_g - \mu_e\cos\varphi)}{a^3} - 3kT\frac{\alpha_g - \alpha_e}{a^3} + \frac{3}{2}\frac{\alpha_g - \alpha_e}{a^3}\frac{I_u I_v}{I_u + I_v} \tag{7}$$

In the case of solute molecules showing both absorption and emission electronic spectra, the dipole moments in the electronic states responsible for the electronic transition and the angle between them can be computed based on similar relations written for both types of electronic spectra [38–41]. In the case in which the solute molecules are active only in absorption spectra, the variational method [42], based on the McRae hypothesis [28] that the molecular polarizability does not vary in the visible photon absorption process, can be applied.

The combined spectral and computational analyses of the molecular parameters help us to obtain supplemental information about the excited states of the molecular structures, contributing to the development of the quantum mechanical calculations for the excited states.

2. Materials and Methods

Four cycloimmonium ylids, pyridinium-dicarbethoxy-methylid (PDCM)—$C_{12}H_{15}O_4N_1$, iso-quinolinium dicarbethoxy methylid (iQDCM)—$C_{16}H_{17}O_4N_1$, pyridinium-acetyl-benzoyl methylid (PABM)—$C_{15}H_{13}O_2N_1$ and benzo-[f]-quinolinium acetyl-benzoyl-methylid (BQABM)—$C_{23}H_{17}O_2N_1$, are considered in this paper. The structural features of the studied cycloimmonium ylids are given in Scheme 1. Two ylids (PDCM and PABM) have pyridine as a common heterocycle. Two pairs of ylids have the same substituents at their carbanions (PDCM and iQDCM, respectively, BQABM and PABM).

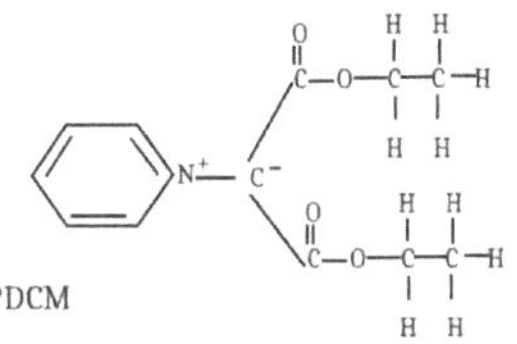

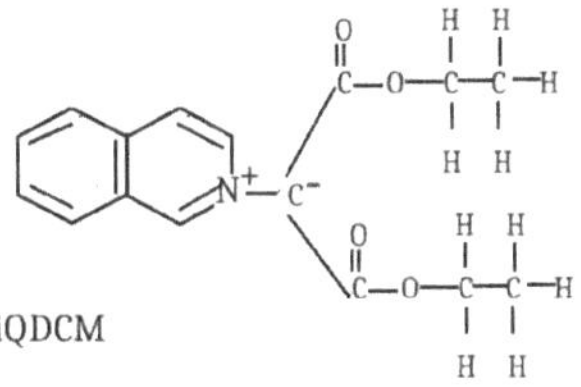

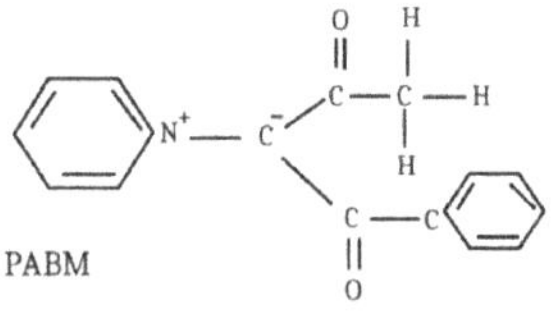

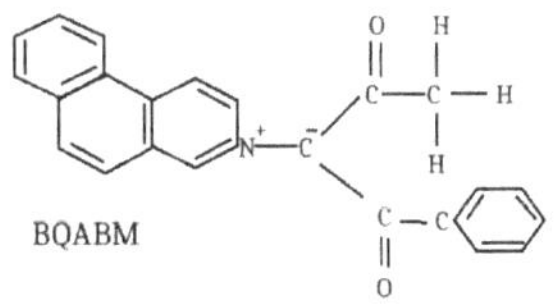

Scheme 1. Structural features of the studied cycloimmonium ylids.

The optimized geometries of the studied ylids, in a common format, are provided in Supplementary Material.

The studied substances were prepared as described in [2] in Alexandru Ioan Cuza University Organic Chemistry Labs. Their purity was assessed by spectral (NMR and IR) and chemical (elemental) means.

The spectral grade solvents were obtained from Merck Company, Darmstadt, Germany, and Sigma-Aldrich, Burlington, MA, United States (headquarters) and used without any purification.

The solvent parameters were taken from [43].

The solvent parameters used in this paper are listed in Table 1, where $E_T(30)$ and E_T^N are from [27].

Table 1. Solvent parameters.

No.	Solvent	$E_T(30)$	E_T^N	Z(kcal/mol)	$f(\varepsilon)$	$f(n)$	π^*	α	β
1	Dioxane	36.0	0.164	64.55	0.286	0.300	0.55	0.00	0.37
2	Benzene	34.3	0.111	59.6	0.299	0.295	0.59	0.00	0.10
3	o-Xylene	33.1	0.074	-	0.302	0.292	0.41	0.00	0.11
4	Toluene	34.5	0.099	-	0.302	0.297	0.54	0.00	0.11
5	Anisole	44.3	0.198	60.8	0.524	0.300	0.73	0.00	0.32
6	Chloroform	39.1	0.259	63.4	0.552	0.267	0.69	0.20	0.10
7	n-Butyl acetate	38.5	0.241	-	0.577	0.240	0.46	0.00	0.45
8	Chlorobenzene	36.8	0.188	62.0	0.605	0.307	0.71	0.00	0.07
9	Ethyl acetate	38.1	0.238	60.5	0.625	0.228	0.55	0.00	0.45
10	Dichloromethane	40.7	0.309	64.3	0.727	0.256	0.82	0.20	0.10
11	Benzyl alcohol	50.4	0.608	82.0	0.804	0.311	0.98	0.60	0.52
12	Cyclohexanol	49.6	0.509	76.5	0.824	0.276	0.45	0.66	0.84
13	n-Butyl alcohol	49.7	0.586	77.5	0.833	0.242	0.47	0.84	0.84
14	Isobutyl alcohol	47.1	0.552	80.6	0.852	0.237	0.40	0.69	0.84
15	n-Propyl alcohol	50.7	0.617	78.2	0.866	0.240	0.52	0.84	0.90
16	Acetone	42.2	0.355	65.5	0.868	0.222	0.62	0.08	0.48
17	Ethanol	51.9	0.654	80.8	0.895	0.221	0.86	0.86	0.75
18	Methanol	55.4	0.762	84.4	0.909	0.203	0.60	0.98	0.66
19	Water	63.1	1	94.6	0.964	0.206	1.09	1.17	0.47
20	Formamide	55.8	0.775	80.5	0.973	0.267	0.97	0.97	0.71

The visible electronic absorption band was recorded with a Specord UV–Vis Carl Zeiss Jena spectrophotometer with a data acquisition system.

The optimized structure and the molecular descriptors of cycloimmonium ylids in the electronic ground state, in vacuum, were established with the Spartan'14 program [44,45]. The density function theory (DFT) was used with the B3LYP method (Becke's three-parameter functional using the Lee–Yang–Parr correlation functional) in combination with the 6-31G* basis set [46].

Statistical analysis was carried out using the Origin 9 program using fitting methods (linear fit and multiple linear regressions). In order to obtain a better correlation between the spectral data and solvent parameters, the aberrant points were eliminated.

3. Results and Discussion

3.1. Computational Results

The optimized structures of the studied methylids are illustrated in Figures 3–6.

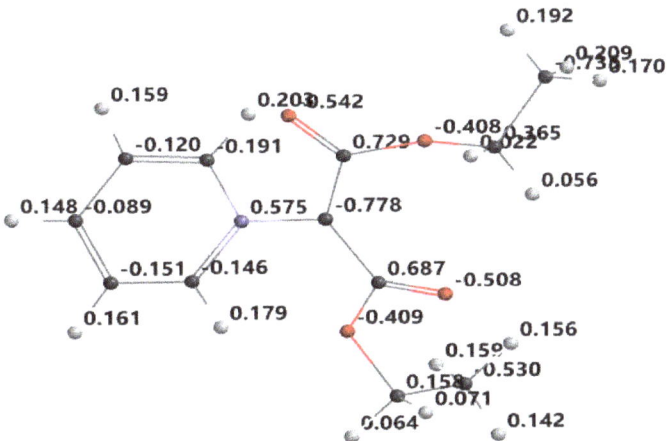

Figure 3. Optimized structure and atomic electrostatic charge for PDCM.

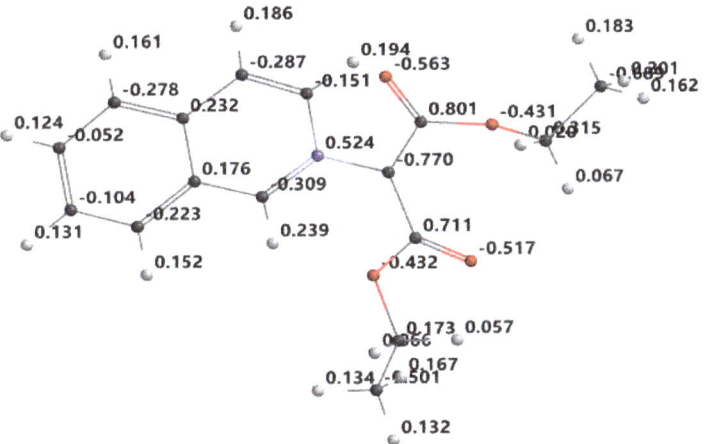

Figure 4. Optimized structure and atomic electrostatic charge for iQDCM.

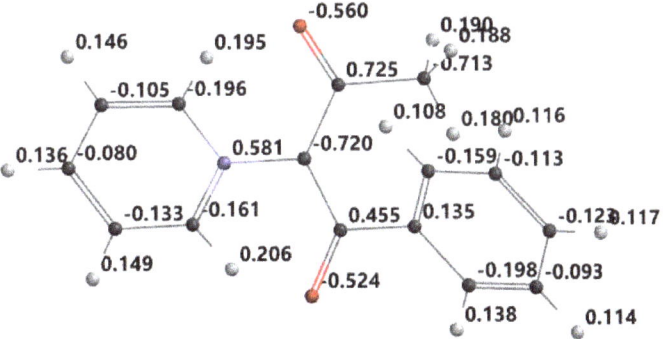

Figure 5. Optimized structure and atomic electrostatic charge for PABM.

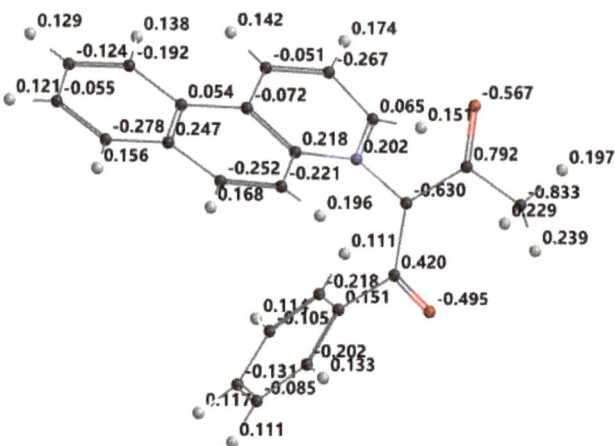

Figure 6. Optimized structure and atomic electrostatic charge for BQABM.

From the Figures 3–6, it can be seen that the ylid covalent bond is polarized; the electronic charge is shifted to carbanion from nitrogen. In this way, the studied methylids have a zwitterionic and basic nature in their ground electronic state. One can see that the covalent bond C=O is also polarized due to the high electronegativity of oxygen. Pronounced charge separation on the covalent bond C=O due to high electronegativity of oxygen can be seen in Figures 3–6. The smallest charge separation is on the C=O bond neighboring the benzene ring in PABM (Figure 5). This fact favors the ability of the studied methylids to participate in specific interactions with protic molecules [3,5,7].

Figures 7–10 show the transitions between HOMO and LUMO orbitals with intramolecular charge transition (ICT) from the carbanion towards the heterocycle for all studied cycloimmonium ylids.

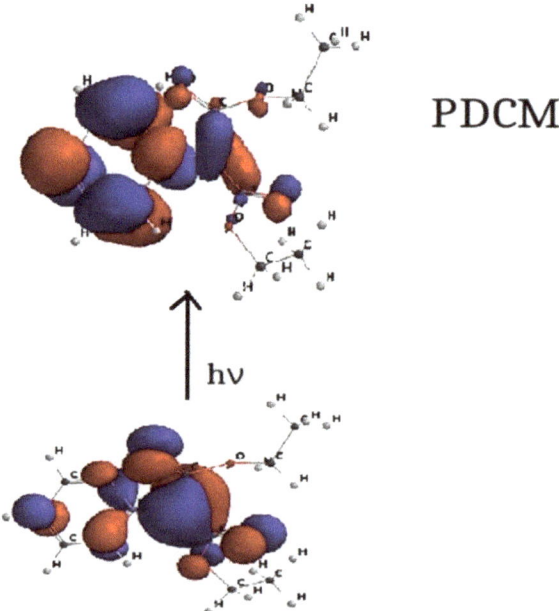

Figure 7. HOMO–LUMO orbital transition for PDCM.

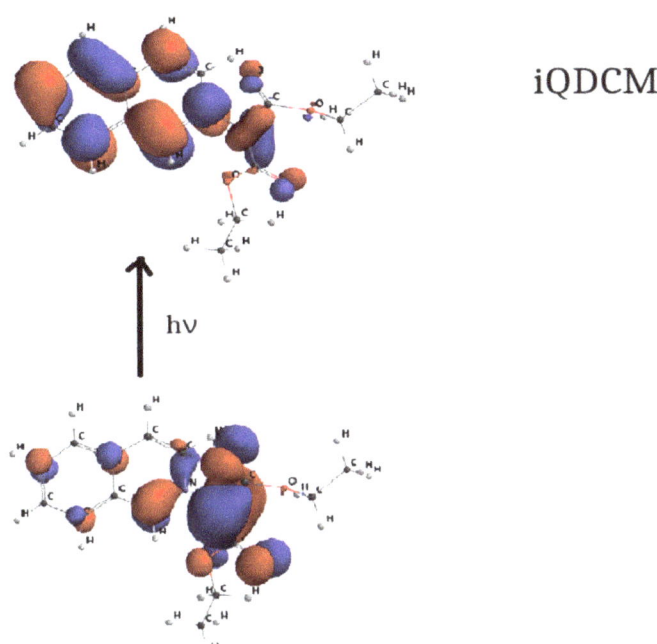

Figure 8. HOMO–LUMO orbital transition for iQDCM.

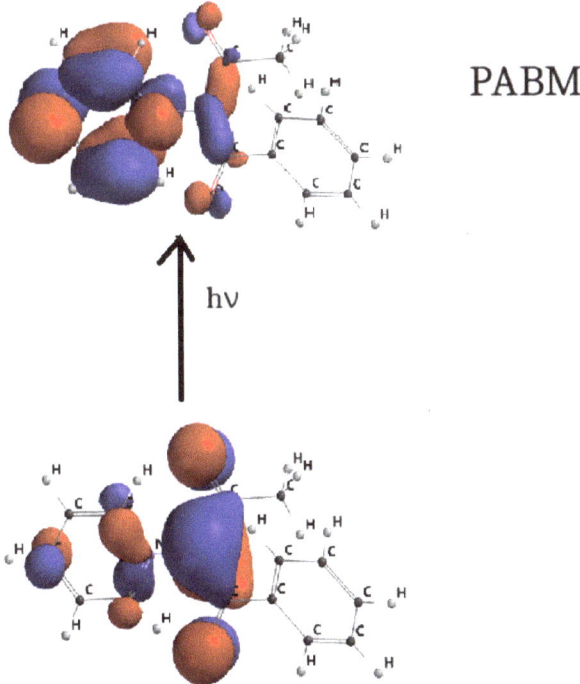

Figure 9. HOMO–LUMO orbital transition for PABM.

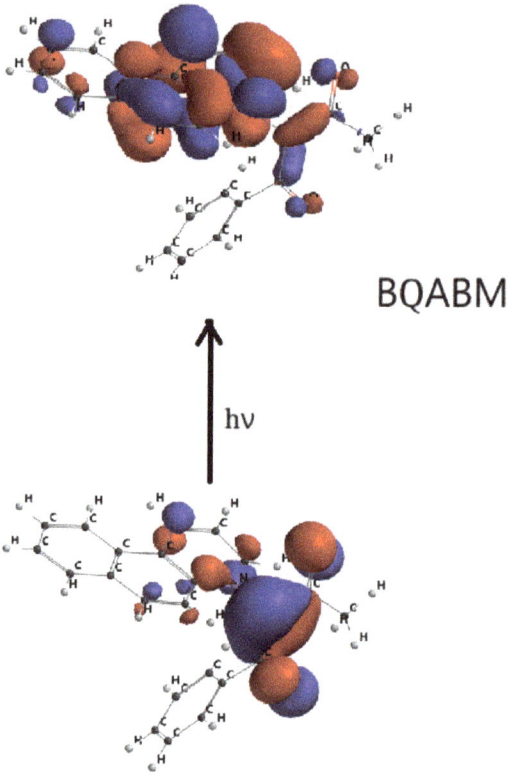

Figure 10. HOMO–LUMO orbital transition for BQABM.

In the ground electronic state (HOMO), the studied methylids are characterized by a high electronic charge on the carbanion, while in their excited electronic state (LUMO), the non-bonded electrons are shifted towards the heterocycle. So, the electric dipole moment of the considered molecules decreases [6,7,23,24] in the process of the visible photon absorption. This results in the solvation energy of the studied methylids being higher in the ground electronic state compared with the first excited-state solvation energy.

The molecular descriptors obtained in quantum mechanical analysis for the considered cycloimmonium ylids are listed in Table 2. The results from Table 2 correspond to gaseous state of the studied molecules.

From Table 2, it can be seen that studied molecules are dipolar and polarizable. According to Koopmans's theorem [47] the values with the changed sign of HOMO and LUMO energies are the ionization potential and electron affinity of the studied methylids. The relatively small values of these parameters show that the cycloimmonium ylids can be used as precursors in chemical reactions.

The HOMO–LUMO gap (Table 2) shows the high chemical reactivity and biological activity of the studied molecules.

The HBA and HBD count shows that the methylids do not donate protons to the solvent molecules (null value of HBD count) but can receive protons from protic solvents. There are three places (HBA count is 3) in methylids where the hydrogen bond can be realized.

The data from Table 2 were corroborated with the spectral results used in this study for calculating the excited dipole moments for the studied molecules.

Table 2. Molecular descriptors of the studied methylids computed by Spartan '14.

Properties \ Ylid	PDCM	iQDCM	PABM	BQAM
Total energy (au)	−829.93	−975.57	−784.5894	−1091.84
Dipole moment (Debye)	7.40	8.34	3.24	8.64
Polarizability (Å^3)	60.50	64.74	61.20	69.66
Solvation energy (kJ/mol)	−49.38	−48.11	−39.36	−54.42
Weight (amu)	237.255	287.315	239.274	339.394
Volume (Å^3)	244.96	296.33	254.49	356.49
Area (Å^2)	268.89	315.88	264.40	352.67
EHOMO (eV)	−5.34	−5.25	−5.38	−5.17
ELUMO (eV)	−2.18	−2.40	−1.94	−2.50
$|\Delta E| = |\text{EHOMO-ELUMO}|$ (eV)	3.16	2.85	3.44	2.67
HBA count	3	3	3	3
HBD count	0	0	0	0

3.2. Spectral Results

The experimental wavenumbers (expressed in cm^{-1}) at the maximum of the visible electronic absorption bands of the considered methylids are listed in Table 3.

Table 3. Wavenumbers at the maximum of the electronic absorption band for the studied methylids.

No.	Solvent	ν (cm^{-1})			
		PDCM	iQDCM	PABM	BQABM
1	Dioxane	22,900	21,607	22,830	21,050
2	Benzene	22,550	21,116	23,400	20,600
3	o-Xylene	22,450	21,120	23,850	21,350
4	Toluene	22,720	21,460	23,750	20,250
5	Anisole	23,040	21,725	23,400	21,180
6	Chloroform	23,280	22,040	24,630	21,500
7	n-Butyl acetate	23,020	21,450	24,280	21,620
8	Chlorobenzene	22,950	21,270	23,800	21,200
9	Ethyl acetate	23,300	21,516	24,330	20,880
10	Dichloromethane	23,160	21,970	24,570	21,310
11	Benzyl alcohol	24,770	23,405	26,020	23,040
12	Cyclohexanol	24,530	23,070	26,000	23,950
13	n-Butyl alcohol	24,550	23,265	25,980	24,100
14	Isobutyl alcohol	24,700	23,270	26,300	24,160
15	n-Propyl alcohol	24,950	23,260	26,450	24,250
16	Acetone	23,450	21,970	24,480	21,900
17	Ethanol	24,970	23,685	26,760	24,160
18	Methanol	25,950	23,960	27,500	24,200
19	Water	25,450	24,180	27,620	24,520
20	Formamide	25,190	24,100	27,020	23,070

The data from Table 3 show that the visible band of cycloimmonium ylids is sensitive to the solvent nature; it shifts to blue when the ylids are passed from non-polar/aprotic solvents to polar/protic ones, due to the increase in the orientation and specific interactions in the ground electronic state.

One can establish linear dependence between the energy at the maximum of the visible band of cycloimmonium ylids and the empirical parameter introduced by Kosower for the solvent polarity (see Equation (8) and Table 4).

$$E_{max}\left(\frac{\text{kcal}}{\text{mol}}\right) = MZ\left(\frac{\text{kcal}}{\text{mol}}\right) + N \tag{8}$$

Table 4. Regression coefficients for Equation (8).

Molecule	M ± ΔM	N ± ΔN	R	Standard Deviation	Nr
PDCM	0.31 ± 0.02	46.8 ± 1.6	0.93	0.8	16
iQDCM	0.27 ± 0.02	45 ± 1.5	0.92	0.9	17
PABM	0.36 ± 0.02	46.4 ± 2.1	0.91	1.2	16
BQABM	0.44 ± 0.04	32.9 ± 2.6	0.92	1.2	15

The slope M, in relation to (8), indicates the strength of the molecular interactions of the methylid with the solvent and the cut at the origin, N (kcal/mol), approximates the energy at the maximum of the visible absorption band when it is recorded in the vaporous state of the spectrally active molecule. From Table 4, it can be seen that cycloimmonium ylids with symmetrically substituted carbanion are less sensitive to the solvent action compared to those with asymmetric carbanion. The cut at the origin in Equation (8) decreases with the increase in the benzene rings in the ylid molecule. In the last column of Table 4, the number Nr of the points used in statistical analysis after the elimination of the aberrant points is specified.

The dependence of the type (8) is illustrated in Figure 11. The very good linear dependence between $E_{max}\left(\frac{\text{kcal}}{\text{mol}}\right)$ and the Kosower empirical polarity is given due to the similar mechanism determining the visible band appearance for the cycloimmonium ylids and for the complex considered by Kosower.

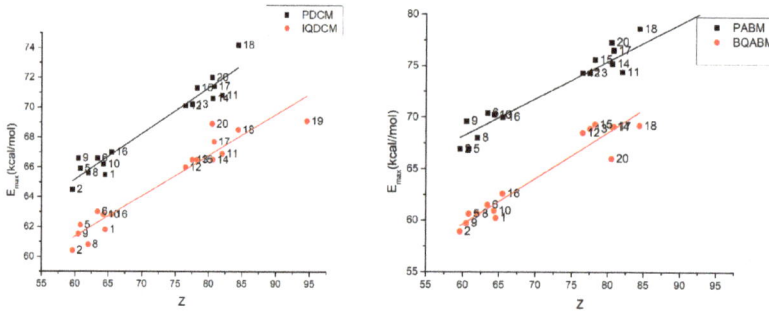

Figure 11. Energy at the maximum of the absorption band versus Z solvent parameter for the studied molecules.

The standard molecule considered by Reichardt in obtaining the solvent polarity, betaine 1 dye (see Figure 2a), has a chemical structure more similar to that of the cycloimmonium ylids. It is normal to obtain a good linear dependence between the energy at the maximum of the visible band of cycloimmonium ylids and the parameters of the $E_T(30)$ empirical scale, from Equation (9), as shown by the data of Table 5.

$$E_{max}\left(\frac{kcal}{mol}\right) = CE_T(30) + D \qquad (9)$$

Table 5. Regression coefficients for Equation (9).

Molecule	$C \pm \Delta C$	$D \pm \Delta D$	R	Standard Deviation	N
PDCM	0.39 ± 0.03	51.1 ± 1.2	0.91	0.86	19
iQDCM	0.35 ± 0.02	48.6 ± 1.1	0.91	0.88	20
PABM	0.45 ± 0.03	52.5 ± 1.2	0.95	0.93	18
BQABM	0.54 ± 0.04	40.6 ± 2	0.90	1.3	16

In the last column of Table 5, the number N of the points used in statistical analysis after the elimination of the aberrant points is specified.

Very good correlation between the positions of the visible band of cycloimmonium ylids and those of the Reichardt' standard dye is illustrated in Figure 12. This fact can be explained by the similitude between the electronic transitions in the visible photon absorption process for both types of molecules.

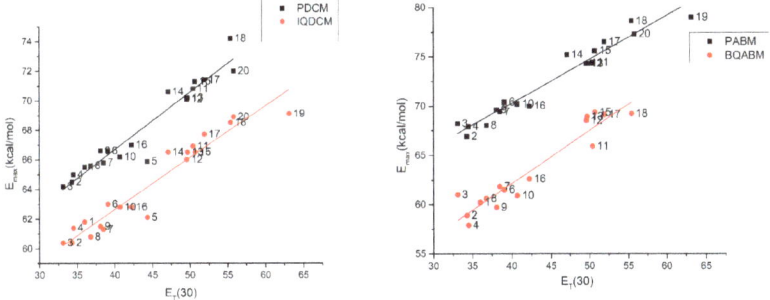

Figure 12. Energy at the maximum of the absorption band versus the $E_T(30)$ parameter for the studied molecules.

The mono-parameter empirical scales of solvents proposed by Kosower and by Reichardt categorize the solvents regarding their influence on the ITC visible band of cycloimmonium ylids but do not establish relations between the correlation coefficients of Equations (8) or (9) and the molecular descriptors computed for the cycloimmonium ylids.

By using dependences of the type (5), in which the theoretical results are combined with the empirical ones, the correlation coefficients resulting from theoretical means express their dependence on the molecular descriptors of the spectrally active molecules by equations of the types (6) and (7), allowing new estimations to be obtained for the excited states of the studied molecules.

In our attempt to use an equation of the type (5) in statistical analysis of the spectral data, the equation of the type (10) was obtained, showing an important role of the universal interactions (described by the term $C_1 f(\varepsilon)$) and of the hydrogen bond (described by the term $C_2 \alpha$), in which the methylids accept protons from the protic molecules of the solvents. An equation of the type (10) suggests the basic nature of cycloimmonium ylids and their dipolar characteristics in orientation interactions.

$$\nu = \nu_0 + C_1 f(\varepsilon) + C_2 \alpha \qquad (10)$$

In the last column of Table 6, the number N of the points used in statistical analysis after the elimination of the aberrant points is specified.

Table 6. Regression coefficients for Equation (10).

Ylid	$v_0 \pm \Delta v_0$	$C_1 \pm \Delta C_1$	$C_2 \pm \Delta C_2$	R	N
PDCM	$22{,}335 \pm 237$	1134 ± 451	1949 ± 249	0.94	20
iQDCM	$21{,}101 \pm 191$	746 ± 361	2136 ± 200	0.96	20
PABM	$22{,}969 \pm 290$	1656 ± 550	2607 ± 304	0.95	20
BQABM	$20{,}153 \pm 395$	1724 ± 723	2794 ± 382	0.94	18

The correlation coefficients obtained in statistical analysis based on relation (5) and the data of Tables 1 and 3, when the parameters without real importance (f(n) and β) were eliminated, are given in Table 6.

The values computed by Equation (10) and the experimental data referring to the visible electronic absorption band of the considered cycloimmonium ylids are plotted in Figures 13 and 14.

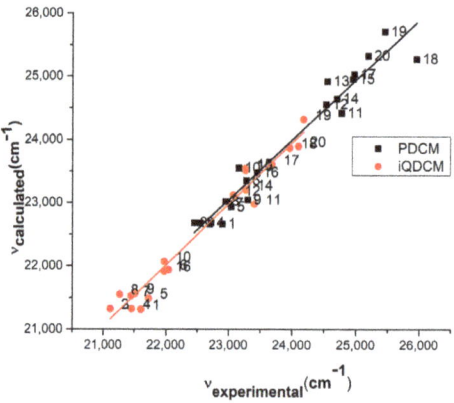

Figure 13. Experimental vs. computed values of the wavenumbers in maximum of the visible electronic absorption band of PDCM and iQDCM molecules.

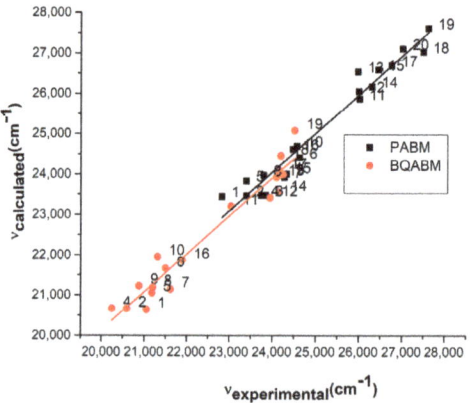

Figure 14. Experimental vs. computed values of the wavenumbers in maximum of the visible electronic absorption band of PABM and BQABM molecules.

There is a very good correlation between computed and experimental wavenumbers at the maximum of the absorption band: the slopes are nearly 1 (0.95 for PDCM, 0.97 for iQDCM, 0.96 for PABM, 0.94 for BQABM), and the regression coefficient Adj R-square is over 0.94 for all studied molecules.

Using the correlation coefficients from Table 6, the contribution of each type of interaction was computed for the four cycloimmonium ylids under study. The term $C_1 f(\varepsilon)$ describes the strength of the universal interactions, while the term $C_2 \alpha$ describes the contribution of the specific interactions. Table 7a,b show the results of calculation for the contribution of these interactions to the total spectral shift recorded in the respective solvent, in cm^{-1}, as well as in percentages.

Table 7. (a). Contribution of universal interactions $C_1 f(\varepsilon)$ (cm^{-1}) and of the specific interactions $C_2 \alpha$ (cm^{-1}) to the total spectral shift recorded in each solvent. The numbers correspond to the solvents from Table 1. (b). Contribution of universal interactions $C_1 f(\varepsilon)$ (cm^{-1}) and of the specific interactions $C_2 \alpha$ (cm^{-1}) to the total spectral shift recorded in each solvent. The numbers correspond to the solvents from Table 1.

(a)				
	PDCM		iQDCM	
No.	$C_1 f(\varepsilon)$; (P%)	$C_2 \alpha$; (P%)	$C_1 f(\varepsilon)$; (P%)	$C_2 \alpha$; (P%)
1	323; (100)	0	215; (100)	0
2	338; (100)	0	224; (100)	0
3	341; (100)	0	227; (100)	0
4	341; (100)	0	227; (100)	0
5	592; (100)	0	393; (100)	0
6	624; (61.5)	390; (38.5)	414; (49.2)	428; (50.8)
7	652; (100)	0	433; (100)	0
8	684; (100)	0	454; (100)	0
9	706; (100)	0	469; (100)	0
10	852; (68.6)	390; (31.4)	545; (56.0)	428; (44.0)
11	909; (43.7)	1170; (56.3)	603; (32.0)	1284; (68.0)
12	931; (41.9)	1287; (58.1)	618; (30.4)	1412; (69.6)
13	941; (36.5)	1638; (63.5)	625; (25.8)	1798; (74.2)
14	963; (41.7)	1346; (58.3)	639; (30.2)	1477; (69.8)
15	979; (37.4)	1638; (62.6)	650; (26.6)	1798; (73.4)
16	981; (86.3)	156; (13.7)	651; (79.2)	171; (20.8)
17	1011; (37.6)	1677; (62.4)	671; (26.7)	1840; (73.3)
18	1027; (35.0)	1911; (65.0)	682; (24.5)	2097; (75.5)
19	1089; (32.3)	2282; (67.7)	723; (22.4)	2504; (77.6)
20	1099; (36.7)	1892; (63.3)	730; (26.0)	2076; (74.0)
(b)				
	PABM		BQABM	
No.	$C_1 f(\varepsilon)$; (P%)	$C_2 \alpha$; (P%)	$C_1 f(\varepsilon)$; (P%)	$C_2 \alpha$; (P%)
1	475; (100)	0	492; (100)	0
2	496; (100)	0	514; (100)	0
3	501; (100)	0	519; (100)	9
4	501; (100)	0	519; (100)	0

Table 7. Cont.

5	870; (100)	0	901; (100)	0
6	916; (63.7)	522; (36.3)	949; (63)	558; (37)
7	958; (100)	0	992; (100)	0
8	1004; (100)	0	1041; (100)	0
9	1036; (100)	0	1075; (100)	0
10	1207; (69.8)	522; (30.2)	1250; (69.1)	558; (30.9)
11	1335; (46.0)	1566; (54.0)	1382; (45.2)	1674; (54.8)
12	1368; (44.3)	1723; (55.7)	1417; (43.5)	1841; (56.5)
13	1384; (38.7)	2192; (61.3)	1433; (38.0)	2344; (62.0)
14	1414; (44.0)	1801; (56.0)	1465; (43.2)	1925; (56.8)
15	1438; (39.6)	2192; (60.4)	1490; (38.9)	2344; (61.1)
16	1441; (87.3)	209; (12.7)	1493; (87.0)	223; (13.0)
17	1486; (39.8)	2245; (60.2)	1539; (39.0)	2399; (61.0)
18	1509; (37.1)	2558; (62.9)	1563; (36.4)	2743; (63.6)
19	1600; (34.4)	3054; (65.6)	1658; (33.7)	3264; (66.3)
20	1615; (39.0)	2531; (61.0)	1674; (38.2)	2706; (61.8)

The data of Table 7a,b show that the universal orientation-induction interactions are prevalent in non-protic solvents, while the specific interactions of the type hydrogen bonds become very important in the protic solvents, even more strong than those universal.

By using the values of C_1 (cm^{-1}) from Table 6 and relations (6) and (7), the excited-state dipole moment of the studied cycloimmonium ylids can be estimated (in the limits of the variational model [7,23,42]. The values of the molecular descriptors from Table 2 were used in estimating the excited dipole moment of the studied molecules.

The results of estimating the excited dipole moment of the studied molecules are listed in Table 8. The values of the molecular radius, a, of cycloimmonium ylids were computed using the values of the molecular area and volume computed by Spartan '14.

Table 8. Excited- and ground-state dipole moments of cycloimmonium ylids.

Ylid	a (Å)	μ_g(D)	μ_e(D)	$\alpha_g\left(\text{Å}^3\right)$	ΔE(cm^{-1})
PDCM	2.733	7.40	7.08	60.50	25,469
iQDCM	2.814	8.34	8.14	64.74	22,988
PABM	2.888	3.24	1.61	61.20	27,747
BQABM	3.033	8.64	8.07	69.66	29,602

We mention that the estimation of the excited-state dipole moment was conducted using the values of the ground electronic state of the dipole moment and polarizability for each molecule considered in its vaporous phase (Table 2).

From the data of Table 8, it can be seen that the smallest dipole moment in the ground state corresponds to PABM, a molecule with a uni-ring heterocycle and with an asymmetrically substituted carbanion. It is smaller than the ground-state dipole moment of PDCM, a cycloimmonium ylid with the same heterocycle due to the presence of four oxygen atoms in PDCM compared with PABM containing only two oxygen atoms. Oxygen is known as an electronegative atom able to increase the electronic charge delocalization. The ground dipole moment of PABM is smaller than that of BQABM due to the structure of the heterocycle in the last ylid. It can be seen that the charge separation on the C=O bond neighboring the benzene ring is very small (see Figure 5).

From Table 8, it can be seen that the electric dipole moment of all cycloimmonium ylids under study is diminished by the visible photon absorption in accordance with the shift of this absorption band to blue when the solvent polarity increases. The largest decrease by excitation was obtained for PABM. It is possible that the carbanion asymmetry determines its sp3 hybridization and facilitates the electronic charge shift towards the heterocycle.

In Table 8, the values ΔE of the distance between the ground and the first excited level (LUMO-HOMO) expressed in cm^{-1} are also listed, and it can be observed that the $n - \pi^*$ transition is realized from the ground state towards one non-occupied level situated below the LUMO.

4. Conclusions

By corroborating the computational and spectral data regarding the chosen four cycloimmonium ylids, one can observe the dipolar and polarizable characteristics of these molecules.

The two empirical scales, proposed by Kosower and by Reichardt, categorize the solvents regarding their global effects on the studied cycloimmonium ylids.

Results of the present study showed very good correspondence between the wavenumbers at the maximum of the electronic bands determined by the same mechanism (based on the electron charge transfer) both for the studied molecules and for the standard molecules used in the definition of the mono-parameter polarity scales.

The multi-parameter solvent scales can be used in determining the nature of molecular interactions and the contribution of each type of interaction to the total spectral shifts of the electronic bands.

Corroborating the theoretical results with the empirical treatment of the solvent effect on the electronic absorption bands of the spectrally active molecules, as cycloimmonium ylids, one can obtain information about the excited-state dipole moments.

The excited-state dipole moment of each studied cycloimmonium ylid is smaller than that in its ground electronic state, proving the shift of electronic charge towards the heterocycle in the visible absorption process.

Supplementary Materials: The following supporting information can be downloaded at: https://www.mdpi.com/article/10.3390/liquids4010009/s1.

Author Contributions: Conceptualization, D.O.D. and D.B.; Methodology, D.O.D.; Software, A.V. and D.B.; Validation, D.B. and D.O.D.; Formal Analysis, A.V.; Investigation, D.O.D. and A.V.; Data Curation, A.V.; Writing—Original Draft Preparation, D.O.D. and A.V.; Writing—Review & Editing, D.O.D. and D.B.; Visualization, A.V. and D.B.; Supervision, D.O.D. and D.B.; Project Administration, D.O.D. All authors have equal contribution to this study. All authors have read and agreed to the published version of the manuscript.

Funding: This research received no external funding.

Data Availability Statement: Data are contained within the article and supplementary materials.

Acknowledgments: We acknowledge the efforts of Professor Christian Reichardt in developing the experimental and theoretical studies regarding solvent effects in Organic Chemistry; for the proposal and enrichment of two solvent polarity scales, E_(T) (30) and E_T^N, based on betaine dyes for about 400 simple and complex liquids; and for the great number of important applications in biology, techniques and fundamental science induced by his research. His understanding of liquid complexity and his perseverance, tenacity, devotion and modesty inspire the young generation and greatly contribute to scientific knowledge and productivity.

Conflicts of Interest: The authors declare no conflicts of interest.

References

1. Johnson, A.W. *Ylid Chemistry*; Academic Press: New York, NY, USA, 1966.
2. Zugravescu, I.; Petrovanu, M. *N-ylid Chemistry*; McGraw Hill: New York, NY, USA, 1976.
3. Dorohoi, D.O. Electronic spectra of N-ylids. *J. Mol. Struct.* **2004**, *704*, 31–43. [CrossRef]

4. Dorohoi, D.O.; Dascalu, C.F.; Teslaru, T.; Gheorghies, L.V. Electronic absorption spectra of two 3-aryl-pyridazinium-2,4,6-picryl-benzoyl-methylids. *Spectrosc. Lett.* **2012**, *45*, 383–391. [CrossRef]
5. Dorohoi, D.O.; Dimitriu, D.G.; Dimitriu, M.; Closca, V. Specific interactions in N-ylid solutions, studied by nuclear magnetic resonance and electronic spectroscopy. *J. Mol. Struct.* **2013**, *1044*, 79–86. [CrossRef]
6. Morosanu, A.C.; Gritco-Todirascu, A.; Creanga, D.E.; Dorohoi, D.O. Computational and solvatochromic study of pyridinium-acetyl-benzoyl-methylid (PABM). *Spectrochim. Acta Mol. Biomol. Spectrosc.* **2018**, *189*, 307–315. [CrossRef] [PubMed]
7. Dorohoi, D.O.; Creanga, D.E.; Dimitriu, D.G.; Morosanu, A.C.; Gritco-Todirascu, A.; Mariciuc, G.G.; Melniciuc, N.P.; Ardelean, E.; Cheptea, C. Computational and Spectral Means for Characterizing the Intermolecular Interactions in Solutions and for Estimating Excited State Dipole Moment of Solute. *Symmetry* **2020**, *12*, 1299. [CrossRef]
8. Babusca, D.; Morosanu, A.C.; Dimitriu, D.G.; Dorohoi, D.O.; Cheptea, C. Spectroscopic and quantum mechanical study of molecular interactions of iso-quinolinium ylids in polar solutions. *Mol. Cryst. Liq. Cryst.* **2020**, *698*, 87–97. [CrossRef]
9. Leontie, L.; Olariu, I.; Rusu, G. On the charge transport in some new carbanion disubstituted ylides in thin films. *Mater. Chem. Phys.* **2003**, *80*, 506–511. [CrossRef]
10. Leontie, L.; Roman, M.; Branza, F.; Podaru, C.; Rusu, I.G. Electrical and optical properties of some new synthetized ylids in thin films. *Synth. Mat.* **2009**, *159*, 642–648. [CrossRef]
11. Pawda, A. *1,3-Dipolar Cycloaddition Reactions Chemistry*; Wiley Interscience: New York, NY, USA, 1984.
12. Petrovanu, M.; Zugravescu, I. *Cycloaddition Reactions*; Romanian Academic Publishing House: Bucuresti, Romania, 1997.
13. Bejan, V.; Moldoveanu, C.; Mangalagiu, I.I. Ultrasound assisted reactions of steroid analogous of anticipated biological activities. *Ultrason. Sonochem.* **2009**, *16*, 312–315. [CrossRef] [PubMed]
14. Saxena, H.O.; Faridi, U.; Kumar, J.; Luqman, S.; Darokar, M.; Shanker, K.; Chanotiya, C.S.; Gupta, M.; Negi, A.S. Synthesis of chalcone derivatives on steroidal framework and their anticancer activities. *Steroids* **2007**, *72*, 892–900. [CrossRef] [PubMed]
15. Wu, Y.; Batist, G.; Zamfir, L. Identification of a novel steroid derivative, NSC12983, as a paclitaxel-like tubulin assembly promoter by 3-D virtual screening. *Anti-Cancer Drug Design* **2001**, *16*, 129. [PubMed]
16. Ungureanu, M.; Mangalagiu, I.; Grosu, G.; Petrovanu, M. Antimicrobial activity of some new pyridazine derivatives. *Ann. Pharm. Fr.* **1997**, *55*, 69–72. [PubMed]
17. Caprosu, M.; Butnariu, R.M.; Mangalagiu, I. Synthesis and antimicrobial activity of some new pyridazine derivatives. *Heterocycles* **2005**, *65*, 1871–1873. [CrossRef]
18. Mantu, D.; Luca, M.C.; Moldoveanu, C.; Zbancioc, G.; Mangalagiu, I.I. Synthesis and Antituberculosis Activity of some new Pyridazine Derivatives. Part II. *Eur. J. Med. Chem.* **2010**, *45*, 5164–5168. [CrossRef]
19. Seifi, M.; Bahonar, S.; Ebrahimipour, S.Y.; Simpson, J.; Dusek, M. Combination of pyridinium and isoquinolinium ylides with phenylisocyanate and isothiocyanates: Synthesis, characterisation, and X-ray crystal structures of mesoionic monosubstituted 3-oxo-propanamides or thioamides. *Aust. J. Chem.* **2015**, *68*, 1577–1582. [CrossRef]
20. Sar, S.; Guha, S.; Prabakar, T.; Maiti, D.; Sen, S. Blue light-emitting diode-mediated in situ generation of pyridinium and isoquinolinium ylides from aryl diazoesters: Their application in the synthesis of diverse dihydroindilozine. *J. Org. Chem.* **2021**, *86*, 11736–11747. [CrossRef] [PubMed]
21. Pop, V.; Dorohoi, D.; Holban, V. Molecular interactions in binary solutions of 4-amino-phthalimide and 3-p-cumyl-pyridazinium-acetyl-benzoyl-methylid. *Spectrochim. Acta Part A Mol. Spectrosc.* **1994**, *50*, 2281–2289. [CrossRef]
22. Dulcescu, M.M.; Stan, C.; Dorohoi, D.O. Spectral study of intermolecular interactions in water-ethanol solutions of some carbanion disubstituted ylids. *Rev. Roum. Chim.* **2010**, *55*, 403–408. Available online: http://ucb.icf.ro/rrch/ (accessed on 15 June 2023).
23. Dorohoi, D.O.; Partenie, D.H.; Morosanu, A.C. Specific and universal interactions in Benzo-[f]-Quinolinium Acetyl-Benzoyl Methylid (BQABM) solutions; excited state dipole moment of BQABM. *Spectrochim. Acta Part A Mol. Biomol. Spectrosc.* **2019**, *213*, 184–191. [CrossRef] [PubMed]
24. Dulcescu-Oprea, M.M.; Morosanu, A.C.; Dimitriu, D.G.; Gritco-Todirascu, A.; Dorohoi, D.O.; Cheptea, C. Solvatochromic study of pyridinium acetyl benzoyl methylid (PABM) in ternary protic solvents. *J. Mol. Struct.* **2021**, *1227*, 129539. [CrossRef]
25. Dimroth, K.; Reichardt, C.; Siepmann, T.; Bohlmann, F. Über Pyridinium-N-phenol-betaine und ihre Verwendung zur Charakterisierung der Polarität von Lösungsmitteln. *Justus Liebigs Ann. Chem.* **1963**, *661*, 1–37. [CrossRef]
26. Reichardt, C. Pyridinium-N-phenolate betaine dyes as empirical indicators of solvent polarity: Some new findings. *Pure Appl. Chem.* **2008**, *80*, 1415–1432. [CrossRef]
27. Reichardt, C. *Solvents and Solvent Effects in Organic Chemistry*, 3rd ed.; Wiley-VCH: Weinheim, Germany, 2003.
28. McRae, E.G. Theory of solvent influence on the molecular electronic spectra. Frequency shifts. *J. Phys. Chem.* **1957**, *61*, 562–572. [CrossRef]
29. Bakhshiev, N.G. *Spectroscopy of Intermolecular Interactions*; Nauka: Leningrad, Russian Federation, 1972. (In Russian)
30. Abe, T. Theory of solvent influence on the molecular electronic spectra. Frequency shifts. *Bull. Chem. Soc. Jpn.* **1965**, *38*, 1314–1318. [CrossRef]
31. Kamlet, M.J.; Abboud, J.I.; Abraham, M.; Taft, R.W. Linear solvation energy relationships 23. A comprehensive collection of the solvatochromic parameters π^*, α, β and some methods for simplifying the generalized solvatochromic equation. *J. Org. Chem.* **1983**, *48*, 2877–2887. [CrossRef]
32. Catalán, J.; Hopf, H. Empirical Treatment of the Inductive and Dispersive Components of Solute−Solvent Interactions: The Solvent Polarizability (SP) Scale. *Eur. J. Org. Chem.* **2004**, *2004*, 4694–4702. [CrossRef]

33. Brooker, L.G.S.; Keyes, G.H.; Heseltine, D.W. Color and Constitution. X.[1] Absorption of the Merocyanines. *J. Am. Chem. Soc.* **1951**, *73*, 5332–5350. [CrossRef]
34. Kosower, E.M. The Effect of Solvent on Spectra. II. Correlation of Spectral Absorption Data with Z-Values. *J. Am. Chem. Soc.* **1958**, *80*, 3261–3267. [CrossRef]
35. Streitwieser, A.; Heathcoch, C.H.; Kosower, E.M. *Introduction to Organic Chemistry*, 4th ed.; Maxwell MacMillan International: New York, NY, USA, 1992; pp. p. 678 and essay 4 after p. 621.
36. Griffiths, T.R.; Puch, D.C. Solvent polarity studies. Part I. New Z values and relationships with other solvent polarity scales. *J. Solut. Chem.* **1979**, *8*, 247–258. [CrossRef]
37. Sıdır, İ.; Sıdır, Y.G. Investigation on the interactions of E -4-methoxycinnamic acid with solvent: Solvatochromism, electric dipole moment and pH effect. *J. Mol. Liq.* **2017**, *249*, 1161–1171. [CrossRef]
38. Kawski, A. On the Estimation of Excited-State Dipole Moments from Solvatochromic Shifts of Absorption and Fluorescence Spectra. *Z. Nat. A* **2002**, *57*, 255–262. [CrossRef]
39. Kawski, A.; Bojarski, P. Comments on the determination of excited state dipole moment of molecules using the method of solvatochromism. *Spectrochim. Acta Part A Mol. Biomol. Spectrosc.* **2011**, *82*, 527–528. [CrossRef] [PubMed]
40. Gahlaut, R.; Tewari, N.; Bridhkoti, J.P.; Joshi, N.K.; Joshi, H.C.; Pant, S. Determination of ground and excited states dipole moments of some naphthols using solvatochromic shift method. *J. Mol. Liq.* **2011**, *163*, 141–146. [CrossRef]
41. Zakerhamidi, M.S.; Moghadam, M.; Ghanadzadeh, A.; Hosseini, S. Anisotropic and isotropic solvent effects on the dipole moment and photophysical properties of rhodamine dyes. *J. Lumin.* **2012**, *132*, 931–937. [CrossRef]
42. Dorohoi, D.O. Excited state molecular parameters determined by spectral means. *Ukr. J. Phys.* **2018**, *63*, 701–707. [CrossRef]
43. Available online: https://www.stenutz.eu./chem.solv26.php (accessed on 22 June 2023).
44. Spartan'14 for Windows, Macintosh and Linux. In *Tutorial and User's Guide*; Wavefunction Inc.: Irvine, CA, USA, 2010.
45. Young, D. *Computational Chemistry*; Wiley Interscience: Sebastopol, CA, USA, 2001.
46. Lin, C.Y.; George, M.W.; Gill, P.M.W. E.D.F.2: A density functional for predicting molecular vibrational frequencies. *Aust. J. Chem.* **2004**, *57*, 365–370. [CrossRef]
47. Koopmans, T. Über die Zuordnung von Wellenfunktionen und Eigenwerten zu den einzelnen elektronen eines Atom. *Physica* **1934**, *1*, 104–113. [CrossRef]

Disclaimer/Publisher's Note: The statements, opinions and data contained in all publications are solely those of the individual author(s) and contributor(s) and not of MDPI and/or the editor(s). MDPI and/or the editor(s) disclaim responsibility for any injury to people or property resulting from any ideas, methods, instructions or products referred to in the content.

Article

Use of DFT Calculations as a Tool for Designing New Solvatochromic Probes for Biological Applications

Cynthia M. Dupureur

Department of Chemistry & Biochemistry, University of Missouri St. Louis, St. Louis, MO 63121, USA; cdup@umsl.edu

Abstract: The intramolecular charge transfer behavior of push–pull dyes is the origin of their sensitivity to environment. Such compounds are of interest as probes for bioimaging and as biosensors to monitor cellular dynamics and molecular interactions. Those that are solvatochromic are of particular interest in studies of lipid dynamics and heterogeneity. The development of new solvatochromic probes has been driven largely by the need to tune desirable properties such as solubility, emission wavelength, or the targeting of a particular cellular structure. DFT calculations are often used to characterize these dyes. However, if a correlation between computed (dipole moment) and experimentally measured solvatochromic behavior can be established, they can also be used as a design tool that is accessible to students. Here, we examine this correlation and include case studies of the effects of probe modifications and conformation on dipole moments within families of solvatochromic probes. Indeed, the ground state dipole moment, an easily computed parameter, is correlated with experimental solvatochromic behavior and can be used in the design of new environment-sensitive probes before committing resources to synthesis.

Keywords: fluorescence; solvatochromism; DFT

Citation: Dupureur, C.M. Use of DFT Calculations as a Tool for Designing New Solvatochromic Probes for Biological Applications. *Liquids* **2024**, *4*, 148–162. https://doi.org/10.3390/liquids4010007

Academic Editors: William E. Acree, Jr., Franco Cataldo and Enrico Bodo

Received: 24 November 2023
Revised: 5 January 2024
Accepted: 29 January 2024
Published: 4 February 2024

Copyright: © 2024 by the author. Licensee MDPI, Basel, Switzerland. This article is an open access article distributed under the terms and conditions of the Creative Commons Attribution (CC BY) license (https://creativecommons.org/licenses/by/4.0/).

1. Introduction

Fluorescence has become one of the most common means of molecular detection. Sensitivity, ease-of-use, and the broad array of fluorescence techniques have fueled these developments. Fluorescent molecules that exhibit differential emission properties in response to environment, e.g., viscosity and/or solvent polarity, are solvatochromic and thus particularly efficient probes [1].

For many years, cellular membranes, composed primarily of amphipathic phospholipids, were principally thought to serve as boundaries and barriers for cells. However, in the past few decades, we have come to better understand a remarkable dynamic complexity. The existence of lipid rafts, or mobile domains, that are distinct in their composition and properties from the surrounding lipid has been known for some time [2,3]. Lipids containing more unsaturated phospholipids and cholesterol tend to be more rigid and less hydrated and therefore more nonpolar [4,5], with lower local viscosity (Lo; Figure 1). Conversely, regions with more saturated phospholipids and less cholesterol have more mobility, are more polar, and exhibit higher local viscosity (Ld).

Thus, the structures of cellular membranes lend themselves well to the application of solvatochromic probes. There are a number of well-known fluorescent probes of lipid order. These are well discussed in recent reviews [1,4], and some are discussed below.

What has driven interest in these probes is a developing understanding of the role of lipid order and lipid dynamics in a variety of important cellular processes. One of these is cellular stress, in the form of starvation or oxidative stress. Dioxaborine- and Nile Red- based dyes have been recently applied to map the effects of these types of stresses on cellular lipid structures [6,7]. Characterizing changes in lipid order as a result of apoptosis, or programmed cell death, is another area of strong interest. Here, Nile

Red and hydroxyflavones have been applied [8,9], and this area of probe application has been reviewed recently [10]. Another cellular process being probed with solvatochromic lipid dyes is viral entry. Laurdan and DiO are two probes that have been applied in this area [11–14].

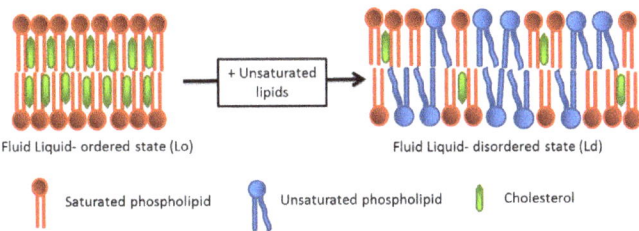

Figure 1. Lo and Ld states of phospholipids. From Ref. [5] with permission.

1.1. Desirable Properties in a Solvatochromic Probe

Advantageous excitation and emission wavelength ranges are of primary concern in choosing a probe application. This would include ranges that do not overlap with other emission signals in a cell (350–550 nm; [15–17]). This is the primary driving force for the development of red emitting probes, e.g., 600–750 nm [18–20]. Also important are properties that affect sensitivity. An ideal, sensitive fluorescent probe should have appropriate solubilities, high extinction coefficients (>30,000 M^{-1} cm^{-1}), high quantum yields in the medium of interest (>50%), and sufficient photostability for the purpose [21]. Often, a delicate balance of all of these properties is required for the application of a probe, and sometimes, it is necessary to compromise on one or more of these elements to achieve an experimental goal. For example, some of the most commonly used solvatochromic probes (laurdan and NR12S) do not have the most desirable photostability [16].

The structural features that lend themselves to solvatochromism include extended π systems. Second, molecules that respond to solvent polarity typically have large dipole moments and, more precisely, a large difference between the ground and excited state dipole moments. Structurally, this translates to uneven distribution of electrons in the molecule, usually facilitated by the presence of polar groups, more specifically, electron donating and electron withdrawing groups separated by an extended system of conjugation through which electrons can easily travel via push–pull or intramolecular charge transfer (ICT) behavior [22]. If a more polar solvent stabilizes the probe in its excited state (larger probe dipole moment) to a greater extent than it stabilizes the ground state (smaller probe dipole moment), positive solvatochromism results [23]. This is exhibited as a red (bathochromic) shift in optical spectra (Figure 2). If, instead, a more polar solvent stabilizes the ground state of the probe more than the excited state, negative solvatochromism results (not shown). In this case, the transition is of higher energy and the optical spectrum exhibits a blue or hypsochromic shift [23].

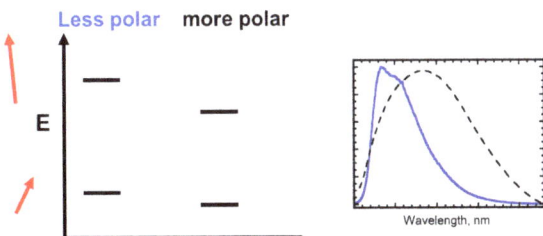

Figure 2. Positive solvatochromism. When moving to a solvent of greater polarity, the excited state dipole moment (red) is larger than the ground state dipole moment (red), resulting in a red shift (right, dashed).

1.2. The Problem of Design

The early history of applying solvatochromic probes to the study of lipids was serendipitous, that is, dyes in the Nile Red family were known to be sensitive to solvent polarity [18]. However, more recently the optimal properties of a solvatochromic dye have received more direct attention. Older dyes have been used as scaffolding and modified for specific purposes. Much of the design of solvatochromic probes stems from specific needs such as solubility in a lipid environment [21], facile synthetic routes, and a general understanding that useful push–pull dyes require electron donating and withdrawing groups separated by extended conjugation. Occasionally, functional groups are added because they are thought to optimize for specific properties such as organelle targeting [7], photostability [16], or solubility [24]. The somewhat meandering history of lipid probe literature does not typically document extensive rational design. Indeed, there is very little literature on the rational design of these dyes [24–26]. Desirable behaviors of these dyes are discussed, but usually in hindsight.

While there are a number solvatochromic dyes widely used for biological imaging, it is generally acknowledged that due to the complexity of properties that require balance for an application, there is an ongoing need to develop new dyes [1]. For synthetic chemists, this provides an attractive challenge. However, not all researchers possess this skill or have ready access to it. It would be advantageous to have a means of assessing the potential of new structures for solvatochromic behavior before committing the resources for synthesis and characterization.

Density functional theory (DFT) calculations are a common feature of papers that describe the synthesis and characterization of new solvatochromic probes. These calculations are readily accessible to even the most junior researchers via programs like Spartan and Gaussian, and these calculations provide a number of molecular parameters that are correlated with good solvatochromic behavior. The first are the energies of HOMO and LUMO, from which the difference is easily computed. Small HOMO LUMO gaps (e.g., 3–5 eV; [27,28]) facilitate electronic transitions and are common among solvatochromic probes. Secondly, maps of the HOMO and LUMO in ground and excited states illustrate the movement of electron density of the molecule upon excitation. Another important visual is the electrostatic potential (ESP) map, which illustrates charge distribution. When dramatic, these latter images can clearly show the potential for intramolecular charge transfer (ICT) or push–pull behavior, a hallmark feature of solvatochromic probes. See Figure 3 for examples of these types of DFT data for a solvatochromic probe.

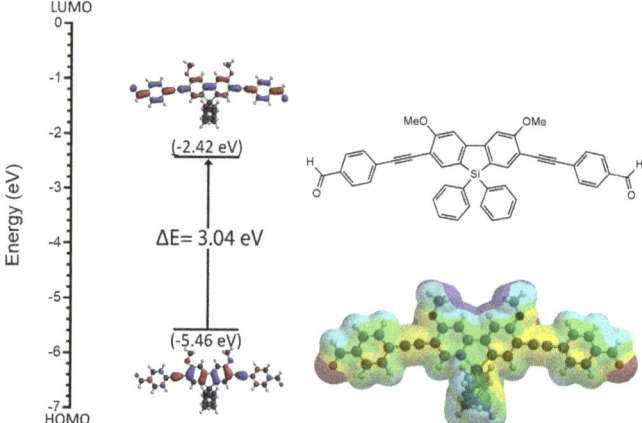

Figure 3. Example of DFT data for solvatochromic probe MF5. ESP map (lower right) is for the ground state in polar solvent. Adapted from [29]. Red indicates areas of highest electron density, and blue the lowest.

Although the landscape of solvatochromic probe design is complex, it would be desirable and supportive to be able to compute a single, accessible parameter that is correlated with experimental solvatochromic behavior. Transition dipole moments ($\mu_e - \mu_g$) are of course informative. A common range for these values for solvatochromic probes is 5–14 D, but these are often computed from the slopes of Lippert–Mataga plots and thus are derived from experimental data [26,30,31]. Exited state dipole moments are seldom reported [30]. Computed ground state dipole moments are sometimes reported and are often large for solvatochromic probes [31–33]. Discussed here is the potential of DFT, in particular, ground state dipole moments, to contribute to the design process, rather than as a form of characterization presented with synthesis. First, reported data on a few series of known solvatochromic dyes are compiled to determine the extent to which a computed dipole moment correlates with experimentally determined solvatochromic behavior. Then, we will examine, in retrospect, how modifications to known families of dye structures can affect the ground state dipole moment and electrostatic potential maps. The effect of conformation on computed dipole moments will also be examined. Then, we will explore the application of DFT in the design of metallafluorenes as lipid probes. All of these investigations are accessible to students and can provide excellent training in molecular properties and probe design.

2. Methods

2.1. DFT Calculations

Routine dipole moments were computed using Spartan '18 (Wavefunction, Irvine, CA, USA). The probe structures were first energy minimized. Using the density functional basis set B3LYP-6-31G*, the equilibrium geometry was calculated in the ground state in polar solvent (DMF). Output includes the dipole moment. To systematically search for low energy conformers (conformer distribution), the density functional sets wB97X-V and 6-311+G were applied.

2.2. Determination of E_{T30} and Lippert–Mataga Slopes

The Stokes shift ($\Delta \bar{\nu}$) is calculated as follows:

$$\Delta \bar{\nu} = \bar{\nu}_A - \bar{\nu}_F \tag{1}$$

where $\bar{\nu}_A$ is the λ_{max} of the absorption spectra and $\bar{\nu}_F$ is the λ_{max} of the emission spectra in wavenumbers (cm^{-1}).

Small variations in the forms of solvent polarizability are often applied [32,34–36], and this can lead to some variability. To normalize that, Stokes shift data from the literature were plotted vs. a form of the solvent polarizability function $f_1(\epsilon, \eta)$, defined as follows [37]:

$$f_1(\epsilon, \eta) = \left(\frac{\epsilon - 1}{2\epsilon + 1} - \frac{\eta^2 - 1}{2\eta^2 + 1} \right) \tag{2}$$

where (ϵ) is the dielectric constant and (η) is the refractive index of the solvent. Polarity functions were either obtained from the literature [34,38] or computed from solvent reference data [39]. Table 1 features a summary of solvent data used in this study. Published Stokes shift data were plotted vs. both the above polarizability function $f_1(\epsilon, \eta)$ and vs. E_{T30} [23] using Kaleidagraph 3.51 software (Synergy Software, Reading, PA, USA).

Table 1. Solvent data used in this study [a].

Solvent	Dielectric Constant [b] ϵ	Refractive Index [b] η	E_{T30} [c]	$f_1(\epsilon, \eta)$ [d]
Water	80.1	1.3330	63.1	0.3217
Glycol	37.0	1.4385	56.3	0.2719
Methanol	32.7	1.3284	55.4	0.3086

Table 1. Cont.

Solvent	Dielectric Constant [b] ϵ	Refractive Index [b] η	E_{T30} [c]	$f_1(\epsilon,\eta)$ [d]
Ethanol	24.5	1.3614	51.9	0.2911
Dichloromethane	8.93	1.4241	40.7	0.2172
Tetrahydrofuran	7.58	1.4072	37.4	0.2096
Ethyl acetate	6.00	1.3724	38.1	0.1993
Toluene	2.38	1.4969	33.9	0.01350
Dioxane	2.22	1.4224	36.0	0.02164
Carbon tetrachloride	2.24	1.4601	32.4	0.01400
Acetonitrile	37.5	1.3441	45.6	0.30500
Dimethylformamide	36.7	1.4305	43.2	0.27440
DMSO	46.7	1.4783	45.1	0.26340
Chloroform	4.81	1.4458	39.1	0.14700
Cyclohexane	2.02	1.4262	30.9	-0.001600
n-hexane	1.88	1.3749	31.0	-0.0014
Acetone	20.7	1.3587	42.2	0.2842
Benzene	2.27	1.5011	34.3	0.001700
Diethyl ether	4.33	1.3524	34.5	0.16760

[a] Used to compute Lippert–Mataga and E_{T30} slopes from published Stokes shift data for Table 2. [b] Taken from Ref. [39]. [c] Taken from Ref. [23]. [d] Taken from Refs. [34,38] or computed from reference data as described in the Section 2.

3. Results

3.1. Experimental Assessments of Solvatochromism

The quickest and most entertaining way to assess the solvatochromism of a probe is to observe solutions in various solvents under a UV lamp (Figure 4). Solvatochromism is usually formally assessed by measuring the Stokes shift, that is, the difference in wavenumbers between the excitation and emission maxima ($\Delta \bar{\nu} = \bar{\nu}_A - \bar{\nu}_F$), in a range of solvents of differing polarity. The absorbance (or excitation) maximum of a probe can sometimes respond to solvent polarity, but in general, emission peak maxima are more sensitive to this property.

tol Ac$_2$O DCM DMSO CHCl$_3$ MeCN EtOH MeOH H$_2$O

Figure 4. Solvatochromism of MF5 in various solvents. Imaged upon excitation at 395 nm. H$_2$O refers to 10 mM Tris, pH 8.

These solvent polarities are typically represented by either a form of a solvent polarizability function dependent on the solvent dielectric constant and refractive index [34,38] or by Reichardt's E_{T30} series [23]. Such data for solvents used in this study are summarized in Table 1 (Section 2).

Lippert–Mataga plots of the Stokes shift vs. the former function (Figure 5) remain in general use [37,40], although variations in the function are not uncommon [32,34–36]. In addition, theoretical variations in the treatment of solvent behavior have emerged from Bakshiev [38,41,42] and Kawski, Chamma, and Vaillet (KCV) [31,43] that are sometimes reported in addition to (or in lieu of) Lippert–Mataga plots [34,38,40]. All of these variations can be reflected in the quality of the correlation with the Stokes shift behavior of solvatochromic probes [35,38,40]. Further, probes can often have specific solvent interactions that can reduce the quality of the correlation depending on how well the theory aligns with solvent behavior.

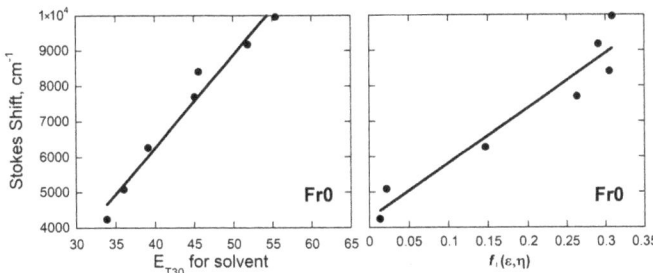

Figure 5. Lippert–Mataga (**left**) and E_{T30} (**right**) plot for the solvatochromic dye Fr0. Plots were generated with data obtained from Ref. [44] as described in Section 2.

Depending on the application, authors may or may not generate these plots, even though it is easily rendered from reported Stokes shift data and solvent reference data.

Published Lippert–Mataga slopes for solvatochromic probes are in excess of 3000 [40,43,45,46]. Another version of the Stokes shift and solvent behavior correlation uses the Reichardt E_{T30} series and is also common [26,32,47,48]. Here, large slopes of 100–600, either positive or negative, are indicative of significant solvatochromism. Those with little to no solvatochromic behavior (like many of the metallafluorenes) have single digit slopes.

3.2. Solvatochromic Behavior vs. Dipole Moment

To the author's knowledge, there is no published source for a collection of solvatochromic data of various probes. There are two goals: One is to initiate a growing list of these data. The other is to assess a possible correlation between the computed ground state dipole moment and the experimental data for published solvatochromic probes. To that end, Stokes shift data for a collection of solvatochromic probes were gleaned from published data and used to prepare E_{T30} and Lippert–Mataga plots to generate slopes that relate experimental spectral data with a measure of solvent polarizability. Because the primary interest here is understanding how modifications to a known probe affect the ground state dipole moment, the primary focus is on prodan, Nile Red, and fluorene-based probes and their derivatives. To add strength to the examination of a possible correlation between experimental and computational data, a few other known solvatochromic probes have been included.

Even the gathering of published experimental data and the computation of slopes was informative: The number and identify of solvents used were quite variable across the literature. Indeed, in the author's experience, solubility can be a factor. Scatter in the plots of experimental data vs. solvent property data is not unusual and is typically attributed to solvent interactions [34,45]. A more sophisticated understanding of this scatter or uncertainty is that the commonly used mathematical treatments of solvent behavior may not always accurately account for properties and interactions that can affect the behavior of solvatochromic probes. To further complicate the analysis, it is not uncommon for poor correlations to be published and even interpreted. Noise in these plots is usually attributed to solvent interactions [34,45]. To reduce the introduction of noise from the experimental plots into the secondary correlation of slope with computed dipole moments, only slopes with correlations in excess of 0.8 are reported.

The resulting Lippert–Mataga and E_{T30} slopes and ground state dipole moments appear in Table 2. E_{T30} slopes range from 320 to single digits. Correspondingly, the highest Lippert–Mataga slopes are close to 17,000, while the lowest ones are near 100. These spreads provide substantial dynamic range to explore subsequent correlations. For some probes explored here, there was no identifiable correlation between published Stokes shift data and solvent polarizability. This was noted for either Lippert–Mataga or E_{T30} for Nile Red, 2APMC, and 2BME (Table 2), but was also true of other probes for which there was no discernible correlation for either Lippert–Mataga or E_{T30} (**A1–A3** [24]). Large computed ground state dipole moments range from 8 to 37 D, but lower values are also represented. For a few probes, a range of dipole moments is tabulated. This is related to conformation, which is discussed in more detail below.

Table 2. Data for some solvatochromic probes.

Probe	Structure	Lippert–Mataga Slope	E_{T30} Slope	Dipole Moment, D [a]	Ref.
Nile Red		4083	c	8.0 [b]	[21,36]
NR12S		12,860	200	14–37 [d]	[21]
Prodan		3881 [b]	127	7.7	[49]
C laurdan		6334	82	6.9	[50]
PA		7256 [b]	154	6.0	[51]
PK		9503	95	4.8	[51]
7AMC		11,840	307	9.2	[52]
FR0		18380	267	9.2	[44]
FR1/PP3		14,170	280	13.4	[26]
FR2/PP6		14,880	212	6.4	[26]

Table 2. Cont.

Probe	Structure	Lippert–Mataga Slope	E_{T30} Slope	Dipole Moment, D [a]	Ref.
FR3		12,880	200	6.4–12.2 [d]	[16]
FR4		16,720	265	2.2–13.5 [d]	[16]
FR8		13,440	320	9.56	[44]
2,5APMC		[c]	88	4.5 [b]	[40]
2BME		11,640	[c]	2.0–7.8 [e]	[45]
MF1		293	4.6	3.0	[29]
MF2		157	2.5	3.2	[29]
MF3		124	2.1	2.0	[29]
MF4		253	4.3	2.2	[29]
MF5		6500	126	8.8	[29]

[a] Computed in polar solvent for this work using Spartan unless otherwise indicated. [b] Literature values. [c] No meaningful correlations of solvent data with Stokes shift (R < 0.8). [d] Dipole moment varies with manual bond rotation. [e] Dipole moment of lowest energy conformers explored using Spartan.

3.3. Assessing a Correlation between Experimental Spectral Data and Computed Dipole Moments

To more clearly explore the correlation between the ground state dipole moment and experimental data via Lippert–Mataga and E_{T30} slopes, these data were plotted against one another (Figure 6A,B). Given the nature of the data, some scatter is expected in this secondary plot. In spite of that, however, there is a convincing correlation between the computed ground state dipole moment and both slopes for this series of solvatochromic probes.

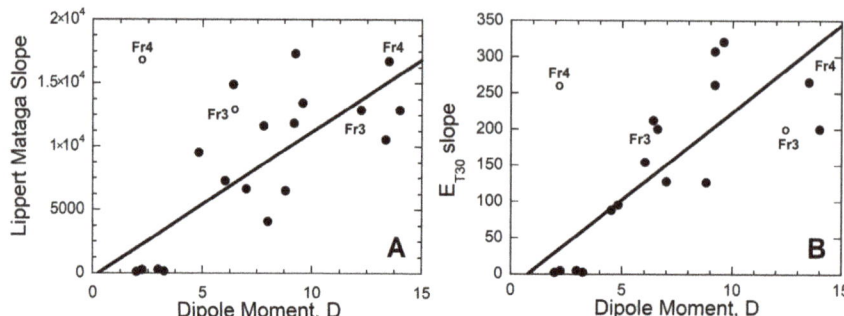

Figure 6. Correlations between experimental measures of solvatochromism and computed ground state dipole moment. (**A**) Lippert–Mataga and (**B**) E_{T30}. Data for the probes were taken from Table 2.

A couple of outlying points provide an opportunity to explore some possible reasons for this weaker correlation. One possibility is that there is noise in the primary plot that determines the slope and might be traceable to errors or choice of solvents. Another more intriguing possibility is related to the importance of bond angles and conformation on the computed dipole moment for probes. While some of the probes in Table 2 do not have rotatable bonds that significantly affect the distribution of polar atoms, a number do have important points of conformational variability. This issue is explored in more detail below.

3.4. Effects of Probe Modifications and Conformation on Dipole Moment and Electrostatic Potential

Here, we explore issues of modification, conformation, and design, using families of solvatochromic probes as examples.

3.4.1. Impact of Probe Modification: Fluorene Series

The conjugated polycyclic structure of fluorene makes it an ideal aromatic core for solvatochromic probes. Indeed, fluorene-based dyes have been investigated as membrane probes and show promising quantum yields and photostability [16,21,53–55]. The structural diversity of these probes provide an opportunity to examine the effect on the ground state dipole moment (Table 2). In Figure 7, two examples are shown illustrating how modifications to **Fr0** affect both the direction and magnitude of the ground state dipole moment and could therefore affect the solvatochromic response. Modifications made at the **Fr0** carbonyl, in particular, the introduction of additional polar atoms, have a significant impact on the dipole moment. These data from this series illustrate that modifying a known solvatochromic probe can affect the dipole moment and, by extension, the solvatochromic behavior either in an advantageous or disadvantageous way.

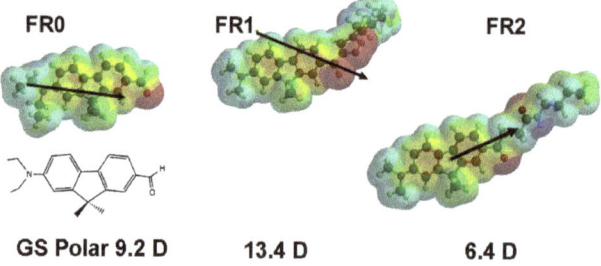

Figure 7. Effect of modifications to fluorene on computed ground state dipole moment in polar solvent as observed from electrostatic potential (ESP) diagrams obtained from DFT calculations, as described in Section 2. Red indicates areas of highest electron density and blue the lowest.

3.4.2. Impact on Conformation: Fluorene and Nile Red Series

It is reasonable to expect that when the spatial relationships among polar atoms in the molecule are changed, the dipole moment will also change. Therefore, it is essential to explore conformational aspects when using DFT for design purposes. Discussed here are two approaches.

In an initial computation on **Fr3**, with no attention to conformation, the ground state dipole moment was 6.37 D, which led to an outlying point (open circle in Figure 6). However, exploring a different orientation for the heterocyclic ring and the rest of the sidechain yielded a higher dipole moment (12.23 D) that aligns better with the other correlated data. A similar pattern was observed for **Fr4** and alternate conformations.

Nile Red remains one of the most commonly used solvatochromic probes [1,17,18,21,56]. This dye was known many decades before its application to the study of lipid dynamics and has a large reported ground state dipole moment (8 D; [31]); Table 1).

A number of Nile Red derivatives have been prepared with goals in mind that include membrane solubility [21,24] and organelle targeting [56]. How is the dipole moment affected by these designs? The most commonly used of these is NR12S [21]. With the chain extending out from the polycyclic core, the computed dipole moment is much larger, 36.6 D (Figure 8, Table 1). If, instead, the chain is rotated over the polycyclic core, the dipole moment is closer to that of Nile Red. However, the electron distribution is visibly altered by the addition of the chain, and this is reflected in the direction of the dipole moments of these molecules.

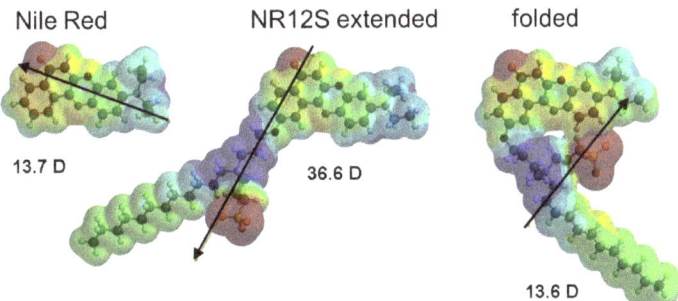

Figure 8. Impact of substitution and conformation on the computed ground state dipole moment of Nile Red as observed from electrostatic potential (ESP) diagrams obtained from DFT calculations, as described in the Methods section. A single bond moving the chain was rotated manually in Spartan. Red indicates areas of highest electron density and blue the lowest.

Spartan and other similar software packages do offer a systematic conformational search, which allow for the exploration of a large number of conformers without input. A number of solvatochromic probes, including NR12S, have a large number of rotatable bonds, which extends the computation time dramatically and must be factored into investigations. To illustrate this option here, an automatic conformer distribution search was conducted on 2BME (Table 2), which has limited points of bond rotation (and therefore more manageable computation times). Interestingly, the reported dipole moment of 2 D falls well outside the correlation [45]. However, the distribution search yielded a higher dipole moment (7.8 D) in these low energy conformers, which fell well into correlation with the spectral data. Where possible, exploration of conformation is therefore highly recommended, bearing in mind the effect of a large number of rotatable bonds on computation time.

Finally, it is important to note that molecular conformations might not necessarily be known in the environment upon which the probe is reporting. However, it is also possible that dipole moments that align with the correlation might correspond to a dominant bond angle/conformation in the medium in which the probe is applied. Collectively, these data suggest a possibly useful correlation between experimental and computed solvatochromic behavior. And while commonly used solvatochromic probes are represented here, a continued expansion of the library would inform the correlation.

3.4.3. Use of DFT in Probe Design: 2,7-Disubstituted Metallafluorenes

The 2,7 positions of fluorene are synthetically accessible locations for extending the conjugation, as well as for providing a means of influencing electron movement for ICT behavior and possibly engineering solvatochromic behavior. There are a few examples of 2,7 substitution via alkene linkers [57]. These are indeed solvatochromic, but as shown above, flexibility can impact electron distribution. This is easily managed with an alkynyl linker [58]. This was then followed by a series of papers that expanded the library of 2,7 substituents [59–62].

We have assessed a small library of these compounds (MFs) via both experimental and theoretical approaches. First, this substituent does influence spectral behavior (ϵ, λ_{max}, and quantum yield), which indicates that these substitutions can be an effective means of tuning these properties [29]. We showed that a few 2,7-disubstituted metallafluorenes can detect detergents and stain cells [63,64].

More recently, we noted that 2,7-benzaldehyde substitution results in an impressive ground state dipole moment (9 D), visible solvatochromic behavior, and competitive Lippert–Mataga and E_{T30} slopes (6500 and 126, respectively) [29]. In contrast, MFs with other electron withdrawing substituents have low dipole moments (e.g., 3 D) and are not solvatochromic. However, since solvent spectral data are seldom reported for nonsolvatochromic compounds, these data serve as important controls for the correlations.

This general structure also provides a wealth of opportunities for exploring the design of solvatochromic probes, and DFT is especially useful in exploring structure space without committing resources for synthesis. As an example, a structure–dipole moment relationship exploration was conducted via DFT calculations on a series of MFs related to **MF5**. As summarized in Figure 9, the methoxy groups are absolutely critical to the magnitude of the dipole moment; substituting the central atom with C decreased the dipole moment, while removing the phenyl groups increased it. Thus, even within the fluorene core there are opportunities to increase the dipole moment and hence solvatochromism.

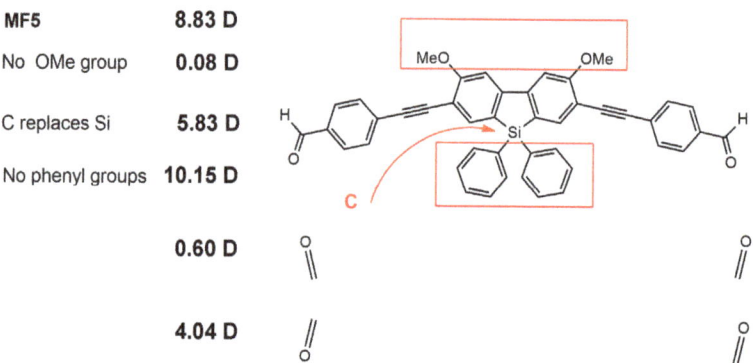

MF5	8.83 D
No OMe group	0.08 D
C replaces Si	5.83 D
No phenyl groups	10.15 D
	0.60 D
	4.04 D

Figure 9. Structure–dipole moment relationships for MF5. Ground state dipole moments were computed via DFT as described in Section 2. Red boxes highlight groups removed.

Further, consistent with the conformational variations explored above for other probes, dipole moment varies with the rotation of the carbonyl; even the preferred conformation cannot be easily known in the environment.

Finally, one can take this exercise further and design new molecules with high dipole moments. For example, benzoxazole or benzimidazole derivatives are fluorescent [65–67]. 2,7-disubstitution of sila- or germafluorene with these groups form chimeric probes which increase the computed ground state dipole moment in polar solvent to 15 and 13 D, respectively [68]. These are excellent leads for increasing the solvatochromic behavior of metallafluorenes.

4. Conclusions

Derivatization of existing probes is often conducted with a goal to increase desirable properties such as solubility and emission wavelength. However, such modifications can have an impact on the ground state dipole moment and could affect solvatochromic behavior. Through the analysis of published Stokes shift data for a few series of known solvatochromic probes, an informative correlation between the computed ground state dipole moment and experimental solvatochromic behavior has been established. Molecular conformations accessible via bond rotation can also affect the ground state dipole moment and thus the correlation, which is an important factor in conducting DFT calculations for this purpose. Expanding the library of computed data for other solvatochromic probes would be needed to assess the greater generality of the correlation. Finally, while DFT calculations are a common form of probe characterization, they can also provide powerful design tools for the development of new solvatochromic probes, as well as a way to enlist the efforts of junior scientists to contribute productively to probe development.

Funding: This research received no external funding.

Data Availability Statement: The datasets generated during and/or analyzed during the current study are available from the corresponding author on reasonable request.

Conflicts of Interest: The author holds and has filed patents that are relevant to the content of this article: US Patent 20220162235 and provisional patent 63/511,403.

Abbreviations

DFT: density functional theory; HOMO, highest occupied molecular orbital; LUMO, lowest unoccupied molecular orbital; MF, metallafluorene.

References

1. Klymchenko, A.S. Solvatochromic and Fluorogenic Dyes as Environment-Sensitive Probes: Design and Biological Applications. *Acc. Chem. Res.* **2017**, *50*, 366–375. [CrossRef]
2. Sezgin, E.; Levental, I.; Mayor, S.; Eggeling, C. The mystery of membrane organization: Composition, regulation and roles of lipid rafts. *Nat. Rev. Mol. Cell Biol.* **2017**, *18*, 361–374. [CrossRef]
3. Levental, I.; Levental, K.R.; Heberle, F.A. Lipid Rafts: Controversies Resolved, Mysteries Remain. *Trends Cell Biol.* **2020**, *30*, 341–353. [CrossRef]
4. Klymchenko, A.S.; Kreder, R. Fluorescent probes for lipid rafts: From model membranes to living cells. *Chem. Biol.* **2014**, *21*, 97–113. [CrossRef] [PubMed]
5. Zalba, S.; Ten Hagen, T.L. Cell membrane modulation as adjuvant in cancer therapy. *Cancer Treat. Rev.* **2017**, *52*, 48–57. [CrossRef] [PubMed]
6. Ashoka, A.H.; Ashokkumar, P.; Kovtun, Y.P.; Klymchenko, A.S. Solvatochromic Near-Infrared Probe for Polarity Mapping of Biomembranes and Lipid Droplets in Cells under Stress. *J. Phys. Chem. Lett.* **2019**, *10*, 2414–2421. [CrossRef] [PubMed]
7. Danylchuk, D.I.; Jouard, P.-H.; Klymchenko, A.S. Targeted Solvatochromic Fluorescent Probes for Imaging Lipid Order in Organelles under Oxidative and Mechanical Stress. *J. Am. Chem. Soc.* **2021**, *143*, 912–924. [CrossRef] [PubMed]
8. Darwich, Z.; Klymchenko, A.S.; Kucherak, O.A.; Richert, L.; Mely, Y. Detection of apoptosis through the lipid order of the outer plasma membrane leaflet. *Biochim. Biophys. Acta* **2012**, *1818*, 3048–3054. [CrossRef] [PubMed]
9. Shynkar, V.V.; Klymchenko, A.S.; Kunzelmann, C.; Duportail, G.; Muller, C.D.; Demchenko, A.P.; Freyssinet, J.M.; Mely, Y. Fluorescent biomembrane probe for ratiometric detection of apoptosis. *J. Am. Chem. Soc.* **2007**, *129*, 2187–2193. [CrossRef] [PubMed]
10. Niu, J.; Ma, Y.; Yang, Y.; Lv, H.; Wang, J.; Wang, T.; Liu, F.; Xu, S.; Jiang, Z.; Lin, W. Lighting up the changes of plasma membranes during apoptosis with fluorescent probes. *Coord. Chem. Rev.* **2023**, *476*, 214926. [CrossRef]
11. Yang, S.T.; Kreutzberger, A.J.B.; Kiessling, V.; Ganser-Pornillos, B.K.; White, J.M.; Tamm, L.K. HIV virions sense plasma membrane heterogeneity for cell entry. *Sci. Adv.* **2017**, *3*, e1700338. [CrossRef] [PubMed]
12. Carravilla, P.; Nieva, J.L.; Eggeling, C. Fluorescence Microscopy of the HIV-1 Envelope. *Viruses* **2020**, *12*, 348. [CrossRef] [PubMed]

13. Chojnacki, J.; Waithe, D.; Carravilla, P.; Huarte, N.; Galiani, S.; Enderlein, J.; Eggeling, C. Envelope glycoprotein mobility on HIV-1 particles depends on the virus maturation state. *Nat. Comm.* **2017**, *8*, 545. [CrossRef] [PubMed]
14. Lorizate, M.; Brugger, B.; Akiyama, H.; Glass, B.; Muller, B.; Anderluh, G.; Wieland, F.T.; Krausslich, H.G. Probing HIV-1 membrane liquid order by Laurdan staining reveals producer cell-dependent differences. *J. Biol. Chem.* **2009**, *284*, 22238–22247. [CrossRef]
15. Algar, W.R.; Hildebrandt, N.; Vogel, S.S.; Medintz, I.L. FRET as a biomolecular research tool—Understanding its potential while avoiding pitfalls. *Nat. Methods* **2019**, *16*, 815–829. [CrossRef]
16. Shaya, J.; Collot, M.; Benailly, F.; Mahmoud, N.; Mely, Y.; Michel, B.Y.; Klymchenko, A.S.; Burger, A. Turn-on Fluorene Push-Pull Probes with High Brightness and Photostability for Visualizing Lipid Order in Biomembranes. *ACS Chem. Biol.* **2017**, *12*, 3022–3030. [CrossRef]
17. Martinez, V.; Henary, M. Nile Red and Nile Blue: Applications and Syntheses of Structural Analogues. *Chemistry* **2016**, *22*, 13764–13782. [CrossRef] [PubMed]
18. Greenspan, P.; Mayer, E.P.; Fowler, S.D. Nile red: A selective fluorescent stain for intracellular lipid droplets. *J. Cell Biol.* **1985**, *100*, 965–973. [CrossRef]
19. Wang, Y.; Chen, J.; Di, C.; Hu, Y.; Munyemana, J.C.; Shu, Y.; Wang, J.-H.; Qiu, H. A novel colorimetric and red-emitting fluorescent probe based on benzopyrylium derivatives for selective detection and imaging of SO_2 derivatives in cells and zebrafish. *Dye. Pigment.* **2023**, *212*, 111129. [CrossRef]
20. Liu, F.; Wang, Z.; Wang, W.; Luo, J.-G.; Kong, L. Red-Emitting Fluorescent Probe for Detection of γ-Glutamyltranspeptidase and Its Application of Real-Time Imaging under Oxidative Stress in Cells and In Vivo. *Anal. Chem.* **2018**, *90*, 7467–7473. [CrossRef]
21. Kucherak, O.A.; Oncul, S.; Darwich, Z.; Yushchenko, D.A.; Arntz, Y.; Didier, P.; Mely, Y.; Klymchenko, A.S. Switchable nile red-based probe for cholesterol and lipid order at the outer leaflet of biomembranes. *J. Am. Chem. Soc.* **2010**, *132*, 4907–4916. [CrossRef]
22. Misra, R.M.; Bhattacharyya, S.P. *Intramolecular Charge Transfer: Theory and Applications*, 1st ed.; Wiley-VCH.: Weinheim, Germany, 2018.
23. Reichardt, C. Solvatochromic Dyes as Solvent Polarity Indicators. *Chem. Rev.* **1994**, *94*, 2319–2358. [CrossRef]
24. Sun, R.; Wan, W.; Jin, W.; Bai, Y.; Xia, Q.; Wang, M.; Huang, Y.; Zeng, L.; Sun, J.; Peng, C.; et al. Derivatizing Nile Red fluorophores to quantify the heterogeneous polarity upon protein aggregation in the cell. *Chem. Commun.* **2022**, *58*, 5407–5410. [CrossRef]
25. Liu, L.; Lei, Y.; Zhang, J.; Li, N.; Zhang, F.; Wang, H.; He, F. Rational Design for Multicolor Flavone-Based Fluorophores with Aggregation-Induced Emission Enhancement Characteristics and Applications in Mitochondria-Imaging. *Molecules* **2018**, *23*, 2290. [CrossRef] [PubMed]
26. Shaya, J.; Fontaine-Vive, F.; Michel, B.Y.; Burger, A. Rational Design of Push-Pull Fluorene Dyes: Synthesis and Structure-Photophysics Relationship. *Chemistry* **2016**, *22*, 10627–10637. [CrossRef] [PubMed]
27. Tarai, A.; Huang, M.; Das, P.; Pan, W.; Zhang, J.; Gu, Z.; Yan, W.; Qu, J.; Yang, Z. ICT and AIE Characteristics Two Cyano-Functionalized Probes and Their Photophysical Properties, DFT Calculations, Cytotoxicity, and Cell Imaging Applications. *Molecules* **2020**, *25*, 585. [CrossRef] [PubMed]
28. Li, L.; Xu, Y.; Chen, Y.; Zheng, J.; Zhang, J.; Li, R.; Wan, H.; Yin, J.; Yuan, Z.; Chen, H. A Family of Push-Pull Bio-Probes for Tracking Lipid Droplets in Living Cells with the Detection of Heterogeneity and Polarity. *Anal. Chim. Acta* **2020**, *1096*, 166–173. [CrossRef]
29. Jarrett-Noland, S.; McConnell, W.; Braddock-Wilking, J.; Dupureur, C. Solvatochromic Behavior of 2,7-Disubstituted Sila- and Germafluorenes. *Chemosensors* **2023**, *11*, 160. [CrossRef]
30. Mes, G.; de Jong, B.; van Ramesdonk, H.; Verhoeven, J.; Warman, J.; de Haas, M.; Horsman-van den Dool, L. Excited-State Dipole Moment and Solvatochromism of Highly Fluorescent Rod-Shaped Bichromophoric Molecules. *J. Am. Chem. Soc.* **1984**, *106*, 6524–6528. [CrossRef]
31. Kawski, A.; Bojarski, P.; Kuklinski, B. Estimation of ground- and excited-state dipole moments of Nile Red dye from solvatochromic effect on absorption and fluorescence spectra. *Chem. Phys. Lett.* **2008**, *463*, 410–412. [CrossRef]
32. Vequi-Suplicy, C.C.; Coutinho, K.; Lamy, M.T. Electric dipole moments of the fluorescent probes Prodan and Laurdan: Experimental and theoretical evaluations. *Biophys. Rev.* **2014**, *6*, 63–74. [CrossRef] [PubMed]
33. Mukherjee, S.; Chattopadhyay, A.; Samanta, A.; Soujanya, T. Dipole Moment Change of NBD Group upon Excitation Studied Using Solvatochromic and Quantum Chemical Approaches: Implications in Membrane Research. *J. Phys. Chem.* **1994**, *98*, 2809–2812. [CrossRef]
34. Patil, O.; Ingalgondi, P.; Mathapati, G.; Gounalli, S.; Sankarappa, T.; Hanagodimath, S. Ground and Excited State Dipole Moments of a Dye. *J. Appl. Phys.* **2016**, *8*, 55–59.
35. Gulseven Sidir, Y.; Sidir, I. Solvent effect on the absorption and fluorescence spectra of 7-acetoxy-6-(2,3-dibromopropyl)-4,8-dimethylcoumarin: Determination of ground and excited state dipole moments. *Spectrochim. Acta A Mol. Biomol. Spectrosc.* **2013**, *102*, 286–296. [CrossRef] [PubMed]
36. Yablon, D.G.; Schilowitz, A.M. Solvatochromism of Nile Red in nonpolar solvents. *Appl. Spectrosc.* **2004**, *58*, 843–847. [CrossRef] [PubMed]
37. Vazquez, M.E.; Blanco, J.B.; Imperiali, B. Photophysics and biological applications of the environment-sensitive fluorophore 6-N,N-dimethylamino-2,3-naphthalimide. *J. Am. Chem. Soc.* **2005**, *127*, 1300–1306. [CrossRef] [PubMed]

38. Karthik, C.; Manjuladevi, V.; Gupta, R.; Kumari, S. Solvatochromism of tricycloquinazoline based disk-shaped liquid crystal: A potential molecular probe for fluorescence imaging. *RSC Adv.* **2015**, *5*, 84592–84600. [CrossRef]
39. Lide, D.R. *Handbook of Chemistry and Physics*, 87th ed.; Lide, D.R., Ed.; CRC Press: Boca Raton, FL, USA, 2006.
40. Thipperudrappa, J. Analysis of solvatochromism of a biologically active ketocyanine dye using different solvent polarity scales and estimation of dipole moments. *Int. J. Life Sci. Pharm. Res.* **2014**, *4*, 1–11.
41. Bakhshiev, N.G. Universal Intermolecular Interactions and Their Effect on the Position the Electronic Spectra of Molecules in Two Component Solutions. *Opt. Spektrosk.* **1964**, *16*, 821–832.
42. Chamma, A.; Viallet, P. Determination du moment dipolaire d'une molecule dans un etat excite singulet. *Comptes Rendus Acad. Des. Sci.* **1970**, *27*, 1901–1904.
43. Kumari, R.; Varghese, A.; George, L. Estimation of Ground-State and Singlet Excited-State Dipole Moments of Substituted Schiff Bases Containing Oxazolidin-2-One Moiety through Solvatochromic Methods. *J. Fluor.* **2017**, *27*, 151–165. [CrossRef]
44. Kucherak, O.A.; Didier, P.; Mély, Y.; Klymchenko, A.S. Fluorene Analogues of Prodan with Superior Fluorescence Brightness and Solvatochromism. *J. Phys. Chem. Lett.* **2010**, *1*, 616–620. [CrossRef]
45. Renuka, C.G.; Shivashankar, K.; Boregowda, P.; Bellad, S.S.; Muregendrappa, M.V.; Nadaf, Y.F. An Experimental and Computational Study of 2-(3-Oxo-3H-benzo[f] chromen-1-ylmethoxy)-Benzoic Acid Methyl Ester. *J. Solut. Chem.* **2017**, *46*, 1535–1555. [CrossRef]
46. Aaron, J.-J.; Buna, M.; Parkanyi, C.; Antonious, M.S.; Tine, A.; Cisse, L. Quantitative Treatment of the Effect of Solvent on the Electronic Absorption and Fluorescence Spectra of Substituted Coumarins: Evaluation of the First Excited Singlet-State Dipole Moments. *J. Fluor.* **1995**, *5*, 337–347. [CrossRef]
47. Cha, S.; Choi, M.G.; Jeon, H.R.; Chang, S.K. Negative Solvatochromism of Merocyanine Dyes: Application as Water Content Probes for Organic Solvents. *Sens. Actuators B Chem.* **2011**, *157*, 14–18. [CrossRef]
48. Arathi, A.S.; Mallick, S.; Koner, A.L. Tuning Aggregation-Induced Emission of 2,3-Napthalimide by Employing Cyclodextrin Nanocavities. *ChemistrySelect* **2016**, *1*, 3535–3540. [CrossRef]
49. Pandey, A.; Rai, R.; Pal, M.; Pandey, S. How polar are choline chloride-based deep eutectic solvents? *Phys. Chem. Chem. Phys.* **2014**, *16*, 1559–1568. [CrossRef] [PubMed]
50. Kim, H.M.; Choo, H.J.; Jung, S.Y.; Ko, Y.G.; Park, W.H.; Jeon, S.J.; Kim, C.H.; Joo, T.; Cho, B.R. A two-photon fluorescent probe for lipid raft imaging: C-laurdan. *Chembiochem* **2007**, *8*, 553–559. [CrossRef] [PubMed]
51. Niko, Y.; Kawauchi, S.; Konishi, G. Solvatochromic pyrene analogues of Prodan exhibiting extremely high fluorescence quantum yields in apolar and polar solvents. *Chemistry* **2013**, *19*, 9760–9765. [CrossRef]
52. Giordano, L.; Shvadchak, V.V.; Fauerbach, J.A.; Jares-Erijman, E.A.; Jovin, T.M. Highly Solvatochromic 7-Aryl-3-hydroxychromones. *J. Phys. Chem. Lett.* **2012**, *3*, 1011–1016. [CrossRef] [PubMed]
53. Zhang, H.; Fan, J.; Dong, H.; Zhang, S.; Xu, W.; Wang, J.; Gao, P.; Peng, X. Fluorene-Derived Two-Photon Fluorescent Probes for Specific and Simultaneous Bioimaging of Endoplasmic Reticulum and Lysosomes: Group-Effect and Localization Hua. *J. Mater. Chem. B* **2013**, *1*, 5450–5455. [CrossRef]
54. Chen, R.F.; Fan, Q.L.; Liu, S.J.; Zhu, R.; Pu, K.Y.; Huang, W. Fluorene and Silafluorene Conjugated Copolymer: A New Blue Light-Emitting Polymer. *Synth. Met.* **2006**, *156*, 1161–1167. [CrossRef]
55. Yang, T.; Zuo, Y.; Zhang, Y.; Gou, Z.; Wang, X.; Lin, W. Novel Fluorene-Based Fluorescent Probe with Excellent Stability for Selective Detection of SCN- and Its Applications in Paper-Based Sensing and Bioimaging. *J. Mat. Chem. B* **2019**, *7*, 4649–4654. [CrossRef]
56. Niko, Y.; Klymchenko, A.S. Emerging Solvatochromic Push-Pull Dyes for Monitoring the Lipid Order of Biomembranes in Live Cells. *J. Biochem.* **2021**, *170*, 163–174. [CrossRef] [PubMed]
57. Auvray, M.; Bolze, F.; Clavier, G.; Mahuteau-Betzer, F. Silafluorene as a promising core for cell-permeant, highly bright and two-photon excitable fluorescent probes for live-cell imaging. *Dye. Pigment.* **2021**, *187*, 109083–109091. [CrossRef]
58. Li, L.; Xu, C.; Li, S. Synthesis and photophysical properties of highly emissive compounds containing a dibenzosilole core. *Tet. Lett.* **2010**, *51*, 622–624. [CrossRef]
59. Hammerstroem, D.W.; Braddock-Wilking, J.; Rath, N.P. Synthesis and characterization of luminescent 2,7-disubstituted silafluorenes. *J. Organomet. Chem.* **2016**, *813*, 110–118. [CrossRef]
60. Hammerstroem, D.W.; Braddock-Wilking, J.; Rath, N.P. Luminescent 2,7-disubstituted germafluorenes. *J. Organomet. Chem.* **2017**, *830*, 196–202. [CrossRef]
61. Germann, S.; Jarrett, S.J.; Dupureur, C.M.; Rath, N.P.; Gallaher, E.; Braddock-Wilking, J. Synthesis of Luminescent 2-7 Disubstituted Silafluorenes with alkynyl-carbazole, -phenanthrene, and -benzaldehyde substituents. *J. Organomet. Chem.* **2020**, *927*, 121514. [CrossRef]
62. Braddock-Wilking, J.; Dupureur, C.; Germann, S.; Spikes, H. Luminescent Silafluorene and Germafluorene Compounds. U.S. Patent 20220162235, 26 May 2022.
63. Spikes, H.J.; Jarrett-Noland, S.J.; Germann, S.M.; Braddock-Wilking, J.; Dupureur, C.M. Group 14 Metallafluorenes as Sensitive Luminescent Probes of Surfactants in Aqueous Solution. *J. Fluor.* **2021**, *31*, 961–969. [CrossRef]
64. Spikes, H.J.; Jarrett-Noland, S.J.; Germann, S.M.; Olivas, W.; Braddock-Wilking, J.; Dupureur, C.M. Group 14 Metallafluorenes for Lipid Structure Detection and 2 Cellular Imaging. *Chem. Proc.* **2021**, *5*, 83–88. [CrossRef]

65. Xiong, J.F.; Li, J.X.; Mo, G.Z.; Huo, J.P.; Liu, J.Y.; Chen, X.Y.; Wang, Z.Y. Benzimidazole derivatives: Selective fluorescent chemosensors for the picogram detection of picric acid. *J. Org. Chem.* **2014**, *79*, 11619–11630. [CrossRef] [PubMed]
66. Reiser, A.; Leyshon, L.; Saunders, D.; Mijovic, M.; Bright, A.; Bogie, J. Fluorescence of Aromatic Benzoxazole Derivatives. *J. Am. Chem. Soc.* **1972**, *94*, 2414–2421. [CrossRef]
67. Barwiolek, M.; Wojtczak, A.; Kozakiewicz, A.; Babinska, M.; Tafelska-Kaczmarek, A.; Larsen, E.; Szlyk, E. The synthesis, characterization and fluorescence properties of new benzimidazole derivatives. *J. Lum.* **2019**, *211*, 88–95. [CrossRef]
68. Dupureur, C.; Jarrett-Noland, S.; McConnell, W.; Gnawali, G.; Germann, S.; Braddock-Wilking, J. Chimeric Environment Sensitive Fluorescent Probes. Provisional Patent Application 63/511,403, 30 June 2023.

Disclaimer/Publisher's Note: The statements, opinions and data contained in all publications are solely those of the individual author(s) and contributor(s) and not of MDPI and/or the editor(s). MDPI and/or the editor(s) disclaim responsibility for any injury to people or property resulting from any ideas, methods, instructions or products referred to in the content.

Communication

Solvent Polarity/Polarizability Parameters: A Study of Catalan's SPP^N, Using Computationally Derived Molecular Properties, and Comparison with π^* and $E_T(30)$

W. Earle Waghorne

UCD School of Chemistry, University College Dublin, D04 V1W8 Dublin, Ireland; earle.waghorne@ucd.ie

Abstract: Catalan's SPP^N, a measure of solvent polarity/polarizability has been analysed in terms of molecular properties derived from computational chemistry. The results show that SPP^N correlates positively with the molecular dipole moment and quadrupolar amplitude and negatively with the molecular polarizability. These correlations are shared with Kamet and Taft's π^* and Reichardt and Dimroth's $E_T(30)$. Thus, one can associate the solvent polarity with non-specific interactions involving the permanent charges on solvent molecules. It is also noted that the opposite correlations, all three parameters increasing with increasing solvent polarity but decreasing with increasing solvent polarizability, creates an ambiguity in their use, for example, in linear free energy relationships.

Keywords: solvent parameters; polarity; polarizability; $E_T(30)$; Catalan SPP; computational chemistry

1. Introduction

Understanding and predicting the effects of changes in solvent on chemical processes are among the classical problems of solution chemistry. Thus, virtually all thermodynamic, kinetic and spectroscopic properties of chemical systems are sensitive to changes in the solvent, some changing by orders of magnitude.

The use of linear free energy relationships to explore or predict the effect of changes in solvent on chemical processes is well established. The principle is straightforward and one writes:

$$SP = \sum c_i P_i \tag{1}$$

where SP is some property, such as the log of the solubility of a solute, the P_i are experimental parameters reflecting properties of the solvent (or solute) and that c_i are coefficients reflecting the response of the solute (or solvent). In effect each $c_i P_i$ term represents the effect of a different intermolecular interaction on SP.

Generally the parameters P_i are derived from experiments that are designed to isolate a particular interaction. In general P_i values represent specific, acid–base interactions: (1) the basicity of the solvent (Kamlet and Taft's β [1], Catalan's SB [2], Gutmann's DN [3]); (2) the acidity of the solvent (Kamlet and Taft's α [4], Catalan's SA [5], Gutmann's AN [3]) or non-specific interactions that are collected together in parameters that are measures of the polarity/polarizability of the solvent (Kamlet and Taft's π^* [6], Catalan's SPP [7], Reichardt and Dimroth's $E_T(30)$ [8]).

The range of experimental parameters raises several interesting questions, including (1) the extent to which these parameters simply reflect properties of the solvent molecules, (2) which molecular properties contribute to the experimental parameters and (3) whether different parameters, for basicity say, reflect the same molecular properties.

These have been explored in a series of papers where experimental parameters have been correlated with a set of molecular properties derived from computational chemistry. These have shown that measures of solvent basicity (β, DN [9] and SB [10]) reflect the partial charge on the most negative atom of the solvent molecule and the energy of the donor orbital, while measures of the solvent acidity (α [11], AN and SA [10]) reflect only the partial charge on the most positive hydrogen atom of the solvent molecule.

In the present paper, a similar analysis of Catalan's SPP^N, a measure of solvent polarity and polarizability, is reported and the results compared to those reported previously for Kamlet and Tafts's π^* and Reichardt and Dimroth's $E_T(30)$ [12,13].

2. Methods and Materials

2.1. Computational Details

The computational methods have been discussed in some detail previously [9,11–13]. Since the calculated molecular properties can depend on the method used, calculations were carried out using both the Hartree–Fock and density functional (B3LYP functional) methods using the 6-311+g(3df,2p) basis set. For the molecular dipole moments, quadrupolar amplitudes and polarizabilities, plots of the values calculated using the two methods against each other are linear and so the choice of method is immaterial [9]. This isn't the case for orbital energies, for which such plots show scatter around a straight line [9]. The situation for partial atomic charges is more complex, since these are not quantum mechanical observables and require a model that ascribes the electronic charge density to the individual atoms. Partial charges based on Mullican's model are highly dependent on the basis set and appear not to converge as the basis set is increased. Those based on the Hirshfeld [14], Natural bond order [15] and the CM5 [16] model do converge as the basis set is improved [9] and all have been used previously. It was fond that the Hirshfeld and NBO models gave very similar results [12] and, in a recent paper considering Abraham parameters it was recognized that the CM5 model overestimates the partial charges on nitrogen atoms, for amides at least [17]. Thus the analyses reported here use partial charges derived using Hirshfeld's model.

All calculations were carried out using the Gaussian 9 software [18].

2.2. Analyses of Parameters

As in previous work, the normalized parameter, SPP^N in this case, has been considered. The use of normalized scales allows relatively direct comparison of the results of the analyses. Values of SPP^N were taken from [7].

The approach used is a simple [9], multivariable regression of the experimental parameter, P, against molecular descriptors, Q_i, representing different molecular properties. Thus:

$$P = P^0 + \sum a_i Q_i \quad (2)$$

where P^0 is the value of P when all of the $a_i Q_i$ terms are zero and the a_i are the coefficients recovered from the regression. In essence each $a_i Q_i$ term represents the contribution of a particular interaction to P.

The molecular descriptors are normalized and calculated from the calculated molecular properties, q_i, as:

$$Q_i = \frac{q_i - q_i^{min}}{q_i^{max} - q_i^{min}} \quad (3)$$

where q_i^{min} and q_i^{max} are the minimum and maximum values of q_i. With the exception of the orbital energies, the values of q_i^{min} are taken to be zero.

Seven molecular properties: the partial charges on the most negative atom and on the most positive hydrogen atom, the molecular dipole moments, quadrupolar amplitudes (see Appendix A) and polarizabilities and the energies of electron donor and acceptor orbitals are considered. In general, the orbital energies are taken to be those of the highest occupied and lowest unoccupied molecular orbitals; the exception to this is the case of solvents with aromatic groups. This was discussed previously [9] but, briefly, solvents with a common functional group have, for example, similar basicity parameters, whether they are aliphatic or aromatic, although for aromatic solvents, the high energy π-bonds of the aromatic ring are clearly the HOMOs, while for aliphatic solvents the HOMO is associated with the basic functional group. The fact that the basicity parameters are similar argues for the use of the energies of the orbitals at the interacting functional group rather than those of the aromatic ring.

In applying Equation (1), the experimental parameter, P, was initially correlated with all seven molecular descriptors; subsequently, descriptors making negligible contributions (typically $a_i \leq 0.03$) or with standard deviations comparable to the value of the coefficient, were excluded and the correlation repeated with the remaining descriptors. Since both the calculated properties and experimental parameters for solvents with a common functional group are similar, the distributions of these can't be represented by statistics based on the normal distribution. Thus, the decision as to whether a descriptor was making a significant contribution to P was decided on the basis of the increase in the standard deviation between the calculated and experimental values of P, when the descriptor was excluded from the correlation.

3. Results

Catalan's SPP^N is based on the differences in the solvatochromism of 2-(dimethylamino)-7-nitrofluorine and its homomorph, 2-fuoro-7-nitrofluorine (Figure 1), which show similar solvatochromism with that of 2-(dimethylamino)-7-nitrofluorine being substantially larger. This assumes that specific interactions, such as hydrogen bonding to the NO_2 moiety will be removed by the subtraction, leaving only effects from solvent polarity and polarizability.

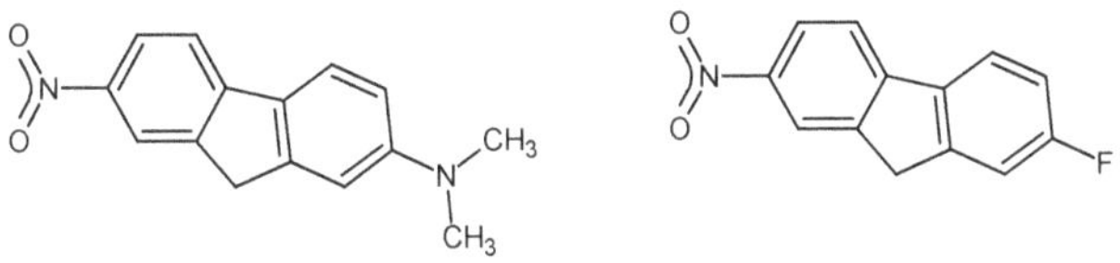

Figure 1. Structures of 2-(dimethylamino)-7-nitrofluorine and 2-fuoro-7-nitrofluorine.

In the analysis of SPP^N polychlorinated alkanes, flexible esters and pyridine were outliers and were excluded from the analysis. The analyses showed that only three solvent molecular properties: the dipole moment, quadrupolar amplitude and polarizability, showed significant correlations with SPP^N. The coefficients recovered from the analyses are listed in Table 1. Also listed in Table 1 are the corresponding coefficients for Kamlet, Abboud and Taft's π and Reichardt and Dimroth's normalized $E_T^N(30)$. The SPP^N values calculated using the coefficients in Table 1 and solvent descriptors, Q_i, recovered from properties calculated using the density functional method are compared with experimental values in Figure 2 (the dashed lines are one standard deviation above and below the line representing perfect agreement); the plot using properties recovered from Hartree–Fock calculations is similar. The values for pyridine (brown circle below the line) and the polychloroalkanes (grey triangles) are included in Figure 2.

The values for the esters provide a cautionary example for the use of computational methods. For rigid molecules the molecular properties recovered from computational calculations are essentially determined by the method and basis set but for flexible molecules the properties may also vary depending on the assumed molecular structure. This was discussed previously [12] with regard to the variation of calculated properties of carboxylic acids with rotation around the C–OH bond of the acid group. It was found that the dipole moments and quadrupolar amplitudes of the acids depended strongly on the conformation but that the other properties did not. It was also found that all molecular properties were relatively insensitive to coiling of the alkyl chain of longer chain carboxylic acids.

Table 1. Comparison of the a_i values for Catalan SPP, Kamlet, Abboud and Taft's π^* and Reichardt and Dimroth's $E_T^N(30)$.

	SPP^N		$\pi^{*\ b}$		$E_T^N(30)\ ^c$	
	DF [a]	HF	DF	HF	DF	HF
Intercept	0.63 ± 0.02	0.63 ± 0.02	0.03 ± 0.04	0.01 ± 0.05	0.05 ± 0.02	0.05 ± 0.03
a_{pol}	−0.14 ± 0.04	−0.18 ± 0.03	−0.51 ± 0.07	−0.58 ± 0.07	−0.24 ± 0.04	−0.32 ± 0.05
a_{d-pole}	0.35 ± 0.02	0.34 ± 0.03	0.63 ± 0.04	0.63 ± 0.04	0.19 ± 0.02	0.18 ± 0.03
a_{q-pole}	0.39 ± 0.05	0.49 ± 0.06	0.68 ± 0.07	0.74 ± 0.07	0.19 ± 0.04	0.21 ± 0.04
$a_{E_{Donor}}$			0.40 ± 0.05	0.45 ± 0.05		
$a_{q+}\ ^d$					0.58 ± 0.02	0.61 ± 0.03
$\sigma\ ^b$	0.05	0.05	0.11	0.11	0.07	0.07

[a] DF and HF indicate analyses based on properties calculated using density functional and Hartree–Fock calculation methods, respectively; uncertainties are standard deviations. [b] Ref. [12]; [c] Ref. [13]; [d] Charges calculated using Hirshfeld's model [14].

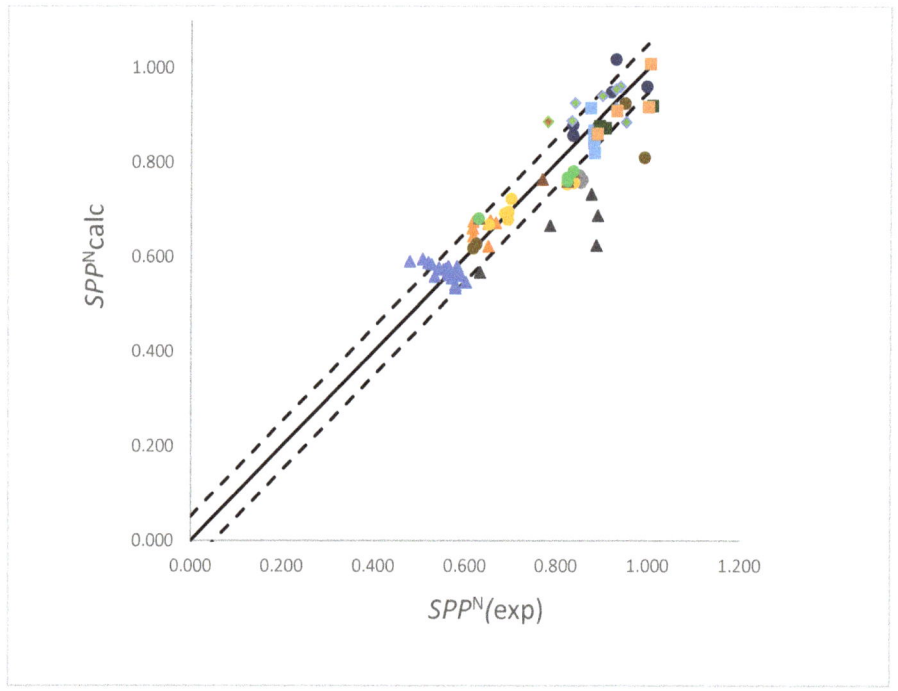

Figure 2. Comparison of the experimental SPP^N values and values calculated using the coefficients in Table 1 (for density functional calculations). The symbols represent: blue triangles, alkanes; dark grey triangles, chloro-alkanes; orange triangles, aromatics; yellow circles, ethers; green circles, nitriles; light grey circles, alcohols; brown circles, amines and pyridine; dark blue circles, esters; light blue squares, ketones; grey diamonds, amides; beige squares, S=O and P=O compounds. The solid line represents perfect agreement and the dashed lines are 1 standard deviation above and below this.

In the case of the esters, the O=CR−O moiety will be planar while rotation around the C−O bond is relatively free. The variations in the calculated molecular properties with rotation around the C−O bond are shown, as normalized molecular descriptors, in Table 2;

also shown are the corresponding SPP values calculated using the descriptors and the coefficients reported in Table 1.

Table 2. Solvent descriptors [a] for esters with different molecular geometries.

O=C−O−R Dihedral Angle	$Q_{d\text{-pole}}$	$Q_{q\text{-pole}}$	Q_{pol}	$Q_{E_{Donor}}$	$Q_{E_{Acceptor}}$	Q_{q-}	Q_{q+}	SPP_{calc}
			Propyl Formate–SPP = 0.815					
0°	0.77	0.45	0.35	0.61	0.32	0.40	0.06	0.97
60° [b]	0.66	0.34	0.35	0.60	0.37	0.38	0.06	0.91
120° [b]	0.47	0.26	0.35	0.59	0.37	0.38	0.06	0.83
180° [b]	0.40	0.33	0.35	0.59	0.29	0.41	0.06	0.84
Average	*0.58*	*0.35*	*0.35*	*0.60*	*0.34*	*0.39*	*0.06*	*0.89*
			methyl acetate–SPP = 0.785					
0°	0.78	0.20	0.27	0.66	0.29	0.42	0.08	0.91
60° [b]	0.67	0.20	0.27	0.66	0.32	0.40	0.08	0.88
120° [b]	0.46	0.32	0.27	0.67	0.32	0.41	0.08	0.86
180° [b]	0.33	0.34	0.27	0.64	0.25	0.42	0.08	0.84
Average	*0.56*	*0.26*	*0.27*	*0.65*	*0.29*	*0.41*	*0.08*	*0.87*
			ethyl acetate–SPP = 0.795					
0°	0.79	0.30	0.35	0.67	0.29	0.43	0.08	0.93
180°	0.36	0.38	0.35	0.65	0.25	0.42	0.08	0.84
			propyl acetate–SPP = 0.795					
0°	0.80	0.48	0.42	0.68	0.28	0.43	0.08	0.97
180°	0.38	0.40	0.42	0.66	0.25	0.42	0.08	0.84
			butyl acetate–SPP = 0.784					
0°	0.81	0.61	0.50	0.68	0.28	0.43	0.08	1.00
180°	0.37	0.48	0.50	0.66	0.24	0.42	0.08	0.85
			methyl salicylate–SPP = 0.836					
0°	0.19	0.70	0.65	0.86	0.47	0.40	0.22	0.85
60° [c]	0.27	0.61	0.64	0.85	0.42	0.40	0.24	0.84
120° [c]	0.34	0.53	0.64	0.85	0.42	0.40	0.25	0.83
180° [c]	0.44	0.51	0.65	0.87	0.48	0.37	0.19	0.84
Average	*0.31*	*0.59*	*0.64*	*0.86*	*0.45*	*0.39*	*0.23*	*0.84*

[a] $0 \leq Q_i \leq 1$; calculated from molecular properties using Equation (2); molecular properties from density functional calculations, Q_{q-} and Q_{q+} are calculated using Hirshfeld's model; [b] from single point calculations following rotation of the O=C−O−R dihedral angle of the optimized 0° structure; [c] 60° and 120° results are from single point calculations; angles are rotation of the benzene ring with respect to the O=CO−C plane; 0° corresponds to the OH group of the benzene ring being on the same side as the COC oxygen and 180° to the OH being on the side of the C=O.

Again, it can be seen that, while the variations are much smaller than those for the carboxylic acids, it is only the dipole moments and quadrupolar amplitudes that vary significantly with rotation around the C−O bond while the other molecular properties are relatively constant.

Results for methylsalicylate are included in Table 2, since rotation of the benzene ring progressively moves the O−H in the 2 position on the benzene ring away from the C−O−C oxygen to the C=O oxygen. In this case, the dipole moment essentially doubles between the 0° and 180° rotamers but, in the calculation of SPP^N, this is almost exactly compensated by the corresponding increase in the quadrupolar amplitude.

4. Discussion

Like SPP^N, both π^* and $E_T^N(30)$ show positive dependences on the solvent dipole moments and quadrupolar amplitudes and a negative dependence on the solvent molecule's polarizability. Each, however, shows an additional contribution.

In the case of $E_T^N(30)$, the additional contribution is from the charge on the most positive hydrogen atom on the solvent molecule and reflects hydrogen bonding at the pendant oxygen atom of the betaine dye used to define the scale. Of course, this is well known and was used by Kamlet and Taft to define their α scale of hydrogen bond acidity [4].

The reason for dependence of π^* on the energy of the electron donor orbital of the solvent molecule is less clear. Both Catalan and Kamlet and Taft considered the possibility of hydrogen bonding to the NO_2 group on the aromatic probe molecules, but if this were the case one would expect a correlation with the charge on the most negative atom of the solvent molecule, which is the dominant contribution to measures of hydrogen bond basicity (Kamlet and Taft's β [9,12], Gutmann's donor number [9], Abraham's B [17] and Catalan's SB [10]).

Leaving these aside, it is interesting that all three parameters show positive correlations with the solvent's molecular dipole moment and quadrupolar amplitude and a negative correlation with the solvent molecule's polarizability.

Since these are intended as measures of a solvent's polarity and polarizability, it is reasonable to associate the polarity with the dipole moment and quadrupolar amplitude. That is, the polarity is related to the intensity of the permanent charges imbedded in the solvent.

The basis of solvent parameters based on solvatochromism is that the equilibrium solvation of the ground state interacts with the excited state of the probe (By the Franck–Condon principle, the time scale of the electronic transition is vastly less than the time scale of molecular motions, so that, initially, the excited state interacts with the ground state solvation shells.) and so, if, for example. the excited state of the probe has greater charge separation than the ground state (This situation leads to an inverse relationship between the solvent parameter and the energy of the electronic transition; that is, the larger the energy gap the smaller the solvent parameter. The reverse situation, where the ground state has greater charge separation leads to a direct relationship; that is, to parameters that increase as the transition energy increases.), it will interact more strongly with the surrounding solvation shells, lowering the excitation energy.

It is implicit in the design of all based on solvatochromism, that all solute–solvent interactions will be stronger with the form of the solute with the greatest charge separation. This assumption is generally found to be correct; thus, basicity scales are positively correlated with the partial charge on the most negative atom and the energy of the donor orbital [9,10,12,17] and acidity parameters are positively correlated with the partial charge on the most positive hydrogen atom [10,11].

As is clear from Table 1, the polarity/polarizability parameters show positive correlations with the dipole moment and quadrupolar amplitude, measures of the intensity of charges imbedded in the surrounding solvent, but negative correlations with the molecular polarizability of the solvent. That is, increasing polarizability leads to decreases in SPP^N, π^* and $E_T^N(30)$.

The expectation that polarizability should correlate positively with these experimental parameters essentially arises from the assumption that charge–induced dipole interactions will dominate. However, solvent molecules may also interact with the probe molecules by London forces, in which case the stronger interaction is likely to be with the form with the lower charge separation, the ground state in the above case, where the excited state has the greater dipole moment. The observed negative correlation between SPP^N, π^* and $E_T^N(30)$ the polarizability of the solvent molecules, argues that this latter interaction dominates.

Whatever the explanation, SPP^N, π^* and $E_T^N(30)$ all show the same negative correlation with the solvent molecular polarizability, which makes their use in linear free energy relationships problematic. Thus, solvents with equal values of SPP^N, for example, can have quite different polarizabilities, compensated by differences in the dipole moments and/or quadrupolar amplitudes.

5. Conclusions

The analysis of SPP^N shows that, like Kamlet and Taft's π^* and Reichardt's $E_T^N(30)$, shows positive correlations with the molecular dipole moment and quadrupolar amplitude of solvent molecules and a negative correlation with the molecular polarizability.

Since these parameters are measures of the solvent polarity and polarizability it is reasonable to ascribe the solvent polarity to the permanent charges on the solvent molecules. In the approach used here, the dipole moment and quadrupolar amplitude are used as measures of the intensity of these charges. Thus, the solvent polarity is determined by the permanent charges in the solvent molecule and acts by non-specific interactions of these permanent charges with permanent charges on the solute. In contrast, solvent polarizability is determined by the polarizability of the individual solvent molecules and involves interactions with the non-polar parts of the solute through London forces.

It seems surprising that these experimental parameters reflect London rather than the expected charge/dipole–induced dipole interactions. However, the experimental parameters are based largely on probe molecules containing one or more aromatic rings, generally one for π^* [6], two for SPP^N [7] and six for $E_T^N(30)$ [8], and so are likely to interact well through London forces.

The observation that the polarity and polarizability act on the experimental parameters in opposite directions, increasing polarity increasing the values of the parameter while increasing polarizability decreases the values creates an obvious difficulty in the use of these parameters in linear free energy relationships, for example.

Funding: This research received no external funding.

Data Availability Statement: Data are contained within the article.

Conflicts of Interest: The author declares no conflict of interest.

Appendix A

The quadrupolar amplitude is calculated as $A = \sqrt{\sum q_{ij} q_{ij}}$ $I = x, y, z$ and $j = x, y, z$, where the q_{ij} are the components of the traceless quadrupole. Complex charge distributions, such as those of polyatomic molecules, are commonly represented by a series of superimposed point objects. The first is a point charge (the net charge), which is a scalar quantity, the second is the dipole, which is a vector and the third is the quadrupole, which is a tensor. Just as the dipole has no net charge, the quadrupole has no net moment. The dipole moment and quadrupolar amplitude are used here simply as quantitative measures of the scale of charge centers imbedded in the bulk solvent, the "intensity" of embedded charges. The simplest way to see the necessity for both the dipolar and quadrupolar contributions is to consider CO_2, which, despite having partial charges on the O and C atoms has a zero dipole moment, but a nonzero quadrupolar amplitude.

References

1. Kamlet, M.J.; Taft, R.W. The solvatochromic comparison method. I. The β-scale of solvent hydrogen-bond acceptor (HBA) basicities. *J. Am. Chem. Soc.* **1976**, *98*, 377–383. [CrossRef]
2. Catalán, J.; Díaz, C.; López, V.; Pérez, P.; de Paz, J.-L.G.; Rodriguez, J.G. A generalized solvent basicity scale: The solvatochromism of 5-nitroindoline and its homomorph 1-methyl-5-nitroindoline. *Liebigs Ann.* **1996**, *1996*, 1785–1794. [CrossRef]
3. Gutmann, V. Empirical parameters for donor and acceptor properties of solvents. *Electrochim. Acta* **1976**, *21*, 661–670. [CrossRef]
4. Taft, R.W.; Kamlet, M.J. The solvatochromic comparison method. 2. The α-scale of solvent hydrogen-bond donor (HBD) acidities. *J. Am. Chem. Soc.* **1976**, *98*, 2886–2894. [CrossRef]

5. Catalán, J.; Díaz, C. A generalized solvent acidity scale: The solvatochromism of *o-tert*-butylstilbazolium betaine dye and its homomorph *o,o'*-di-tert-butylstilbazolium betaine dye. *Liebigs Ann.* **1997**, *1997*, 1941–1949. [CrossRef]
6. Kamlet, M.J.; Abboud, J.L.; Taft, R.W. The solvatochromic comparison method. 6. The π^* scale of solvent polarities. *J. Am. Chem. Soc.* **1977**, *99*, 6027–6038. [CrossRef]
7. Catalán, J.; López, V.; Pérez, P.; Martín-Villamil, R.; Rodriguez, J.G. Progress towards a generalized solvent polarity scale: The solvatochromism of 2-(dimethylamino)-7-nitrofluorene and its homomorph 2-fluoro-7-nitrofluorene. *Liebigs Ann.* **1995**, *1995*, 241–252. [CrossRef]
8. Dimroth, K.; Reichardt, C.; Siepmann, T.; Bohlmann, F. Über pyridinium-*N*-phenol-betaine und ihre verwendung zur charakterisierung der polarität von lösungsmitteln. *Justus Liebigs Ann. Chem.* **1963**, *661*, 1–37. [CrossRef]
9. Waghorne, W.E.; O'Farrell, C. Solvent basicity, a study of Kamlet–Taft β and Gutmann *DN* values using computationally derived molecular properties. *J. Solution Chem.* **2018**, *47*, 1609–1625. [CrossRef]
10. Waghorne, W.E. Solvent Acidity and Basicity Scales: Analysis of Catalan's SB and SA scales and Gutmann's Acceptor Number and Comparison with Kamlet and Taft's β and α Solvent Scales and Abraham's B and A Solute Scales. 2024, in preparation. 2024; in preparation.
11. Waghorne, E. A Study of Abraham's effective hydrogen bond acidity and polarity/polarizability parameters, *A* and *S*, using computationally derived molecular properties. *J. Solution Chem.* **2023**. [CrossRef]
12. Waghorne, W.E. A study of Kamlet–Taft β and π^* scales of solvent basicity and polarity/polarizability using computationally derived molecular properties. *J. Solution Chem.* **2020**, *49*, 466–485. [CrossRef]
13. Waghorne, W.E. A study of the Reichardt $E_T^N(30)$ parameter using solvent molecular properties derived from computational chemistry and consideration of the Kamlet and Taft α scale of solvent hydrogen bond acidities. *J. Solution Chem.* **2020**, *49*, 1360–1372. [CrossRef]
14. Hirshfeld, F.L. Bonded-atom fragments for describing molecular charge densities. *Theor. Chim. Acta* **1977**, *44*, 129–138. [CrossRef]
15. Foster, J.P.; Weinhold, F. Natural hybrid orbitals. *J. Amer. Chem. Soc.* **1980**, *102*, 7211–7218. [CrossRef]
16. Marenich, A.V.; Jerome, S.V.; Craner, C.J.; Truhlar, D.G. Charge Model 5: An extension of Hirshfeld population analysis for the accurate description of molecular interactions in gaseous and condensed phases. *J. Chem. Theory Comput.* **2012**, *6*, 527–541. [CrossRef] [PubMed]
17. Waghorne, W.E. A study of the Abraham effective solute hydrogen bond basicity parameter using computationally derived molecular properties. *J. Solution Chem.* **2023**, *51*, 1133–1147. [CrossRef]
18. Scuseria, G.E.; Robb, M.A.; Cheeseman, J.R.; Scalmani, G.; Barone, V.; Mennucci, B.; Petersson, G.A.; Nakatsuji, H.; Caricato, M.; Li, X.; et al. *Gaussian*, 9th ed.; Gaussian, Inc.: Wallingford, CT, USA, 2009.

Disclaimer/Publisher's Note: The statements, opinions and data contained in all publications are solely those of the individual author(s) and contributor(s) and not of MDPI and/or the editor(s). MDPI and/or the editor(s) disclaim responsibility for any injury to people or property resulting from any ideas, methods, instructions or products referred to in the content.

Communication

Effective Recognition of Lithium Salt in (Choline Chloride: Glycerol) Deep Eutectic Solvent by Reichardt's Betaine Dye 33

Manish Kumar, Abhishek Kumar and Siddharth Pandey *

Department of Chemistry, Indian Institute of Technology Delhi, Hauz Khas, New Delhi 110016, India
* Correspondence: sipandey@chemistry.iitd.ac.in; Tel.: +91-11-26596503; Fax: +91-11-26581102

Abstract: Deep eutectic solvents (DESs) have emerged as novel alternatives to common solvents and VOCs. Their employment as electrolytes in batteries has been an area of intense research. In this context, understanding changes in the physicochemical properties of DESs in the presence of Li salts becomes of utmost importance. Solvatochromic probes have the potential to gauge such changes. It is reported herein that one such UV–vis molecular absorbance probe, Reichardt's betaine dye 33, effectively manifests changes taking place in a DES Glyceline composed of H-bond accepting salt choline chloride and H-bond donor glycerol in a 1:2 molar ratio, as salt LiCl is added. The lowest energy intramolecular charge–transfer absorbance band of this dye exhibits a 17 nm hypsochromic shift as up to 3.0 molal LiCl is added to Glyceline. The estimated E_T^N parameter shows a linear increase with the LiCl mole fraction. Spectroscopic responses of betaine dye 33, N,N-diethyl-4-nitroaniline and 4-nitroaniline are used to assess empirical Kamlet–Taft parameters of dipolarity/polarizability (π^*), H-bond-donating acidity (α) and H-bond-accepting basicity (β) as a function of LiCl concentration in Glyceline. LiCl addition to Glyceline results in an increase in α and no change in π^* and β. It is proposed that the added lithium interacts with the oxygen of the –OH functionalities on the glycerol rendering of the solvent with increased H-bond-donating acidity. It is observed that pyrene, a popular fluorescence probe of solvent polarity, does respond to the addition of LiCl to Glyceline, however, the change in pyrene response starts to become noticeable only at higher LiCl concentrations ($m_{\text{LiCl}} \geq 1.5$ m). Reichardt's betaine dye is found to be highly sensitive and versatile in gauging the physicochemical properties of DESs in the presence of LiCl.

Keywords: deep eutectic solvent; Reichardt's betaine dye 33; Glyceline; Lithium chloride; pyrene

Citation: Kumar, M.; Kumar, A.; Pandey, S. Effective Recognition of Lithium Salt in (Choline Chloride: Glycerol) Deep Eutectic Solvent by Reichardt's Betaine Dye 33. *Liquids* **2023**, *3*, 393–401. https://doi.org/10.3390/liquids3040024

Academic Editors: Franco Cataldo and William E. Acree, Jr.

Received: 23 June 2023
Revised: 12 September 2023
Accepted: 21 September 2023
Published: 28 September 2023

Copyright: © 2023 by the authors. Licensee MDPI, Basel, Switzerland. This article is an open access article distributed under the terms and conditions of the Creative Commons Attribution (CC BY) license (https://creativecommons.org/licenses/by/4.0/).

1. Introduction

Deep eutectic solvents (DESs) have emerged as viable alternatives not only to toxic organic solvents but also to ionic liquids [1–7]. While many commonly used organic solvents are hazardous to the immediate environment and belong to the class of volatile organic compounds (VOCs), recent toxicity reports are not favorable as far as common ionic liquids are concerned [7,8]. Escalating costs associated with the manufacture of many organic solvents combined with the complexity of synthesis and purification of most ionic liquids further restrict the use of these solvent media in science and technology today [7,8]. A DES, in this context, affords a solubilizing media that is mostly non-toxic and inexpensive. DESs can be prepared by simple mixing of two judiciously selected constituents. There have been many discoveries in the types of DESs depending upon the constituents but the most important are type III DESs. They consist of an H-bond donor (HBD) and an H-bond acceptor (HBA), that are inexpensive, non-toxic, and easily acquired. After mixing the two constituents, the melting point of the resulting mixture is usually much lower than the melting points of each of the constituents resulting in a liquid state of matter under ambient conditions. Among several classes of DESs proposed in the recent literature, the ones prepared by mixing a common ammonium salt, such as HBA with a suitable HBD, are perhaps the most investigated so far [3,5,9]. Specifically, the DESs constituted of choline chloride as the HBA and one of the HBDs, namely urea, glycerol, ethylene glycol, and malonic acid, are the initial DESs in this class that were reported around two decades ago [3,5,9]. Applications of these DESs in various strata of science and technology have been growing ever since [2,3].

One of the major areas of application of DESs is in electrochemistry, where DESs have shown potential as worthy electrolytes for batteries [10,11]. As a consequence, salt-added DESs have become a solvent system subject to rigorous investigation of late. In this context, investigations of potential uses for DESs in Li-ion batteries have naturally emerged. Changes in the physicochemical properties of the DESs due to the presence of Li salt have subsequently become an active area of research. Understanding of the solvation and dynamics of Li salts within DESs is being pursued by researchers worldwide.

Solvatochromic probe behavior within a Li salt-added DES system can reveal changes in the physicochemical properties of the milieu due to the addition of the Li salt; it also reveals information on solute solvation and dynamics in the process [12–14]. Information gained from the responses of spectroscopic probes can be useful in understanding reactivity, separation, extraction, and electrochemistry involving solutes with similar functionalities. We have found that Reichardt's betaine dye 33 (structure provided in Figure 1), which is known to manifest dipolarity/polarizability along with the H-bond donating (HBD) acidity of the solubilizing medium, is effectively able to gauge the consequences of adding LiCl to the DES constituted of choline chloride (ChCl) and glycerol (Gly) in a 1:2 molar ratio named Glyceline. The use of betaine dyes to obtain physicochemical changes and solute solvation affords a simple and effective way to obtain insights to such complex systems.

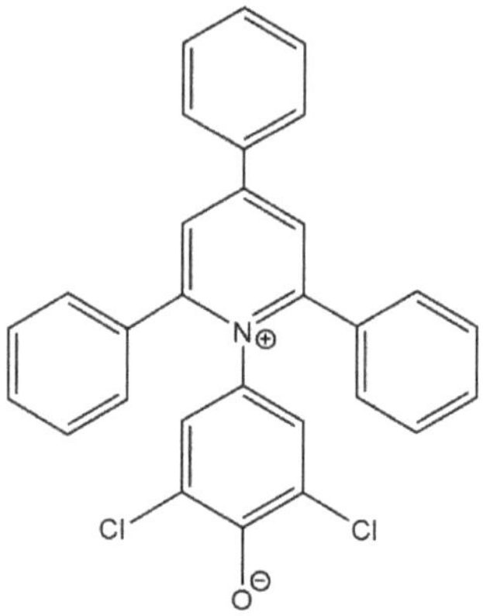

Figure 1. Structure of Reichardt's betaine dye 33.

2. Materials and Methods

Glycerol ($\geq$99.5%), choline chloride ($\geq$99.0%) and LiCl with >99% (by mass) purity were purchased from Sigma-Aldrich (St. Louis, MO, USA) and stored in an Auto Secador desiccator cabinet. 2,6-Dichloro-4-(2,4,6-triphenylpyridinium-1-yl)phenolate (betaine dye 33) was purchased in the highest available purity from Fluka ($\geq$99%, HPLC grade). 4-Nitroaniline (NA) and N,N-diethyl-4-nitroaniline (DENA) were purchased in the highest purity from Spectrochem Co., Ltd. (Mumbai, India). and Frinton Laboratories, respectively. Pyrene [$\geq$99.0% (GC), puriss for fluorescence] was obtained in highest purity from Sigma-Aldrich Co.

The calculated amount of glycerol and choline chloride was transferred to a glass vial and weighed using an analytical balance with a precision of $\pm$0.1 mg. The components were mixed thoroughly to obtain a homogeneous solution and subjected to vacuum for approximately 6 h. As per the requirement, a pre-calculated amount of LiCl was added to this solution and mixed over a magnetic stirrer at 60 °C until all of the LiCl was dissolved, and a homogeneous solution was obtained. Stock solution of all of the probes was prepared by dissolving the required amount in ethanol in a pre-cleaned amber glass vial and stored at 4 ± 1 °C to retard any photochemical reaction. An appropriate amount of the probe solution from the stock was transferred to the 1 cm path length quartz cuvette. Ethanol was evaporated using a gentle stream of high purity nitrogen gas to achieve the desired final concentration of the probe. A pre-calculated amount of LiCl was added—Glyceline DES was directly added to the cuvette and the solution was thoroughly mixed. The final concentrations of DENA, NA, betaine dye 33 and pyrene were ~20, 20, 50 and 10 µM, respectively. A Perkin-Elmer Lambda 35 double-beam spectrophotometer with variable bandwidth was used for the acquisition of the UV–vis molecular absorbance spectra of DENA, NA and betaine dye 33. Steady-state fluorescence spectra of pyrene (λ_{ex} = 337 nm) were acquired on an Edinburgh Instruments Ltd. (Livingston, UK) spectrofluorimeter (FLS1000-SS-S) with STGM325-X grating excitation and STGM325-M grating emission monochromators with a 450 W Xe arc lamp as the excitation source, a single cell TEC holder and a Red PMT as the detector. All spectra were duly corrected by subtracting the spectral responses from suitable blanks prior to data analysis. Data analysis was performed using SigmaPlot v14.5 software.

3. Results and Discussion

UV–vis molecular absorbance spectra of Reichardt's betaine dye 33 dissolved in LiCl-added Glyceline under ambient conditions are presented in Figure 2A (the maximum molal concentration of LiCl in the system was m_{LiCl} = 3.0 mol·kg^{-1}, which corresponds to mole fraction χ_{LiCl} = 0.24). A careful examination of the spectra reveals that the lowest energy absorbance band of the dye shows systematic monotonic hypsochromic shift as LiCl is added to Glyceline. It is well-established that 2,6-diphenyl-4-(2,4,6-triphenylpyridinium-1-yl)phenolate (Reichardt's betaine dye 30) exhibits an unusually high solvatochromic band shift; the lowest energy intramolecular charge-transfer absorption band of betaine dye 30 is hypsochromically-shifted by ca. 357 nm in going from relatively nonpolar diphenyl ether (λ_{max}~810 nm) to water (λ_{max}~453 nm) [15–17]. It is established that the negative solvatochromism of betaine dye 30 originates from the differential solvation of its highly polar equilibrium ground-state and the less polar first Franck–Condon excited-state with increasing solvent polarity [15–17]. There is a considerable charge transfer from the phenolate to the pyridinium part of the zwitterionic molecule. Because of its zwitterionic nature the solvatochromic probe behavior of betaine dye 30 is strongly affected by the HBD acidity of the solvent; H-donating solvents stabilize the ground-state more than the excited-state. The empirical scale of solvent 'polarity', $E_T(30)$ for betaine dye 30, is defined as the molar transition energy of the dye traditionally in kcal·mol^{-1} at room temperature and normal pressure according to the expression $E_T(30)$ = 28591.5/λ_{max} in nm [15–17]. However, in the present work a derivative of betaine dye 30, 2,6-dichloro-4-(2,4,6-triphenylpyridinium-1-yl)phenolate (betaine dye 33), is used to investigate LiCl-added Glyceline system due to it having certain advantages over betaine dye 30. The low solubility of betaine dye 30 in many H-bonded solvent systems renders it unsuitable for our investigations. Betaine dye 33, on the other hand, has no such problems due to inherent structural differences with betaine dye 30. For historical reasons, it has been related to number 33, and the lowest energy absorbance transition of this dye [i.e., $E_T(33)$] is calculated the same way $E_T(30)$ is calculated [17].

Thus, from the absorbance spectra of betaine dye 33 presented in Figure 2A, the corresponding $E_T(33)$ are estimated and converted into E_T^N using Equations (1) and (3):

$$E_T(30) = 0.9953(\pm 0.0287) \times E_T(33) - 8.1132(\pm 1.6546) \quad (1)$$

R = 0.9926, standard error of estimate = 0.8320, n = 20

$E_T(30)$ was obtained from $E_T(33)$ (i.e., Equation (1)) by acquiring the lowest energy UV–vis absorbance band for both the dyes in 20 different solvents, and performing linear regression analysis between the two.

$$E_T^N = \frac{[E_T(30)_{SOLVENT} - E_T(30)_{TMS}]}{[E_T(30)_{WATER} - E_T(30)_{TMS}]} \quad (2)$$

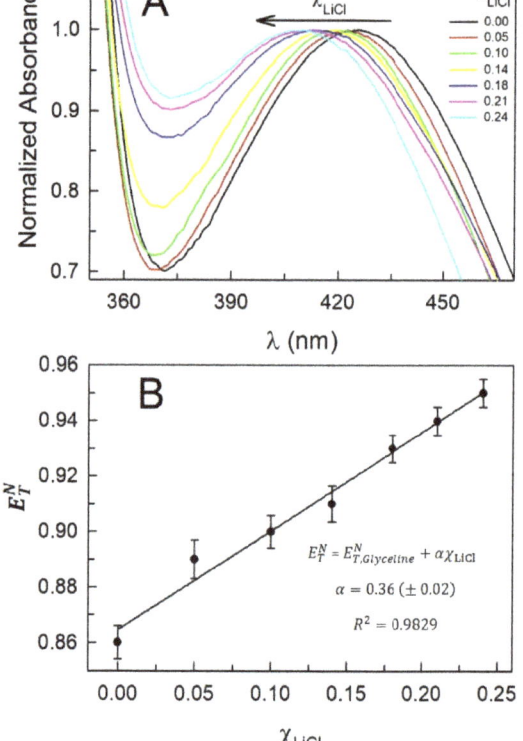

Figure 2. Absorbance spectra of Reichardt's betaine dye 33 [50 µM] in Glyceline and LiCl-added Glyceline under ambient conditions (panel **A**) and variation in E_T^N with a mole fraction of LiCl (χ_{LiCl}) (panel **B**). The solid straight line is the best fit obtained from the linear regression analysis. Error in E_T^N is $\leq \pm 0.007$.

Here, TMS stands for tetramethylsilane. From $E_T(30)_{WATER}$ = 63.1 kcal·mol^{-1} and $E_T(30)_{TMS}$ = 30.7 kcal·mol^{-1}, we obtain

$$E_T^N = \frac{[E_T(30)_{SOLVENT} - 30.7]}{32.4} \quad (3)$$

E_T^N is easier to conceive as it is dimensionless and varies between 0 for TMS (extreme non-polar) and 1 for water (extreme polar) [17]. Table 1 lists the lowest energy absorbance maxima ($\lambda_{max,33}$) of betaine dye 33 along with the estimated E_T^N for the LiCl-added Glyceline system. A hyposchromic shift of 17 nm is observed in going from no LiCl to 3.0 m of LiCl in Glyceline which transforms to an increase in E_T^N from 0.86 to 0.95. It is convenient to note that a plot of E_T^N versus χ_{LiCl} exhibits good linear behavior ($R^2 > 0.98$) with a slope of 0.36 ($\pm$0.02) (Figure 2B). Thus, it is concluded that as LiCl is added to Glyceline, the dipolarity/polarizabilty and/or HBD acidity of the system increases; and the increase is effectively manifested in the spectral response of the Reichardt's betaine dye 33 in a linear manner with the mole fraction of LiCl.

Table 1. Absorbance maxima for Reichardt's betaine dye 33 ($\lambda_{max,33}$), DENA ($\lambda_{max,DENA}$) and NA ($\lambda_{max,NA}$) and corresponding estimated Kamlet–Taft empirical solvent parameters, at different mole fractions of LiCl (χ_{LiCl}) in Glyceline under ambient conditions. Error in λ_{max} are $\leq \pm 0.5$ nm. Error in E_T^N is $\leq \pm 0.007$ and errors in α, β and π^* are $\leq \pm 0.005$.

m_{LiCl} (mol·kg^{-1})	χ_{LiCl}	$\lambda_{max,33}$ (nm)	E_T^N	$\lambda_{max,DENA}$ (nm)	π^*	$\lambda_{max,NA}$ (nm)	β	α
0.0	0.00	426	0.86	422	1.21	385	0.43	0.85
0.5	0.05	422	0.89	422	1.21	385	0.43	0.89
1.0	0.10	419	0.90	422	1.21	385	0.43	0.92
1.5	0.14	417	0.91	422	1.21	385	0.43	0.94
2.0	0.18	413	0.93	422	1.21	385	0.43	0.98
2.5	0.21	411	0.94	422	1.21	385	0.43	1.00
3.0	0.24	409	0.95	422	1.21	385	0.43	1.02

Whether the increase in E_T^N upon addition of LiCl to the DES Glyceline is due to the increase in the dipolarity/polarizabilty or the HBD acidity or both is explored by assessing empirical Kamlet–Taft solvatochromic indicators of solvent dipolarity/polarizability (π^*), HBD acidity (α), and HBA basicity (β) [18–22]. The π^* is estimated from the absorption maximum (ν_{DENA}, in kK) of DENA, a non-hydrogen bond donor solute, using [18,19]:

$$\pi^* = 8.649 - 0.314\nu_{DENA} \quad (4)$$

and then α was estimated from $E_T(30)$ and π^* values [18,20].

$$\alpha = [E_T(30) - 14.6(\pi^* - 0.23\delta) - 30.31]/16.5 \quad (5)$$

The δ parameter in Equation (5) is a "polarizability correction term" equal to 0.0 for nonchlorinated aliphatic solvents, 0.5 for polychlorinated aliphatics, and 1.0 for aromatic solvents [21]. Finally, β values are determined from the enhanced solvatochromic shift of NA relative to its homomorph DENA, $-\Delta\nu$(DENA–NA)/kK [18,22]:

$$\beta = -0.357\nu_{NA} - 1.176\pi^* + 11.12 \quad (6)$$

Interestingly, the UV–vis absorbance spectra of both DENA and NA, respectively, do not show any statistically meaningful variation upon addition of up to 3.0 m LiCl to the DES Glyceline (Figure 3A). Based on Equations (4) and (6), this subsequently reflects in no change in the β and the π^* parameters as LiCl is added to Glyceline (Table 1). The parameter α, which depends on the E_T parameter along with π^* (Equation (5)), does increase with an increasing concentration of LiCl in Glyceline (Figure 3B).

The Kamlet–Taft empirical parameters for solvent polarity (π^*, α, and β) clearly indicate the surprising outcome that, as LiCl is added to DES Glyceline, dipolarity/polarizability of the medium does not change, nor does the H-bond accepting basicity—the medium acquires more H-bond donating acidity [parameter α increases linearly with increasing χ_{LiCl} within the system with a slope = 0.70 ($\pm$0.02)]. Within Glyceline, it is reported that the Cl$^-$ of ChCl are involved in H-bonding with the –OH functionalities of glycerol that in turn contribute to DES formation [23]. We believe that added Li$^+$ preferentially combines with the oxygen of the –OH functionalities of glycerol thus rendering the HBD acidity of the medium to increase [24]. The diminished HBA basicity due to this is compensated by the presence of additional Cl$^-$ of the LiCl. Since both added Li$^+$ and Cl$^-$ are involved in various H-bonding within the system, diminishing their charges, no effective increase in dipolarity/polarizability is observed.

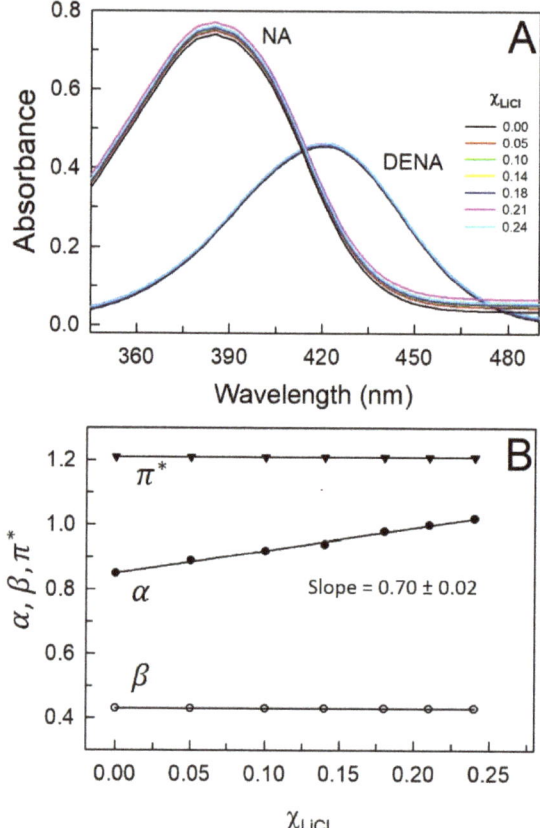

Figure 3. UV–visible absorbance spectra of N,N-diethyl-4-nitroaniline (DENA, 20 µM) and 4-nitroaniline (NA, 20 µM) (panel **A**) and variation in α, β and π^* with mole fraction of LiCl (χ_{LiCl}) within Glyceline (panel **B**) under ambient conditions. Errors in α, β and π^* are $\leq \pm 0.005$.

Since fluorescence polarity probes are known for their higher sensitivity, we employed pyrene as one of such probes to assess the effect of LiCl addition on DES Glyceline. Fluorescence emission spectra of pyrene is constituted of five vibronic bands with band 1-to-band 3 intensity ratio (Py I_1/I_3) increases monotonically with increasing dipolarity of the pyrene cybotactic region [25–29]. Emission spectra of pyrene in LiCl-added Glyceline is acquired at five different temperatures in the range 298.15 to 358.15 K (representative spectra are shown in Figure 4A).

The estimated Py I_1/I_3 at different χ_{LiCl} are plotted in Figure 4C. A careful examination of the data reveals that statistically meaningful changes in Py I_1/I_3 start to appear only above 1.5 m LiCl—for $m_{LiCl} < 1.5$ m, the pyrene probe is not able to manifest polarity changes in the system as LiCl is added to Glyceline. Betaine dye 33 response, however, could effectively reflect the changes in the medium at very low LiCl concentrations as well. It is interesting to note that at higher LiCl concentrations, the Py I_1/I_3 decreases suggesting a decrease in the dipolarity of the pyrene cybotactic region in the presence of LiCl. We again invoke the explanation given above that both Li$^+$ and Cl$^-$ tie up with the charged species present in the solution thus lowering the dipolarity of the medium—this lowering in dipolarity may be overshadowed by the increased HBD acidity that becomes reflected in the response of the betaine dye 33. Further support for this is afforded by the variation in Py I_1/I_3 of the LiCl-added Glyceline as the temperature is increased

(Figure 4B shows pyrene emission spectra at two different temperatures). Figure 4D depicts the clear decrease in Py I_1/I_3 as the temperature of LiCl-added Glyceline system is increased—the decrease is observed to be linear. This observation is akin to the decrease in static dielectric constants (ε) of several liquids, including several DESs and ionic liquids, as the temperature is increased. Also, similar observations were reported for the LiCl-added ChCl:Urea and glycerol, respectively, as well as LiTf$_2$N-added 1-ethyl-3-methylimidazolium bis(trifluoromethylsulfonyl)imide ([C$_2$C$_1$im][Tf$_2$N]) at similar temperatures [30–32].

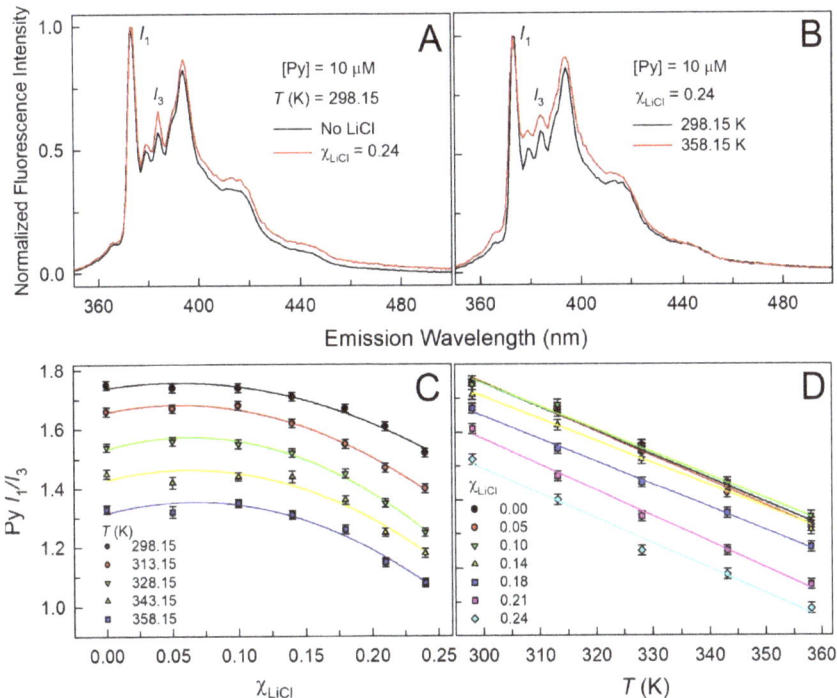

Figure 4. Fluorescence emission spectra of pyrene within Glyceline in the absence and presence of LiCl (χ_{LiCl} = 0.24) at 298.15 K (panel **A**) and at 298.15 K and 358.15 K at χ_{LiCl} = 0.24 (panel **B**). Band 1 to 3 emission intensity ratio of pyrene (Py I_1/I_3) within LiCl-added Glyceline for temperatures ranging between 298.15 and 358.15 K (panel **C**) and for different LiCl mole fractions at all investigated temperatures (panel **D**). Error in Py I_1/I_3 is $\leq \pm 0.02$.

4. Conclusions

Reichardt's betaine dye 33 is able to effectively manifest the changes taking place in DES Glyceline as LiCl salt is added. The response of the betaine dye 33 in concert with responses from DENA and NA (to obtain empirical Kamlet–Taft parameters) affords a scenario where it is clear that as LiCl is added to Glyceline, the HBD acidity of the medium increases with little or no change in the dipolarity/polarizability and HBA basicity. The interaction of Li species with the oxygens of the –OH functionalities of glycerol imparts increased HBD acidity to the medium with other interactions compensating for each other in such a manner that there is little or no increase in dipolarity/polarizability and HBA basicity. The fluorescence probe pyrene is able to reflect the decrease in the dipolarity but only at higher LiCl concentrations ($m_{LiCl} \geq 1.5$ m). A decrease in dipolarity with increasing temperature, however, is amply manifested through the pyrene response. The sensitivity and versatility of Reichardt's betaine dye in effective gauging changes in the physicochemical properties of the liquid medium is amply demonstrated.

Author Contributions: Conceptualization, methodology, visualization, writing—original draft, M.K.; data collection, A.K.; writing—review and editing, supervision, resources, project administration, S.P. All authors have read and agreed to the published version of the manuscript.

Funding: This work was generously supported by the Council of Scientific and Industrial Research, EMR-II (CSIR-EMR-II), Government of India, through a grant to Siddharth Pandey [grant number 01(3043)/21/EMR-II].

Data Availability Statement: Not applicable.

Acknowledgments: Manish Kumar would like to acknowledge the Council of Scientific and Industrial Research (CSIR) of the Government of India for his Senior Research Fellowship (SRF).

Conflicts of Interest: The authors declare no conflict of interest.

References

1. Wagle, D.V.; Zhao, H.; Baker, G.A. Deep eutectic solvents: Sustainable media for nanoscale and functional materials. *Acc. Chem. Res.* **2014**, *47*, 2299–2308. [CrossRef] [PubMed]
2. Hansen, B.B.; Spittle, S.; Chen, B.; Poe, D.; Zhang, Y.; Klein, J.M.; Horton, A.; Adhikari, L.; Zelovich, T.; Doherty, B.W.; et al. Deep Eutectic Solvents: A Review of Fundamentals and Applications. *Chem. Rev.* **2021**, *121*, 1232–1285. [CrossRef] [PubMed]
3. Zhang, Q.; De Oliveira Vigier, K.; Royer, S.; Jérôme, F. Deep eutectic solvents: Syntheses, properties and applications. *Chem. Soc. Rev.* **2012**, *41*, 7108–7146. [CrossRef] [PubMed]
4. LaRocca, M.M.; Baker, G.A.; Heitz, M.P. Assessing rotation and solvation dynamics in ethaline deep eutectic solvent and its solutions with methanol. *J. Chem. Phys.* **2021**, *155*, 034505. [CrossRef]
5. Smith, E.L.; Abbott, A.P.; Ryder, K.S. Deep Eutectic Solvents (DESs) and Their Applications. *Chem. Rev.* **2014**, *114*, 11060–11082. [CrossRef]
6. Sahu, S.; Banu, S.; Sahu, A.K.; Kumar, B.P.; Mishra, A.K. Molecular-level insights into inherent heterogeneity of maline deep eutectic system. *J. Mol. Liq.* **2022**, *350*, 118478. [CrossRef]
7. Płotka-Wasylka, J.; De la Guardia, M.; Andruch, V.; Vilková, M. Deep Eutectic Solvents vs. Ionic Liquids: Similarities and Differences. *Microchem. J.* **2020**, *159*, 105539. [CrossRef]
8. Flieger, J.; Flieger, M. Ionic liquids toxicity-benefits and threats. *Int. J. Mol. Sci.* **2020**, *21*, 6267. [CrossRef]
9. Abbott, A.P.; Boothby, D.; Capper, G.; Davies, D.L.; Rasheed, R.K. Deep Eutectic Solvents Formed between Choline Chloride and Carboxylic Acids: Versatile Alternatives to Ionic Liquids. *J. Am. Chem. Soc.* **2004**, *126*, 9142–9147. [CrossRef]
10. Cruz, H.; Jordão, N.; Branco, L.C. Deep Eutectic Solvents (DESs) as Low-Cost and Green Electrolytes for Electrochromic Devices. *Green Chem.* **2017**, *19*, 1653–1658. [CrossRef]
11. Li, Q.; Chen, J.; Fan, L.; Kong, X.; Lu, Y. Progress in Electrolytes for Rechargeable Li-Based Batteries and Beyond. *Green Energy Environ.* **2016**, *1*, 18–42. [CrossRef]
12. Acree, W.E., Jr.; Wilkins, D.C.; Tucker, S.A.; Griffin, J.M.; Powell, J.R. Spectrochemical Investigations of Preferential Solvation. 2. Compatibility of Thermodynamic Models versus Spectrofluorometric Probe Methods for Tautomeric Solutes Dissolved in Binary Mixtures. *J. Phys. Chem.* **1994**, *98*, 2537–2544. [CrossRef]
13. Rai, R.; Pandey, S. Solvatochromic probe response within ionic liquids and their equimolar mixtures with tetraethylene glycol. *J. Phys. Chem. B* **2014**, *118*, 11259–11270. [CrossRef] [PubMed]
14. Nunes, R.; Nunes, N.; Elvas-Leitão, R.; Martins, F. Using Solvatochromic Probes to Investigate Intermolecular Interactions in 1,4-Dioxane/Methanol/Acetonitrile Solvent Mixtures. *J. Mol. Liq.* **2018**, *266*, 259–268. [CrossRef]
15. Catalán, J.; de Paz, J.L.G.; Reichardt, C. On the Molecular Structure and UV/vis Spectroscopic Properties of the Solvatochromic and Thermochromic Pyridinium-N-Phenolate Betaine Dye B30. *J. Phys. Chem. A* **2010**, *114*, 6226–6234. [CrossRef]
16. Reichardt, C. Polarity of Ionic Liquids Determined Empirically by Means of Solvatochromic Pyridinium N-phenolate Betaine Dyes. *Green Chem.* **2005**, *7*, 339–351. [CrossRef]
17. Reichardt, C. Solvatochromic Dyes as Solvent Polarity Indicators. *Chem. Rev.* **1994**, *94*, 2319–2358. [CrossRef]
18. Kamlet, M.J.; Abboud, J.L.M.; Abraham, M.H.; Taft, R.W. Linear Solvation Energy Relationships. 23. A Comprehensive Collection of the Solvatochromic Parameters, Pi*, Alpha, and Beta, and Some Methods for Simplifying the Generalized Solvatochromic Equation. *J. Org. Chem.* **1983**, *48*, 2877–2887. [CrossRef]
19. Kamlet, M.J.; Abboud, J.L.; Taft, R.W. The Solvatochromic Comparison Method. 6. The π^* Scale of Solvent Polarities. *J. Am. Chem. Soc.* **1977**, *99*, 6027–6038. [CrossRef]
20. Taft, R.W.; Kamlet, M.J. The Solvatochromic Comparison Method. 2. The α-Scale of Solvent Hydrogen-Bond Donor (HBD) Acidities. *J. Am. Chem. Soc.* **1976**, *98*, 2886–2894. [CrossRef]
21. Taft, R.W.; Abboud, J.L.M.; Kamlet, M.J. Solvatochromic comparison method. 20. Linear solvation energy relationships. 12. The d.delta. term in the solvatochromic equations. *J. Am. Chem. Soc.* **1981**, *103*, 1080–1086. [CrossRef]
22. Kamlet, M.J.; Taft, R.W. The Solvatochromic Comparison Method. I. The β-Scale of Solvent Hydrogen-Bond Acceptor (HBA) Basicities. *J. Am. Chem. Soc.* **1976**, *98*, 377–383. [CrossRef]

23. Makris, D.P.; Lalas, S. Glycerol and Glycerol-Based Deep Eutectic Mixtures as Emerging Green Solvents for Polyphenol Extraction: The Evidence so Far. *Molecules* **2020**, *25*, 5842. [CrossRef] [PubMed]
24. Yu, D.; Troya, D.; Korovich, A.G.; Bostwick, J.E.; Colby, R.H.; Madsen, L.A. Uncorrelated Lithium-Ion Hopping in a Dynamic Solvent–Anion Network. *ACS Energy Lett.* **2023**, *8*, 1944–1951. [CrossRef]
25. Dong, D.C.; Winnik, M.A. The Py scale of solvent polarities. *Can. J. Chem.* **1984**, *62*, 2560–2565. [CrossRef]
26. Pandey, S.; Baker, S.N.; Pandey, S.; Baker, G.A. Fluorescent Probe Studies of Polarity and Solvation within Room Temperature Ionic Liquids: A Review. *J. Fluoresc.* **2012**, *22*, 1313–1343. [CrossRef]
27. Street, K.W., Jr.; Acree, W.E., Jr. Experimental Artifacts and Determination of Accurate Py Values. *Analyst* **1986**, *111*, 1197–1201. [CrossRef]
28. Street, K.W., Jr.; Acree, W.E., Jr.; Fetzer, J.C.; Shetty, P.H.; Poole, C.F. Polycyclic Aromatic Hydrocarbon Solute Probes. Part V: Fluorescence Spectra of Pyrene, Ovalene, Coronene, and Benzo[ghi]perylene Dissolved in Liquid Alkylammonium Thiocyanate Organic Salts. *Appl. Spectrosc.* **1989**, *43*, 1149–1153. [CrossRef]
29. Tucker, S.A.; Cretella, L.E.; Waris, R.; Street, K.W., Jr.; Acree, W.E., Jr.; Fetzer, J.C. Polycyclic Aromatic Hydrocarbon Solute Probes. Part VI: Effect of Dissolved Oxygen and Halogenated Solvents on the Emission Spectra of Select Probe Molecules. *Appl. Spectrosc.* **1990**, *44*, 269–273. [CrossRef]
30. Dhingra, D.; Pandey, A.; Pandey, S. Pyrene Fluorescence To Probe a Lithium Chloride-Added (Choline Chloride + Urea) Deep Eutectic Solvent. *J. Phys. Chem. B* **2019**, *123*, 3103–3111. [CrossRef]
31. Kumar, M.; Anjali; Dhingra, D.; Yadav, A.; Pandey, S. Effect of lithium salt on fluorescence quenching in glycerol: A comparison with ionic liquid/deep eutectic solvent. *Phys. Chem. Chem. Phys.* **2022**, *24*, 459–467. [CrossRef] [PubMed]
32. Kadyan, A.; Pandey, S. Florescence Quenching within Lithium Salt-Added Ionic Liquid. *J. Phys. Chem. B* **2018**, *122*, 5106–5113. [CrossRef] [PubMed]

Disclaimer/Publisher's Note: The statements, opinions and data contained in all publications are solely those of the individual author(s) and contributor(s) and not of MDPI and/or the editor(s). MDPI and/or the editor(s) disclaim responsibility for any injury to people or property resulting from any ideas, methods, instructions or products referred to in the content.

Communication

Prediction of Paracetamol Solubility in Binary Solvents Using Reichardt's Polarity Parameter Combined Model

Elaheh Rahimpour [1,2] and Abolghasem Jouyban [1,3,*]

[1] Pharmaceutical Analysis Research Center, Tabriz University of Medical Sciences, Tabriz 51656-65811, Iran; rahimpour_e@yahoo.com
[2] Infectious and Tropical Diseases Research Center, Tabriz University of Medical Sciences, Tabriz 51636-39888, Iran
[3] Faculty of Pharmacy, Near East University, North Cyprus, Mersin 10, Nicosia P.O. Box 99138, Turkey
* Correspondence: ajouyban@hotmail.com

Abstract: The objective of this research is to propose a general model utilizing the solvatochromic polarity of electronic transition energy (ET) of the Reichardt indicator to predict paracetamol solubility in the solvent mixtures. In order to model validation, the available ET (30) values of nine aqueous mixtures obtained from existing literature sources were utilized. The trained model yielded a relatively accurate estimation of paracetamol solubility in the investigated systems.

Keywords: paracetamol; solubility prediction; Reichardt indicator; binary mixtures

1. Introduction

Paracetamol, known as N-acetyl-p-aminophenol, is highly valued for its analgesic and antipyretic properties in the treatment of various conditions such as fever, headache, arthritis, neuralgia, post-surgical pain, and providing palliative care to advanced cancer patients [1]. While it is mostly administered as a tablet, other forms such as intravenous preparations, suppositories, and solutions are also available in the market [2]. For efficient drug absorption, it must be in an aqueous solution form at the absorption site. The improved aqueous solubility of drugs or drug candidates can increase their bioavailability, reduce their dosage, and ultimately enhance their efficacy. Therefore, the aqueous solubility of any drug candidate is a crucial physicochemical property essential for its successful development. This aspect of drug development is often limited by poor solubility, and, as a result, it is crucial to determine drug candidate solubility as early as possible. There is considerable interest in the development of models that accurately predict aqueous solubility directly from a chemical structure [3]. In the case of the low aqueous solubility of a drug, addition of a permissible organic solvent, cosolvency, is an appropriate solution. Cosolvency helps the formulation scientists to dissolve the desired amount of the drug in a given volume of the liquid formulation. In some cases, i.e., in injectable solution, there is a volume restriction problem too. More solubilizing cosolvent with a lower toxicity and less side effects is more favorable. Desolubilization of a drug is also required where recrystallization is the aim of the experiments. In these cases, the drug and the related compounds are dissolved in a good solvent; usually an organic solvent and an anti-solvent is added to the mixtures to induce crystallization process. These practical applications reveal the importance of solubility data in binary solvent mixtures. Despite the experimental determination of the solubility in cosolvent + water mixtures, there are some models to calculate the solubility in mixed solvent systems. These models facilitate the process of data usage in industrial applications. The extended Hildebrand solubility approach of Martin [4], mixture response surface [5], the combined nearly ideal binary solvent/Redlich–Kister equation [6], the log-linear model of Yalkowsky [7], the modified Wilson model [8], phenomenological model [9], fluctuation theory [10], the excess free

Citation: Rahimpour, E.; Jouyban, A. Prediction of Paracetamol Solubility in Binary Solvents Using Reichardt's Polarity Parameter Combined Model. *Liquids* **2023**, *3*, 512–521. https://doi.org/10.3390/liquids3040032

Academic Editor: Enrico Bodo

Received: 29 October 2023
Revised: 4 December 2023
Accepted: 8 December 2023
Published: 14 December 2023

Copyright: © 2023 by the authors. Licensee MDPI, Basel, Switzerland. This article is an open access article distributed under the terms and conditions of the Creative Commons Attribution (CC BY) license (https://creativecommons.org/licenses/by/4.0/).

energy approach [11], the Jouyban–Acree model [12], and Kamlet–Abboud–Taft-linear solvation energy relationship [13,14] were the well-known reported mathematical models for solubility prediction in cosolvency mixtures. One of the commonly used models that have demonstrated accurate predictions of solubility is the Jouyban–Acree model, which is dependent on both temperature and solvent compositions [12]. Beyond their general forms, these models can be customized by introducing the chemical and physical properties of solvent and solute into their parameters. Some of these parameters that exhibit quantitative structure–property relationships (QSPRs) are the Hansen [15] and Catalan [16] solubility parameters, Abraham solvation parameters and solvatochromic polarity parameters (e.g., electronic transition energy (ET) or Reichardt's polarity). In continuation of our previous works in combining the QSPR parameters with the Jouyban–Acree model, this study seeks to suggest a combined Jouyban–Acree model with Reichardt's polarity parameter that can predict and correlate paracetamol solubility in the cosolvency systems. To achieve this, data on paracetamol solubility along with ET 30 values in different cosolvency systems were gathered from the literature and utilized to develop a comprehensive model capable of predicting paracetamol solubility accurately. We used paracetamol as a model drug in this work, since a very wide range of solubility data in cosolvent + water mixtures are available for this drug.

2. Computational Methods

Until now, the solubility pattern of paracetamol has been studied in the binary aqueous mixtures of ethanol [17], 1-propanol [18], 2-propanol [19], polyethylene glycol 200 (PEG 200), PEG 400 [20], propylene glycol (PG) [21], PEG 600, N-methyl pyrrolidone (NMP) [22], methanol [23], carbitol [24], 1,4-dioxane [25], acetonitrile [26], dimethylformamide (DMF), and dimethylsulfoxide (DMSO) [27] and non-aqueous binary mixtures of NMP + PEG 600 [22], PEG 600 + PG [28], PEG 200 + ethanol, PEG 400 + ethanol, PEG 600 + ethanol [29], PG + ethanol [30] and ethyl acetate + ethanol [31]. The solubility data in the mole fraction unit were used for the studied computations. ET (30) values for the binary aqueous mixtures of ethanol, methanol, PG, 2-propanol, 1-propanol, acetonitrile, DMF, DMSO, and 1,4-dioxane were obtained from a reference by using interpolation for the desired co-solvent mass fraction [32]. However, ET (30) values for other mixtures were not available and excluded from the computations. Furthermore, the Abraham solubility parameters for the investigated solvents were taken from a reference [33]. It is obvious that some of the mentioned cosolvents, like methanol, DMF or 1,4-dioxane, are highly toxic and could not be used in the preparation of oral/parenteral/topical pharmaceutical formulations. We included these cosolvents in our study to show the capability of the proposed model to cover various cosolvent + water systems. In addition, these binary solvent mixtures could be used in other industrial applications such as crystallization, preparation of the nanoparticles, etc.

The investigated model in this work was the Jouyban–Acree model as the most precise cosolvency model available, it depicts the correlation between the solubility of a solute and both the temperature and the solvent composition. In binary cosolvency systems at different temperatures, the Jouyban–Acree model can be expressed in a general form as follows [12]:

$$\ln x_{m,T} = w_c \cdot \ln x_{c,T} + w_w \cdot \ln x_{w,T} + \frac{w_c w_w}{T} \sum_{i=0}^{2} J_i (w_c - w_w)^i \qquad (1)$$

where $x_{m,T}$, $x_{c,T}$ and $x_{w,T}$ denote the solubility of the solute in the solvent mixtures, cosolvent and water at temperature T/K; w_c, and w_w are the mass fractions of mono solvents 1 (the cosolvents (c) in this work), and 2 (water (w) in this work) in the absence of the solute; and J_i terms are the model coefficients representing the two-body (d-d, d-c, c-c, d-w, w-w (d = drug)) and three-body (d-d-d, d-d-c, d-c-c, d-c-d, c-c-c, d-d-w, d-w-w, d-w-d, w-w-w, d-w-c) interactions in the solute saturated mixture solution [12]. One can integrate the

Jouyban–Acree model with certain parameters such as Reichardt's polarity parameter to analyze the characteristics of solvents with regard to their physicochemical properties. By including these values in Equation (1) for a given solute, the combined model can be obtained as

$$\ln x_{m,T} = w_c \cdot \ln x_{c,T} + w_w \cdot \ln x_{w,T} + \left(\frac{w_c w_w}{T}\right)\left(J_1' + J_2' \cdot E_{m,T}^N\right) \\ + \left(\frac{w_c w_w (w_c - w_w)}{T}\right)\left(J_3' + J_4' \cdot E_{m,T}^N\right) + \left(\frac{w_c w_w (w_c - w_w)^2}{T}\right)\left(J_5' + J_6' \cdot E_{m,T}^N\right) \quad (2)$$

where J' terms are the model parameters and $E_{m,T}^N$ is the ET (30) values for the desired binary mixtures. The symbols used in prior models remain unchanged in this case. The model constants in Equation (2) are determined through a no-intercept least square analysis.

To investigate the capability of Reichardt's polarity parameter for improving the solubility prediction power of the Jouyban–Acree model, the obtained results were compared with the Jouyban–Acree model combined with Abraham solvation parameters. For this purpose, Equation (3) as a simplified model for one solute was used:

$$\ln x_{m,T} = w_c \cdot \ln x_{c,T} + w_w \cdot \ln x_{w,T} + \left(\frac{w_c w_w}{T}\right)\left(\begin{array}{l}J_1'' + J_2''(c_c - c_w)^2 + J_3''(e_c - e_w)^2 + J_4''(s_c - s_w)^2 \\ + J_5''(a_c - a_w)^2 + J_6''(b_c - b_w)^2 + J_7''(v_c - v_w)^2 + J_8''(a_c b_c - a_w b_w)^2\end{array}\right) \\ + \left(\frac{w_c w_w (w_c - w_w)}{T}\right)\left(\begin{array}{l}J_9'' + J_{10}''(c_c - c_w)^2 + J_{11}''(e_c - e_w)^2 + J_{12}''(s_c - s_w)^2 \\ + J_{13}''(a_c - a_w)^2 + J_{14}''(b_c - b_w)^2 + J_{15}''(v_c - v_w)^2 + J_{16}''(a_c b_c - a_w b_w)^2\end{array}\right) \\ + \left(\frac{w_c w_w (w_c - w_w)^2}{T}\right)\left(\begin{array}{l}J_{17}'' + J_{18}''(c_c - c_w)^2 + J_{19}''(e_c - e_w)^2 + J_{20}''(s_c - s_w)^2 \\ + J_{21}''(a_c - a_w)^2 + J_{22}''(b_c - b_w)^2 + J_{23}''(v_c - v_w)^2 + J_{24}''(a_c b_c - a_w b_w)^2\end{array}\right) \quad (3)$$

Solvent coefficients, namely c, e, s, a, b and v, exhibit variation based on the type of solvent being analyzed. The phase's affinity to interact with solutes via polarizability-based interactions is expressed as e, whereas s quantifies the dipolarity/polarity of the solvent phase. Hydrogen-bond acidity and basicity of the solvent phase are designated as a and b coefficients. Additionally, v represents the overall dispersion interaction energy between the solvent phase and the solute. Also, J'' terms are the model parameters.

To determine accuracy, the mean relative deviation (MRD) is employed and computed via the following formula:

$$MRD\% = \frac{100}{NDP}\sum\left(\frac{|Calculated\ solubity\ value - Observed\ solubity\ value|}{Observed\ solubity\ value}\right) \quad (4)$$

The formula involves NDP, which represents the quantity of data points in every set. The definition of the MRD is very similar to that of the relative standard deviation (RSD) for the repeated experiments. One could directly compare the numerical values of the MRDs with the RSD values for experimental measurements, where the ideal model should provide MRD% close to RSD values. The RSD for repeated paracetamol solubility data using the same chemicals and the same instruments and procedures varied from 3.3% to 17.0% and as a general rule; with a lower solubility, a larger RSD is obtained [34]. Concerning the paracetamol solubility data reported from different laboratories, the overall RSD varied from 17.6% to 21.1% [35]. In order to demonstrate the predictive ability of the models in question, a leave-one-solvent-system-out method was utilized for cross-validation. During each analysis, one data set was omitted from the training process and the trained model was then used to predict its corresponding solubility.

3. Results and Discussion

The experimental paracetamol solubility data in binary aqueous mixtures of ethanol, methanol, PG, 2-propanol, 1-propanol, acetonitrile, DMF, DMSO, and 1,4-dioxane were used to train Equations (1)–(3). In the first step, the Jouyban–Acree model and its combined form with Reichardt's polarity parameter were used for each binary system data correlating, individually. The MRD% values for these computations are given in Table 1. As can be seen, MRD% values for all solubility systems were less than 15%, showing the reliability of data for fitting to the mathematical model. Furthermore, the low MRD% values being obtained separately for each cosolvency system is an initial criterion for including them in the generation of a general model.

Table 1. MRDs% for solubility of paracetamol in the aqueous binary systems at various temperatures for Equations (1) and (2).

No.	Solvent Mixtures	T (K)	MRDs (±SD)%	
			Equation (1)	Equation (2)
1	Ethanol + water	293.2	6.2 ± 7.1	4.9 ± 6.4
		298.2	3.2 ± 2.8	1.6 ± 1.6
		303.2	3.3 ± 3.0	2.0 ± 1.7
		308.2	3.9 ± 3.3	1.4 ± 1.4
		313.2	4.3 ± 4.7	4.2 ± 2.9
2	PG + water	293.2	2.3 ± 3.0	2.3 ± 3.0
		298.2	2.3 ± 2.1	2.3 ± 2.1
		303.2	1.9 ± 3.1	1.9 ± 3.1
		308.2	2.3 ± 2.3	2.3 ± 2.3
		313.2	3.3 ± 2.6	3.3 ± 2.5
3	Methanol + water	298.2	0.6 ± 0.5	0.6 ± 0.4
4	1,4-Dioxane + water	293.2	13.8 ± 13.1	10.9 ± 9.0
		298.2	7.9 ± 9.0	6.1 ± 4.7
		303.2	8.1 ± 6.9	5.9 ± 3.9
		308.2	9.7 ± 8.7	6.9 ± 6.4
		313.2	11.3 ± 9.5	8.6 ± 7.6
5	1-Propanol + water	293.2	7.9 ± 6.1	8.1 ± 5.0
		298.2	4.0 ± 4.0	3.3 ± 2.2
		303.2	3.6 ± 3.0	0.7 ± 0.9
		308.2	6.8 ± 5.6	6.6 ± 3.6
		313.2	7.6 ± 6.6	7.2 ± 4.7
6	Acetonitrile + water	293.2	11.6 ± 19.0	5.3 ± 10.2
		298.2	10.4 ± 13.6	3.0 ± 5.9
		303.2	9.1 ± 8.6	2.3 ± 2.4
		308.2	8.5 ± 6.3	3.6 ± 3.9
		313.2	8.1 ± 7.4	5.5 ± 6.5
7	DMSO + water	298.2	12.8 ± 16.7	11.6 ± 15.2
		303.2	6.0 ± 6.5	3.8 ± 4.5
		308.2	5.3 ± 5.8	3.1 ± 4.2
		313.2	10.4 ± 12.3	10.1 ± 11.8
8	DMF + water	298.2	8.3 ± 9.6	5.7 ± 8.1
		303.2	4.9 ± 6.4	1.9 ± 2.5
		308.2	4.9 ± 4.3	1.7 ± 2.3
		313.2	6.8 ± 7.6	5.3 ± 7.2
9	2-Propanol + water	293.2	8.7 ± 7.3	9.0 ± 7.0
		298.2	5.5 ± 4.4	5.2 ± 4.8
		303.2	2.6 ± 2.4	1.8 ± 1.5
		308.2	4.7 ± 4.3	4.3 ± 3.3
		313.2	9.2 ± 6.5	9.0 ± 6.1

Another point in Table 1 was the low value of MRD% for the combined form of the Jouyban–Acree model with Reichardt's polarity parameters compared with the Jouyban–Acree model. The Jouyban–Acree model, in its general form, is not influenced by the characteristics and properties of either the solute or solvent. Despite this, factors such as solute ionization in solvent mixtures, solubilization/desolubilization capacity, density, dielectric constant, and physical/chemical stability can impact solubility. These parameters can be described in ET (30) values reported for the solvent mixtures.

The next step was the correlation of all data for the generation of a general model for solubility prediction. The trained version of the combined form of the Jouyban–Acree model with Reichardt's polarity parameters for the paracetamol solubility prediction in aqueous solvent mixtures was as:

$$\ln x_{m,T} = w_c \cdot \ln x_{c,T} + w_w \cdot \ln x_{w,T} + \left(\frac{w_c w_w}{T}\right)\left(5922.694 - 75.549 E_{m,T}^N\right) \\ + \left(\frac{w_c w_w (w_c - w_w)}{T}\right)\left(6900.277 - 135.008 E_{m,T}^N\right) + \left(\frac{w_c w_w (w_c - w_w)^2}{T}\right)\left(8395.463 - 156.285 E_{m,T}^N\right) \quad (5)$$

It is important to highlight that the statistical significance of all the model constants was confirmed through t-test analysis at a probability level of <0.1. The back-calculated solubility data, comprising 422 data points, showed an overall MRD% of 37.6%. Table 2 displays the MRD% values calculated for paracetamol solubility data in different solvent mixtures at varying temperatures, using Equation (5). For the trained model, the lowest predicted solubility data deviation (MRD = 4.3%) can be observed for a solvent mixture of 2-propanol and water at a temperature of 303.2 K. Conversely, the highest deviation (MRD = 139.0%) occurs for a solvent mixture of PG and water at a temperature of 293.2 K.

One can remove J'_1, J'_3, and J'_5 from Equation (2) to reach below model with $J'_i E_{m,T}^N$ parameters.

$$\ln x_{m,T} = w_c \cdot \ln x_{c,T} + w_w \cdot \ln x_{w,T} \\ + \left(\frac{w_c w_w}{T}\right)\left(35.147 E_{m,T}^N\right) + \left(\frac{w_c w_w (w_c - w_w)}{T}\right)\left(7.948 E_{m,T}^N\right) \quad (6)$$

The overall MRD% for back-calculated data with this trained equation is 46.9% which does not have significant difference with Equation (5) demonstrating Equation (6) with three parameters can be used instead of Equation (5) with six parameters.

The effectiveness of Reichardt's polarity parameter in improving the solubility prediction accuracy of the Jouyban–Acree model was examined by comparing the results with those obtained from the Jouyban–Acree model that was combined with Abraham solvation parameters. Abraham solvation parameters are a set of empirical coefficients that include multiple parameters that represent different molecular interactions such as polarizability, dipolarity/polarity, hydrogen-bond acidity, basicity, and dispersion. Each parameter contributes to a different aspect of solvation, creating a more accurate representation of the overall behavior of the solvent. The use of multiple parameters in Abraham solvation parameters, as well as their flexibility and applicability to a wider range of solvents, offers advantages in predicting solvation behavior over other solubility parameters.

The trained form of Equation (3) for the paracetamol solubility in nine included aqueous binary systems is

$$\ln x_{m,T} = w_c \cdot \ln x_{c,T} + w_w \cdot \ln x_{w,T} + \left(\frac{w_c w_w}{T}\right)\left(\begin{array}{l}-785.592(c_c - c_w)^2 + 711.412(e_c - e_w)^2 + 295.216(s_c - s_w)^2 \\ +163.222(a_c - a_w)^2 - 55.665(b_c - b_w)^2 + 253.856(v_c - v_w)^2 - 12.758(a_c b_c - a_w b_w)^2\end{array}\right) \\ + \left(\frac{w_c w_w (w_c - w_w)}{T}\right)\left(\begin{array}{l}11591.238 - 999.834(c_c - c_w)^2 - 22533.610 e_c - e_w)^2 - 650.844(s_c - s_w)^2 \\ -280.764(a_c - a_w)^2 - 118.233(b_c - b_w)^2 + 842.913(v_c - v_w)^2\end{array}\right) \\ + \left(\frac{w_c w_w (w_c - w_w)^2}{T}\right)\left(\begin{array}{l}9977.706 - 1677.087(c_c - c_w)^2 - 13326.989(e_c - e_w)^2 - 656.926(s_c - s_w)^2 \\ -189.332(a_c - a_w)^2 - 64.860(b_c - b_w)^2 + 343.005(v_c - v_w)^2\end{array}\right) \quad (7)$$

The overall MRD% is 12.4% (Table 2). As can be seen, a relatively high difference was observed for back-calculated MRD% of Equation (7) with 10.0% and Equation (5) with 37.6 for the similar data. A similar trained model was proposed for the solubility of paracetamol in various cosolvent + water mixtures with an overall MRD% of 19.6%, employing the Hansen solubility parameters [35]. However, these differences are normal and the possible reason for this difference in accuracy is the number and nature of parameters used in each model. The Abraham solubility parameter model incorporates multiple parameters that represent different types of molecular interactions, whereas Reichardt's polarity parameter represents only the solvents' relative polarities, which can be less specific to the solute. Another possible reason could be the variant data set used for model validation. The models' performances heavily depend upon the data set used for validation, and any biases in the data set can affect the predictive capability of a model. For example, this difference between the MRD% values of the two models decreased when excluding the PG+ water system with MRD% 10.2% for Equation (7) and 23.4% for Equation (5). Therefore, it is essential to use a diverse data set for validation, including compounds with different chemical structures and properties, to ensure accurate predictions.

Table 2. MRDs% for solubility of paracetamol in the aqueous binary systems at various temperatures for Equations (5) and (7).

No.	Solvent Mixtures	T (K)	MRDs (±SD)% Equation (5)	MRDs (±SD)% Equation (7)
1	Ethanol + water	293.2	21.8 ± 24.2	7.2 ± 6.7
		298.2	15.7 ± 14.1	4.0 ± 3.8
		303.2	14.5 ± 12.7	3.9 ± 4.2
		308.2	14.6 ± 13.4	4.5 ± 4.2
		313.2	9.8 ± 12.1	3.9 ± 4.2
2	PG + water	293.2	139.0 ± 112.4	10.9 ± 10.1
		298.2	125.9 ± 100.9	9.7 ± 6.8
		303.2	125.5 ± 99.2	6.6 ± 5.7
		308.2	122.4 ± 94.5	7.2 ± 7.2
		313.2	126.3 ± 97.7	4.5 ± 8.8
3	Methanol + water	298.2	56.4 ± 51.2	17.9 ± 14.2
4	1,4-Dioxane + water	293.2	8.8 ± 6.4	20.2 ± 16.4
		298.2	10.4 ± 7.8	13.5 ± 11.7
		303.2	11.8 ± 9.7	11.3 ± 12.1
		308.2	15.2 ± 13.7	11.4 ± 11.4
		313.2	16.7 ± 14.5	12.6 ± 11.7
5	1-Propanol + water	293.2	14.3 ± 12.2	18.7 ± 17.1
		298.2	16.6 ± 14.5	17.1 ± 14.4
		303.2	19.9 ± 17.4	16.8 ± 12.8
		308.2	28.2 ± 19.7	16.2 ± 11.3
		313.2	28.3 ± 18.9	14.8 ± 10.4
6	Acetonitrile + water	293.2	43.8 ± 25.9	11.8 ± 19.7
		298.2	41.7 ± 26.1	10.6 ± 14.1
		303.2	39.8 ± 27.6	9.0 ± 8.9
		308.2	39.3 ± 27.0	8.1 ± 6.9
		313.2	41.9 ± 26.2	7.8 ± 7.0
7	DMSO + water	298.2	32.8 ± 29.2	9.4 ± 10.1
		303.2	34.6 ± 31.3	8.2 ± 10.8
		308.2	36.1 ± 33.1	11.6 ± 14.4
		313.2	37.2 ± 35.1	16.3 ± 18.5
8	DMF + water	298.2	34.4 ± 31.7	7.5 ± 8.1
		303.2	35.2 ± 32.5	5.4 ± 5.2
		308.2	35.9 ± 33.5	5.4 ± 5.2
		313.2	36.8 ± 34.4	7.9 ± 8.1
9	2-Propanol + water	293.2	11.6 ± 9.5	13.5 ± 10.9
		298.2	7.8 ± 7.3	10.2 ± 7.7
		303.2	4.3 ± 3.8	5.6 ± 4.1
		308.2	4.9 ± 3.9	3.4 ± 4.8
		313.2	8.0 ± 6.4	7.1 ± 5.5
	Overall		37.6 ± 54.2	10.0 ± 10.8

It should be noted that even though Reichardt-polarity-parameter combined model gave a higher error percentage compared to the Abraham-solubility-parameter combined model, the error range was still acceptable. As mentioned above, the RSD values for repeated paracetamol solubility determination varied from 17.6 to 21.1% [34]. These observations suggest that Reichardt's polarity parameter can potentially be used as an alternative to Abraham solubility parameters if a less complex model is desired, although it may not provide the same level of accuracy as Abraham solubility parameters.

Cross-validation was employed using the leave-one-solvent system-out method to assess the prediction capabilities of the trained models. A comprehensive report of the cross-validation process for the analyzed models is presented in Table 3. The tabulated results show that the overall $MRDs\%$ increased from 15.3 to 20.2 for the ethanol + water mixture, 127.8 to 177.8 for PG + water, 60.9 to 56.4 for methanol + water, 41.2 to 12.6 for 1,4-dioxane +water, 25.7 to 21.5 for 1-propanol + water, 41.3 to 47.6 for acetonitrile +water, 35.2 to 36.3 for DMSO + water, 35.5 to 37.0 for DMF + water, and 7.3 to 7.8 for the 2-propanol + water system. It can be concluded that the combined form of the Jouyban–Acree model with Reichardt's polarity parameters has an acceptable reliability to predict the paracetamol solubility data in the investigated mixtures. A cross-validation process was also employed for the Jouyban–Acree model combined with Abraham solvation parameters, and the overall $MRD\%$ value increased from 10.0% to 1.1×10^7. As can be seen, the Reichardt-polarity-parameter combined model showed better results compared to the Abraham-solvation-parameters combined model. A possible reason for it can be this fact that the Reichardt-polarity-parameter combined model is relatively simple, requiring only one parameter to predict solubility (the solvent polarity parameter) whereas, the Abraham-solvation-parameters combined model requires multiple parameters, including the hydrogen-bond acidity and basicity, polarizability, and volume parameters. In some cases, having fewer model parameters can make a model less prone to overfitting and better suited to predict solubility, especially if the data set is limited. The performance of the models also depends on the quality and diversity of the training data used to optimize the parameters. Therefore, a more detailed investigation is needed to determine the performance differences between the models.

Table 3. Leave-solvent-system-out cross-validation for the proposed models.

No.	Solvent Mixtures	T (K)	MRDs (±SD)%	
			Equation (5)	Equation (7)
1	Ethanol + water	293.2	27.4 ± 30.0	8.3 ± 6.8
		298.2	20.7 ± 18.9	5.1 ± 5.2
		303.2	19.2 ± 17.3	4.8 ± 5.7
		308.2	19.3 ± 18.0	5.3 ± 5.5
		313.2	14.2 ± 16.4	4.4 ± 4.0
2	PG + water	293.2	193.7 ± 154.9	162.4 ± 199.6
		298.2	176.2 ± 139.5	147.8 ± 179.4
		303.2	175.0 ± 170.0	138.8 ± 164.5
		308.2	170.0 ± 130.4	137.5 ± 164.9
		313.2	174.1 ± 133.9	138.9 ± 166.1
3	Methanol + water	298.2	60.9 ± 55.1	82.5 ± 91.2
4	1,4-Dioxane + water	293.2	38.2 ± 34.3	160.8 ± 190.8
		298.2	40.5 ± 34.3	141.5 ± 169.6
		303.2	41.0 ± 34.8	132.3 ± 165.7
		308.2	42.8 ± 35.1	117.0 ± 142.0
		313.2	43.4 ± 35.0	111.2 ± 132.3

Table 3. Cont.

No.	Solvent Mixtures	T (K)	MRDs (±SD)%	
			Equation (5)	Equation (7)
5	1-Propanol + water	293.2	17.8 ± 16.2	26.2 ± 21.8
		298.2	20.9 ± 19.9	25.6 ± 19.2
		303.2	24.3 ± 21.1	25.6 ± 18.0
		308.2	32.8 ± 23.6	25.6 ± 17.0
		313.2	32.8 ± 22.6	24.2 ± 16.1
6	Acetonitrile + water	293.2	50.1 ± 29.4	$1.4 \times 10^8 \pm 3.5 \times 10^8$
		298.2	48.1 ± 29.9	$1.1 \times 10^8 \pm 2.7 \times 10^8$
		303.2	46.1 ± 31.6	$8.4 \times 10^7 \pm 1.4 \times 10^8$
		308.2	45.8 ± 31.1	$6.0 \times 10^7 \pm 1.4 \times 10^8$
		313.2	48.1 ± 30.0	$4.2 \times 10^7 \pm 1.0 \times 10^8$
7	DMSO + water	298.2	34.7 ± 30.8	72.4 ± 92.7
		303.2	36.4 ± 32.6	70.6 ± 90.1
		308.2	36.4 ± 32.6	68.4 ± 87.3
		313.2	37.7 ± 34.2	65.0 ± 82.4
8	DMF + water	298.2	36.0 ± 32.9	118.5 ± 215.7
		303.2	36.7 ± 33.7	116.8 ± 210.8
		308.2	37.3 ± 34.5	114.1 ± 202.8
		313.2	38.1 ± 35.4	112.5 ± 197.8
9	2-Propanol + water	293.2	12.1 ± 10.2	36.5 ± 32.1
		298.2	8.5 ± 7.9	32.5 ± 26.7
		303.2	4.9 ± 4.3	26.7 ± 20.6
		308.2	5.2 ± 4.1	22.6 ± 17.2
		313.2	8.1 ± 4.1	18.2 ± 15.0
	Overall		50.1	1.1×10^7

4. Conclusions

This research involved the development of a trained model based on Reichardt's polarity parameter to predict paracetamol solubility in cosolvency systems. The use of the Jouyban–Acree model was examined, as well as its combined version with Reichardt's polarity parameter. The effectiveness of Reichardt's polarity parameter in improving the solubility prediction accuracy of the Jouyban–Acree model was examined by comparing the results with those obtained from the Jouyban–Acree model combined with Abraham solvation parameters. Upon analysis, the model was deemed to have a satisfactory level of accuracy in predicting solubilities, as evidenced by the overall *MRDs*% of 37.6.

Author Contributions: Conceptualization, A.J.; Methodology, E.R.; Validation, A.J.; Investigation, E.R.; Data curation, E.R.; Writing—original draft, E.R.; Writing—review & editing, A.J.; Supervision, A.J.; Funding acquisition, A.J. All authors have read and agreed to the published version of the manuscript.

Funding: This work was supported by Research Affairs of Tabriz University of Medical Sciences, Tabriz, Iran under grant number of 67898.

Data Availability Statement: No new data were created or analyzed in this study. Data sharing is not applicable to this article.

Conflicts of Interest: The authors declare no conflict of interest.

References

1. Duncan, C.; Watson, D.; Stein, A. Diagnosis and management of headache in adults: Summary of SIGN guideline. *BMJ* **2008**, *337*, a2329. [CrossRef] [PubMed]
2. Jibril, F.; Sharaby, S.; Mohamed, A.; Wilby, K.J. Intravenous versus oral acetaminophen for pain: Systematic review of current evidence to support clinical decision-making. *Can. J. Hosp. Pharm.* **2015**, *68*, 238. [CrossRef] [PubMed]

3. Llinàs, A.; Glen, R.C.; Goodman, J.M. Solubility challenge: Can you predict solubilities of 32 molecules using a database of 100 reliable measurements? *J. Chem. Inf. Model.* **2008**, *48*, 1289–1303. [CrossRef]
4. Adjei, A.; Newburger, J.; Martin, A. Extended Hildebrand approach: Solubility of caffeine in dioxane–water mixtures. *J. Pharm. Sci.* **1980**, *69*, 659–661. [CrossRef]
5. Ochsner, A.B.; Belloto Jr, R.J.; Sokoloski, T.D. Prediction of xanthine solubilities using statistical techniques. *J. Pharm. Sci.* **1985**, *74*, 132–135. [CrossRef]
6. Acree, W.E., Jr. Mathematical representation of thermodynamic properties: Part 2. Derivation of the combined nearly ideal binary solvent (NIBS)/Redlich-Kister mathematical representation from a two-body and three-body interactional mixing model. *Thermochim. Acta* **1992**, *198*, 71–79. [CrossRef]
7. Yalkowsky, S.H.; Roseman, T.J. Solubilization of drugs by cosolvents. *Tech. Solubilization Drugs* **1981**, *12*, 91–134.
8. Xu, X.; Pinho, S.P.; Macedo, E.A. Activity coefficient and solubility of amino acids in water by the modified Wilson model. *Ind. Eng. Chem. Res.* **2004**, *43*, 3200–3204. [CrossRef]
9. Khossravi, D.; Connors, K.A. Solvent effects on chemical processes, I: Solubility of aromatic and heterocyclic compounds in binary aqueous—Organic solvents. *J. Pharm. Sci.* **1992**, *81*, 371–379. [CrossRef]
10. Ruckenstein, E.; Shulgin, I. Solubility of drugs in aqueous solutions: Part 1. Ideal mixed solvent approximation. *Int. J. Pharm.* **2003**, *258*, 193–201. [CrossRef]
11. Williams, N.; Amidon, G. Excess free energy approach to the estimation of solubility in mixed solvent systems III: Ethanol-propylene glycol-water mixtures. *J. Pharm. Sci.* **1984**, *73*, 18–23. [CrossRef] [PubMed]
12. Jouyban, A.; Acree, W.E., Jr. Mathematical derivation of the Jouyban-Acree model to represent solute solubility data in mixed solvents at various temperatures. *J. Mol. Liq.* **2018**, *256*, 541–547. [CrossRef]
13. Taft, R.W.; Abboud, J.-L.M.; Kamlet, M.J.; Abraham, M.H. Linear solvation energy relations. *J. Solut. Chem.* **1985**, *14*, 153–186. [CrossRef]
14. Kamlet, M.J.; Doherty, R.M.; Abboud, J.-L.M.; Abraham, M.H.; Taft, R.W. Linear solvation energy relationships: 36. Molecular properties governing solubilities of organic nonelectrolytes in water. *J. Pharm. Sci.* **1986**, *75*, 338–349. [CrossRef] [PubMed]
15. Hansn, C.M. *Hansen Solubility Parameters: A User's Handbook*, 2nd ed.; CRC Press; Taylor & Francis Group: Boca Raton, FL, USA, 2007.
16. Catalán, J. Toward a generalized treatment of the solvent effect based on four empirical scales: Dipolarity (SdP, a new scale), polarizability (SP), acidity (SA), and basicity (SB) of the medium. *J. Phys. Chem. B* **2009**, *113*, 5951–5960. [CrossRef]
17. Jiménez, J.A.; Martínez, F. Thermodynamic magnitudes of mixing and solvation of acetaminophen in ethanol+ water cosolvent mixtures. *Rev. Acad. Colomb. Cienc.* **2006**, *30*, 87–99. [CrossRef]
18. Pourkarim, F.; Mirheydari, S.N.; Martinez, F.; Jouyban, A. Solubility of acetaminophen in 1-propanol+ water mixtures at T = 293.2–313.2 K. *Phys. Chem. Liq.* **2020**, *58*, 456–472. [CrossRef]
19. Hojjati, H.; Rohani, S. Measurement and prediction of solubility of paracetamol in water– isopropanol solution. Part 1. Measurement and data analysis. *Org. Process Res. Dev.* **2006**, *10*, 1101–1109. [CrossRef]
20. Ahumada, E.A.; Delgado, D.R.; Martínez, F. Solution thermodynamics of acetaminophen in some PEG 400+ water mixtures. *Fluid Phase Equilibria* **2012**, *332*, 120–127. [CrossRef]
21. Jiménez, J.A.; Martínez, F. Thermodynamic study of the solubility of acetaminophen in propylene glycol+ water cosolvent mixtures. *J. Braz. Chem. Soc.* **2006**, *17*, 125–134. [CrossRef]
22. Soltanpour, S.; Jouyban, A. Solubility of acetaminophen and ibuprofen in polyethylene glycol 600, N-methyl pyrrolidone and water mixtures. *J. Solut. Chem.* **2011**, *40*, 2032–2045. [CrossRef]
23. Muñoz, M.M.; Jouyban, A.; Martínez, F. Solubility and preferential solvation of acetaminophen in methanol+ water mixtures at 298.15 K. *Phys. Chem. Liq.* **2016**, *54*, 515–528. [CrossRef]
24. Shakeel, F.; Alanazi, F.K.; Alsarra, I.A.; Haq, N. Solubilization behavior of paracetamol in Transcutol–water mixtures at (298.15 to 333.15) K. *J. Chem. Eng. Data* **2013**, *58*, 3551–3556. [CrossRef]
25. Bustamante, P.; Romero, S.; Peña, A.; Escalera, B.; Reillo, A. Enthalpy–entropy compensation for the solubility of drugs in solvent mixtures: Paracetamol, acetanilide, and nalidixic acid in dioxane–water. *J. Pharm. Sci.* **1998**, *87*, 1590–1596. [CrossRef]
26. Rahimpour, E.; Agha, E.M.H.; Martinez, F.; Barzegar-Jalali, M.; Jouyban, A. Solubility study of acetaminophen in the mixtures of acetonitrile and water at different temperatures. *J. Mol. Liq.* **2021**, *324*, 114708. [CrossRef]
27. Cysewski, P.; Jeliński, T.; Przybyłek, M.; Nowak, W.; Olczak, M. Solubility Characteristics of Acetaminophen and Phenacetin in Binary Mixtures of Aqueous Organic Solvents: Experimental and Deep Machine Learning Screening of Green Dissolution Media. *Pharmaceutics* **2022**, *14*, 2828. [CrossRef]
28. Soltanpour, S.; Jouyban, A. Solubility of acetaminophen and ibuprofen in polyethylene glycol 600, propylene glycol and water mixtures at 25 °C. *J. Mol. Liq.* **2010**, *155*, 80–84. [CrossRef]
29. Jouyban, A.; Soltanpour, S.; Acree, W.E., Jr. Solubility of acetaminophen and ibuprofen in the mixtures of polyethylene glycol 200 or 400 with ethanol and water and the density of solute-free mixed solvents at 298.2 K. *J. Chem. Eng. Data* **2010**, *55*, 5252–5257. [CrossRef]
30. Jiménez, J.A.; Martínez, F. Temperature dependence of the solubility of acetaminophen in propylene glycol+ ethanol mixtures. *J. Solut. Chem.* **2006**, *35*, 335–352. [CrossRef]

31. Bustamante, P.; Romero, S.; Reillo, A. Thermodynamics of paracetamol in amphiprotic and amphiprotic—Aprotic solvent mixtures. *Pharm. Pharmacol. Commun.* **1995**, *1*, 505–507.
32. Marcus, Y. The use of chemical probes for the characterization of solvent mixtures. Part 2. Aqueous mixtures. *J. Chem. Soc. Perkin Trans. 2* **1994**, 1751–1758. [CrossRef]
33. Acree, W.E., Jr.; Grubbs, L.M.; Abraham, M.H. *Prediction of Partition Coefficients and Permeability of Drug Molecules in Biological Systems with Abraham Model Solute Descriptors Derived from Measured Solubilities and Water-to-Organic Solvent Partition Coefficients*; InTech: New York, NY, USA, 2012; pp. 100–102.
34. Rahimpour, E.; Moradi, M.; Sheikhi-Sovari, A.; Rezaei, H.; Rezaei, H.; Jouyban-Gharamaleki, V.; Kuentz, M.; Jouyban, A. Comparative drug solubility studies using shake-flask versus a laser-based robotic method. *AAPS PharmSciTech* **2023**, *24*, 207. [CrossRef] [PubMed]
35. Hatefi, A.; Jouyban, A.; Mohamadian, E.; Acree, W.E., Jr.; Rahimpour, E. Prediction of paracetamol solubility in cosolvency systems at different temperatures. *J. Mol. Liq.* **2019**, *273*, 282–291. [CrossRef]

Disclaimer/Publisher's Note: The statements, opinions and data contained in all publications are solely those of the individual author(s) and contributor(s) and not of MDPI and/or the editor(s). MDPI and/or the editor(s) disclaim responsibility for any injury to people or property resulting from any ideas, methods, instructions or products referred to in the content.

Article

Reichardt's Dye-Based Solvent Polarity and Abraham Solvent Parameters: Examining Correlations and Predictive Modeling

William E. Acree, Jr. [1,*] and Andrew S. I. D. Lang [2]

[1] Department of Chemistry, University of North Texas, Denton, TX 76203, USA
[2] Department of Computing & Mathematics, Oral Roberts University, Tulsa, OK 74171, USA; alang@oru.edu
* Correspondence: bill.acree@unt.edu

Abstract: The concept of "solvent polarity" is widely used to explain the effects of using different solvents in various scientific applications. However, a consensus regarding its definition and quantitative measure is still lacking, hindering progress in solvent-based research. This study hopes to add to the conversation by presenting the development of two linear regression models for solvent polarity, based on Reichardt's $E_T(30)$ solvent polarity scale, using Abraham solvent parameters and a transformer-based model for predicting solvent polarity directly from molecular structure. The first linear model incorporates the standard Abraham solvent descriptors s, a, b, and the extended model ionic descriptors j^+ and j^-, achieving impressive test-set statistics of $R^2 = 0.940$ (coefficient of determination), MAE = 0.037 (mean absolute error), and RMSE = 0.050 (Root-Mean-Square Error). The second model, covering a more extensive chemical space but only using the descriptors s, a, and b, achieves test-set statistics of $R^2 = 0.842$, MAE = 0.085, and RMSE = 0.104. The transformer-based model, applicable to any solvent with an associated SMILES string, achieves test-set statistics of $R^2 = 0.824$, MAE = 0.066, and RMSE = 0.095. Our findings highlight the significance of Abraham solvent parameters, especially the dipolarity/polarizability, hydrogen-bond acidity/basicity, and ionic descriptors, in predicting solvent polarity. These models offer valuable insights for researchers interested in Reichardt's $E_T(30)$ solvent polarity parameter and solvent polarity in general.

Keywords: Reichardt's dye; $E_T(30)$ solvent polarity parameter; hydrogen bonding; solvatochromism; Abraham model; predictive modeling

1. Introduction

For over 50 years, researchers have employed the term "solvent polarity" to explain variations in chemical reaction rates, spectroscopic properties, and thermophysical characteristics of solute molecules dissolved in different solvents. However, despite its widespread usage, "solvent polarity" lacks a universally accepted definition and quantitative measure that the scientific community has agreed upon. In its broadest interpretation, the term encompasses the entire range of intermolecular interactions, including Coulombic forces, dispersion forces, charge transfer, hydrogen bonding, directional dipole–dipole interactions, and solvophobic effects experienced by molecules and ionic species. It should be noted that interactions leading to changes in the solute's chemical identity through complex formation, oxidation–reduction reactions, protonation, or other structural-altering processes do not fall within the scope of this broad definition of "solvent polarity" [1].

More refined studies have attempted to distinguish and quantify the distinct solvent effects originating from hydrogen-bonding interactions compared to dipole–dipole interactions through spectroscopic and calorimetric measurements. Various empirical scales for solvent polarity and acidity/basicity have been established using spectroscopic probe molecules such as betaine-30 dye, 4-nitroanisole [2], N,N-diethyl-4-nitroaniline [2], pyrene [3], and other large polycyclic aromatic hydrocarbons [4], 2-(N,N-dimethylamino)-7-nitrofluorene [5], Brooker's merocyanine dye [6], and N-ethyl-4-carbethoxypyridinium

iodide [7]. Among these scales, the most notable one is based on the betaine-30 dye, commonly known as Reichardt's Dye, which is used to define the $E_T(30)$ and normalized $\hat{E}_T(30)$ solvent polarity scales [8–10]:

$$E_T(30) \text{ (kcal mol}^{-1}) = 2.8591 \cdot 10^{-3} \cdot \tilde{v}_{max} \tag{1}$$

$$\hat{E}_T(30) = (2.8591 \cdot 10^{-3} \cdot \tilde{v}_{max} - 30.7)/32.4 \tag{2}$$

where $\tilde{v}_{max}$ is the wavenumber of the maximum in the π to π^* absorption band. The above list of spectroscopic probe molecules is not exhaustive but serves to illustrate that numerous organic molecules and ionic species exhibit solvatochromic behavior. A comprehensive compilation of probe molecules can be found elsewhere [2,9].

Calorimetric probe studies [11–14] have been employed with some degree of success in quantifying the strength of hydrogen bonding between a solute and solvent. By measuring the enthalpies of solution for a reference compound that is incapable of forming hydrogen bonds in the given solvent or by measuring the enthalpies of solution for the solute in an "inert" solvent, it is possible to estimate the contributions from non-hydrogen bonding interactions. Alternatively, mathematical expressions derived from semi-theoretical solution models can be used to estimate the non-hydrogen bonding effects. Each estimation method yields a different value for the hydrogen bond enthalpy. There is also no universally accepted method for addressing non-hydrogen bonding contributions, nor is there a single hydrogen-bonding interaction whose strength can be considered a universal reference value.

Quantitative structure–property relationships and linear free energy relationships are occasionally employed to mathematically elucidate the variations in a solute's thermophysical properties across different solvent media. Among the various proposed relationships, the Abraham general solvation parameter model is one of the most widely utilized methods. This model's popularity stems from both its ability to encompass a wide range of solute properties and its foundation in the diverse molecular interactions that govern the specific solute property under investigation. The Abraham model is constructed upon two linear free energy relationships [15–18]. The first equation models the transfer of neutral molecules and ionic species between two condensed phases:

$$SP = c_p + e_p \cdot E + s_p \cdot S + a_p \cdot A + b_p \cdot B + v_p \cdot V + j_p^+ \cdot J^+ + j_p^- \cdot J^- \tag{3}$$

and the second modeling of the transfer of neutral molecules from the gas phase to a condensed phase:

$$SP = c_k + e_k \cdot E + s_k \cdot S + a_k \cdot A + b_k \cdot B + l_k \cdot L \tag{4}$$

where SP represents a "specific property" of solutes within a given solvent, partitioning system, or biological/pharmaceutical process, and where the subscripts p and k distinguish the solvent parameters between the two different transfer systems. In this study, SP corresponds to the logarithm of the solute's water-to-organic solvent partition coefficient (log P) or gas-to-organic solvent partition coefficient (log K). Specifically, it refers to the logarithm of the ratio between the solute's molar solubility in two different solvents or phases: log $(C_{S,organic}/C_{S,water})$ (Equation (3)) and log $(C_{S,organic}/C_{S,gas})$ (Equation (4)). In these equations, $C_{S,organic}$ and $C_{S,water}$ represent the molar solubility of the solute in the organic solvent and water, respectively, while $C_{S,gas}$ denotes the molar gas phase concentration that can be calculated from the solute's vapor pressure. It is important to note that the numerical values of the lowercase equation coefficients in Equations (3) and (4) will vary for each specific process.

The right-hand side of Equations (3) and (4) represent the different types of solute–solvent interactions that are believed to govern the specific solute transfer process being described. Each term quantifies a particular solute–solvent interaction as the product of the solute property (uppercase alphabetic letters) and the complementary solvent properties (lowercase alphabetic letters). It is the lowercase letters that contain valuable chemical information regarding the polarity/polarizability and hydrogen-bonding characteristics of the solvent. The five solvent properties are defined as follows:

- e represents the ability of the solvent to interact with surrounding solvent molecules through electron lone pair interactions;
- s is a measure of the dipolarity/polarizability of the organic solvent;
- a and b refer to the hydrogen-bond acidity and hydrogen-bond basicity of the solvent;
- v and l describe the solvent's dispersion forces and cavity formation, providing the space in which the dissolved solute resides.

The last two terms on the right-hand side of Equation (3) correspond to the interactions between ions and the solubilizing medium. The term $j_p^+ \cdot J^+$ represents cations, while the term $j_p^- \cdot J^-$ represents anions. These terms are utilized to describe zwitterionic compounds, such as amino acids and the betaine-30 dye. When the ionic coefficients j_p^+ and j_p^- are set to zero, Equation (1) reduces to the standard equation for neutral species. That is, when j_p^+ and j_p^- are unavailable, the same set of numerical values (c_p, e_p, s_p, a_p, b_p, and v_p) for a given solvent is employed to describe solute transfer for both neutral and ionic solutes.

The complimentary solute descriptors on the right-hand side of Equations (3) and (4) are defined as follows: E denotes the molar refraction of the given solute in excess of that of a linear alkane having a comparable molecular size; S is a combination of the electrostatic polarity and polarizability of the solute; A and B refer to the respective hydrogen-bond donating and accepting capacities of the dissolved solute; V corresponds to the McGowan molecular volume of the solute calculated from atomic sizes and chemical bond numbers; and L is the logarithm of the solute's gas-to-hexadecane partition coefficient measured at 298.15 °K. The solute descriptors and their calculation from measured experimental data are described in greater detail elsewhere [16,19,20].

In the present study, our primary focus is on the lowercase solvent coefficients of Equation (3) (c, e, s, a, b, v, j$^+$, and j$^-$). Note that since we are exclusively dealing with the coefficients from Equation (3), we will drop the subscript p going forward. These coefficients are determined by experimental partition coefficient data and molar solubility ratios. Our investigation aims to explore potential correlations between these coefficients and normalized $\hat{E}_T(30)$ values.

Past studies have established that the $E_T(30)$ parameter is significantly correlated with:

- The H-bond donating and dipolarity/polarizability parameters of the Kamlet–Taft scale [21,22];
- The solvent acidity and solvent dipolarity parameters of the Catalán scale [22,23].

The aforementioned scales are all based on the solvatochromism of select spectroscopic probe molecules. As such, the numerical values are based on the transition energy corresponding to the promotion of an electron from the ground electronic state to an excited electronic state. Solute transfer between two phases does not involve an electronic transition within the probe molecule.

The primary objective of this study is to explore whether the observed correlations between the $E_T(30)$ parameter and solvent properties established through solute transfer measurements hold true in general. By investigating the relationship between $E_T(30)$ and solvent properties, our objective is to gain insights into the underlying chemistry that influences the determination of $E_T(30)$ values. Additionally, we provide models that enable the prediction of the $E_T(30)$ parameter using Abraham solvent coefficients. These coefficients can be determined through experimental methods or, in the case of organic solvents, predicted using open methods [24].

In addition to the primary objective, this study aims to develop a transformer-based model capable of accurately predicting $\hat{E}_T(30)$ values for a wide range of solvents directly from structure information encoded in the "language" of SMILES. We aim to provide a practical and efficient tool for estimating $\hat{E}_T(30)$ values without requiring extensive experimental measurements or descriptor calculations. Specifically, we will create a predictive model by fine-tuning an explicit ChemBERTa-2 model [25] using $\hat{E}_T(30)$ values as the endpoint. This approach was previously employed to directly predict other physical–chemical endpoints from structure, exhibiting comparable accuracy to conventional machine-learning techniques [26].

2. Materials and Methods

We compiled a comprehensive collection of unique $E_T(30)$ values and their corresponding normalized $\hat{E}_T(30)$ values by conducting an extensive literature search [9,27–30]. In cases where multiple measurements were found for the same solvent, we recorded only the most recent value. This process resulted in a dataset of $E_T(30)$ measurements for 491 solvents, which we have made available as Open Data on Figshare [31]. The dataset contains several types of solvents, including organic mono-solvents with various functional groups, binary and ternary solvent mixtures with multiple volume ratios, complexes, salts, and ionic liquids.

Similarly, we gathered Abraham solvent parameters, including the ionic parameters j^+ and j^- where applicable, from the relevant literature sources [20,24,32–35]. Once again, we retained only the most recent values. This dataset has also been available as Open Data on Figshare [36].

To create a combined measurement-parameter modeling dataset, we initially excluded ten rows with $E_T(30)$ measurements obtained at temperatures greater than 40 °C. The betaine-30 dye exhibits strong thermochromism, and $E_T(30)$ values determined at lower temperatures are always larger than values measured at elevated temperatures [10]. We then merged the two datasets by matching standard INCHIKEYs across tables. The resulting modeling dataset contains 481 $E_T(30)/\hat{E}_T(30)$ measurements, out of which 113 have associated standard Abraham solvent parameters (c, e, s, a, b, v), of which 44 additionally possess the ionic parameters j^+ and j^-. The modeling dataset is included as part of this study's Supplementary Materials.

To address the different amounts of available descriptor information in the subsets of data, we developed three distinct models:

- Model 1: A linear model regressed on data from solvents that had all Abraham solvent parameters (c, e, s, a, b, v, j^+, j^-) available;
- Model 2: A linear model regressed on data from solvents that had the standard set of Abraham solvent parameters (c, e, s, a, b, v) available;
- Model 3: A fine-tuned transformer-based large language model (LLM) trained using the "language" of SMILES.

Models 1 and 2 were optimized using a best subset selection technique with five-fold cross-validation. For Model 3, we fine-tuned a transformer-based large-language model (LLM) ChemBERTa-2 [25], specifically ChemBERTa-77M-MTR, which is openly accessible on Hugging Face [37]. This enabled us to predict $\hat{E}_T(30)$ values directly from the solvent's structural representation written using the SMILES format, bypassing the need for descriptor derivation or calculation. Prior to fine-tuning, we eliminated duplicate rows of mixtures, keeping only the 50:50 (by volume) mixtures, ensuring that each $\hat{E}_T(30)$ value was associated with a unique SMILES string, resulting in an AI modeling subset of 461 solvents. Mixtures in the dataset—alcohol–water mixtures in various volume ratios and ionic liquids—are represented using the standard technique of separating the SMILES strings with a period. For example, an ethanol-water mixture is represented by the SMILES string CO.O.

We began by randomly splitting the dataset into training, validation, and test sets (80:10:10) and trained the model using the Trainer class from the transformer's library with the adamw_torch optimizer. We optimized the performance of the fine-tuned model by closely monitoring the model's performance on the validation set every ten steps during the training process. This approach allowed us to employ the early stopping technique, effectively preventing overfitting without compromising model accuracy. We then used five-fold cross-validation with the determined optimal training parameters to calculate comparable inter-model test-set statistics. Supplementary Materials accompanying this article include the Python code used for fine-tuning the LLM.

We refined Model 3 to enhance its predictive power using an enumerated SMILES technique [38]. This approach effectively increased the number of solvent SMILES/$\hat{E}_T(30)$-value pairs from 461 to 5221. For each of the 461 solvents, we generated 30 random SMILES using open-source Python code provided by Bjerrum [38], which preserved the original chemical structure but introduced random spelling variations of the SMILES strings. Due to the random generation process, duplicate SMILES were common, especially for smaller molecules. To address this, we eliminated duplicate entries, resulting in 5221 unique SMILES representations. These SMILES strings were then matched with their corresponding $\hat{E}_T(30)$ values and placed in the same fold as the solvent from which their alternative SMILES spelling originated. This step was taken to prevent the same chemical structure from appearing in multiple folds, which could potentially distort the results. That is, this approach ensured that the same solvent was not utilized in both the training and testing phases. Finally, we fine-tuned the transformer-based AI language model ChemBERTa-2 on this expanded dataset, using the same procedure and parameters as before, including five-fold cross-validation.

For those interested in learning more about using LLMs and other AI techniques for cheminformatics, we refer the reader to the excellently written tutorials available on DeepChem [39].

3. Results

The base modeling dataset contains the SMILES strings and $E_T(30)/\hat{E}_T(30)$ values of 481 solvents. Within this dataset, we identified three distinct subsets:

- $N = 461$ solvents with unique SMILES strings. In cases where mixtures had multiple ratios with the same SMILES, only the 50:50 volume ratio was retained;
- $N = 113$ solvents that possess all the standard Abraham solvent parameters (c, e, s, a, b, v);
- $N = 44$ solvents with the complete set of Abraham solvent parameters, including j^+ and j^-.

In this section, we provide the results of developing three distinct models specific to the three subsets. These models are presented from the most accurate yet less broadly applicable, to the slightly less accurate but still highly effective, and finally, to the most generally applicable model.

3.1. Predicting E_T-Values Using Complete Abraham Solvent Parameters

Using a best subset linear regression technique on the subset of 44 solvents with known extended Abraham solvent parameters (c, e, s, a, b, v, j^+, j^-), we found that the optimal model is one with five significant descriptors ($p < 0.05$; s, a, b, v, j^+, j^-), with corresponding 5-fold cross-validated test-set statistics of $R^2 = 0.940$ (coefficient of determination), MAE = 0.037 (mean absolute error), and RMSE = 0.050 (Root-Mean-Square Error). The model coefficients are as presented in Equation (5):

$$\hat{E}_T(30) = 0.945 + 0.047 \cdot s + 0.016 \cdot a + 0.103 \cdot b + 0.034 \cdot j^+ + 0.066 \cdot j^- \qquad (5)$$

Betaine-30 is a zwitterionic molecule used to measure $E_T(30)$ values. Despite being overall neutral, it contains both a cation and an anion moiety. Therefore, it is reasonable that the derived model incorporates the two additional ionic descriptors, j^+ and j^-, in addition to the standard solvent polarity associated with the s-parameter and the hydrogen bond donating/accepting-related parameters a and b. Unfortunately, there are relatively few solvents for which these two additional ionic terms have been calculated.

3.2. Predicting ET-Values Using Standard Abraham Solvent Parameters

Employing a best subset linear regression approach again but now on the more significant subset of 113 solvents that possess the six standard Abraham solvent parameters (c, e, s, a, b, v), we identified an optimal model with three significant descriptors ($p < 0.05$; s, a, b). The corresponding five-fold cross-validated test-set statistics for this more generally applicable model are equal to: $R^2 = 0.842$, MAE = 0.085, and RMSE = 0.104. The corresponding model coefficients are presented in Equation (6):

$$\hat{E}_T(30) = 1.090 + 0.051 \cdot s + 0.081 \cdot a + 0.140 \cdot b \tag{6}$$

As with model one, we see the expected polarity-related descriptors and the hydrogen bond donating/accepting-related descriptors a and b. This observation is in accord with previous studies that found a link between the $E_T(30)$ parameter and the solubilizing media's acidity/basicity and dipolarity [21–23]. Unlike previous studies, solvent property equation coefficients in the Abraham model are not deduced from spectroscopic properties but rather from the experimental partition coefficient and solubility ratio data. The spectral properties insofar as the betaine-30 dye molecule is concerned, particularly in the cases where hydrogen-bond formation is possible, are influenced by more than a single dye-solvent type of molecular interaction.

As an information note, Abraham model equation coefficients have been reported for peanut oil and several mono-organic solvents lacking an experimental $\hat{E}_T(30)$ value. This provides the opportunity to illustrate the predictive nature of Model 2. The results of our predictive computations are summarized in the second column of Table 1. While the lack of experimental data prevents a comparison of observed versus predicted values, we note that the calculated $\hat{E}_T(30)$ values based on Model 2 differ only slightly from experimental values of solvents having similar functional groups and molecular structures. For example, the estimated values for isopropyl acetate ($\hat{E}_T(30) = 0.319$), pentyl acetate ($\hat{E}_T(30) = 0.290$), tert-butyl acetate ($\hat{E}_T(30) = 0.315$), and isopropyl myristate ($\hat{E}_T(30) = 0.322$) are comparable to experimental values for other monoester solvents such as methyl acetate ($\hat{E}_T(30) = 0.247$), ethyl acetate ($\hat{E}_T(30) = 0.225$), propyl acetate ($\hat{E}_T(30) = 0.210$), and butyl acetate ($\hat{E}_T(30) = 0.241$). The slightly larger estimated value of $\hat{E}_T(30) = 0.361$ for dimethyl adipate likely arises because of the additional ester functional group [40].

Table 1. Calculated $\hat{E}_T(30)$ values for select organic solvents based on models 2 and 3.

Solvent	Model 2	Model 3
Undecane	0.011	0.052
Hexadecane	0.035	0.090
Methylcyclohexane	0.103	0.073
2,2,4-Trimethylpentane	0.032	0.122
Hexadec-1-ene	0.078	0.056
Deca-1,9-diene	0.202	0.013
1,2-Dimethylbenzene	0.140	0.153
1,3-Dimethylbenzene	0.123	0.153
Ethylbenzene	0.132	0.142
PGDP [a]	0.315	0.201
Isopropyl myristate	0.322	0.140
4-Methylpentan-2-ol	0.466	0.403
2-Ethylhexan-1-ol	0.474	0.563
2,2,2-Trifluoroethanol	0.778	0.941
2-Propoxyethanol	0.455	0.657
2-Isopropoxyethanol	0.448	0.634
3-Methoxybutan-1-ol	0.473	0.591
1-tert-Butoxypropan-2-ol	0.444	0.484
Isopropyl acetate	0.319	0.222
Pentyl acetate	0.290	0.170

Table 1. Cont.

Solvent	Model 2	Model 3
tert-Butyl acetate	0.315	0.210
N-Ethylformamide	0.515	0.748
N,N-Dibutylformamide	0.410	0.359
1-OctadecaZnol	0.446	0.459
N-Methyl-2-piperidone	0.524	0.341
N-Formylmorpholine	0.515	0.314
Peanut oil	0.301	0.257
2-(2-Ethoxyethoxy)ethanol	0.504	0.647
N,N-Dimethylacetamide	0.474	0.592
o-Nitrophenyl octyl ether	0.342	0.217
Dimethyl adipate	0.361	0.250

[a] PGDP is the abbreviation for propylene glycol dipelarginate.

3.3. Predicting ET-Values Directly from Structure by Fine-Tuning a Chemical Foundation Model

By fine-tuning a ChemBERTa-2 model on the largest subset of 461 solvents with unique SMILES and employing five-fold cross-validation, we created a transformer-based model that predicted $\hat{E}_T(30)$ values directly from the structure. The resulting model exhibited satisfactory performance, with the following test-set statistics: $R^2 = 0.808$, MAE = 0.071, RMSE = 0.099.

By employing SMILES enumeration [38], we achieved significant improvements in model performance, as evidenced by the test-set statistics calculated on the same dataset as before. Specifically, we calculated the test-set statistics from the measured and predicted $\hat{E}_T(30)$ values for the 461 original solvents using their original SMILES strings. The resulting test-set statistics were as follows: $R^2 = 0.824$, MAE = 0.066, RMSE = 0.095. The comparison between the measured and enhanced transformer-based model predicted $\hat{E}_T(30)$ values is illustrated in Figure 1.

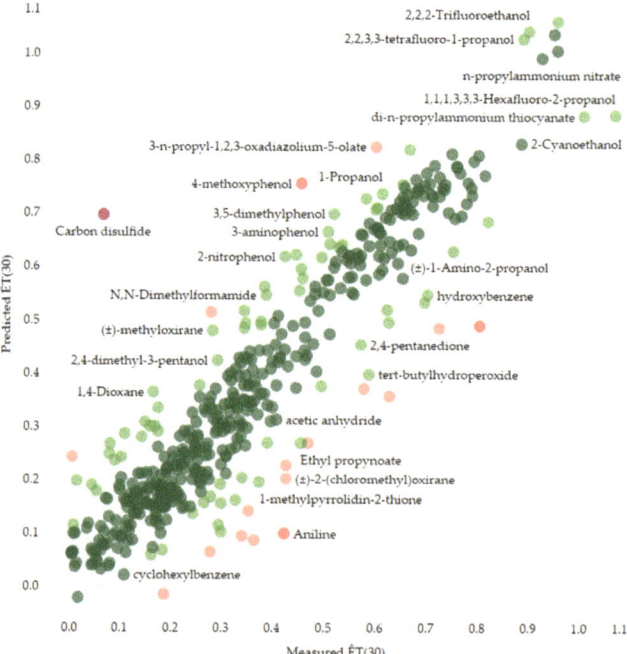

Figure 1. Experimental vs. Model 3-predicted $\hat{E}_T(30)$ values for 461 solvents colored by absolute error.

To demonstrate its capabilities, we present the $\hat{E}_T(30)$ values predicted by Model 3 for several significant sustainable solvents sourced from Bradley et al. [24]. These solvents exhibit a range of $\hat{E}_T(30)$ values, including D-limonene ($\hat{E}_T(30) = 0.035$), 2-methyltetrahydrofuran ($\hat{E}_T(30) = 0.163$), cyclademol ($\hat{E}_T(30) = 0.244$), oleic acid ($\hat{E}_T(30) = 0.342$), geraniol ($\hat{E}_T(30) = 0.477$), propionic acid ($\hat{E}_T(30) = 0.578$), acetic acid ($\hat{E}_T(30) = 0.623$), propylene glycol ($\hat{E}_T(30) = 0.761$), and glycerol ($\hat{E}_T(30) = 0.815$). It is evident that Model 3 is useful in predicting $\hat{E}_T(30)$ values and may be particularly valuable where direct measurement of $\hat{E}_T(30)$ values is difficult. See Table 1 for Model 3-predicted $\hat{E}_T(30)$ values for solvents with Abraham solvent coefficients but no experimentally determined $\hat{E}_T(30)$ values. This allows a comparison between the predictions of Model 2 and Model 3, highlighting the similarity in the obtained results and reinforcing the utility of Model 3.

However, interpreting this model can be challenging because it does not rely on correlations with parameters that encode known chemical information. Some information can be gleaned by examining the attention mechanism of the model [41]. By analyzing the attention patterns, it is possible to gain insights into the model's attentional priorities and understand which words (atoms) or phrases (functional groups) are deemed important for generating specific outputs, possibly discovering new chemistry, see Figure 2 for an example, selected at random.

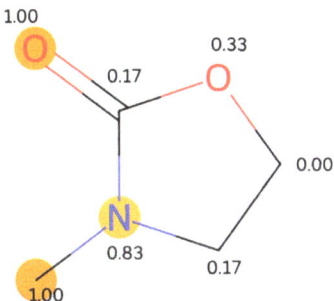

Figure 2. 3-methyl-2-oxazolidinone with scaled attention annotations and highlights averaged over all layers and heads showing how the model has learned to focus on various atoms that are important in determining $E_T(30)$ values.

The general analysis of the attention mechanism of Model 3 is beyond the scope of this paper, though we list it as an area of future research.

During the analysis of model outliers, it was observed that for Model 2, the three ionic liquids present in the modeling dataset ([BMIm]$^+$ BF$_4^-$, [BMIm]$^+$ Tf$_2$N$^-$, [BMIm]$^+$ PF$_6^-$) exhibited the highest absolute errors. These compounds demonstrated underpredicted $\hat{E}_T(30)$ values for Model 2, resulting in a mean absolute error of 0.152. It is worth mentioning that none of these ionic liquids possessed available ionic descriptors, which are likely crucial for achieving more accurate results.

In contrast, Model 3 exhibited a high level of accuracy in predicting the $\hat{E}_T(30)$ values for all three ionic liquids, as evidenced by a low mean absolute error of 0.051. However, an apparent failure of Model 3, illustrated in Figure 1, was observed in the case of carbon disulfide. In this instance, Model 3 significantly overpredicted the $\hat{E}_T(30)$ value, yielding a predicted value of 0.695 compared to the value of 0.065 collected from the literature, resulting in a substantial AE (absolute error) of 0.630. Conversely, Model 2 provided a more reasonable prediction for carbon disulfide, with an AE of 0.129. Model 2 had no significant outliers, likely due to its relatively small chemical space.

4. Discussion

The Abraham solvent parameter-based models, models 1 (s, a, b, j$^+$, j$^-$) and 2 (s, a, b), demonstrate strong predictive capabilities for compounds with measured Abraham solvent parameters. These models also exhibit good explanatory behavior regarding "solvent polarity," as indicated by significant Abraham solvent descriptors such as s (encodes dipolarity/polarizability information) and descriptors a and b (which encode hydrogen-bond acidity/basicity related information). Test-set statistics support this observation, with Model 1 achieving an R^2 of 0.940, MAE of 0.037, and RMSE of 0.050 ($N = 44$), while Model 2 achieves an R^2 of 0.842, MAE of 0.085, and RMSE of 0.104 ($N = 113$).

These findings align with previous studies that have identified correlations between $E_T(30)$ values and hydrogen-bond acidity/basicity and dipolarity/polarizability parameters on different scales, such as the Kamlet–Taft scale [21,22] and the Catalán scale [22,23]. Additionally, we note that other factors like electron lone pair interactions and volume-related dispersion forces, represented by the e and v parameters, respectively, do not significantly contribute to $E_T(30)$ values.

Including ionic descriptors, j^+ and j^-, significantly improves the model's performance, although it is essential to note that the chemical space of the sample is limited. Nonetheless, the poor performance of Model 2 in predicting $E_T(30)$ values for ionic liquids (which lack available j^+ and j^- descriptor values) underscores the importance of these descriptors in determining $E_T(30)$ values in general.

Abraham solvent parameters also have the advantage over some other descriptor-based models in that they can be determined experimentally for most solvents (including mixtures, ionic liquids) and even predicted themselves for mono-solvents using standard machine learning techniques [24].

The optimized version of Model 3 exhibits strong predictive ability (R^2 = 0.824, MAE = 0.066, RMSE = 0.095). It stands out as the most versatile among all our models in that it can be applied to any solvent with an associated SMILES string, enhancing its practical utility.

The predictive applicability of Model 3 will prove useful, particularly in those organic solvents in which the betaine-30 dye has very limited solubility. The betaine-30 dye is insoluble in perfluorohydrocarbons, alkanes, aliphatic ethers, thioethers, amines, and esters having long alkyl chains. For these solvents, "experimental" $E_T(30)$ values are often based on the absorption measurements of a lipophilic penta-*tert*-butyl substituted betaine dye, whose solvatochromic behavior is highly correlated with that of the betaine-30 zwitterionic molecule [30]. However, the interpretability of Model 3 poses challenges. In certain scenarios, it may produce significant errors, as observed with carbon disulfide, or give different predictions for molecules that exist in more than one tautomeric form. Further investigation should focus on analyzing the attention heads of Model 3, as this exploration may unveil new insights into chemistry and offer potential avenues for future research.

Finally, our use of an LLM is non-traditional and stands out in comparison to the standard linear techniques also used in the paper. Our LLM model is designed to seamlessly integrate into various workflows, adding a layer of abstraction that reduces the complexity for end-users. By leveraging ontology, our model benefits from a structured approach, enhancing its ability to understand, interpret, and generate predictions accurately. Agent assistance, on the other hand, enables the model to provide actionable insights and maintain interactive and dynamic communication with users. Knowledge engineering plays a crucial role in our model by facilitating the extraction, structuring, and analysis of domain-specific knowledge, enabling our model to make reliable predictions and offer a deeper understanding of the underlying chemical phenomena.

Supplementary Materials: The following supporting information can be downloaded at: https://www.mdpi.com/article/10.3390/liquids3030020/s1, Dataset S1: $E_T(30)$ Modeling Dataset.csv; Code S2: Fine-tuning ChemBERTa-2 to predict E_T.py.

Author Contributions: Conceptualization, W.E.A.J.; methodology, A.S.I.D.L.; data collection and curation, W.E.A.J. and A.S.I.D.L.; modeling, A.S.I.D.L.; visualizations, A.S.I.D.L.; writing—original draft preparation, W.E.A.J. and A.S.I.D.L.; writing—review and editing, W.E.A.J. and A.S.I.D.L. All authors have read and agreed to the published version of the manuscript.

Funding: This research received no external funding.

Institutional Review Board Statement: Not applicable.

Informed Consent Statement: Not applicable.

Data Availability Statement: The Abraham model solvent coefficient data and the $E_T(30)$ solvent polarity data presented in this study are openly available from Figshare https://figshare.com/articles/dataset/Dataset_Abraham_Model_Solvent_Coefficients/23546670 and https://figshare.com/articles/dataset/Dataset_ET_30_Solvent_Polarity_Parameters/23546709, respectively.

Conflicts of Interest: The authors declare no conflict of interest.

References

1. Reichardt, C. *Solvents and Solvent Effects in Organic Chemistry*, 2nd ed.; VCH: Weinheim, Germany, 1988.
2. Kamlet, M.J.; Abboud, J.L.; Taft, R.W. The solvatochromic comparison method. 6. The π^* scale of solvent polarities. *J. Am. Chem. Soc.* **1977**, *99*, 6027–6038. [CrossRef]
3. Dong, D.C.; Winnik, M.A. The Py scale of solvent polarities. *Can. J. Chem.* **1984**, *62*, 2560–20655. [CrossRef]
4. Acree, W.E.; Tucker, S.A.; Fetzer, J.C. Fluorescence Emission Properties of Polycyclic Aromatic Compounds in Review. *Polycycl. Aromat. Compd.* **1991**, *2*, 75–105. [CrossRef]
5. Catalán, J. On the E_T (30), π^*, P_y, S', and SPP Empirical Scales as Descriptors of Nonspecific Solvent Effects. *J. Org. Chem.* **1997**, *62*, 8231–8234. [CrossRef] [PubMed]
6. Brooker, L.G.S.; Craig, A.C.; Heseltine, D.W.; Jenkins, P.W.; Lincoln, L.L. Color and constitution. XIII. Merocyanines as solvent property indicators. *J. Am. Chem. Soc.* **1965**, *87*, 2443–2450. [CrossRef]
7. Kosower, E.M. The effect of solvent on spectra. I. A new empirical measure of solvent polarity-Z-values. *J. Am. Chem. Soc.* **1958**, *80*, 3253–3260. [CrossRef]
8. Reichardt, C.; Harbusch-Gornert, E. Pyridinium N-phenoxide betaines and their application for the characterization of solvent polarities. X. Extension, correction, and new definition of the ET solvent polarity scale by application of a lipophilic penta-tert-butyl-substituted pyridinium N-phenoxide betaine dye. *Liebigs Ann. Chem.* **1983**, *1983*, 721–743. [CrossRef]
9. Reichardt, C. Solvatochromic Dyes as Solvent Polarity Indicators. *Chem. Rev.* **1994**, *94*, 2319–2358. [CrossRef]
10. Reichardt, C. Polarity of ionic liquids determined empirically by means of solvatochromic pyridinium N-phenolate betaine dyes. *Green Chem.* **2005**, *7*, 339–351. [CrossRef]
11. Arnett, E.M.; Mitchell, E.J.; Murty, T.S.S.R. Basicity. Comparison of hydrogen bonding and proton transfer to some Lewis bases. *J. Am. Chem. Soc.* **1974**, *96*, 3875–3891. [CrossRef]
12. Catalán, J.; Gomez, J.; Couto, A.; Laynez, J. Toward a solvent basicity scale: The calorimetry of the pyrrole probe. *J. Am. Chem. Soc.* **1990**, *112*, 1678–1681. [CrossRef]
13. Catalán, J.; Gomez, J.; Saiz, J.L.; Couto, A.; Ferraris, M.; Laynez, J. Calorimetric quantification of the hydrogen-bond acidity of solvents and its relationship with solvent polarity. *J. Chem. Soc. Perkin Trans. 2 Phys. Org. Chem.* **1995**, *1*, 2301–2305. [CrossRef]
14. Rakipov, I.T.; Petrov, A.A.; Akhmadeev, B.S.; Varfolomeev, M.A.; Solomonov, B.N. Thermodynamic of dissolution and hydrogen bond of the pyrrole, N-methylpyrrole with proton acceptors. *Thermochim. Acta* **2016**, *640*, 19–25. [CrossRef]
15. Abraham, M.H. Scales of hydrogen bonding: Their construction and application to physicochemical and biochemical processes. *Chem. Soc. Rev.* **1993**, *22*, 73–83. [CrossRef]
16. Abraham, M.H.; Ibrahim, A.; Zissimos, A.M. The determination of sets of solute descriptors from chromatographic measurements. *J. Chromatogr. A* **2004**, *1037*, 29–47. [CrossRef]
17. Abraham, M.H.; Acree, W.E., Jr. Equations for the Transfer of Neutral Molecules and Ionic Species from Water to Organic phases. *J. Org. Chem.* **2010**, *75*, 1006–1015. [CrossRef]
18. Abraham, M.H.; Acree, W.E., Jr. Solute Descriptors for Phenoxide Anions and Their Use to Establish Correlations of Rates of Reaction of Anions with Iodomethane. *J. Org. Chem.* **2010**, *75*, 3021–3026. [CrossRef]
19. Abraham, M.H.; Acree, W.E., Jr. Descriptors for the Prediction of Partition Coefficients and Solubilities of Organophosphorus Compounds. *Sep. Sci. Technol.* **2013**, *48*, 884–897. [CrossRef]
20. Abraham, M.H.; Acree, W.E., Jr. Solvation Descriptors for Zwitterionic α-Aminoacids; Estimation of Water-Solvent Partition Coefficients, Solubilities, and Hydrogen-Bond Acidity and Hydrogen-Bond Basicity. *ACS Omega* **2019**, *4*, 2883–2892. [CrossRef]
21. Marcus, Y. The properties of organic liquids that are relevant to their use as solvating solvents. *Chem. Soc. Rev.* **1993**, *22*, 409–419. [CrossRef]
22. Spange, S.; Lienert, C.; Friebe, N.; Schreiter, K. Complementary interpretation of ET(30) polarity parameters of ionic liquids. *Phys. Chem. Chem. Phys.* **2020**, *22*, 9954–9966. [CrossRef] [PubMed]
23. Spange, S.; Weiss, N.; Schmidt, C.H.; Schreiter, K. Reappraisal of Empirical Solvent Polarity Scales for Organic Solvents. *Chem. Methods* **2021**, *1*, 42–60. [CrossRef]
24. Bradley, J.-C.; Abraham, M.H.; Acree, W.E., Jr.; Lang, A.S.I.D. Predicting Abraham model solvent coefficients. *Chem. Cent. J.* **2015**, *9*, 1–10. [CrossRef] [PubMed]

25. Ahmad, W.; Simon, E.; Chithrananda, S.; Grand, G.; Ramsundar, B. ChemBERTa-2: Towards Chemical Foundation Models. *arXiv* **2022**, arXiv:2209.01712. [CrossRef]
26. Lang, A.S.I.D.; Chong, W.C.; Wörner, J.H.R. Fine-Tuning ChemBERTa-2 for Aqueous Solubility Prediction. *Ann. Chem. Sci. Res.* **2023**, *4*, 1–3. [CrossRef]
27. Buhvestov, U.; Rived, F.; Rafols, C.; Bosch, E.; Roses, M. Solute-solvent and solvent-solvent interactions in binary solvent mixtures. Part 7. Comparison of the enhancement of the water structure in alcohol-water mixtures measured by solvatochromic indicators. *J. Phys. Org. Chem.* **1998**, *11*, 185–192. [CrossRef]
28. Reichardt, C. Pyridinium N-phenoxide betaine dyes and their application to the determination of solvent polarities. Part XXX. Pyridinium-N-phenolate betaine dyes as empirical indicators of solvent polarity: Some new findings. *Pure Appl. Chem.* **2008**, *80*, 1415–1432. [CrossRef]
29. Cerón-Carrasco, J.P.; Jacquemin, D.; Laurence, C.; Planchat, A.; Reichardt, C.; Sraidi, K. Determination of a Solvent Hydrogen-Bond Acidity Scale by Means of the Solvatochromism of Pyridinium-N-phenolate Betaine Dye 30 and PCM-TD-DFT Calculations. *J. Phys. Chem. B* **2014**, *118*, 4605–4614. [CrossRef]
30. Cerón-Carrasco, J.P.; Jacquemin, D.; Laurence, C.; Planchat, A.; Reichardt, C.; Sraidi, K. Solvent polarity scales: Determination of new ET(30) values for 84 organic solvents. *J. Phys. Org. Chem.* **2014**, *27*, 512–518. [CrossRef]
31. Lang, A.S.I.D.; Acree, W.E., Jr. Dataset: ET(30) Solvent Polarity Parameters. *Figshare* **2023**. [CrossRef]
32. Jiang, B.; Horton, M.Y.; Acree, W.E., Jr.; Abraham, M.H. Ion-specific equation coefficient version of the Abraham model for ionic liquid solvents: Determination of coefficients for tributylethylphosphonium, 1-butyl-1-methylmorpholinium, 1-allyl-3-methylimidazolium and octyltriethylammonium cations. *Phys. Chem. Liq.* **2017**, *55*, 358–385. [CrossRef]
33. Sinha, S.; Yang, C.; Wu, E.; Acree, W.E., Jr. Abraham Solvation Parameter Model: Examination of Possible Intramolecular Hydrogen-Bonding Using Calculated Solute Descriptors. *Liquids* **2022**, *2*, 131–146. [CrossRef]
34. Longacre, L.; Wu, E.; Yang, C.; Zhang, M.; Sinha, S.; Varadharajan, A.; Acree, W.E., Jr. Development of Abraham Model Correlations for Solute Transfer into the tert-Butyl Acetate Mono-Solvent and Updated Equations for Both Ethyl Acetate and Butyl Acetate. *Liquids* **2022**, *2*, 258–288. [CrossRef]
35. Varadharajan, A.; Sinha, S.; Xu, A.; Daniel, A.; Kim, K.; Shanmugam, N.; Wu, E.; Yang, C.; Zhang, M.; Acree, W.E., Jr. Development of Abraham Model Correlations for Describing Solute Transfer into Transcutol Based on Molar Solubility Ratios for Pharmaceutical and Other Organic Compounds. *J. Solut. Chem.* **2023**, *52*, 70–90. [CrossRef]
36. Lang, A.S.I.D.; Acree, W.E., Jr. Dataset: Abraham Model Solvent Coefficients. *Figshare* **2023**. [CrossRef]
37. DeepChem (DeepChem). Hugging Face. 2022. Available online: https://huggingface.co/DeepChem (accessed on 19 June 2023).
38. Bjerrum, E.J. SMILES Enumeration as Data Augmentation for Neural Network Modeling of Molecules. *arXiv* **2017**, arXiv:1703.07076.
39. Tutorials (DeepChem). 2023. Available online: https://deepchem.io/tutorials/the-basic-tools-of-the-deep-life-sciences/ (accessed on 19 June 2023).
40. Shanmugam, N.; Zhou, A.; Motati, R.; Yao, E.; Kandi, T.; Longacre, L.; Benavides, D.; Motati, S.; Acree, W.E., Jr. Development of Abraham Model Correlations for Dimethyl Adipate from Measured Solubility Data of Nonelectrolyte Organic Compounds. *Phys. Chem. Liq.* **2023**, in press. [CrossRef]
41. Vig, J. A Multiscale Visualization of Attention in the Transformer Model. In Proceedings of the 57th Annual Meeting of the Association for Computational Linguistics: System Demonstrations, Florence, Italy, 28 July–2 August 2019; Association for Computational Linguistics: Florence, Italy, 2019; pp. 37–42. [CrossRef]

Disclaimer/Publisher's Note: The statements, opinions and data contained in all publications are solely those of the individual author(s) and contributor(s) and not of MDPI and/or the editor(s). MDPI and/or the editor(s) disclaim responsibility for any injury to people or property resulting from any ideas, methods, instructions or products referred to in the content.

Perspective

Polarity of Aqueous Solutions

Pedro P. Madeira [1], Luisa A. Ferreira [2], Vladimir N. Uversky [3] and Boris Y. Zaslavsky [2,*]

[1] Centro de Investigacao em Materiais Ceramicos e Compositos, Department of Chemistry, University of Aveiro, 3810-193 Aveiro, Portugal; p.madeira@ua.pt
[2] Cleveland Diagnostics, 3615 Superior Ave., Cleveland, OH 44114, USA; luisa.ferreira@clevelanddx.com
[3] Department of Molecular Medicine and Byrd Alzheimer's Research Institute, Morsani College of Medicine, University of South Florida, Tampa, FL 33612, USA; vuversky@usf.edu
* Correspondence: boris.zaslavsky@clevelanddx.com

Abstract: This short review describes the expansion of the solvatochromic approach utilizing water-soluble solvatochromic dyes to the analysis of solvent features of aqueous media in solutions of various compounds. These solvent features (polarity/dipolarity, hydrogen bond donor ability (HBD acidity), and hydrogen bond acceptor ability (HBA basicity)) vary depending on the nature and concentration of a solute. Furthermore, the solvent features of water (the solvent dipolarity/polarizability and hydrogen bond donor ability) in solutions of various compounds describe multiple physicochemical properties of these solutions (such as the solubility of various compounds in aqueous solutions, salting-out and salting-in constants for polar organic compounds in the presence of different inorganic salts, as well as water activity, osmotic coefficients, surface tension, viscosity, and the relative permittivity of aqueous solutions of different individual compounds) and are likely related to changes in the arrangement of hydrogen bonds of water in these solutions.

Keywords: water; solvent properties; hydrogen-bonds; solubility; physicochemical properties of aqueous solutions

1. Introduction

Unusual and important physical and chemical properties of liquid water, such as surface tension and high permittivity, are based on its ability to form hydrogen bonds. The strength of these bonds depends, in part, on the mutual positions of the interacting molecules. For that and possibly other reasons, the conventional use of thermodynamics in many studies of liquid water properties may be viewed as debatable or at least requiring more complex models. The conventional use of thermodynamics is based on the ideal gas laws, but this simple approximation seems rather questionable in the case of liquid water.

Conclusions based on various thermodynamic properties of aqueous electrolyte solutions could not be used to explain the effects of salts on the structure of water. The attempts to use Flory–Huggins theory (see in [1]) to explain liquid–liquid phase separation have been inadequate, especially for aqueous two-phase systems [2]. For liquid–liquid phase separation to occur in water, the necessary and sufficient condition is the emergence of an interfacial tension. The interfacial tension between organic solvent and water two-phase systems is well-known [3] to increase with increasing dissimilarity of the two solvents. However, in any of these organic solvent–water systems, the value of interfacial tension is not affected by the increase in the organic solvent concentration. On the other hand, in aqueous two-phase systems formed by two polymers or a single polymer and salt, it is well-established experimentally [4–7] that the interfacial tension increases with increasing concentrations of the polymers, i.e., with increasing differences between the polymer concentrations in the two phases.

According to the recently reported [8] model of phase separation in aqueous mixtures of two polymers, the phase diagram may be described in terms of the polymers' effects on the solvent features of water. The different properties of the coexisting phases in aqueous two-phase systems are successfully described by different solvent properties of water in the phases [9–11]. It should be emphasized that the Flory–Huggins theory of the incompatibility of polymers in solution considers the solvent solely as a diluent of unfavorable contacts between polymers and cannot be used to explain the liquid–liquid phase separation in aqueous media. This is unsurprising, as the theory was stated [12] as inapplicable to polar systems. Therefore, attempts to use this theory for the explanation of liquid–liquid phase separation (LLPS) in biological systems seem to be counterproductive.

It is well-established that the high overall concentration of biological macromolecules, occupying up to 40% of the cellular volume and typical for different biological systems [13,14], may influence different properties of proteins and nucleic acids in vivo, such as conformational stability, folding mechanisms, aggregation propensity, interactions with partners, etc. This effect was initially ascribed to the restriction of the volume accessible to a query protein or nucleic acid by the excluded volume effects induced by macromolecular components. Additional weak or "soft" nonspecific interactions between the target protein or nucleic acid and surrounding macromolecules have been suggested to explain numerous experimental observations [15–19]. We showed [20] that most of the crowding effects can be attributed to the effects of proteins or nonionic crowding polymers on the solvent properties of water. The same effects are also displayed by various polyols and other osmolytes [21].

The solvent properties of the aqueous solutions of individual phase-forming polymers describe the binodal line for such systems much better [8]. The two-phase distribution of solutes, from small organic compounds to large proteins, depends on the solvent properties of the two aqueous phases and not on the type and concentrations of the phase-forming polymers [22,23].

The debate between those accepting the salt classification as water structure-making and structure-breaking [24] and its opponents [25,26] remains unsettled. The issue is important because water structure is commonly ignored in molecular biology even though it affects protein folding and LLPS leading to the formation of membrane-less organelles and impacting multiple other biological processes. Computer simulations and predictions of protein structures usually ignore the presence of water in protein solutions, while the results of such simulations are often considered [27] as experimental evidence. Our attempts to produce a suitable theoretical model for phase separation in aqueous two-phase systems were only partially successful [28]. It has been shown [28] that interfacial tension in aqueous two-phase systems may be described by the linear combination of differences between the solvent properties of the coexisting phases.

We found that such proteins as heat-shock protein HspB6 [29], plant stress dehydrins [30], and crystallins affect the solvent properties of water in their solutions as strongly as other solutes. For such studies to succeed it must be proven, first, that there are no direct protein interactions with the solvatochromic dyes [29,30]. Most of the proteins are inclined to bind aromatic compounds; therefore, only a few proteins have been studied so far. Serum albumin has been shown by NMR [31,32] to alter the hydrogen bond donor acidity of water. These studies need to be extended to explore various chaperones and other proteins.

The discovery [33] of disordered proteins with enlarged water-exposed surfaces indicated, at least indirectly, the role of water in the regulation of protein functions but did not hinder computer modeling of the protein characteristics.

The role of different water properties in various tissues [34] as a factor in the distribution of drugs and viruses is generally neglected because the distribution is often viewed solely as a function of binding to specific receptors or other tissue-specific structures. The possibility of water properties in the vicinity of the receptor altering local drug concentration and thus affecting the equilibrium of the drug–receptor binding is commonly ignored.

All of the above considerations allow us to hypothesize that the emergence of interfacial tension and resulting liquid–liquid phase separation in biological systems may be the consequence of different effects of phase-forming proteins and RNAs on the properties of water in the cytoplasm, nucleoplasm, mitochondrial matrix, or stroma of chloroplasts. The further studies of interfacial tension might be more productive for gaining deeper insights into the molecular mechanism of liquid–liquid phase separation in biology than those of unarguably important structural details of the macromolecules participating in and/or driving such phase separation.

Physicochemical properties of aqueous solutions, such as water activity, osmotic coefficient, surface tension, relative permittivity, and viscosity, are well-known to differ in solutions of various compounds depending on the compound nature and concentration. Water in all these solutions is generally viewed as a media maintaining essentially the same bulk properties. It established, however, that solubility of various compounds change quite significantly in aqueous solutions of different salts [35], polyols [36], and other osmolytes [37], amino acids [38,39], and polymers [40] due to possible rearrangement of the hydrogen bonds.

Different solvents are often classified according to their polarity [41] generally considered an overall measure of all specific and non-specific solute–solvent interactions (electrostatic, dipole–dipole, H-bonding). There is a large number of different polarity scales based on different probes and spectroscopic techniques (NMR, IR, UV/visible absorption and emission spectroscopy) [41]. According to Ab Rani et al. [42], there is no absolute correct measure of polarity; different polarity values provide different estimates for the same solvent. There is no useful concept of "right" or "wrong". The usefulness of an empirical polarity scale is its ability to describe and predict solvent dependent phenomena.

The set of multi-parameter polarity scales pioneered by Kamlet and Taft [43–45] includes three separate scales. One is based on the ability of the solvent to serve as a hydrogen bond donor (HBD) acidity (α) [45], the other based on the ability to serve as a hydrogen bond acceptor (HBA) basicity (β) [44], and the scale based on the ability of the solvent to participate in dipole–dipole interactions (dipolarity/polarizability, π^*) [43]. The combination of these three parameters describes the ability of a given solvent to participate in solute–solvent interactions much better than any single parameter.

In our studies we used a set of three solvatochromic dyes; 4-nitroanisole for estimation of dipolarity/polarizability of aqueous media, π^*-value; 4-nitrophenol for the solvent HBA basicity, β-value; and Reichardt's carboxylated betaine dye sodium {2,6-diphenyl-4-[4-(4-carboxylato-phenyl)-2,6-diphenylpyridinium-1-yl]}phenolate} for solvent HBD acidity, α-value.

Solvent features of aqueous media in solutions of over 60 various solutes including inorganic salts [46,47], free amino acids [47,48], small organic compounds [49–51], polymers [38,52–54], and a few proteins [52,54] were estimated.

It has been established that the solvent features estimated for aqueous solutions of over 60 different compounds are linearly related to each other as reported in [39]. This relationship may be described as follows:

$$\pi^*_{ij} = k_{\pi^*j} + k_{\alpha j}\alpha_{ij} + k_{\beta j}\beta_{ij}, \tag{1}$$

where i denotes the solute concentration; j denotes the particular solute; and k_{π^*j}, $k_{\alpha j}$, and $k_{\beta j}$ are constant coefficients.

It was found [39] that the above coefficients are linearly interrelated, as illustrated graphically in Figure 1 for all examined compounds:

$$k_{\pi^*j} = 1.096_{\pm 0.002} - 1.235_{\pm 0.002}k_{\alpha j} - 0.5956_{\pm 0.0003}k_{\beta j}, \tag{2}$$
$$N = 61; r^2 = 0.99991; SD = 0.014; F = 3{,}369{,}994,$$

where N—number of solutes; r^2—correlation coefficient; SD—standard deviation; F—variation ratio.

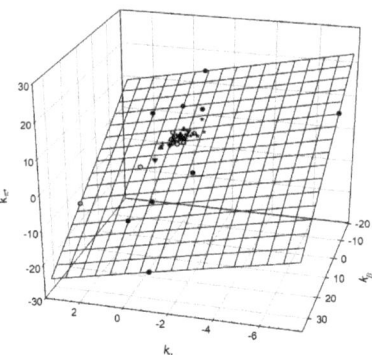

Figure 1. Linear interrelationship between the solvent properties of the solutes (data from [39]).

The above relationship seems to be generally applicable, and it confirms that the aqueous solutions of various compounds can be viewed as different solvents of the same aqueous nature. It also indicates that dipole–dipole interactions in water depend on hydrogen bonding. The permanent dipole moment of water is known to be 1.85 D in the gas phase, while it increases to 2.9 ± 0.6 D [55] in the liquid phase. This difference is generally attributed to polarization of water molecules due to the hydrogen bonding in liquid water. Theoretical consideration and discussion of this effect may be found in [53]. Hence, it seems reasonable to suggest that all three solvent features—π^*, α, and β—in water depend on properties of hydrogen bonds, and that they already embody dispersion forces.

2. Solvent Features Characterize the Physicochemical Properties of Aqueous Solutions

The solubility of various compounds in aqueous solutions often depends on the nature and concentration of co-solutes exemplified by salting-out and salting-in effects induced by inorganic salts [46].

The solubility of different amino acids and their derivatives vary with the concentration of co-solutes glucose, sucrose, and other polyols [56] in the co-solute-specific manner [57]. Similar effects were observed in the presence of urea [58]. All the solubility data in the presence of different co-solutes have been found to correlate strongly with the solvent features, dipolarity/polarizability, π^*, and hydrogen bond donor ability, α, of water in the corresponding co-solute solution [39].

Carbohydrates have been shown to exert a long-range influence on water solvation dynamics [59] and alter the dielectric properties of solutions [60].

Certain free amino acids, such as arginine and lysine, enhance solubility of various compounds [59,61,62]. The solubility of gluten and coumarin in arginine solutions [61], for instance, is strongly associated with the solvent dipolarity/polarizability, π^*, and HBD acidity, α, of water in arginine solutions [63]. This contradicts the assumption [63] that the increased solubility of these two solutes is solely due to their direct interactions with arginine. Similarly, the solubility of a series of alkyl gallates in aqueous solutions of arginine and lysine has been shown [64] to be influenced by the differential effects of the two amino acids on the water's dipolarity/polarizability, π^*. This finding contradicts the notion [39] that solubility enhancement occurs through direct interactions with or hydrophobic interactions with lysine.

The aqueous solubility of alanine, valine, leucine, and isoleucine decrease while those of phenylalanine and tryptophan increase in the presence of 3 to 7 wt.% PEG-20,000 in water [56]. Additionally, the solubility of sucrose in aqueous solutions of PEG-200 and PEG-400 decreases with increasing PEG concentration [65]. Conversely, highly lipophilic compounds generally exhibit increased solubility in water with a higher concentration of PEG-400 [58]. The addition of PEG commonly precipitates proteins from aqueous solutions, with PEG's precipitating efficiency escalating with the polymer's molecular weight [66].

The solubility of alkyl bromides in aqueous solutions of nonionic polymers increases linearly with increasing polymer concentration in solutions of PVP-24,000 and PEG-6000 [67]. Solubility of trimethoprim increases tenfold in 70% wt. solution of PVP-40,000 [40], while that of naproxen increases six-fold in the presence of 10% wt. PVP-25,000 [63].

It is well-established that certain physicochemical properties of aqueous solutions of various compounds in the presence of different salts are linearly related. As an example, the optical rotation of different enantiomeric amino acids and glucose in the presence of various sodium salts [68,69] are linearly related. The solubility of amino acids in aqueous solutions of urea [39] and sorbitol [40] are linearly related as well. The so-called Collander solvent regression relationship between logarithms of partition coefficients of compounds of the same chemical nature in different organic solvent–water biphasic systems [70] is well-known and is widely used [71].

It has been recently shown that specific physicochemical properties, such as the salting-out and salting-in constants for polar organic compounds in the presence of different inorganic salts [72] he same interrelationships may be observed between logarithms of distribution coefficients of drugs in octanol–buffer systems with buffers of various ionic compositions [73]. Logarithms of partition coefficients of different solutes in aqueous two-phase systems of various ionic compositions may be described by similar three-dimensional linear relationships [43].

These linear relationships may be described as follows:

$$Y_1 = k_0 + k_1 Y_2 + k_2 Y_3, \qquad (3)$$

where Y_1, Y_2, and Y_3 are solute properties in aqueous solutions in the presence of solutes 1, 2, and 3; and k_0, k_1, and k_2 are constants.

Analysis of water activity, osmotic coefficients, surface tension, viscosity, and relative permittivity of aqueous solutions of different individual compounds at the same concentrations has shown [43] that the above relationship is generally applicable. The observed [43] relationships typically hold for significant ranges of concentrations—up to 70 mass% in polymer solutions or up to 7 M in solutions of small compounds. Similar relationships are observed for all physicochemical properties. It is also valid for describing different physicochemical properties of the same compound.

Various types of water–water interactions probably define all the physicochemical properties of aqueous solutions; hence, these properties are interrelated.

All physicochemical properties of aqueous solutions of different compounds may be described via a linear combination of the solvent dipolarity/polarizability and hydrogen bond donor ability.

3. Solvent Properties of Aqueous Solutions Are Due to Rearrangement of Hydrogen Bonds

Attenuated total reflection Fourier transform infrared spectroscopy (ATR-FTIR) is one of the most readily available experimental methods for the analysis of the rearrangement of H-bonds in aqueous solutions of various compounds [74–76].

Analysis of arrangement of H-bonds in aqueous solutions is based on inspecting the OH stretching band. This band may be viewed as composed of several components, represented by a set of Gaussian curves. Each Gaussian curve is assigned to water molecules existing in different H-bonded environments [74,75,77–79]. These assignments are based on the rather uncertain models of water structure. The Gaussian curves positioned at lower optical frequencies are generally assigned to water molecules forming strong, ice-like, hydrogen bonds, while those at higher frequencies are assigned to water molecules in an environment with weaker and/or distorted hydrogen bonds. Fitting the OH stretching band in water and all the aqueous solutions of various compounds with one to five Gaussian components showed [80] that the satisfactory fit was always obtained with exactly four components positioned at 3080 cm^{-1}, 3230 cm^{-1}, 3400 cm^{-1}, and 3550 cm^{-1} in agreement with the data reported by Kitadai, et al. [76]. The optimal fit to experimentally observed OH stretching band in pure water and all solutions examined so far was obtained [80] with four Gaussian curves assigned as follows: (I) 3080 cm^{-1}—water molecules with four tetrahedrally arranged hydrogen bonds; (II) 3230 cm^{-1}—water molecules with four distorted hydrogen bonds; (III) 3400 cm^{-1}—water molecules with loosely arranged four and three hydrogen bonds; and (IV) 3500 cm^{-1}—water molecules with three, two, and single hydrogen bonds. This assignment is only a rough approximation of the complex hydrogen bond network existing in water [81] based on the internally consistent empirical measurements. The corresponding subpopulations or clusters of water with different properties may be distributed throughout, and the ratio of these subpopulations/clusters, existing in pure water, may vary in solutions of different solutes.

Analysis of the OH stretching band in aqueous solutions is performed by decomposition of the band into four Gaussian components and the estimation of the relative percentage area for each component as the function of the solute concentration. The area sum of all four Gaussian distributions is normalized to unity to better indicate their relative contributions, i.e., fractions of different water subpopulations.

The relative contributions of the four components estimated from analysis of ATR-FTIR spectra and the solvent features of aqueous solutions of inorganic salts, urea, trimethylamine N-oxide, and several nonionic polymers show that these features are strongly correlated with the relative fractions of certain solute specific water subpopulations. Each correlation for solutions of every examined compound j at the ith concentration may be described in the general form as follows:

$$SF_{ji}(\pi^*_{ji}; \alpha_{ji}; \beta_{ji}) = k_{0j} + k_{mj}A_{ji}^m + k_{nj}A_{ji}^n, \tag{4}$$

where SF_{ji} is a solvent feature (dipolarity/polarizability, π^*, solvent HBD acidity, α, solvent HBA basicity, β); A_{ji} is the fraction of water subpopulation (I, II, III, or IV), denoted as I and n (I, II, III, or IV correspondingly); subscript I denotes the solute concentration; subscript j denotes the solute; and k_{0j}, k_{mj}, and k_{nj} are constants.

While the relative contributions of Gaussian components corresponding to the fractions of the water subpopulations I, II, III, and IV may not be totally independent, an analysis of the cross-terms in the covariance matrix suggests that this set of four appears to be both necessary and sufficient, and was demonstrated [80] to be superior to a fit with either three or five Gaussian components. The observed correlations described by Equation (4) imply that the particular water subpopulations or clusters cannot be localized within the hydration layer of a solute because solvatochromic probes are too big to fit in such a layer.

The only other rational explanation seems to be that the relative amounts of the water subpopulations/clusters in bulk water change in the presence of a solute. This explanation agrees with the fact that, typically, all three solvent features for solutions of a given solute may be described by Equation (3) with both fractions of two solute-specific water subpopulation A_{ji}^m and A_{ji}^n, or with either A_{ji}^m and A_{ji}^n. As an example, the solvent dipolarity/polarizability, π^*, in aqueous solutions of NaSCN may be described as follows:

$$\pi^*_I = 2.34_{\pm 0.03} - 2.62_{\pm 0.06}a_i^{II},$$
$$N = 7; r^2 = 0.9976; SD = 0.004; F = 1973, \tag{5a}$$

the solvent acidity as follows:

$$\alpha_I = 25.1_{\pm 5.7} - 19.8_{\pm 5.2}A_i^{II} - 130.2_{\pm 29.1}A_i^{IV},$$
$$N = 7; r^2 = 0.9901; SD = 0.006; F = 199, \tag{5b}$$

and the solvent basicity as follows:

$$\beta_I = 0.4_{\pm 0.01} - 1.08_{\pm 0.08}A_i^{IV},$$
$$N = 7; r^2 = 0.9709; SD = 0.001; F = 167, \tag{5c}$$

where A_i^{II} is the fraction water subpopulation II (3230 cm^{-1}); and A_i^{VI} is the fraction water subpopulation IV (3550 cm^{-1}) in aqueous solutions of NaSCN at concentration I; all the other parameters are as defined above.

Qualitatively similar relationships were valid for aqueous solutions of all nine solutes examined [80].

By comparing the relationship of different components of the OH stretching band with previously reported solvent features of water, we have confirmed that such relationships may describe the origin of the solute effects on the solvent properties of water.

4. Conclusions and Future Directions

Analysis of solvent properties of aqueous solutions of small organic compounds, such as protective osmolytes, provides information leading to explanation of important biological phenomena. One example is the so-called volume exclusion effect, where it was recently established that the initial theoretical considerations cannot describe the increased protein stability or function change. There is no experimental evidence that has been suggested to explain the observed disagreements between the theory and experimental observations; multiple soft nonspecific interactions have not been shown to exist. The results of the solvatochromic analysis of the osmolyte solutions, however, provided experimental evidence that the explanation is in the changes in the solvent properties of water.

Future applications of the described approach may lead to improvement in our understanding of multiple biological processes, from the function of chaperones and protein folding to the mechanism of liquid–liquid phase separation involved in the formation of biological condensates and the development of new aqueous two-phase systems for better clinical testing of various diseases.

Author Contributions: Conceptualization, B.Y.Z.; validation, P.P.M., L.A.F., V.N.U., and B.Y.Z.; formal analysis, P.P.M., L.A.F., V.N.U., and B.Y.Z.; investigation, P.P.M., L.A.F., V.N.U., and B.Y.Z.; data curation, P.P.M., L.A.F., V.N.U., and B.Y.Z.; writing—original draft preparation, P.P.M., L.A.F., V.N.U., and B.Y.Z.; writing—review and editing, P.P.M., L.A.F., V.N.U., and B.Y.Z.; supervision, B.Y.Z. All authors have read and agreed to the published version of the manuscript.

Funding: This research received no external funding.

Data Availability Statement: Data are contained within the article.

Acknowledgments: The authors are thankful to German Todorov for fruitful discussions.

Conflicts of Interest: Authors Luisa A. Ferreira and Boris Y. Zaslavsky were employed by the company Cleveland Diagnostics. The remaining authors declare that the research was conducted in the absence of any commercial or financial relationships that could be construed as a potential conflict of interest. The funders had no role in the design of the study; in the collection, analyses, or interpretation of data; in the writing of the manuscript; or in the decision to publish the results.

References

1. Guan, Y.; Lilley, T.H.; Treffry, T.E. A new excluded volume theory and its application to the coexistence curves of aqueous polymer two-phase systems. *Macromolecules* **1993**, *26*, 3971–3979. [CrossRef]
2. Zaslavsky, B.Y.; Ferreira, L.A.; Uversky, V.N. Driving Forces of Liquid-Liquid Phase Separation in Biological Systems. *Biomolecules* **2019**, *9*, 473. [CrossRef]
3. Boudh-Hir, M.-E.; Mansoori, G.A. Theory for interfacial tension of partially miscible liquids. *Phys. A Stat. Mech. Its Appl.* **1991**, *179*, 219–231. [CrossRef]
4. Atefi, E.; Mann, J.A., Jr.; Tavana, H. Ultralow interfacial tensions of aqueous two-phase systems measured using drop shape. *Langmuir* **2014**, *30*, 9691–9699. [CrossRef] [PubMed]
5. Bamberger, S.; Seaman, G.V.; Sharp, K.; Brooks, D.E. The effects of salts on the interfacial tension of aqueous dextran poly (ethylene glycol) phase systems. *J. Colloid Interface Sci.* **1984**, *99*, 194–200. [CrossRef]
6. Forciniti, D.; Hall, C.; Kula, M. Interfacial tension of polyethyleneglycol-dextran-water systems: Influence of temperature and polymer molecular weight. *J. Biotechnol.* **1990**, *16*, 279–296. [CrossRef]
7. Ryden, J.; Albertsson, P.-Å. Interfacial tension of dextran—Polyethylene glycol—Water two—Phase systems. *J. Colloid Interface Sci.* **1971**, *37*, 219–222. [CrossRef]

8. Ferreira, L.A.; Uversky, V.N.; Zaslavsky, B.Y. Modified binodal model describes phase separation in aqueous two-phase systems in terms of the effects of phase-forming components on the solvent features of water. *J. Chromatogr. A* **2018**, *1567*, 226–232. [CrossRef]
9. da Silva, N.R.; Ferreira, L.A.; Madeira, P.P.; Teixeira, J.A.; Uversky, V.N.; Zaslavsky, B.Y. Analysis of partitioning of organic compounds and proteins in aqueous polyethylene glycol-sodium sulfate aqueous two-phase systems in terms of solute-solvent interactions. *J. Chromatogr. A* **2015**, *1415*, 1–10. [CrossRef]
10. Ferreira, L.A.; Fan, X.; Madeira, P.P.; Kurgan, L.; Uversky, V.N.; Zaslavsky, B.Y. Analyzing the effects of protecting osmolytes on solute–water interactions by solvatochromic comparison method: II. Globular proteins. *RSC Adv.* **2015**, *5*, 59780–59791. [CrossRef]
11. Madeira, P.P.; Bessa, A.; Teixeira, M.A.; Alvares-Ribeiro, L.; Aires-Barros, M.R.; Rodrigues, A.E.; Zaslavsky, B.Y. Study of organic compounds-water interactions by partition in aqueous two-phase systems. *J. Chromatogr. A* **2013**, *1322*, 97–104. [CrossRef] [PubMed]
12. Tompa, H. *Polymer Solutions*; Butterworth Science Publications: London, UK, 1956.
13. Minton, A.P. Influence of macromolecular crowding upon the stability and state of association of proteins: Predictions and observations. *J. Pharm. Sci.* **2005**, *94*, 1668–1675. [CrossRef]
14. Zhou, H.X.; Rivas, G.; Minton, A.P. Macromolecular crowding and confinement: Biochemical, biophysical, and potential physiological consequences. *Annu. Rev. Biophys.* **2008**, *37*, 375–397. [CrossRef] [PubMed]
15. Minton, A.P. Explicit Incorporation of Hard and Soft Protein-Protein Interactions into Models for Crowding Effects in Protein Mixtures. 2. Effects of Varying Hard and Soft Interactions upon Prototypical Chemical Equilibria. *J. Phys. Chem. B* **2017**, *121*, 5515–5522. [CrossRef]
16. Ando, T.; Yu, I.; Feig, M.; Sugita, Y. Thermodynamics of Macromolecular Association in Heterogeneous Crowding Environments: Theoretical and Simulation Studies with a Simplified Model. *J. Phys. Chem. B* **2016**, *120*, 11856–11865. [CrossRef]
17. Hoppe, T.; Minton, A.P. Incorporation of Hard and Soft Protein-Protein Interactions into Models for Crowding Effects in Binary and Ternary Protein Mixtures. Comparison of Approximate Analytical Solutions with Numerical Simulation. *J. Phys. Chem. B* **2016**, *120*, 11866–11872. [CrossRef] [PubMed]
18. Gnutt, D.; Gao, M.; Brylski, O.; Heyden, M.; Ebbinghaus, S. Excluded-volume effects in living cells. *Angew. Chem. Int. Ed.* **2015**, *54*, 2548–2551. [CrossRef]
19. Wang, Y.; Sarkar, M.; Smith, A.E.; Krois, A.S.; Pielak, G.J. Macromolecular crowding and protein stability. *J. Am. Chem. Soc.* **2012**, *134*, 16614–16618. [CrossRef]
20. Ferreira, L.A.; Madeira, P.P.; Breydo, L.; Reichardt, C.; Uversky, V.N.; Zaslavsky, B.Y. Role of solvent properties of aqueous media in macromolecular crowding effects. *J. Biomol. Struct. Dyn.* **2016**, *34*, 92–103. [CrossRef]
21. Ferreira, L.A.; Uversky, V.N.; Zaslavsky, B.Y. Role of solvent properties of water in crowding effects induced by macromolecular agents and osmolytes. *Mol. Biosyst.* **2017**, *13*, 2551–2563. [CrossRef]
22. Madeira, P.P.; Reis, C.A.; Rodrigues, A.E.; Mikheeva, L.M.; Zaslavsky, B.Y. Solvent properties governing solute partitioning in polymer/polymer aqueous two-phase systems: Nonionic compounds. *J. Phys. Chem. B* **2010**, *114*, 457–462. [CrossRef] [PubMed]
23. Madeira, P.P.; Bessa, A.; de Barros, D.P.; Teixeira, M.A.; Alvares-Ribeiro, L.; Aires-Barros, M.R.; Rodrigues, A.E.; Chait, A.; Zaslavsky, B.Y. Solvatochromic relationship: Prediction of distribution of ionic solutes in aqueous two-phase systems. *J. Chromatogr. A* **2013**, *1271*, 10–16. [CrossRef]
24. Marcus, Y. Effect of ions on the structure of water: Structure making and breaking. *Chem. Rev.* **2009**, *109*, 1346–1370. [CrossRef] [PubMed]
25. Ball, P.; Hallsworth, J.E. Water structure and chaotropicity: Their uses, abuses and biological implications. *Phys. Chem. Chem. Phys.* **2015**, *17*, 8297–8305. [CrossRef]
26. Zhang, Y.; Cremer, P.S. Interactions between macromolecules and ions: The Hofmeister series. *Curr. Opin. Chem. Biol.* **2006**, *10*, 658–663. [CrossRef] [PubMed]
27. Ball, P. Water as an active constituent in cell biology. *Chem. Rev.* **2008**, *108*, 74–108. [CrossRef]
28. Titus, A.R.; Ferreira, L.A.; Belgovskiy, A.I.; Kooijman, E.E.; Mann, E.K.; Mann, J.A., Jr.; Meyer, W.V.; Smart, A.E.; Uversky, V.N.; Zaslavsky, B.Y. Interfacial tension and mechanism of liquid-liquid phase separation in aqueous media. *Phys. Chem. Chem. Phys.* **2020**, *22*, 4574–4580. [CrossRef]
29. Ferreira, L.A.; Gusev, N.B.; Uversky, V.N.; Zaslavsky, B.Y. Effect of human heat shock protein HspB6 on the solvent features of water in aqueous solutions. *J. Biomol. Struct. Dyn.* **2018**, *36*, 1520–1528. [CrossRef]
30. Ferreira, L.A.; Walczyk Mooradally, A.; Zaslavsky, B.; Uversky, V.N.; Graether, S.P. Effect of an Intrinsically Disordered Plant Stress Protein on the Properties of Water. *Biophys. J.* **2018**, *115*, 1696–1706. [CrossRef]
31. Titus, A.R.; Madeira, P.P.; Ferreira, L.A.; Belgovskiy, A.I.; Mann, E.K.; Mann, J.A., Jr.; Meyer, W.V.; Smart, A.E.; Uversky, V.N.; Zaslavsky, B.Y. Arrangement of Hydrogen Bonds in Aqueous Solutions of Different Globular Proteins. *Int. J. Mol. Sci.* **2022**, *23*, 11381. [CrossRef]
32. Madeira, P.P.; Passos, H.; Gomes, J.; Coutinho, J.A.P.; Freire, M.G. Alternative probe for the determination of the hydrogen-bond acidity of ionic liquids and their aqueous solutions. *Phys. Chem. Chem. Phys.* **2017**, *19*, 11011–11016. [CrossRef]
33. Uversky, V.N. Natively unfolded proteins: A point where biology waits for physics. *Protein Sci.* **2002**, *11*, 739–756. [CrossRef]
34. Tanford, C. The hydrophobic effect and the organization of living matter. *Science* **1978**, *200*, 1012–1018. [CrossRef] [PubMed]
35. Collins, K.D. The behavior of ions in water is controlled by their water affinity. *Q Rev. Biophys.* **2019**, *52*, e11. [CrossRef]

36. Politi, R.; Sapir, L.; Harries, D. The impact of polyols on water structure in solution: A computational study. *J. Phys. Chem. A* **2009**, *113*, 7548–7555. [CrossRef] [PubMed]
37. Perry, J.M.; Kanasaki, Y.N.; Karadakov, P.B.; Shimizu, S. Mechanism of dye solubilization and de-aggregation by urea. *Dye. Pigment.* **2021**, *193*, 109530. [CrossRef]
38. Arakawa, T.; Uozaki, M.; Hajime Koyama, A. Modulation of small molecule solubility and protein binding by arginine. *Mol. Med. Rep.* **2010**, *3*, 833–836. [CrossRef]
39. Hirano, A.; Kameda, T.; Arakawa, T.; Shiraki, K. Arginine-assisted solubilization system for drug substances: Solubility experiment and simulation. *J. Phys. Chem. B* **2010**, *114*, 13455–13462. [CrossRef]
40. Garekani, H.A.; Sadeghi, F.; Ghazi, A. Increasing the aqueous solubility of acetaminophen in the presence of polyvinylpyrrolidone and investigation of the mechanisms involved. *Drug Dev. Ind. Pharm.* **2003**, *29*, 173–179. [CrossRef]
41. Reichardt, C.; Welton, T. *Solvents and Solvent Effects in Organic Chemistry*; John Wiley & Sons: Hoboken, NJ, USA, 2011.
42. Ab Rani, M.A.; Brant, A.; Crowhurst, L.; Dolan, A.; Lui, M.; Hassan, N.H.; Hallett, J.P.; Hunt, P.A.; Niedermeyer, H.; Perez-Arlandis, J.M.; et al. Understanding the polarity of ionic liquids. *Phys. Chem. Chem. Phys.* **2011**, *13*, 16831–16840. [CrossRef]
43. Kamlet, M.J.; Abboud, J.L.; Taft, R.W. Solvatochromic Comparison Method. 6. Pi-Star Scale of Solvent Polarities. *J. Am. Chem. Soc.* **1977**, *99*, 6027–6038. [CrossRef]
44. Kamlet, M.J.; Taft, R.W. Solvatochromic Comparison Method. 1. Beta-Scale of Solvent Hydrogen-Bond Acceptor (Hba) Basicities. *J. Am. Chem. Soc.* **1976**, *98*, 377–383. [CrossRef]
45. Taft, R.W.; Kamlet, M.J. Solvatochromic Comparison Method. 2. Alpha-Scale of Solvent Hydrogen-Bond Donor (Hbd) Acidities. *J. Am. Chem. Soc.* **1976**, *98*, 2886–2894. [CrossRef]
46. Ferreira, L.A.; Loureiro, J.A.; Gomes, J.; Uversky, V.N.; Madeira, P.P.; Zaslavsky, B.Y. Why physicochemical properties of aqueous solutions of various compounds are linearly interrelated. *J. Mol. Liq.* **2016**, *221*, 116–123. [CrossRef]
47. Badyal, Y.; Saboungi, M.-L.; Price, D.; Shastri, S.; Haeffner, D.; Soper, A. Electron distribution in water. *J. Chem. Phys.* **2000**, *112*, 9206–9208. [CrossRef]
48. Kemp, D.D.; Gordon, M.S. An interpretation of the enhancement of the water dipole moment due to the presence of other water molecules. *J. Phys. Chem. A* **2008**, *112*, 4885–4894. [CrossRef]
49. Lakshmi, T.; Nandi, P. Effects of sugar solutions on the activity coefficients of aromatic amino acids and their N-acetyl ethyl esters. *J. Phys. Chem.* **1976**, *80*, 249–252. [CrossRef]
50. Gekko, K. Mechanism of polyol-induced protein stabilization: Solubility of amino acids and diglycine in aqueous polyol solutions. *J. Biochem.* **1981**, *90*, 1633–1641. [CrossRef]
51. Nozaki, Y.; Tanford, C. The Solubility of Amino Acids and Related Compounds in Aqueous Urea Solutions. *J. Biol. Chem.* **1963**, *238*, 4074–4081. [CrossRef]
52. Heyden, M.; Brundermann, E.; Heugen, U.; Niehues, G.; Leitner, D.M.; Havenith, M. Long-range influence of carbohydrates on the solvation dynamics of water—Answers from terahertz absorption measurements and molecular modeling simulations. *J. Am. Chem. Soc.* **2008**, *130*, 5773–5779. [CrossRef]
53. Arakawa, T.; Kita, Y.; Koyama, A.H. Solubility enhancement of gluten and organic compounds by arginine. *Int. J. Pharm.* **2008**, *355*, 220–223. [CrossRef] [PubMed]
54. Malmberg, C.G.; Maryott, A.A. Dielectric constants of aqueous solutions of dextrose and sucrose. *J. Res. Natl. Bur. Stand.* **1950**, *45*, 299–303. [CrossRef]
55. Ferreira, L.A.; Uversky, V.N.; Zaslavsky, B.Y. Effects of amino acids on solvent properties of water. *J. Mol. Liq.* **2019**, *277*, 123–131. [CrossRef]
56. Sasahara, K.; Uedaira, H. Solubility of amino acids in aqueous poly (ethylene glycol) solutions. *Colloid Polym. Sci.* **1993**, *271*, 1035–1041. [CrossRef]
57. Tinjacá, D.A.; Muñoz, M.M.; Rahimpour, E.; Jouyban, A.; Martínez, F.; Acree, W.E., Jr. Solubility and apparent specific volume of sucrose in some aqueous polyethylene glycol mixtures at 298.2 K. *Pharm. Sci.* **2018**, *24*, 163–167. [CrossRef]
58. Rytting, E.; Lentz, K.A.; Chen, X.-Q.; Qian, F.; Venkatesh, S. Aqueous and cosolvent solubility data for drug-like organic compounds. *AAPS J.* **2005**, *7*, E78–E105. [CrossRef] [PubMed]
59. Atha, D.H.; Ingham, K.C. Mechanism of precipitation of proteins by polyethylene glycols. Analysis in terms of excluded volume. *J. Biol. Chem.* **1981**, *256*, 12108–12117. [CrossRef]
60. Nayak, A.K.; Panigrahi, P.P. Solubility enhancement of etoricoxib by cosolvency approach. *ISRN Phys. Chem.* **2012**, *2012*, 820653. [CrossRef]
61. Okubo, T.; Chen, S.-X.; Ise, N. Solubility of alkyl bromides in aqueous polymer solutions. *Bull. Chem. Soc. Jpn.* **1973**, *46*, 397–400. [CrossRef]
62. Guptat, R.; Kumar, R.; Singla, A. Enhanced dissolution and absorption of trimethoprim from coprecipitates with polyethylene glycols and polyvinylpyrrolidone. *Drug Dev. Ind. Pharm.* **1991**, *17*, 463–468. [CrossRef]
63. Bettinetti, G.; Mura, P.A.; Liguori, A.U.; Bramanti, G.; Giordano, F. Solubilization and interaction of naproxen with polyvinyl-pyrrolidone in aqueous solution and in the solid state. *IL FARMACO. EDIZIONE PRATICA* **1988**, *43*, 331–343. [PubMed]
64. Ariki, R.; Hirano, A.; Arakawa, T.; Shiraki, K. Arginine increases the solubility of alkyl gallates through interaction with the aromatic ring. *J. Biochem.* **2011**, *149*, 389–394. [CrossRef]

65. Arakawa, T.; Timasheff, S.N. Mechanism of poly(ethylene glycol) interaction with proteins. *Biochemistry* **1985**, *24*, 6756–6762. [CrossRef] [PubMed]
66. Sharma, V.; Singh, J.; Gill, B.; Harikumar, S. SMEDDS: A novel approach for lipophilic drugs. *Int. J. Pharm. Sci. Res.* **2012**, *3*, 2441.
67. Zaslavsky, B.Y.; Ferreira, L.A.; Uversky, V.N. Biophysical principles of liquid–liquid phase separation. In *Droplets of Life*; Academic Press: Cambridge, MA, USA, 2023; pp. 3–82.
68. Rossi, S.; Lo Nostro, P.; Lagi, M.; Ninham, B.W.; Baglioni, P. Specific anion effects on the optical rotation of alpha-amino acids. *J. Phys. Chem. B* **2007**, *111*, 10510–10519. [CrossRef] [PubMed]
69. Lo Nostro, P.; Ninham, B.W.; Milani, S.; Fratoni, L.; Baglioni, P. Specific anion effects on the optical rotation of glucose and serine. *Biopolymers* **2006**, *81*, 136–148. [CrossRef]
70. Collander, R.; Lindholm, M.; Haug, C.M. The partition of organic compounds between higher alcohols and water. *Acta Chem. Scand* **1951**, *5*, 774–780. [CrossRef]
71. Hansch, C.; Leo, A. *Exploring QSAR.: Fundamentals and Applications in Chemistry and Biology*; American Chemical Society: Washington, DC, USA, 1995; Volume 1.
72. Ferreira, L.; Chervenak, A.; Placko, S.; Kestranek, A.; Madeira, P.; Zaslavsky, B. Responses of polar organic compounds to different ionic environments in aqueous media are interrelated. *Phys. Chem. Chem. Phys.* **2014**, *16*, 23347–23354. [CrossRef]
73. Ferreira, L.A.; Chervenak, A.; Placko, S.; Kestranek, A.; Madeira, P.P.; Zaslavsky, B.Y. Effect of ionic composition on the partitioning of organic compounds in octanol–buffer systems. *RSC Adv.* **2015**, *5*, 20574–20582. [CrossRef]
74. Masuda, K.; Haramaki, T.; Nakashima, S.; Habert, B.; Martinez, I.; Kashiwabara, S. Structural change of water with solutes and temperature up to 100 degrees C in aqueous solutions as revealed by attenuated total reflectance infrared spectroscopy. *Appl. Spectrosc.* **2003**, *57*, 274–281. [CrossRef]
75. Chen, Y.; Zhang, Y.-H.; Zhao, L.-J. ATR-FTIR spectroscopic studies on aqueous $LiClO_4$, $NaClO_4$, and $Mg(ClO_4)_2$ solutions. *Phys. Chem. Chem. Phys.* **2004**, *6*, 537–542. [CrossRef]
76. Kitadai, N.; Sawai, T.; Tonoue, R.; Nakashima, S.; Katsura, M.; Fukushi, K. Effects of ions on the OH stretching band of water as revealed by ATR-IR spectroscopy. *J. Solut. Chem.* **2014**, *43*, 1055–1077. [CrossRef]
77. Pavelec, J.; DiGuiseppi, D.; Zavlavsky, B.Y.; Uversky, V.N.; Schweitzer-Stenner, R. Perturbation of water structure by water-polymer interactions probed by FTIR and polarized Raman spectroscopy. *J. Mol. Liq.* **2019**, *275*, 463–473. [CrossRef]
78. Kataoka, Y.; Kitadai, N.; Hisatomi, O.; Nakashima, S. Nature of hydrogen bonding of water molecules in aqueous solutions of glycerol by attenuated total reflection (ATR) infrared spectroscopy. *Appl. Spectrosc.* **2011**, *65*, 436–441. [CrossRef]
79. Guo, Y.-C.; Li, X.-H.; Zhao, L.-J.; Zhang, Y.-H. Drawing out the structural information about the first hydration layer of the isolated Cl− anion through the FTIR-ATR difference spectra. *J. Solut. Chem.* **2013**, *42*, 459–469. [CrossRef]
80. da Silva, N.; Ferreira, L.A.; Belgovskiy, A.I.; Madeira, P.P.; Teixeira, J.A.; Mann, E.K.; Mann, J.A., Jr.; Meyer, W.V.; Smart, A.E.; Chernyak, V.Y.; et al. Effects of different solutes on the physical chemical properties of aqueous solutions via rearrangement of hydrogen bonds in water. *J. Mol. Liq.* **2021**, *335*, 116288. [CrossRef]
81. Brini, E.; Fennell, C.J.; Fernandez-Serra, M.; Hribar-Lee, B.; Luksic, M.; Dill, K.A. How Water's Properties Are Encoded in Its Molecular Structure and Energies. *Chem. Rev.* **2017**, *117*, 12385–12414. [CrossRef]

Disclaimer/Publisher's Note: The statements, opinions and data contained in all publications are solely those of the individual author(s) and contributor(s) and not of MDPI and/or the editor(s). MDPI and/or the editor(s) disclaim responsibility for any injury to people or property resulting from any ideas, methods, instructions or products referred to in the content.

MDPI
St. Alban-Anlage 66
4052 Basel
Switzerland
www.mdpi.com

Liquids Editorial Office
E-mail: liquids@mdpi.com
www.mdpi.com/journal/liquids

Disclaimer/Publisher's Note: The statements, opinions and data contained in all publications are solely those of the individual author(s) and contributor(s) and not of MDPI and/or the editor(s). MDPI and/or the editor(s) disclaim responsibility for any injury to people or property resulting from any ideas, methods, instructions or products referred to in the content.